Recent Advances in Bioinformatics

Recent Advances in Bioinformatics

Edited by Christina Marshall

SYRAWOOD
PUBLISHING HOUSE

New York

Published by Syrawood Publishing House,
750 Third Avenue, 9th Floor,
New York, NY 10017, USA
www.syrawoodpublishinghouse.com

Recent Advances in Bioinformatics
Edited by Christina Marshall

© 2019 Syrawood Publishing House

International Standard Book Number: 978-1-68286-674-0 (Hardback)

Cataloging-in-Publication Data

Recent advances in bioinformatics / edited by Christina Marshall.
    p. cm.
Includes bibliographical references and index.
ISBN 978-1-68286-674-0
1. Bioinformatics. 2. Biology--Data processing. 3. Computational biology.
I. Marshall, Christina.
QH324.2 .R43 2019
570--dc21

TABLE OF CONTENTS

PREFACE

It is often said that books are a boon to mankind. They document every progress and pass on the knowledge from one generation to the other. They play a crucial role in our lives. Thus I was both excited and nervous while editing this book. I was pleased by the thought of being able to make a mark but I was also nervous to do it right because the future of students depends upon it. Hence, I took a few months to research further into the discipline, revise my knowledge and also explore some more aspects. Post this process, I begun with the editing of this book.

Bioinformatics undertakes the development of software systems and methods for biological data analysis and storage. Statistics, computation and other bioinformatics tools are fundamental to the development of analytical models which are used to study biological data accumulated during processes like genome sequencing and annotation. The aim of this book is to present researches that have transformed this discipline and aided its advancement. Chapters in this book cover a number of crucial theories and concepts while also discussing their practical applications. It includes some of the vital pieces of work being conducted across the world on various topics related to bioinformatics. The extensive content of this book provides the readers with a thorough understanding of the subject. Researchers and students in this field will be assisted by this book.

I thank my publisher with all my heart for considering me worthy of this unparalleled opportunity and for showing unwavering faith in my skills. I would also like to thank the editorial team who worked closely with me at every step and contributed immensely towards the successful completion of this book. Last but not the least, I wish to thank my friends and colleagues for their support.

Editor

Interpretive time-frequency analysis of genomic sequences

Hamed Hassani Saadi[1], Reza Sameni[1] and Amin Zollanvari[2*]

Abstract

Background: Time-Frequency (TF) analysis has been extensively used for the analysis of non-stationary numeric signals in the past decade. At the same time, recent studies have statistically confirmed the non-stationarity of genomic non-numeric sequences and suggested the use of non-stationary analysis for these sequences. The conventional approach to analyze non-numeric genomic sequences using techniques specific to numerical data is to convert non-numerical data into numerical values in some way and then apply time or transform domain signal processing algorithms. Nevertheless, this approach raises questions regarding the relative magnitudes under numeric transforms, which can potentially lead to spurious patterns or misinterpretation of results.

Results: In this paper, using the notion of interpretive signal processing (ISP) and by redefining correlation functions for non-numeric sequences, a general class of TF transforms are extended and applied to non-numerical genomic sequences. The technique has been successfully evaluated on synthetic and real DNA sequences.

Conclusion: The proposed framework is fairly generic and is believed to be useful for extracting quantitative and visual information regarding local and global periodicity, symmetry, (non-) stationarity and spectral color of genomic sequences. The notion of interpretive time-frequency analysis introduced in this work can be considered as the first step towards the development of a rigorous mathematical construct for genomic signal processing.

Keywords: Genomic signal processing, Time-frequency analysis, Interpretive signal processing

Background

The application of signal processing techniques in genomics has found a great deal of attention and applications in the past decade [1–4]. Nevertheless, an important class of analytical tools in signal processing that have not been yet fully formulated in genomics is the class of joint time-frequency (TF) distributions and transforms. These are powerful mathematical tools with various applications in signal processing [5, 6].

The major advantage of TF transforms and distributions over conventional Fourier analysis is to simultaneously retrieve the temporal (or spatial) and frequency domain structure of non-stationary data. In other words, while the temporal evolution of the frequency contents of a signal is lost in the conventional spectral estimation using the Fourier analysis, the TF gives a detailed view of such information for non-stationary signals. At the same time, several studies have statistically confirmed the non-stationarity of genomic sequences and suggested the use of non-stationary analysis for these sequences [7–9]. There could be many potential applications by applying the TF transforms to genomic sequences. Nevertheless, the first step in this line of work is to be able to apply these transforms to non-numerical genomic sequences.

The conventional approach to analyze genomic sequences using techniques specific to numerical data is to convert non-numerical genomic data into numerical values in some way and then apply time or transform domain signal processing algorithms to the resulting numeric series [10, 11]. Despite the promising results achieved by these methods, the procedure of converting

*Correspondence: amin.zollanvari@nu.edu.kz
[2]Department of Electrical and Electronic Engineering, Nazarbayev University, Astana, Kazakhstan
Full list of author information is available at the end of the article

genomic data into numerical data has been the bottleneck for these techniques—there is no concrete one-to-one map between non-numeric data and the numeric domain. Moreover, the process of converting non-numeric data to numeric values, can have misleading outcomes. For instance, the genomic alphabet (A, G, C and T) *is not an ordered set*. However, in mapping this alphabet to real values, the sequence is implicitly mapped to an ordered set, which raises expectations regarding their relative magnitudes under numeric transforms, resulting in misinterpretation of the processing results.

In [12], we introduced the notion of *interpretive signal processing* (ISP) as a novel approach for extending signal processing algorithms to non-numerical data. ISP can be seen as a subset of the general notion of *sequential pattern mining*, which has become a prominent topic in sequence data mining during recent years. In ISP, instead of coding non-numerical strings into numerical ones, the basic idea is to use the interpretations of conventional signal processing algorithms to reconstruct similar techniques that are directly applicable to non-numerical data. The notion of ISP is fairly general and may be used in various applications.

In this study, we employ ISP for applying a general class of TF transforms to analyze genomic sequences. As in real valued signals, the advantage of TF analysis over basic Fourier analysis is that it provides a means of analyzing both local and global patterns within a non-stationary sequence. Therefore, using TF transforms, local (yet significant) events which are commonly dominated by averaging in classical Fourier analysis can be identified within a sequence. As a proof-of-principle, we use the ISP-inspired TF analysis for detecting periodicities in coding regions and also detecting repetitive sub-sequences, also known as *tandem repeats*. The length and period of such sequences have important biological implications and several methods have been presented in the past for detecting these sub-sequences [11, 13, 14].

The rest of the paper is organized as follows. We first review the basics of a general class of time-frequency transforms. Then the general idea of ISP is illustrated with a simple example. The extension of time-frequency transforms for non-numeric genomic data is presented next, followed by some concluding remarks and future perspective of the work.

Methods
Joint time-frequency analysis
We first review some general concepts of bilinear TF analysis that are later extended to non-numerical data.

Correlation is a primary concept of great value in most signal processing algorithms. The *instantaneous cross-correlation* of the signals $x(t)$ and $y(t)$ is defined

$$r_{xy}(t, \tau) = x\left(t - \frac{\tau}{2}\right) y^*\left(t + \frac{\tau}{2}\right) \qquad (1)$$

which for real-valued signals is a simple measure of similarity of $x(t - \tau/2)$ and $y(t + \tau/2)$. In fact, in (1), the multiplication operator is used to measure the similarity of its operands. We will later show how this product can be replaced by the values of the similarity matrix of the genomic "alphabets".

By summing $r_{xy}(t, \tau)$ (or integrating for continuous-time signals), the *cross-correlation function* is achieved.

$$R_{xy}(\tau) = \sum_{t=-\infty}^{\infty} r_{xy}(t, \tau) \qquad (2)$$

This function is a measure of the *average similarity* of the two signals with τ-samples of time lag. Alternatively, by summing $r_{xy}(t, \tau)$ over τ, a measure of *signal symmetry* is achieved.

$$s_{xy}(t) = \sum_{\tau=-\infty}^{\infty} r_{xy}(t, \tau) \qquad (3)$$

Taking the Fourier transform of $R_{xy}(\tau)$ with respect to τ, results in the *cross-spectrum*.

$$S_{xy}(f) = \mathscr{F}_{\tau \to f}\{R_{xy}(\tau)\} \qquad (4)$$

where $\mathscr{F}(\cdot)$ represents the Fourier transform.

The *cross-Wigner-Ville* is the Fourier transforms of $r_{xy}(t, \tau)$ with respect to τ.

$$WV_{xy}(t, f) = \mathscr{F}_{\tau \to f}\{r_{xy}(t, \tau)\} \qquad (5)$$

The Wigner-Ville (WV) transform can be interpreted as the time-variant extension of (4), which is specifically useful for the spectral study of non-stationary signals.

The last TF transform that we introduce is the *ambiguity function* (AF), defined as follows

$$AF_{xy}(\eta, \tau) = \mathscr{F}_{t \to \eta}\{r_{xy}(t, \tau)\} \qquad (6)$$

The ambiguity function is basically a *time-frequency correlation* with a maximum at the origin [5]. It has been shown that the ambiguity function can be used to discriminate signals with different spectral color and temporal correlation [5]. The relationship between the instantaneous correlation and the other bilinear transforms is summarized in Fig. 1.

Due to the *bilinear form* of the WV transform (containing the product of $x(\cdot)$ and $y(\cdot)$), undesired cross-terms appear in the time-frequency plane. The cross-terms can be attenuated by filtering the WV transform using a TF *kernel* $\phi(\cdot, \cdot)$, which results in the following general form

$$\rho_{xy}(t, f) = WV_{xy}(t, f) * *\phi(t, f) \qquad (7)$$

where $**$ is the two-dimensional convolution operator. Equation (5) is the most general form of a general class of TF transforms (or TF *distributions*) known as the *Cohen*

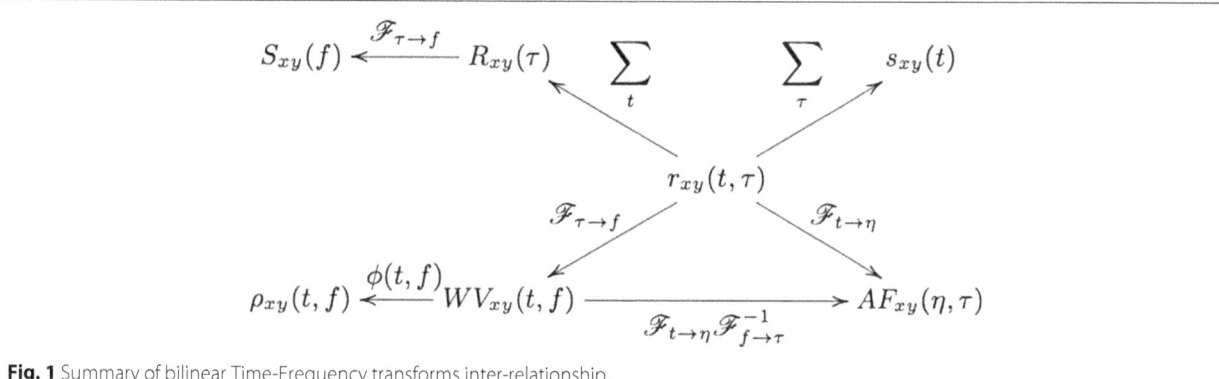

Fig. 1 Summary of bilinear Time-Frequency transforms inter-relationship

class [15]. The properties of these distributions are controlled by $\phi(t,f)$. According to Fig. 1, $\phi(t,f)$ may also be applied to the TF transform in the Ambiguity plane, where it takes a multiplicative form rather than two-dimensional convolution.

Equations (1)–(7) can be calculated for a single signal by setting $y(t) = x(t)$, which gives the similarity of a signal with its time-lagged variants in the time and frequency domain.

To illustrate the application of the introduced TF transforms, we consider a sample segment of the signal $x(t) = y(t)$ consisting of a chirp and two Gaussian signals [16], shown in Fig. 2. The results of (1)–(6) calculated for this signal are shown in Fig. 2. To understand the significance of TF analysis, compare the spectrum of the signal in Fig. 2g with the WV distribution of this signal in Fig. 2f. As we can see, the evolution in time of the frequency content of the signal is totally lost in Fig. 2g. However, one can trace variations in the spectral content as a function of time (sample) in Fig. 2f. For example, the chirp is represented by a relatively wide-band signal, whose (normalized) frequency content is decreasing from 0.4 to 0.1 around time points 20 to 200.

Interpretive signal processing

All practical signal processing algorithms have some intuitive interpretation besides their mathematical formulation. Let us illustrate the idea with a simple example: it is well known that the *inner product* of discrete-time real signals $x(t)$ and $y(t)$ is mathematically defined as follows

$$\langle x(t), y(t) \rangle = \sum_{t=-\infty}^{\infty} x(t)y(t) \tag{8}$$

In (8), whenever $x(t)$ and $y(t)$ have the same sign (both positive or negative), a positive value is added to the summation; while when they differ in sign a negative value is added. Therefore, for zero-mean signals (which have both positive and negative values), if the inner product

is close to zero, one can conclude that the two signals do not have a similar pattern, while a great absolute value of the inner product is an indication of average "co-variance" or similarity of the two signals. In fact, the multiplication operator in (8), provides a measure of point-wise similarity, while the summation gives the average behavior of this similarity. This basic interpretation has led to the definition of a hand full of other measures of signal co-variance. For instance, one may subtract the mean values of $x(t)$ and $y(t)$ to centralize the data (when the mean values do not convey information), or to normalize it by the square roots of the energies of $x(t)$ and $y(t)$, in order to make the inner product dimensionless and to normalize it between -1 and +1. For certain applications, researchers have replaced the point-wise product of $x(t)y(t)$ by other measures of similarity, like $\text{sign}(x(t)y(t))$ (cf. [17]). We can see that while the "inner product as a measure of signal similarity" is a common property of various forms of these definitions, employing the appropriate form hinges on the application.

Herein, we refer to the procedure of reforming signal processing algorithms based on their interpretation, as *interpretive signal processing* (ISP). In [12], we used the notion of ISP to apply *matched filters* in genomic signal processing. We show how this procedure helps us reformulate the Cohen class of time-frequency transforms for genomic sequence data.

Extending the time-frequency analysis to genomic sequences

As shown in Fig. 1, the core of all bilinear transforms is the instantaneous cross- or auto-correlation. In order to extend TF transforms to non-numeric genomic sequence data, *we propose to replace the product of $x(\cdot)$ and $y(\cdot)$ in $r_{xy}(t, \tau)$ with the similarity matrix entries of genomic sequences*. This idea is based on the interpretation of the product as a measure of similarity. Since a DNA sequence consists of four nucleotides, Adenine, Cytosine, Guanine, and Thymine, denoted by A, C, G, and T, respectively,

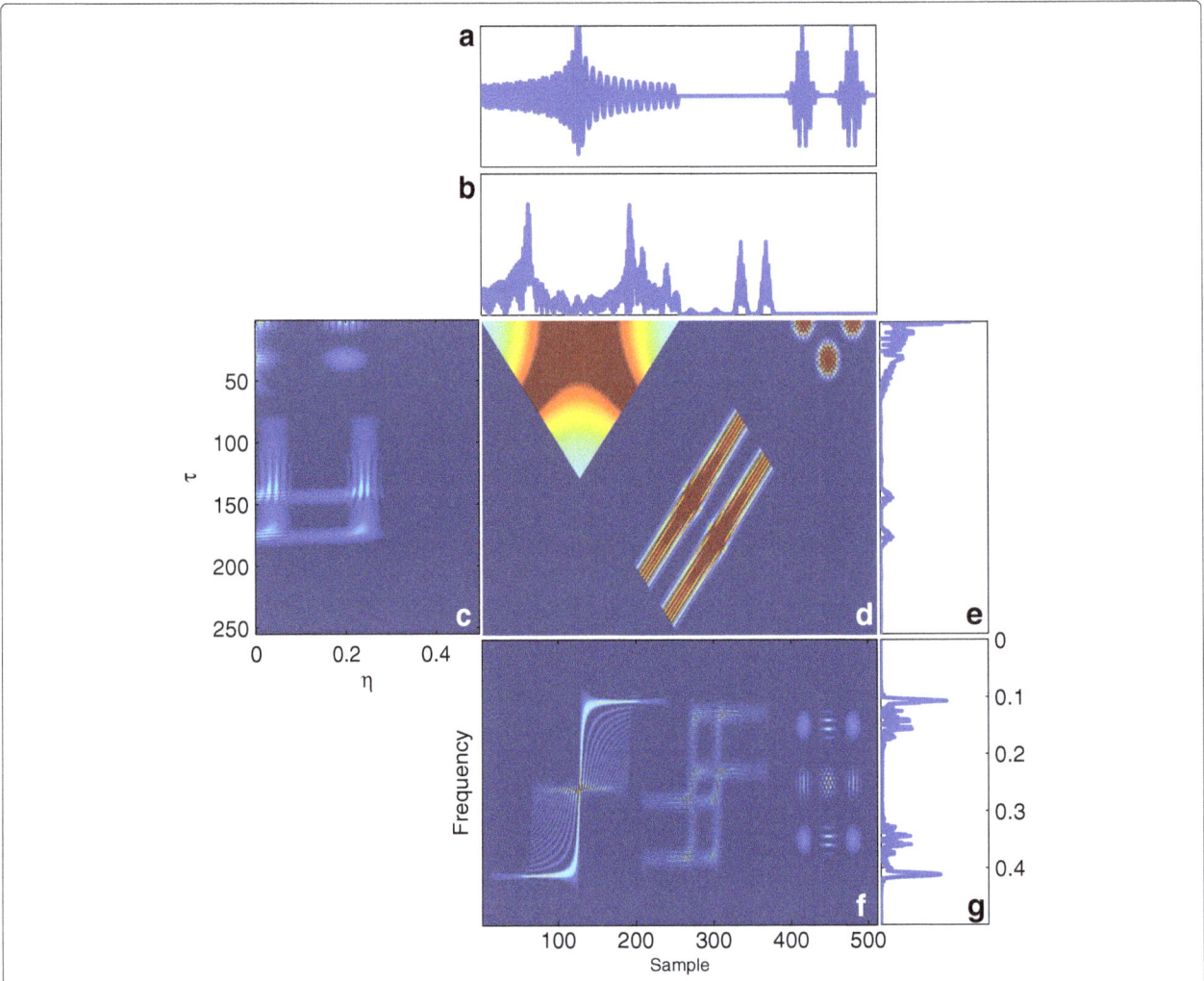

Fig. 2 a A sample signal $x(t)$, **b** $S_{xx}(t)$ showing the local symmetry of $x(t)$, **c** the ambiguity plane, **d** the instantaneous auto-correlation, illustrating the similarity of $x(t)$ with its time lag **e** the auto-correlation function with its maximum peak at $\tau = 0$ **f** the WV transform showing the spectral properties of $x(t)$ versus spatial samples **g** the spectrum of $x(t)$

a possible choice of the similarity matrix is the identity matrix represented as

$$
S = \begin{array}{c} \\ A \\ C \\ G \\ T \end{array} \begin{pmatrix} A & C & G & T \\ 1 & 0 & 0 & 0 \\ 0 & 1 & 0 & 0 \\ 0 & 0 & 1 & 0 \\ 0 & 0 & 0 & 1 \end{pmatrix} \tag{9}
$$

which indicates that each of the DNA nucleotides only resembles itself. In practice, based on experimental statistics, a bioinformatician may choose non-zero values for the off-diagonal entries or values smaller than one for the diagonal entries, indicating the probability of base-pair mutation at a specific locus. Moreover, in order to analyze a specific nucleotide and neglect others, all irrelevant entries of the matrix can be set to zero, which

results in selective frequency or selective pattern analysis for DNA sequences. For proteins sequences, one may use BLOSUM62 and PAM250 matrices [18].

Since all bilinear transforms contain the product of two terms, the proposed approach is an indirect means of mapping non-numeric sequences to numeric values, which is guaranteed to serve as a similarity measure (by definition), and does not suffer from the *ordering* issue in previous mapping techniques (as noted in the introduction).

Results

Case studies

Before applying the method to real DNA and protein sequences, let us consider a synthetic sequence for illustration.

A synthetic DNA sequence

For illustration, consider the following synthetic periodic DNA sequence, with period 4 (ACGT) and length 1000.

$$x_0[n] = \cdots \text{ACGTACGTACGTACGT} \cdots \qquad (10)$$

Real DNA sequences are never fully periodic. In order to make the sequence more realistic, we add some random changes (noise) to the sequence, by random substitutions of some nucleotides.

$$x[n] = \cdots \text{AGGTTCGTACGAACCT} \cdots \qquad (11)$$

Using the trivial similarity matrix in (9), Eqs. (2)–(6) can be calculated for this nucleotide sequence using (1). The results are summarized in Fig. 3.

Part (e) in Fig. 3 shows the WV plane for the noisy synthetic sequence in (11). We can clearly see that the pseudo-periodicity of the sequence has led into a horizontal line at 0.25 (normalized frequency) in this figure, which

is equivalent to a periodicity of $1/0.25 = 4$ samples. Also, this pseudo-periodicity causes a peak in part (f), which shows the global spectral properties of the signal. It has been shown that for stationary and temporally correlated signals (i.e., a colored spectrum), most of the ambiguity function's energy is spread in the τ direction around $\eta = 0$ [5]. This explains the ambiguity function form of our synthetic periodic sequence, which is stationary over time. More examples will be shown for real sequences in the next section. The effect of the correlation lag τ is seen in Fig. 3c, where we can see that due to the periodicity of our synthetic sequence, correlations exist between near (small τ) and far samples (large τ).

A real DNA sequence case study

The proposed framework has been tested on several DNA sequences. As a first case study, we apply the method to a real DNA sequence adopted from the National Center

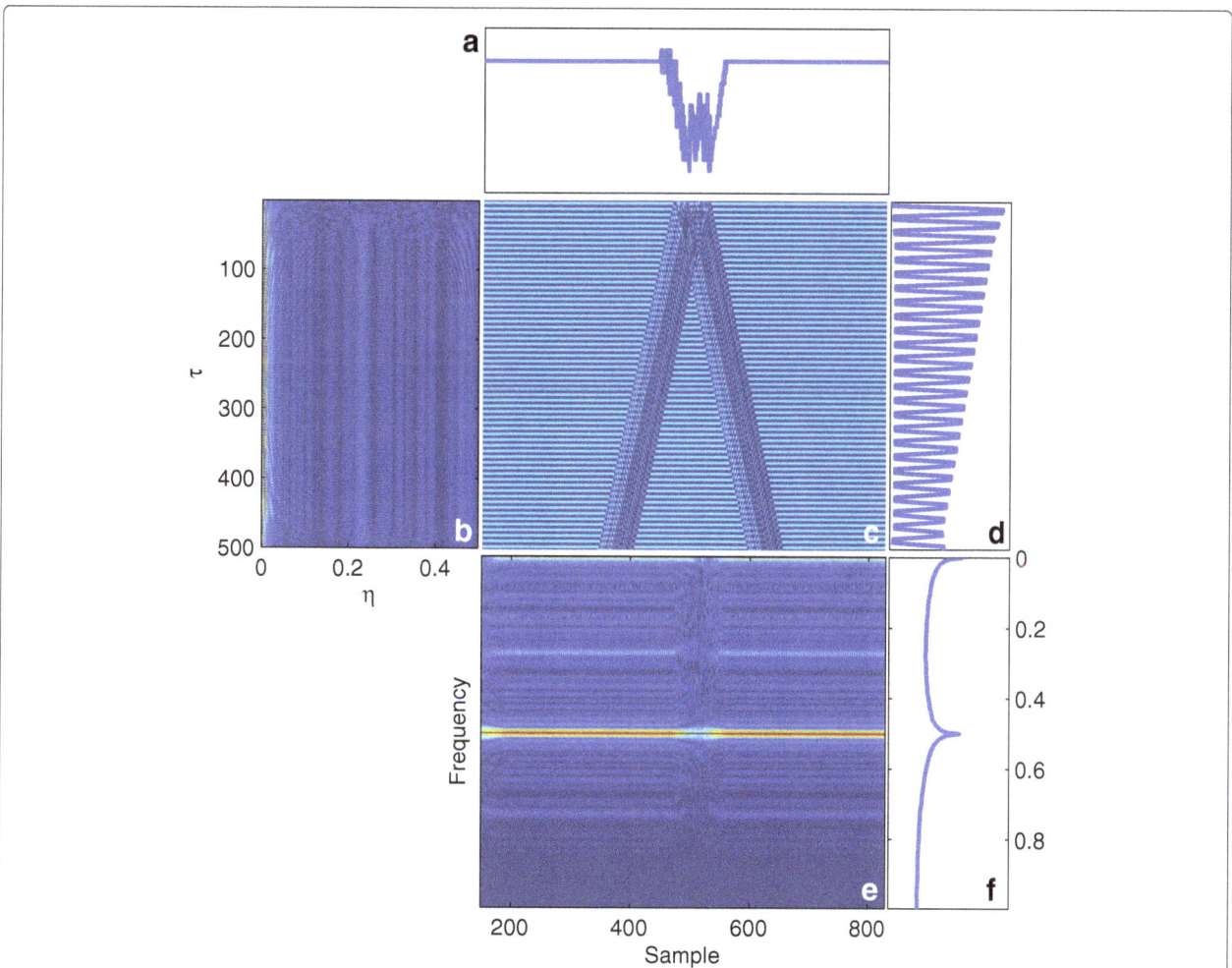

Fig. 3 Results of (1)–(6) for the synthetic DNA sequence in (11). **a** The symmetry function using (3). **b** The ambiguity function using (6), showing the stationarity and spectral color of the sequence. **c** The instantaneous auto-correlation using (1), representing the similarity of the sequence with its time lag nucleotides. **d** The auto-correlation of the sequence using (2), with a maximum peak at $\tau = 0$. **e** The WV transform using (5), indicating the time-frequency properties of the sequence. **f** The spectrum using (4), which shows the global spectral properties for the sequence

for Biotechnology Information (NCBI) with the accession number FJ807392.1 [19]. Figure 4 illustrates Eqs. (1)–(6) for this real DNA sequence as well as a randomly generated DNA sequence.

For comparison, the results in Fig. 4A can be compared with similar results obtained from a totally random synthetic DNA sequence of the same length in Fig. 4B. It is seen that there is no specific structure in the time-frequency transforms of the random sequence, while there are clear structures indicating local periodicities, nonstationarities and spectral color in the real DNA sequence.

According to NCBI, FJ807392.1 is a *Helice tientsinensis microsatellite TJH03 sequence*, which is a repetitive sequence with 282 base nucleotide pairs. Figure 5 is a zoom-in of the WV plane for this sequence, from which its repetitive sub-sequences can be well seen and detected at 0.25 and 0.02 normalized frequencies, corresponding to both short term and long term periodicities in the sequence.

Moreover, due to the repetitive structure of this sequence, it can be considered a stationary and colored sequence, which explains the ambiguity plane structure in Fig. 4A part b, which is concentrated around $\eta = 0$

and spread in the direction of τ. Therefore, the proposed ambiguity plane can be used to study the spectral and stationarity properties of DNA sequences. This is especially useful for feature extraction and classification of DNA and protein sequences.

Identification of protein coding DNA regions
As a second case study, the proposed method is compared with a well known method called *indicator sequences*, for analyzing DNA sequences [1, 20]. Accordingly, indicator sequences of a DNA sequence are four binary sequences corresponding to the four different nucleotides. Each sample in the indicator sequence specifies the presence of the nucleotide at that position. The following is an example of a DNA sequence and its indicator sequences:

DNA sequence: A G C C T G A
Indicator sequence $u_A[n]$: 1 0 0 0 0 0 1
Indicator sequence $u_C[n]$: 0 0 1 1 0 0 0   (12)
Indicator sequence $u_G[n]$: 0 1 0 0 0 1 0
Indicator sequence $u_T[n]$: 0 0 0 0 1 0 0

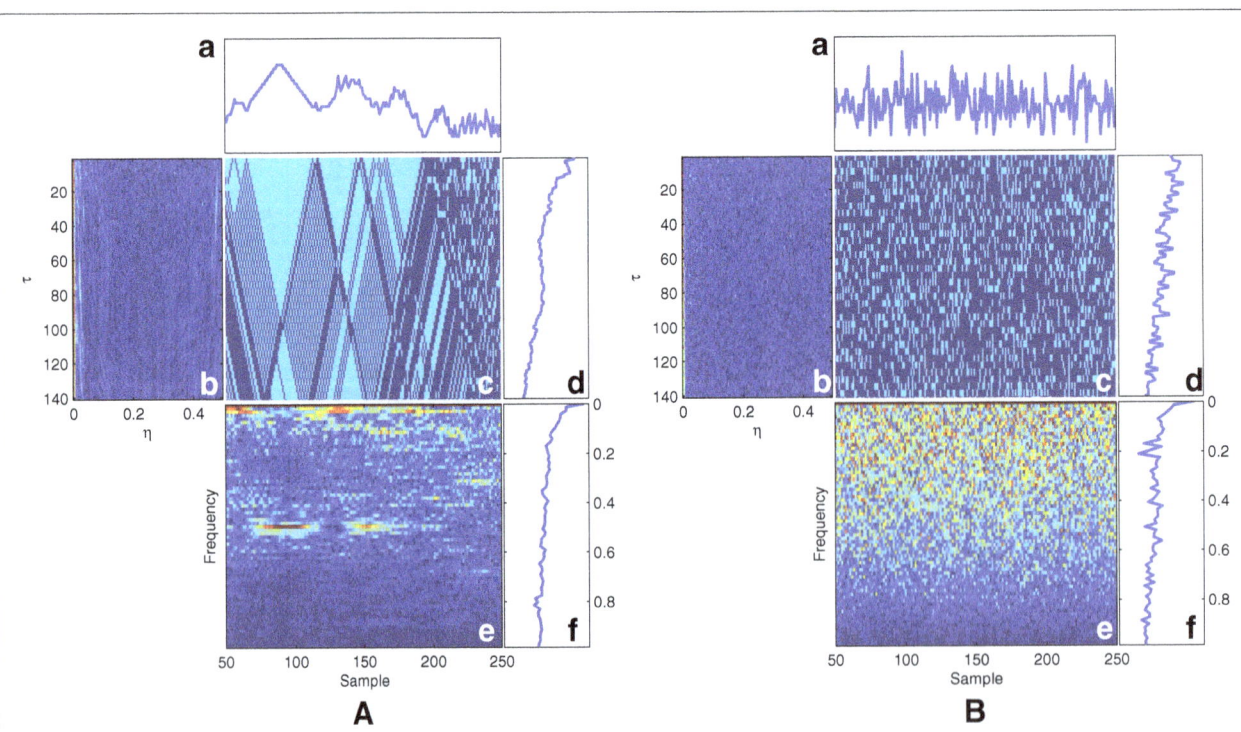

Fig. 4 Results of (1)–(6) for **A** the DNA sequence with the accession number FJ807392.1 and **B** a totally random synthetic DNA sequence. The identity matrix is used as the similarity matrix for computing these functions. In each figure: **a** The symmetry function from (3). **b** The ambiguity function from (6) showing the stationarity and spectral color of the sequence. **c** The instantaneous auto-correlation using (1) representing the similarity of the sequence with its time lag nucleotides. **d** The auto-correlation function of the sequence using (2), with a maximum peak at $\tau = 0$. **e** The WV transform using (5), indicating the time-frequency properties of the sequence. **f** The spectrum using (4), which shows the global spectral properties for the sequence

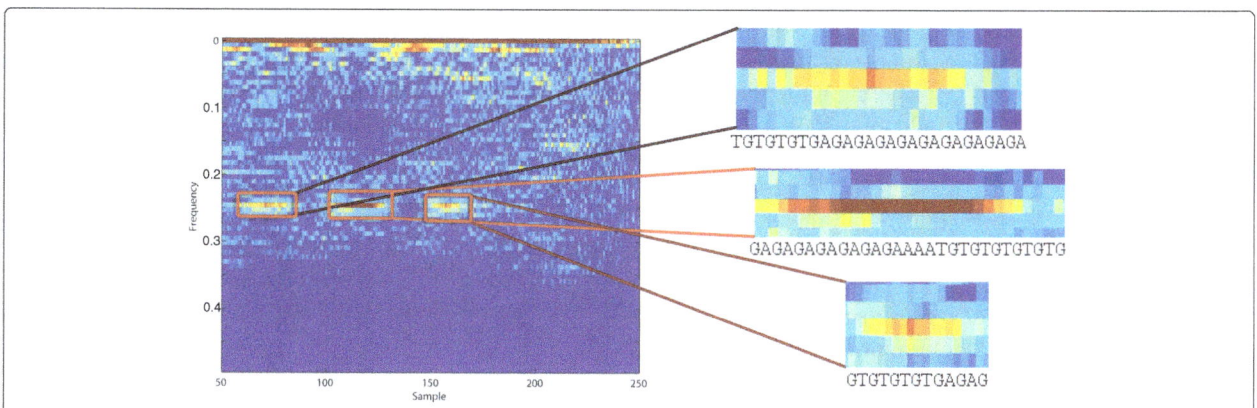

Fig. 5 The WV transform for the DNA sequence in Fig. 4A and its corresponding DNA sub-sequence. Note the local repetitive structure in the DNA sequence and its corresponding pattern in the WV plane

According to [20], the indicator sequence can be used to define the DNA spectrum, as follows:

$$S[k] = |U_A[k]|^2 + |U_C[k]|^2 + |U_G[k]|^2 + |U_T[k]|^2 \tag{13}$$

where $U_i[k]$ is the *discrete Fourier transform* (DFT) of $u_i[n]$ ($i = \{A, C, G, T\}$). In [20], it has been empirically shown that for a coding region within a DNA sequence (a region that can be converted to a protein), Eq. (13) has a clear peak at $k = N/3$ where N is the DNA sequence length. While this observation has been referred to in various studies, to the authors' knowledge, no mathematical explanation has yet been presented for it. However, using our proposed machinery, one observes a similar periodicity by using the identity similarity matrix (9) and by calculating the spectral function (4). The results of this comparison are shown in Fig. 6.

To illustrate this, we take a DNA sequence with the NCBI accession number NM_001244612.1. This DNA sequence with 4165 base pairs is known to be a coding region for human proteins. Figure 6 shows the

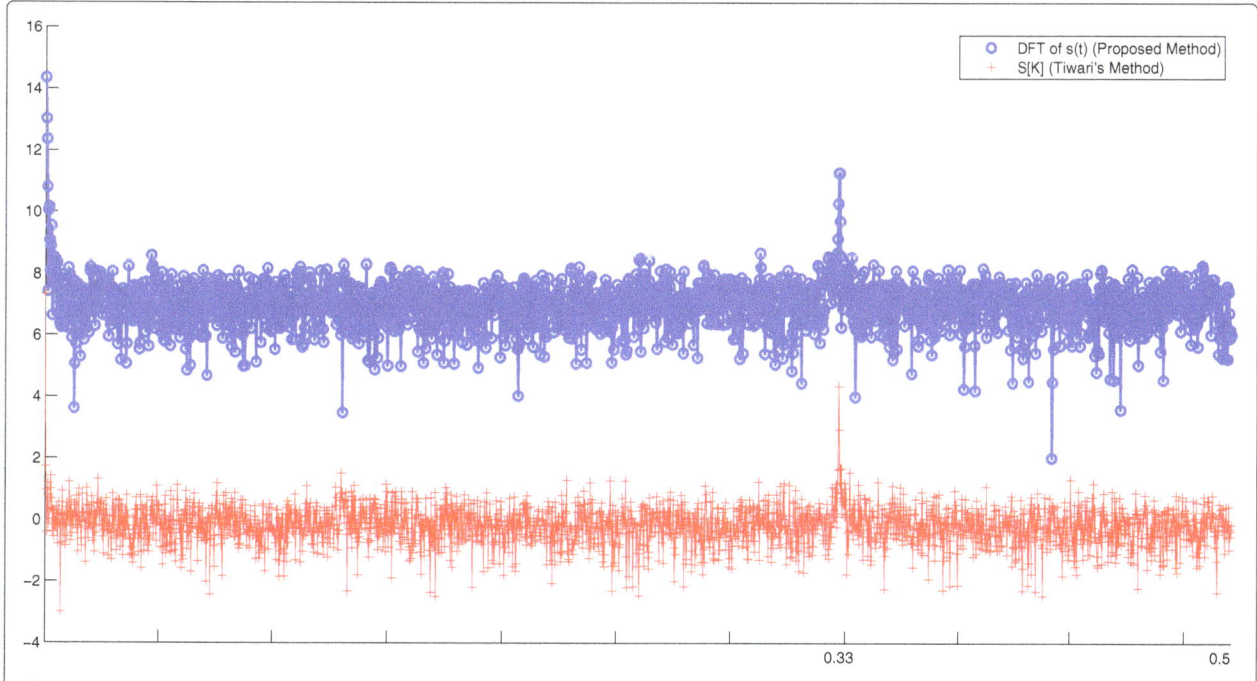

Fig. 6 A comparison between Tiwari's method [20] and the DFT of (3) for a DNA sequence with the accession number NM_001244612.1. Both signals have a peak at $k = N/3$ (for better illustration one of the signals has been shifted up)

DNA spectrum calculated from indicator sequences. The second signal in Fig. 6 is the Fourier transform of the proposed symmetry function (3). The peak for the Fourier transform of (3) reports a periodicity at $N/3$.

Discussion

In our experiments, we show that by defining the instantaneous auto-correlation of DNA sequences using similarity matrices, local and global periodicities in the sequences can be detected by the WV transform in (5). Symmetric sub-sequences can be determined by the symmetry function (3) and the stationarity and spectral properties of sequences can be recognized by the ambiguity function (6). Also, the global spectrum of the sequences can be calculated by the spectrum of the sequences (4) and the global correlation of the sequences can be found by the auto-correlation function (2).

The major advantage of ISP *per se* is to process the non-numerical symbols directly (instead of converting the symbols into numerical values). This property simplifies the interpretation of the output of signal processing algorithms when applied to non-numerical symbols. However, ISP is not always trivial, since the interpretation of mathematical equations is not always straightforward. Moreover, the interpretation of signal processing algorithms is not necessarily unique and in some cases unfeasible. Therefore, in practice ISP can result in algorithms that are only partially applicable to non-numerical data while the remaining parts are left unchanged—as in the TF transforms presented in this work, in which only the instantaneous auto-correlation function was replaced with the similarity matrix of genomic sequences.

Conclusion

In this study, using the notion of interpretive signal processing (ISP), the conventional time-frequency transforms have been extended to analyze non-numerical genomic sequences. Applications of the proposed machinery in determining genome periodicity and detecting tandem repeats were presented using synthetic and real DNA sequences. The results show that the proposed ISP-inspired TF transforms (to which we refer as the interpretive TF analysis) can be useful to analyze genomic sequences.

Other aspects of the proposed interpretive TF analysis that require further work are: 1) investigating other biologically-inspired applications of the proposed machinery; 2) studying different choices of similarity matrices in various applications such as DNA or protein sequence alignment; 3) integrating the proposed machinery in existing sequence analysis toolboxes for extracting further quantitative and visual information from genomic sequences; 4) using the TF representations as features (as an image) and using image classification and clustering techniques for classifying unknown genomic sequences; and 5) extending the hereby proposed notion to higher order spectra (HOS) and higher order time-frequency analysis. These contributions can be considered as a step towards the development of a rigorous mathematical construct for genomic sequence signal processing.

Abbreviations
AF: Ambiguity function; DFT: Discrete Fourier transform; HOS: Higher order spectra; ISP: Interpretive signal processing; TF: Time-frequency; WV transform: Wigner-Ville transform

Funding
Publication of this article was funded by the Nazarbayev University Social Policy Grant (A. Zollanvari).

Authors' contributions
H. H. S. designed and implemented the experiments, and drafted the manuscript. R. S. conceived the study, helped in implementing the experiments, and drafted the manuscript. A. Z. provided the bioinformatics background and helped draft the manuscript; all authors approved the final version of the manuscript.

Competing interests
The authors declare that they have no competing interests.

About this supplement
This article has been published as part of *BMC Bioinformatics* Volume 18 Supplement 4, 2017: Selected original research articles from the Third International Workshop on Computational Network Biology: Modeling, Analysis, and Control (CNB-MAC 2016): bioinformatics. The full contents of the supplement are available online at https://bmcbioinformatics.biomedcentral.com/articles/supplements/volume-18-supplement-4.

Author details
[1] School of Electrical and Computer Engineering, Shiraz University, Shiraz, Iran. [2] Department of Electrical and Electronic Engineering, Nazarbayev University, Astana, Kazakhstan.

References
1. Anastassiou D. Genomic signal processing. Sig Process Mag IEEE. 2001;18(4):8–20.
2. Kakumani R, Ahmad MO, Devabhaktuni V. Comparative genomic analysis using statistically optimal null filters. In: ISCAS. Paris: IEEE; 2010. p. 2235–8.
3. Afreixo V, Ferreira PJSG, Santos D. Fourier analysis of symbolic data: a brief review. Digital Signal Process. 2004;14(6):523–30.
4. Vaidyanathan PP, jun Yoon B. The role of signal-processing concepts in genomics and proteomics. J Frankl Inst. 2004;341:111–35.
5. Flandrin P. Time-frequency/time-scale analysis. San Diego: Academic Press; 1999.
6. Qian S, Chen D. Joint time-frequency analysis: method and application. Upper Saddle River: Prentice Hall; 1996.
7. Bouaynaya N, Schonfeld D. Nonstationary analysis of coding and noncoding regions in nucleotide sequences. IEEE J Sel Top Sig Process. 2008;2(3):357–64.
8. Bouaynaya N, Schonfeld D. Emergence of new structure from nonstationary analysis of genomic sequences. In: IEEE International Workshop on Genomic Signal Processing and Statistics. Phoenix, AZ; 2008. p. 1–4.
9. Zielinski J, Schonfeld NBD, O'Neill W. Time-dependent ARMA modeling of genomic sequences. BMC Bioinforma. 2008;9(Suppl 9):S14.
10. Anastassiou D. Frequency-domain analysis of biomolecular sequences. Bioinformatics. 2000;16(12):1073–81.
11. Sussillo D, Kundaje A, Anastassiou D. Spectrogram analysis of genomes. EURASIP J Appl Sig Process. 2004;2004:29–42.
12. Hassani Saadi H, Sameni R. Using matched filters for similarity search in

genomic data. In: Proceedings of the 16th CSI International Symposium on Artificial Intelligence and Signal Processing (AISP). Shiraz: Iran; 2012. p. 469–72.

13. Buchner M, Janjarasjitt S. Detection and visualization of tandem repeats in DNA sequences. Sig Process IEEE Trans. 2003;51(9):2280–7.

14. Pop PG, Lupu E. DNA repeats detection using BW spectrograms. In: Automation, Quality and Testing, Robotics, 2008. AQTR 2008. IEEE International Conference on. vol. 3. Cluj-Napoca, Romania; 2008. p. 408–12.

15. Cohen L. Time-frequency analysis. Englewood Cliffs: Prentice Hall PTR; 1995.

16. Auger F, Flandrin P, Goncalves P, Lemoine O. Time-frequency toolbox. 1996. Available from: http://tftb.nongnu.org/. Accessed 20 Jan 2017.

17. Theis F, Müller N, Plant C, Böhm C. Robust Second-Order Source Separation Identifies Experimental Responses in Biomedical Imaging In: Vigneron V, Zarzoso V, Moreau E, Gribonval R, Vincent E, editors. Latent Variable Analysis and Signal Separation. vol. 6365 of Lecture Notes in Computer Science. Berlin/Heidelberg: Springer; 2010. p. 466–73.

18. Pevsner J. Bioinformatics and functional genomics. New York: Wiley-Blackwell; 2009.

19. Zhang D, Ding G, Zhang H, Tang B. Isolation characterization of 10 microsatellite markers in Helice tientsinensis (Brachyura: Varunidae). Conserv Genet Resour. 2009;1(1):321–3.

20. Tiwari S, Ramachandran S, Bhattacharya A, Bhattacharya S, Ramaswamy R. Prediction of probable genes by Fourier analysis of genomic sequences. Comput Appl Biosci CABIOS. 1997;13(3):263–70.

2

Automated multigroup outlier identification in molecular high-throughput data using bagplots and gemplots

Jochen Kruppa and Klaus Jung[*] (iD)

Abstract

Background: Analyses of molecular high-throughput data often lack in robustness, i.e. results are very sensitive to the addition or removal of a single observation. Therefore, the identification of extreme observations is an important step of quality control before doing further data analysis. Standard outlier detection methods for univariate data are however not applicable, since the considered data are high-dimensional, i.e. multiple hundreds or thousands of features are observed in small samples. Usually, outliers in high-dimensional data are solely detected by visual inspection of a graphical representation of the data by the analyst. Typical graphical representation for high-dimensional data are hierarchical cluster tree or principal component plots. Pure visual approaches depend, however, on the individual judgement of the analyst and are hard to automate. Existing methods for automated outlier detection are only dedicated to data of a single experimental groups.

Results: In this work we propose to use bagplots, the 2-dimensional extension of the boxplot, to automatically identify outliers in the subspace of the first two principal components of the data. Furthermore, we present for the first time the gemplot, the 3-dimensional extension of boxplot and bagplot, which can be used in the subspace of the first three principal components. Bagplot and gemplot surround the regular observations with convex hulls and observations outside these hulls are regarded as outliers. The convex hulls are determined separately for the observations of each experimental group while the observations of all groups can be displayed in the same subspace of principal components. We demonstrate the usefulness of this approach on multiple sets of artificial data as well as one set of gene expression data from a next-generation sequencing experiment, and compare the new method to other common approaches. Furthermore, we provide an implementation of the gemplot in the package 'gemPlot' for the R programming environment.

Conclusions: Bagplots and gemplots in subspaces of principal components are useful for automated and objective outlier identification in high-dimensional data from molecular high-throughput experiments. A clear advantage over other methods is that multiple experimental groups can be displayed in the same figure although outlier detection is performed for each individual group.

Keywords: Bagplot, Gemplot, High-dimensional data, Outlier, Principal component analysis

---

*Correspondence: klaus.jung@tiho-hannover.de
Institute for Animal Breeding and Genetics, University of Veterinary Medicine
Hannover, Foundation, Bünteweg 17p, D-30559 Hannover, Germany

Background

Modern molecular biology produces high-dimensional data en masse where the number of features is much larger than the sample size of the study or the experiment. Some typical examples are (metric) gene and protein expression data observed with DNA microarrays [1, 2], next-generation sequencing (NGS) [3] or proteomics techniques such as mass spectrometry [4, 5] or 2-D gel electrophoresis [6]. In the case of microarray and proteomics experiments, expression data are generally continuous fluorescence or intensity values [7, 8], while NGS produces expression data as read counts [9] or also as continuous quantities [10]. Other examples of high-dimensional data in molecular biology are methylation levels that can also be observed with microarrays [11] and NGS [12], or data from binding experiments such as chromatin immunoprecipitation (ChIP) [13] in combination with microarrays or NGS, or affinity purification mass spectrometry (AP/MS) [14].

Typical questions analyzed in high-dimensional data focus on the correlation of expression levels with experimental factors or with patient data. For example gene expression data is usually analyzed to detect differentially expressed genes between two levels of an experimental factor (e.g. treatment versus control) or between two patient groups [15–17]. Moreover, high-dimensional expression data is often used to train classifier and regression models to predict therapy outcome [18] or survival [19, 20].

Throughout this work, the term 'observation' is used to specify the measured data of one experimental unit (e.g. one patient), and the term 'sample' means a set of experimental units. Statistical and bioinformatics analyses of the above questions are usually very sensitive to a single observation, i.e. its addition or removal can seriously affect the results. For example when searching for differentially expressed genes or training a classifier model, individual observations can have a strong impact on the ranking of genes or the estimates of the classifier's performance. Therefore, quality control of the raw data often involves their inspection with respect to extreme observations. An extreme observation in an experimental group could either be a) regular but just extreme observation, b) a mislabeled observation or c) the consequence of an incorrect measurement. In the first case robust methods can help to decrease or downweigh the outlier's impact [21–23]. If the last case becomes obvious, the observation can be removed from data analysis. In the case of a mislabeled observation this can be corrected. In either case, the identification of outliers can help to continue with a correct analysis.

The most widely used tools for outlier detection in molecular high-throughput data are hierarchical clustering [24] and principal component analysis [25]. In hierarchical clustering all observations are plotted in a tree, where similar observations appear at branches near to each other and unsimilar observations appear at branches farther away from each other. Principal component analysis (PCA) performs first a dimension reduction so that the high-dimensional data can be represented in a two- or three-dimensional plot while a certain proportion of variance of the original data is maintained. In a PCA plot similar observations group together and unsimilar observations appear again farther away from each other. A useful tool for interactive 3D-visualization of principal components is given by the R-package 'GGobi' [26]. Using hierarchical cluster plots and PCA plots, outliers can be identified by visual inspection which is, however, a subjective decision of the researcher. When using hierarchical clustering, the 'single linkage' approach usually performs best to identify outliers [27].

Besides visual methods, some automated approaches have been proposed, several of them were reviewed by Egan and Morgan [28] and Zimek et al. [29]. Methods for automated outlier detection based on robust PCA were for example proposed by Model et al. [30], Hubert et al. [31] as well as Filzmoser and Todorov [32]. These methods focus, however, only on the data of one experimental group. Because in these approaches, outlier detection and graphical representation are linked, multiple experimental groups can not be displayed in the same plot.

In order to detect outliers individually in multiple experimental groups but to display the results in the same plot we propose an approach that is also based on PCA. In particular, PCA is first performed for the whole data set and bagplots are then used individually for each experimental group to identify outliers in the space of the first two principal components. In addition, we newly present the 'gemplot' - the three-dimensional version of the bagplot - and propose to use this plot for automatically identification of outliers in the space of the first three principal components.

In the methods section, the basic principal of PCA as well as the concepts of bagplot and gemplot are detailed, and the example data is described. In the results section we demonstrate the usefulness of our approach on multiple data sets of artificially constructed principal components and one set of public gene expression data from a kidney cancer study. We close this work with some conclusions.

Methods

In this section, we first describe the basic idea of PCA and detail next the idea of outlier detection by means of boxplots, bagplots and the new gemplot. Finally, we present the example data that we use for illustrating our approach.

Principal component analysis

Genomic data typically follows a high-dimensional setting where the number of molecular features d is much larger than the sample size n. In order to visualize such data, dimension reduction is usually used, for example by means of PCA which has been widely used in genomics data analysis. PCA projects the data into the space of orthogonal principal components. Each component represents a certain proportion of variance of the original data. Thus, in many cases the first two components represent more than 50% of the variance of the original data and are sufficient to display the location of the single observations to each other and to detect extreme observations. Raychaudhuri et al. [33] show microarray data, where the first two components present even more than 90% of the original variance, and Shieh and Hung [34] also point out that a small number of components is usually needed to explain most of the variance in high-dimensional settings. Cangelosi and Goriely [35] show several data sets of gene expression data, where the first three components represent 80% of the original variance. Sharov et al. [36] show gene expression data, where outliers become obvious when plotting the first three components. More precisely, the original $(n \times d)$ data matrix X is transformed into d orthogonal components given by the columns of $Z = X \cdot A$ [37]. It can be shown that the columns of the $(d \times d)$ transformation matrix A are the eigenvectors corresponding to the eigenvalues λ_j $(j = 1, \dots, d)$ of S, where S is the sample covariance matrix of X. Moreover, the eigenvalue λ_j is equal to the variance of the jth principal component. Hence, $V_j = \lambda_j / \sum \lambda_j$ represents the proportion of variance that is explained by the jth component. For the two- and three-dimensional representation of the original data one can than use the first two or three components, respectively, given by the first three columns of Z, i.e. z_1, z_2 and z_3.

Boxplot, bagplot and gemplot

The box-and-whiskers-plot, mostly just called boxplot, is one of the most frequently used graphical tools to display the quantiles of a metric feature. It consists of a box whose lower and upper limits represent the 25- and 75%-quantile of the observed data and a bar within the box represents the median, i.e. the 50%-quantile. Thus the box includes 50% of the data points. In its standard version, the whiskers go from the 75%-quantile to the maximum of the data and from the 25%-quantile to the minimum of the data, respectively. In order to detect outliers the whiskers are often limited by the so called fence, where their maximal length is only allowed to be 1.5 times the interquartile range (the difference between 25- and 75%-quantile). Each observation that is beyond the fence is drawn as a single dot and is regarded as an outlier.

The two-dimensional extension of the boxplot, the bagplot, was first presented in 1999 by Rousseeuw et al. [38]. Instead of a box, a convex hull - called the bag - is contructed so that 50% of the data points in the center of the two-dimensional point cloud are included in this bag. The bag is determined by making use of the concepts of halfspace location depths and depth regions. For each point θ in the two-dimensional space, $ldepth(\theta, (z_1^i, z_2^i))$ is the smallest number of data points (z_1^i, z_2^i), $i = 1, \dots, n$, contained in any closed halfplane with margins through θ. In our case, the coordinates of the data points (z_1^i, z_2^i) are given by the first two principal components. The depth region D_k is then given by the set of all θ with $ldepth \geq k$. If we denote the number of data points in D_k by $\#D_k$, we need to find that k^* for which $\#D_{k^*} \leq \lfloor n/2 \rfloor < \#D_{k^*-1}$. The bag can than be found by interpolating between the sets D_{k^*} by D_{k^*-1}. As center of the data, the depth median T is given by that θ with the highest $ldepth$. If there are multiple such θ, T is given by their center of gravity. Similar to the role of the whiskers in a boxplot a loop is constructed around the bag, where observations outside this loop are regarded as outliers. In order to construct the loop, one first generated the fence by inflating the bag relative to T by a factor of 3. The loop contains all data points within the fence that don't belong to the bag. For the detailed description of the construction of a bagplot we refer to the original work of Rousseeuw et al. [38]. An algorithm for the fast computation of halfspace location depths is for example given in Miller et al. [39].

Rousseeuw et al. already pointed out that the bagplot is, in principle, defined in any dimension. Instead of halfplanes, halfspaces are then used to determine the halfspace location depths. Since it is computationally extreme and also hard to visualize bagplots in higher dimensions, there was not yet an implementation for the three-dimensional case. Here, we present this three-dimensional version for the first time and call it 'gemplot' due to the similarity of its appearance as gemstones. We also provide the software package 'gemPlot' for the R programming environment available at https://github.com/jkruppa/gemPlot. In order to determine the sets D_k in the three-dimensional space, our implementation uses a three-dimensional array that lays a three-dimensional grid across the data points. For each grid point θ, $ldepth(\theta, (z_1^i, z_2^i, z_3^i))$ is calculated. Thus, it is also possible – though time-consuming – to calculate the gemplots in more than three dimensions. A special graphical trick to make the visualization of gemplots possible is the use of transparent colors. Thus, the depth median and the bag ('inner gem') are still visible when drawing the loop ('outer gem'). When the median is not a single point, we found that it is more appropriate not to display the center of gravity but also a further gem which is made up by another convex hull. In our implementation, the gemplot is drawn in an interactive device so that the researcher can rotate the gemplot with the computer mouse and can also zoom in and out. Furthermore, our implementation can also determine outliers in subspaces of more than three dimensions (without visualization), which is however computational very expensive.

Data examples

We demonstrate the practicability of bagplots and gemplots for outlier detection in high-dimensional data on two sets of artificially generated principal components, different data sets of artificial gene expression data, as well as on a data set originating from an RNA-seq experiment in a tumor study.

Artificial data examples

In order to make users familiar with our approach and it's behavior we generated two simple data sets which directly represent the values of three principal components. In both artificial data sets, data for two study groups were generated. In the first example the first three principal components, z_1, z_2 and z_3, were drawn from the same normal distribution within each group. Here, the order to the three components was defined arbitrarily. For study group one the principal components were drawn from $\mathcal{N}(\mu = -3, \sigma = 1)$ and from $\mathcal{N}(\mu = 3, \sigma = 1)$ for the second study group. The samples size for each group was $n = 100$. Since outliers are more frequent in skewed distributions, some principal components in the second example data set were drawn from the exponential distribution. In particular, z_2 in one study group and z_1 and z_3 in the other study group were drawn from $Exp(\lambda = 4)$.

In order to introduce outliers into both examples, we changed the coordinates of arbitrarily chosen observations. In the example with only normally distributed components, we changed the coordinates of observation 3 in the first study group to $(-6, -6, -6)$ and of observation 8

in the second group to $(6, 6, 6)$. In the example with non-normal components, we set the coordinates of observation 73 and 87 in the first group to $(3, 4, 25)$ and $(3, 10, 4)$, respectively. Thus, the outlying character of observation 73 manifests itself in the third component, and that of observation 87 in the second component.

In a further example of artificial data, we wanted to study how outliers are detected by our approach when principal components are derived from high-dimensional expression data. Therefore, we simulated in two different settings expression data from the multivariate normal distribution, $\mathcal{N}_d(\mu, \Sigma)$, for $d = 1000$ genes and $n = 100$ samples. In both settings we used an $(d \times d)$ autoregressive covariance matrix Σ with entries $\sigma_{ij} = \rho^{\tau \cdot |i-j|}$ each representing the covariance between gene i and j ($i, j = 1, \ldots, d$). For $i = j$, we chose $\sigma_{ij} = 1$ for all genes. Setting 1 included data for one experimental group, and with three outliers. The construction of the mean vector μ for the non-outlying observations, and the mean vectors μ_1, μ_2, μ_3 for the outliers is specified in Table 1. The mean vectors for the outliers were constructed so that each outlier represents a different group and appears at one direction of the first three principal components. In setting 2, a second group is included also with three outlying observations. Non-outlying observations and outliers for the second group are constructed by mean vectors κ, κ_1, κ_2, κ_3, also specified in Table 1. In each setting, we varied the parameter τ for the generation of the covariance matrix, as well as the parameter fc (for fold change) that reflects how far away an outlier is from its group (Table 1).

RNA-seq data of kidney tumors and controls

In the above simulated principal components, no within group correlation was given. In real data with multiple experimental groups, it may happen that principal components within a study group are correlated, although the principal components for the whole data are uncorrelated.

Table 1 Construction of simulation parameters for artificial gene expression data from the multivariate normal distribution

Setting	Group 1	Group 2	Outliers	Parameter values
1	$\mu = 0_d$	-	$\mu_1 = fc \otimes J_d$	$fc = 0.5, 0.75$
			$\mu_2 = (fc, -fc)^T \otimes J_{d/2}$	$\rho = 0.75$
			$\mu_3 = (fc, -fc, fc, -fc)^T \otimes J_{d/4}$	$\tau = 0.01, 0.05$
2	$\mu = 0_d$	$\kappa = 5 \cdot J_d$	$\mu_1 = fc \otimes J_d$	$fc = 1.0, 1.5$
			$\mu_2 = (fc, -fc)^T \otimes J_{d/2}$	$\rho = 0.75$
			$\mu_3 = (fc, -fc, fc, -fc)^T \otimes J_{d/4}$	$\tau = 0.05, 0.1., 0.2$
			$\kappa_1 = \kappa + (fc \otimes J_d)$	
			$\kappa_2 = \kappa + ((fc, -fc)^T \otimes J_{d/2})$	
			$\kappa_3 = \kappa + ((fc, -fc, fc, -fc)^T \otimes J_{d/4})$	

Setting 1 represents data for one group with three outliers. Setting 2 represents data for two group, each with three outliers. The last columns shows the simulation parameters that are varied. The mean vectors in group 1 and group 2 for regular observations are given by μ and κ, and those mean vectors for outlying observations are given by μ_1, μ_2, μ_3 and $\kappa_1, \kappa_2, \kappa_3$. In this notation, J_L denotes a vector of ones of length L and $\otimes$ denotes the symbol for the Kronecker product

In the simulated data principal components were generated uncorrelated whithin each group. We therefore considered also one example of real data from an RNA-seq experiment. As will be demonstrated in the analysis of these data, correlation between the components whithin a group can have a strong impact on the outlier detection. The related experiment was performed as part of The Cancer Genome Atlas Project [40] and involved samples from kidney renal clear cell carcinomas. The data set contains expression data of $d = 20531$ genes in $n = 144$ tissue samples from either non-tumor (control) tissue or tumor tissue. The total sample divided into 72 observations from each tissue type. The data is available in the R-package 'SimSeq'. The experiment was originally conducted to find pathways and genes which take part in development of kidney cancer, or renal cell carcinomas (RCC), which are a common group of hemotherapy-resistant cancer types and therefore of special interest for the search of gene mutations [40].

Results

In this section we show the results of outlier detection in the example data sets and the simulation study. In detail, we show the different results that bagplots and gemplots produce and add a comparison with outlier detection by means of hierarchical clustering.

Analysis artificial principal components

In the first example with normally distributed principal components, the outlying observations 3 and 8 were detected in the two-dimensional analysis using a bagplot for each of the two study groups (Fig. 1). In addition, another outlier, observation 64, was detected in the first

group. When the analysis if restocked by the third principal component observation 64 is not charged as an outlier any more, but a new outlier in the second study group, observation 49, appeared.

In the second example, where some principal components were not normally distributed, the vanishing and emergence of outliers when switching to more dimensions becomes even much clearer. In the first (blue labeled) study group only observation 87 is detected as outlier in the bagplot approach (Fig. 2). When turning to the gemplot representation, also observation 73 is be detected. In the second study group (red labeled), observations 65 and 97 are detected as outliers in the bagplot approach but they vanish when using gemplots.

Simulations with artificial gene expression data

Each of the scenarios was simulated in 1000 runs. Setting 1 represented scenarios with only one study group. We first simulated expression data with an overall high correlation between the genes ($\tau = 0.01$, Additional file 1: Figure S1). Depending on how far away the outliers are from the regular observations (fc=1 to fc=4), the number of detected outliers by the different approaches changes. In general, the boxplot approach (1D) detects more outliers than the bagplot approach (2D) which itself detects more outliers than the gemplot approach (3D). When using the 1D method, boxplots for the first three principal components were inspected. For the 2D method, bagplots for PC1 versus PC2, PC1 versus PC3, and PC2 versus PC3 were used. Then, the union of detected outliers was identified for each method. The 3D method included the first three components anyway. All three approaches only detect outliers that are sufficiently distant from the regular

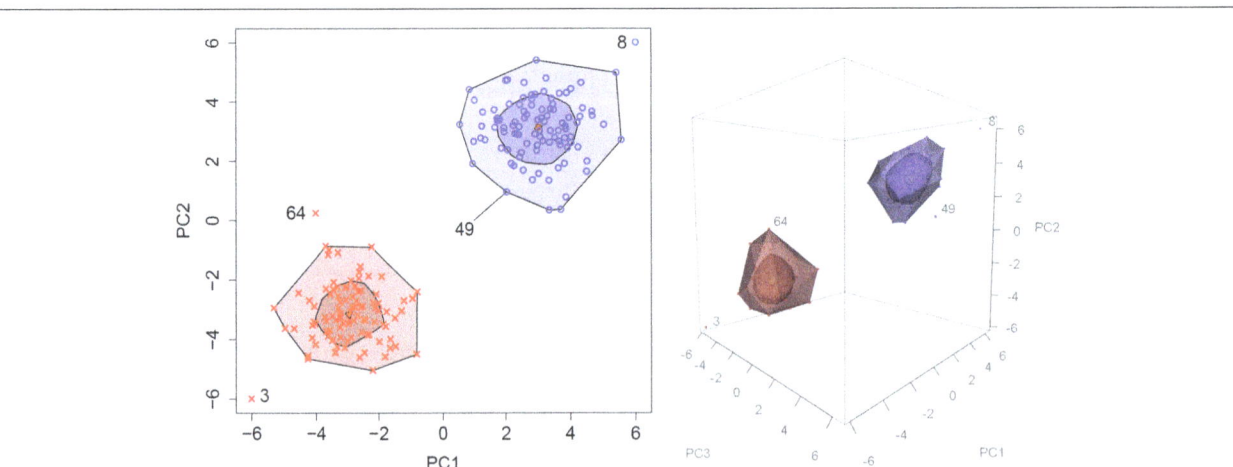

Fig. 1 Bagplots and gemplots under normaly distributed prinicipal components. Bagplots (*left*) and gemplots (*right*) for identifying outliers in the space of the first two or three principal components, respectively. The plots consist of inner bags (or *inner* gemstones) that contain 50% of the samples and an outer loop (or *outer* gemstone). Sample outside the loop or gemstones are flagged as outliers. While multiple experimental groups can be displayed in the same subspace of principal components, outlier detection can be performed separately for each group. Switching from two to three dimensions, new outliers can be found (sample 49) while other sample disappear as outliers (sample 64)

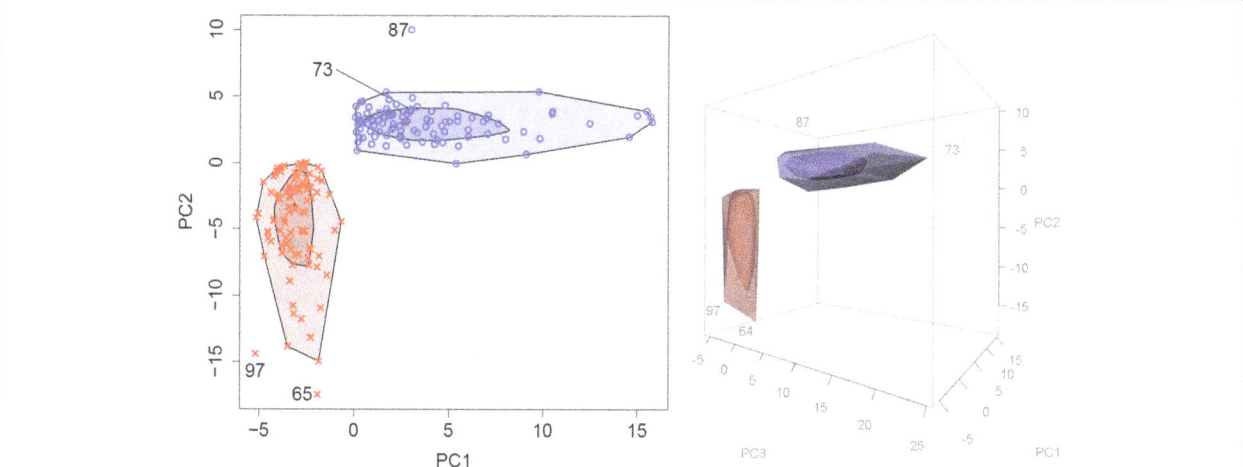

Fig. 2 Bagplots and gemplots under non-normaly distributed prinicipal components. In real data, samples are often non-normaly distributed in some principal components. In the blue labelled group samples follow an exponential distribution on the second and third principal components. Thus, sample 73 becomes only an evident outlier in the three-dimensional gemplot

observations, where the 2D and the 3D method require outliers to be more distant than the 1D methods does. I.e., the boxplot approaches appears to be more sensitive than the bagplot which is more sensitive than the gemplot. On the other hand, boxplot and bagplot produce more false detections than the gemplot approach does, i.e. the gemplot outperforms the other tow approaches with respect to specificity. When the overall correlation between the genes is rather low ($\tau = 0.05$, Additional file 1: Figure S2), the sensitivity of all three approaches increases, while specificity remains nearly the same as with high correlations. The results for scenarios in setting 2, i.e. with two study groups (Additional file 1: Figures S3, S4), are similar to those in setting 1. Again, the highest sensitivity is observed for the 1D approach, while the gemplot approach yields the highest specificity. Likewise,

all three methods have a higher sensitivity when the correlation among genes is low and when outliers are clearly distant from the regular observations. Both factors, show no strong effect on the specificity. All simulations show also that the specificity can be improved by the gemplot approach, even if the variance declared by the thirst principal component is less than 10%.

Analysis of RNA-seq data

For outlier detection we first employed the approach of hierarchical clustering using the single linkage approach (Fig. 3). Normally, observations on branches that form individual clusters or that separate from the remaining cluster tree by an extreme height are subjectively judged as outliers by visual inspection of the cluster tree. Alternatively to visual inspection, one can transfer the heights of

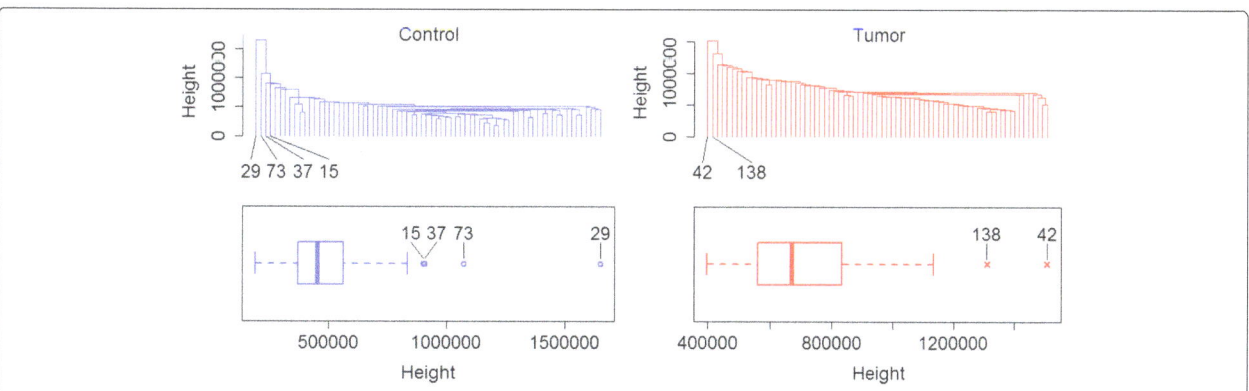

Fig. 3 Hierarchical cluster trees and outlier detection RNA-seq data from a kidney tumor study. Clustering was performed using the 'single linkage' method which is generally recommended to identify outliers. Usually, outliers are selected by subjective judgment of a researcher as samples on branches that separate locally or by conspicuous height from the remaining branches. To select outliers in a more objective way, boxplots on the height of the branches can be used. A disadvantage of hierarchical cluster trees is that outlier detection and arrangement of the tree depend on whether experimental groups are displayed separately or as a whole

the branches to a boxplot representation to identify out-liers in a more objective way. With the latter approach four observations are detected in the control group of the kidney data and two in the tumor group.

When also using boxplots individually on the first three principal components, large number of outliers is detected (Fig. 4). Of these, observation 37 and 73 in the control group were also detected in the clustering approach, while there was no overlap of findings in the tumor group. Here, it becomes also clear that each princi-pal component can uncover a different set of outliers.

When turning to bagplot and gemplot representations of the principal components (Fig. 5), observation 73 is still an outlier in the two-dimensional approach but not in the three-dimensional one. Again, bagplots were used to inspect PC1 versus PC2, PC1 versus PC3, and PC2 versus PC3. Looking at the bagplot of the tumor group the role of correlation between the principal components in this group becomes clear. Observations 74 and 92 are very unlikely to be detected as outliers in boxplots on the first and second principal component. But due to the within-group correlation between the first two compo-nents the shape of the bagplot allows for their detection as outliers. Observation 6 appears not as an outlier in the boxplot approach but is detected by the bagplot as well as by the gemplot. Looking at the 'screeplot' for the PCA on these data, i.e. the proportion of declared variance per principal component, one can see that there is a sharp bend after PC3 (Additional file 1: Figure S5). According

to the 'elbow method', one would argue that the princi-pal components behind PC3 don't provide substantially more information and that they could be omitted for fur-ther analysis. Nevertheless, we run the gemplot approach with the first four components to gain 4% more by adding PC4. In that analysis, no observation was flagged as an outlier. Thus, the number of detected outliers reduces in this example when increasing the number of dimensions from 2D to 3D or 4D, i.e. when increasing the number of principal components. The largest number of outliers was detected by boxplots and smallest number by the gemplots. Since in this example the proportion of vari-ance represented was rather small (PC1: 23%, PC2: 13%, PC3: 5%, PC4: 4%), a final decision on outliers might be critical, here.

In order to demonstrate the impact of outliers we per-formed a differential expression analysis between tumor and control samples using the R-package 'limma' [41]. The functionality of this package allows to downweigh individual observations, and is was shown that this can improve the power of detecting differentially expressed genes [22]. We performed the analysis a) without down-weighing outliers, b) by downweighing the four outliers detected by the bagplot approach and c) with down-weighing the single outlier, observation 6, identified by the gemplot approach. Figure 6 presents smoothed scatterplots of the log foldchanges resulting from anal-ysis with versus analysis without downweighing out-liers as well as scatterplots of FDR-adjusted p-values (i.e. q-values) of the different approaches. The plots show that only one outlier can have a clear impact on the p-values, but only a smaller effect on the fold changes. With four outliers, the effect is even more extreme.

Discussion
In this work we present a new approach for objec-tive and automated outlier detection in molecular high-throughput data using bagplots and gemplots. The approach is useful for a wide range of data, e.g. gene and protein expression data, methylation data, ChIP data or AP/MS data. All these different data types can first be visualized after dimension reduction by principal com-ponent analysis, and bagplots and gemplots can than be applied separately to the observations of each experimen-tal or study group.

Our simulations of gene expression data have shown, that outlier detection with baplots and gemplots on prin-cipal components is less sensitive than using boxplots, however, it also reduces significantly the number of false detections. Although two or three components often rep-resent a large proportion of the original variance, it is recommended that researchers explore the variance pro-portions by scree plots before. A scree plot can help to

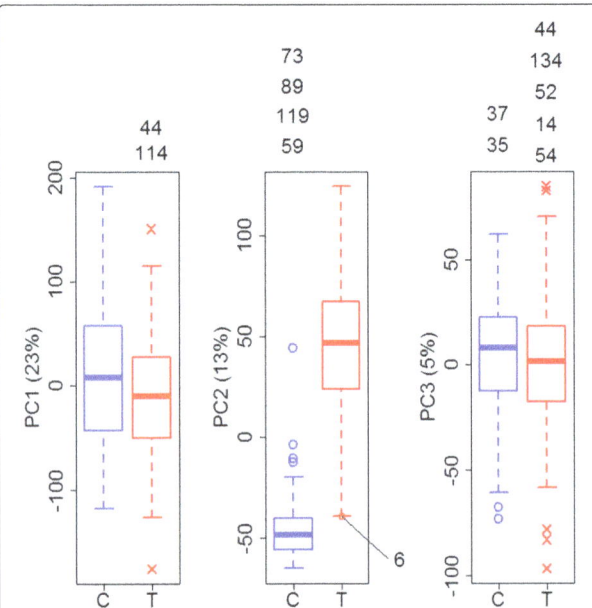

Fig. 4 Boxplots of the first three principal components of the kidney data. Group-specific boxplots to detect outliers on the first three principal components. On each dimension of the principal component space different outliers can be detected

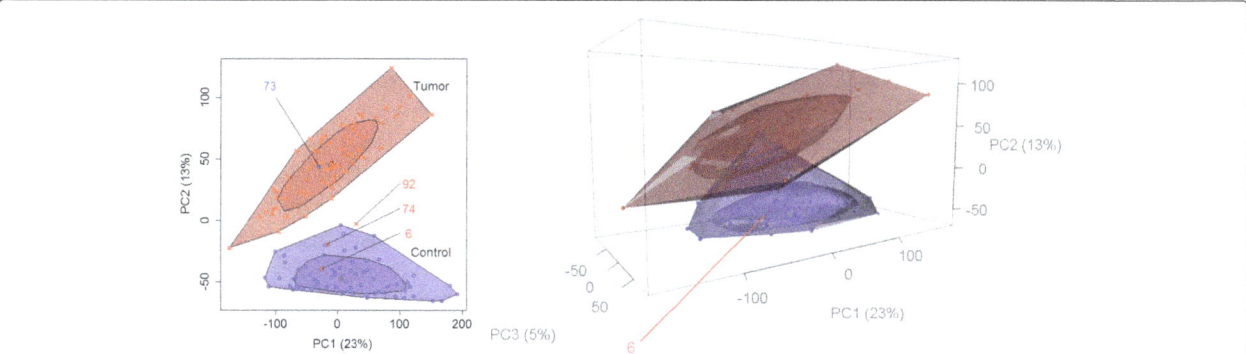

Fig. 5 Bagplots and gemplots for representation of the kidney data. In the control group, observation 73 is detected as outlier by the bagplot appraoch but not vanishes when using gemplots. In the tumor group, three outliers are detected using a bagplot of which only observation 6 remains as an outlier in the gemplot approach

select an appropriate number of components [35]. In the case that the first three components do not represent a large proportion of the original variance, our approach should be applied with cautiousness.

While we observed in the example of the RNA-seq data that there is a tendency to find less outliers when using more dimensions, we would though recommend to explore the data by at least three principal components with gemplots since more of the original variance is explained than by individual principal components.

In particular, our simulations have shown that even if a principal component declares less than 10% of variance, it can contribute to reduce false detections. We observed also in other than the presented data, that gemplots detect less outliers than bagplots or boxplots but this is hard to formalize and depends very much on the shape of the total scatterplot. Thus, an outlier detected by a boxplot of an individual principal component can be fetched back to the set of data points within the fence when adding another principal component. This

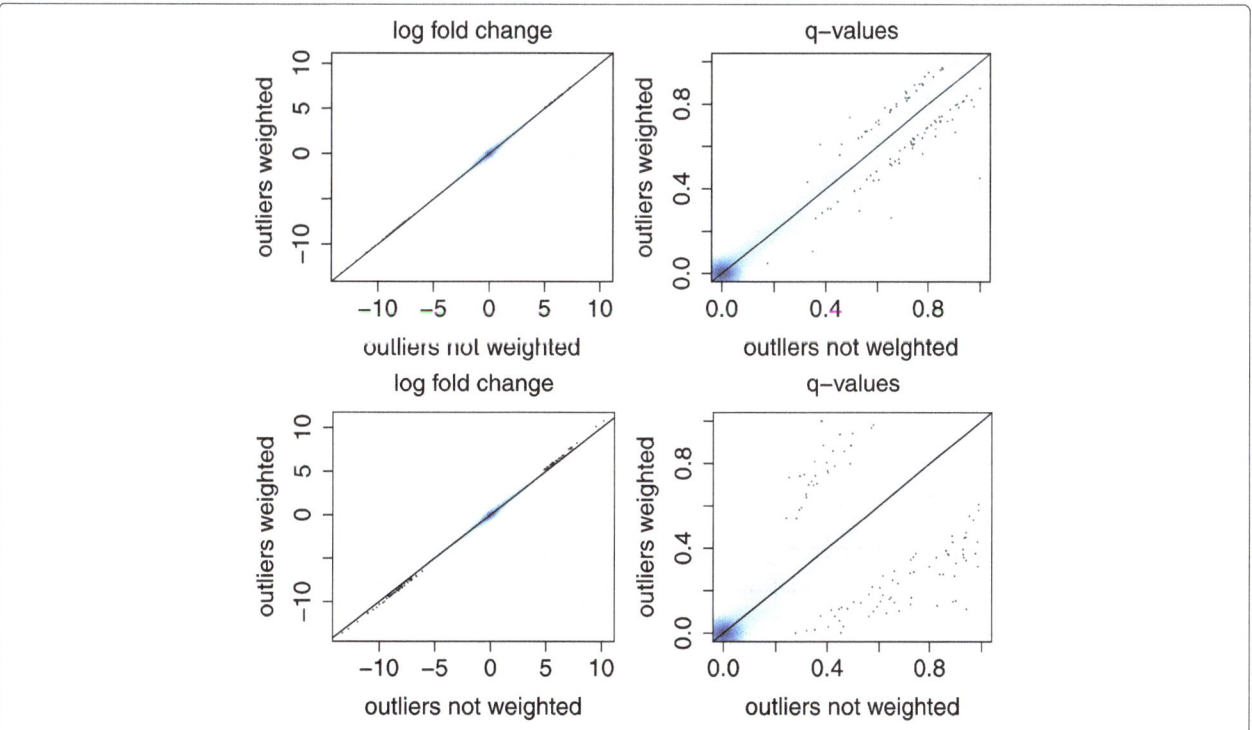

Fig. 6 Fold changes and *p*-values with and without outlier weighting. Comparison of fold changes and FDR-adjusted *p*-values from a differential expression analysis between tumor and control samples of the kidney data. Axes show the result with and without weighting of outliers. The upper plots show the effect of only one outlier, the lower plots show the effect of four outliers

effect is mainly observed when principal components show a within group correlation in a multiple group setting. One could then argue that principal components don't represent the correct directions of the largest variances for that group. Analyzing each group individually by PCA would be more appropriate then. On the other hand, one could argue that a simultaneous analysis of all study groups better reflects the variation of gene expression or other high-dimensional measurements in a biological system as a whole, e.g. the variation of different subtypes in a particular cancer disease. A multigroup PCA would for example be more appropriate if outliers come from intermediate subtypes. In any case, researchers should carefully consider whether a single group or multigroup analysis is more appropriate for their data.

Basically, our R-package 'gemplot' can calculate outlier detection in more than three dimensions, which becomes, however, very slow for more than four components. In contrast to other approaches that are only based on the individual judgement of the researcher, our approach is more objective and provides an automated detection of outliers. A series of experiments can thus be analysed with the same fix criteria for outlier detection. While some other approaches also provide an automated selection of outliers ([30, 31]), these approaches can only cope with one study group since outlier detection and graphical representation are linked. Thus, our approach is specifically useful when multiple experimental groups are to be analyzed in the same space of principal components. In general, the approach can also be used in other subspaces of the data obtained by dimension reduction, e.g. subspaces derived by multidimensional scaling [42] or factor analysis [43].

As pointed out by Lee et al. [44] and by Ma [45], standard PCA can fail to yield consistent estimators of the loading matrix A in high-dimensional settings. Therefore, their approaches for estimating the loadings should also be considered before further exploring the data by bagplots or gemplots.

We also have demonstrated the impact of a small number of outliers on the selection of differentially expressed genes. Similar effects can be assumed for other types of analyses with high-throughput expression data, such as gene set analysis [46–48], classification problems [18–20], and in consequence also for data integration methods of multi-omics data [49, 50]. The concrete handling of outliers detected by our approach depends of course on the specific methods for subsequent analysis. In this regard, another advantage of our approach is that it can be used independently of the methods intended for further analysis.

Conclusion

We present the gemplot as the three-dimensional version of boxplot and bagplot, respectively. We have demonstrated the usability of the gemplot for outlier detection in molecular, high-dimensional data. In contrast to other methods, our approach allows for simultaneous outlier identification in multiple experimental groups. The presented method is less sensitive than other methods – depending on how extreme the outlying data are – but it produces also less false positives.

Additional file

Additional file 1: Supplementary Figures Mean numbers of correct detected outliers and incorrect detected outliers in a simulation with artificial gene expression data. **Figure S1.** Scenario with one study group. The four plots show results when the overall correlation between most genes is high ($\tau = 0.01$). **Figure S2.** Scenario with one study group. The four plots show results when the overall correlation between most genes is low ($\tau = 0.05$). **Figure S3.** Scenario with two study groups. The four plots show results when the overall correlation between most genes is high ($\tau = 0.05$) or low ($\tau = 0.2$) with different fold changes ($fc = 1.00$ or $fc = 1.50$). **Figure S4.** Scenario with two study groups. Continuation of Figure S3 for scenarios where outliers are clearly distant from the regular observations ($fc = 2.00$). **Figure S5.** Screeplot for the principal component analysis of the kidney RNAseq data.

Abbreviations
AP/MS: Affinity purification mass spectrometry; ChIP: Chromatin immunoprecipitation; NGS: Next-generation sequencing; PCA: Principal component analysis; RCC: Renal cell carcinomas

Acknowledgements
None.

Funding
This work was supported by the Niedersachsen-Research Network on Neuroinfectiology (N-RENNT) of the Ministry of Science and Culture of Lower Saxony. The funding body was not involved in the design or conclusion of our study.

Authors' contributions
KJ formulated the idea of using bagplots and gemplots for outlier detection in molecular high-throughput data. Both authors contributed equally in writing the R-package, analyzing the example data and writing the manuscript. Both authors read and approved the final manuscript.

Competing interests
The authors declare that they have no competing interests.

References
1. Schena M, Shalon D, Davis RW, Brown PO. Quantitative Monitoring of Gene Expression Patterns with a Complementary DNA Microarray. Science. 1995;270(5235):467–70.
2. Heller MJ. DNA Microarray Technology: Devices, Systems, and Applications. Ann Rev Biomed Eng. 2002;4(1):129–53.
3. Metzker ML. Sequencing technologies — the next generation. Nat Rev Genet. 2009;11(1):31–46.
4. Aebersold R, Mann M. Mass spectrometry-based proteomics. Nature. 2003;422(6928):198–207.
5. Lenz C, Dihazi H. Introduction to Proteomics Technologies. Stat Anal in Proteomics. 2016;1362:3–27.
6. Görg A, Weiss W, Dunn MJ. Current two-dimensional electrophoresis technology for proteomics. Proteomics. 2004;4(12):3665–85.

7. Yang YH. Normalization for cDNA microarray data: a robust composite method addressing single and multiple slide systematic variation. Nucleic Acids Res. 2002;30(4):15e–15.

8. Coombes KR, Tsavachidis S, Morris JS, Baggerly KA, Hung MC, Kuerer HM. Improved peak detection and quantification of mass spectrometry data acquired from surface-enhanced laser desorption and ionization by denoising spectra with the undecimated discrete wavelet transform. Proteomics. 2005;5(16):4107–17.

9. Anders S, Pyl PT, Huber W. HTSeq–a Python framework to work with high-throughput sequencing data. Bioinformatics. 2014;31(2):166–9.

10. Mortazavi A, Williams BA, McCue K, Schaeffer L, Wold B. Mapping and quantifying mammalian transcriptomes by RNA-Seq. Nat Methods. 2008;5(7):621–28.

11. Bibikova M, Barnes B, Tsan C, Ho V, Klotzle B, Le JM, et al. High density DNA methylation array with single CpG site resolution. Genomics. 2011;98(4):288–95.

12. Hurd PJ, Nelson CJ. Advantages of next-generation sequencing versus the microarray in epigenetic research. Brief Funct Genomic Proteomic. 2009;8(3):174–83.

13. Zhang Y, Liu T, Meyer CA, Eeckhoute J, Johnson DS, Bernstein BE, et al. Model-based Analysis of ChIP-Seq (MACS). Genome Biol. 2008;9(9):R137.

14. Nesvizhskii AI. Computational and informatics strategies for identification of specific protein interaction partners in affinity purification mass spectrometry experiments. Proteomics. 2012;12(10):1639–55.

15. Dudoit S, Yang YH, Callow MJ, Speed TP. Statistical methods for identifying differentially expressed genes in replicated cDNA microarray experiments. Stat Sin. 2002;111–39.

16. Robinson MD, McCarthy DJ, Smyth GK. edgeR: a Bioconductor package for differential expression analysis of digital gene expression data. Bioinformatics. 2009;26(1):139–40.

17. Anders S, Huber W. Differential expression analysis for sequence count data. Genome Biol. 2010;11(10):R106.

18. Alizadeh AA, Eisen MB, Davis RE, Ma C, Lossos IS, Rosenwald A, et al. Distinct types of diffuse large B-cell lymphoma identified by gene expression profiling. Nature. 2000;403(6769):503–11.

19. van de Vijver MJ, He YD, van't Veer LJ, Dai H, Hart AAM, Voskuil DW, et al. A Gene-Expression Signature as a Predictor of Survival in Breast Cancer. N Engl J Med. 2002;347(25):1999–2009.

20. Beer DG, Kardia SLR, Huang CC, Giordano TJ, Levin AM, Misek DE, et al. Gene-expression profiles predict survival of patients with lung adenocarcinoma. Nat Med. 2002;8:816–24.

21. Gottardo R, Raftery AE, Yeung KY, Bumgarner RE. Bayesian Robust Inference for Differential Gene Expression in Microarrays with Multiple Samples. Biometrics. 2006;62:10–18.

22. Ritchie ME, Diyagama D, Neilson J, van Laar R, Dobrovic A, Holloway A, et al. Empirical array quality weights in the analysis of microarray data. BMC Bioinformatics. 2006;7(1):261.

23. Filzmoser P, Maronna R, Werner M. Outlier identification in high dimensions. Comput Stand Data Anal. 2008;52(3):1694–711.

24. Eisen MB, Spellman PT, Brown PO, Botstein D. Cluster analysis and display of genome-wide expression patterns. Proc Natl Acad Sci. 1998;95(25):14863–8.

25. Yeung KY, Ruzzo WL. Principal component analysis for clustering gene expression data. Bioinformatics. 2001;17(9):763–74.

26. Swayne DF, Lang DT, Buja A, Cook D. GGobi: evolving from XGobi into an extensible framework for interactive data visualization. Comput Stat Data Anal. 2003;43(4):423–44.

27. Milligan GW. An examination of the effect of six types of error perturbation on fifteen clustering algorithms. Psychometrika. 1980;45(3):325–42.

28. Egan WJ, Morgan SL. Outlier Detection in Multivariate Analytical Chemical Data. Anal Chem. 1998;70(11):2372–9.

29. Zimek A, Schubert E, Kriegel HP. A survey on unsupervised outlier detection in high-dimensional numerical data. Stat Anal Data Min. 2012;5(5):363–87.

30. Model F, König T, Piepenbrock C, Adorjan P. Statistical process control for large scale microarray experiments. Bioinf. 2002;18:S155—S63.

31. Hubert M, Rousseeuw PJ, Vanden Branden K. ROBPCA: a new approach to robust principal component analysis. Technometrics. 2005;47(1):64–79.

32. Filzmoser P, Todorov V. Robust tools for the imperfect world. Inf Sci. 2013;245:4–20.

33. Raychaudhuri S, Stuart JM, Altman RB. Principal Components Analysis to Summarize Microarray Experiments: Application to Sporulation Time Series. Pac Symp Biocomput. 2000;455–66.

34. Shieh AD, Hung YS. Detecting Outlier Samples in Microarray Data. Stat Appl Genet Mol Biol. 2009;8(1):1–24.

35. Cangelosi R, Goriely A. Component retention in principal component analysis with application to cDNA microarray data. Biol Direct. 2007;2(1):2.

36. Sharov AA, Dudekula DB, Ko MSH. A web-based tool for principal component and significance analysis of microarray data. Bioinformatics. 2005;21(10):2548–9.

37. Rencher AC. Multivariate Statistical Inference and Applications. New York: Wiley; 1998.

38. Rousseeuw PJ, Ruts I, W TJ. The Bagplot: A Bivariate Boxplot. Am Stat. 1999;53(4):.

39. Miller K, Ramaswami S, Rousseeuw P, Sellares J, Souvaine D, Streinu I, et al. Efficient computation of location depth contours by methods of computational geometry. Stat Comput. 2003;13(2):153–62.

40. The Cancer Genome Atlas Research Network. Comprehensive molecular characterization of clear cell renal cell carcinoma. Nature. 2013;499:43–9.

41. Smyth GK. Limma: linear models for microarray data. In: Bioinformatics and computational biology solutions using R and Bioconductor. New York: Springer; 2005. p. 397–420.

42. Hout MC, Papesh MH, Goldinger SD. Multidim Scaling. Wiley Interdisciplinary Reviews: Cognitive Science. 2013;4(1):93–103.

43. Comrey AL, Lee HB. A first course in factor analysis. Hillsdale: Psychology Press; 2013.

44. Lee YK, Lee ER, Park BU. Principal component analysis in very high-dimensional spaces. Stat Sin. 2012;933–56.

45. Ma Z. Sparse principal component analysis and iterative thresholding. Ann Stat. 2013;41(2):772–801.

46. Subramanian A, Tamayo P, Mootha VK, Mukherjee S, Ebert BL, Gillette MA, et al. Gene set enrichment analysis: A knowledge-based approach for interpreting genome-wide expression profiles. Proc Natl Acad Sci. 2005;102(43):15545–50.

47. Goeman JJ, Buhlmann P. Analyzing gene expression data in terms of gene sets: methodological issues. Bioinformatics. 2007;23(8):980–7.

48. Jung K, Becker B, Brunner E, Beissbarth T. Comparison of global tests for functional gene sets in two-group designs and selection of potentially effect-causing genes. Bioinformatics. 2011;27(10):1377–83.

49. Wachter A, Beißbarth T. pwOmics: an R package for pathway-based integration of time-series omics data using public database knowledge: Fig. 1. Bioinformatics. 2015;31(18):3072–4.

50. Montague E, Stanberry L, Higdon R, Janko I, Lee E, Anderson N, et al. MOPED 2.5—An Integrated Multi-Omics Resource: Multi-Omics Profiling Expression Database Now Includes Transcriptomics Data. OMICS: J Integr Biol. 2014;18(6):335–43.

3

A fast and efficient python library for interfacing with the Biological Magnetic Resonance Data Bank

Andrey Smelter[1,2], Morgan Astra[5] and Hunter N. B. Moseley[3,4,5,6*]

Abstract

Background: The Biological Magnetic Resonance Data Bank (BMRB) is a public repository of Nuclear Magnetic Resonance (NMR) spectroscopic data of biological macromolecules. It is an important resource for many researchers using NMR to study structural, biophysical, and biochemical properties of biological macromolecules. It is primarily maintained and accessed in a flat file ASCII format known as NMR-STAR. While the format is human readable, the size of most BMRB entries makes computer readability and explicit representation a practical requirement for almost any rigorous systematic analysis.

Results: To aid in the use of this public resource, we have developed a package called **nmrstarlib** in the popular open-source programming language Python. The **nmrstarlib**'s implementation is very efficient, both in design and execution. The library has facilities for reading and writing both NMR-STAR version 2.1 and 3.1 formatted files, parsing them into usable Python dictionary- and list-based data structures, making access and manipulation of the experimental data very natural within Python programs (i.e. "saveframe" and "loop" records represented as individual Python dictionary data structures). Another major advantage of this design is that data stored in original NMR-STAR can be easily converted into its equivalent JavaScript Object Notation (JSON) format, a lightweight data interchange format, facilitating data access and manipulation using Python and any other programming language that implements a JSON parser/generator (i.e., all popular programming languages). We have also developed tools to visualize assigned chemical shift values and to convert between NMR-STAR and JSONized NMR-STAR formatted files. Full API Reference Documentation, User Guide and Tutorial with code examples are also available.

We have tested this new library on all current BMRB entries: 100% of all entries are parsed without any errors for both NMR-STAR version 2.1 and version 3.1 formatted files. We also compared our software to three currently available Python libraries for parsing NMR-STAR formatted files: PyStarLib, NMRPyStar, and PyNMRSTAR.

Conclusions: The **nmrstarlib** package is a simple, fast, and efficient library for accessing data from the BMRB. The library provides an intuitive dictionary-based interface with which Python programs can read, edit, and write NMR-STAR formatted files and their equivalent JSONized NMR-STAR files. The **nmrstarlib** package can be used as a library for accessing and manipulating data stored in NMR-STAR files and as a command-line tool to convert from NMR-STAR file format into its equivalent JSON file format and vice versa, and to visualize chemical shift values. Furthermore, the nmrstarlib implementation provides a guide for effectively JSONizing other older scientific formats, improving the FAIRness of data in these formats.

Keywords: Biological Magnetic Resonance Bank, Nuclear magnetic resonance, NMR-STAR, JSON, nmrstarlib, Python

---

* Correspondence: hunter.moseley@uky.edu
[3]Department of Molecular and Cellular Biochemistry, University of Kentucky, Lexington, KY 40356, USA
[4]Markey Cancer Center, University of Kentucky, Lexington, KY 40356, USA
Full list of author information is available at the end of the article

Background

The Biological Magnetic Resonance Data Bank (BMRB) is a free, publicly-accessible repository of data on peptides, proteins, and nucleic acids obtained through NMR Spectroscopy [1], that is part of the worldwide Protein Databank (wwPDB) [2]. It currently consists of more than 11,000 individual NMR-STAR file entries, containing a wide range of NMR spectral data, experimental details, and biochemical data collected from thousands of biological samples. The NMR-STAR format is based on the Self-defining Text Archival and Retrieving (STAR) flat file database format [3], with some modifications specific to the BMRB. STAR provides a hierarchical dictionary structure for storing arbitrary data. In NMR-STAR, the format specifies top-level dictionaries called "saveframes", which are used to categorize the data and meta-data about the experiment. Inside each saveframe is an arbitrarily number of key-value pairs and tables of records (loops). The key-value pairs store a single piece of information under a descriptive variable name. Each loop stores a table of records, each record containing a set of values representing individual fields in the record. There are currently two active versions of the BMRB: version 2.1 and version 3.1. While they both use the same NMR-STAR format at the most general level, the layout of the data in the two formats is different.

Python is a free, open-source scripting language which runs on all major operating systems [4, 5]. It is designed to facilitate the development and maintenance of simple, efficient, and readable code. Python has object-oriented programming facilities and includes several high-level data structure objects in its standard library. Among these are the dictionary, a data structure implemented via the `dict` class that stores data as a set of key-value pairs (specific mappings between keys and values). The `OrderedDict` class is identical to the `dict` class except that the order of inserted keys-value pairs is remembered. This is particularly useful for categorical data with sequential relationships. The dictionary data structure is the most straightforward mechanism for representing and using data from NMR-STAR files, which have a nested, mostly dictionary-like structure

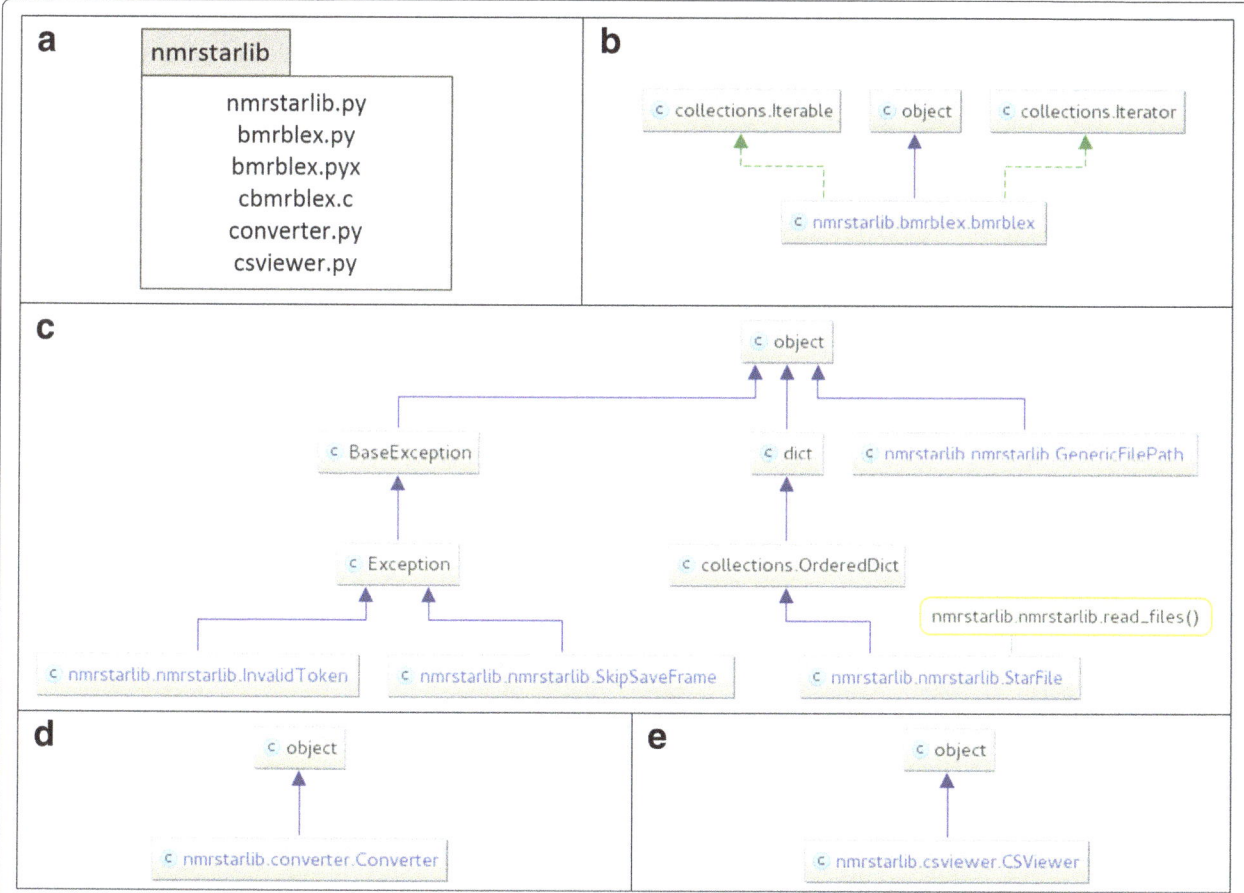

Fig. 1 Organization of the nmrstarlib package version 1.1.0. **a** UML package diagram of the nmrstarlib library; **b** UML class diagram of the bmrblex.py (bmrblex.pyx) module; **c** UML class diagram of the nmrstarlib.py module; **d** UML class diagram of the converter.py module; **e** UML class diagram of the csviewer.py module

themselves. However, to our knowledge no NMR-STAR parsing library using this design exists. The newest major version of Python (version 3.0.0), was initially released on 2008-12-03, however many software libraries and utilities written in Python still use Python version 2.x exclusively. As Python version 3.1 brings many substantial improvements over Python 2.x (including the addition of the `OrderedDict` class, which was later back-ported to Python version 2.7 [6]). As of Python version 3.5 `OrderedDict` is implemented in C which makes it much faster than the Python 2.7 implementation of `OrderedDict`. Moreover in Python 3.6, the `dict` data structure implementation becomes ordered by default and `dict` and `OrderedDict` are more efficient than in any previous versions of Python. While we provide support for Python 2.7 for use by legacy code, we believe that researchers will prefer libraries and tools written in latest version of Python in order to develop maintainable codebases, especially as Python version 2.x becomes less supported over time. Moreover, Python version 2.7 will no longer be maintained after Spring of 2020 [7]. Two publically available Python libraries for parsing NMR-STAR format files PyStarLib [8] and NMRPyStar [9] both require Python version 2.7. PyNMRSTAR [10] works with both major versions of Python (2.7 and 3.3+).

Implementation

The `nmrstarlib` package consists of several modules: `nmrstarlib.py`, `bmrblex.py`, `converter.py`, and `csviewer.py` (Fig. 1a). The `nmrstarlib` module (Fig. 1c) provides the `StarFile` class, which implements a nested Python dictionary/list representation of a BMRB NMR-STAR file. Once a NMR-STAR formatted file is processed into a `StarFile` object, experimental data can be accessed directly from the `StarFile` object, using bracket accessors as with any regular Python `dict` object. The `nmrstarlib` module relies on the `bmrblex` module (Fig. 1b) for processing of tokens. The `bmrblex` module provides the `bmrblex` generator – BMRB lexical analyzer (parser). We provide two versions of the `bmrblex` module: a pure Python version (`bmrblex.py`) and a Python + C extension (`bmrblex.py`, `cbmrblex.c`) for faster performance. The compiled C extensions are implemented in the Cython programming language [11], which we will call the Cython implementaion. If the Cython implementation of `bmrblex` fails for any reason, the library will use the Python implementation, ensuring that the library always works.

The library creates an internal representation of the NMR-STAR format as a nesting of `OrderedDict` objects with the top-level object `StarFile` inheriting from the

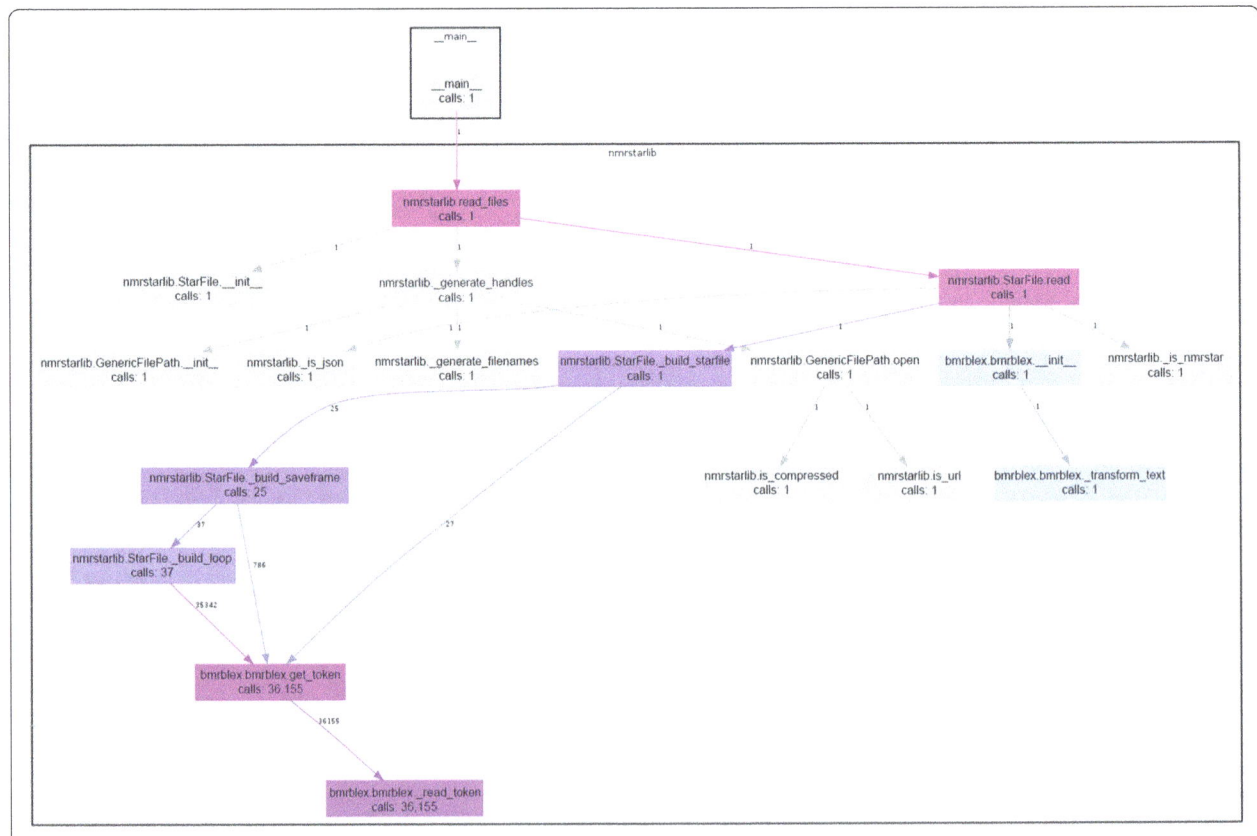

Fig. 2 Diagram showing what function calls are made during the process of StarFile object creation

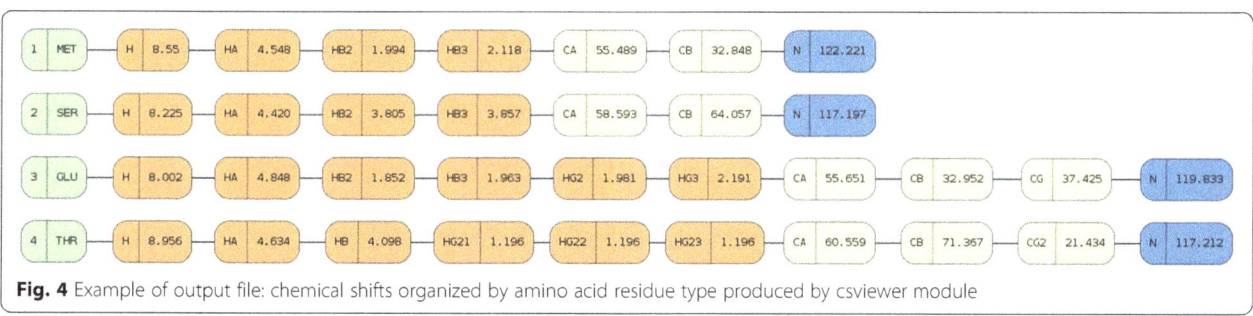

a

```
data_336

save_entry_information
   _Entry.Sf_category        entry_information
   _Entry.Sf_framecode       entry_information
   _Entry.ID        336
   _Entry.Title
;
1H-NMR studies of structural homologies between the heme
environments in horse cytochrome c and in cytochrome c-
552 from Euglena gracilis
;
   _Entry.Type        macromolecule
   _Entry.Version_type        update
   _Entry.Submission_date     1995-07-31
   _Entry.Accession_date      1996-04-12
   _Entry.Last_release_date         .
   _Entry.Original_release_date     .
   _Entry.Origination         BMRB
   _Entry.NMR_STAR_version    3.1.1.61
   _Entry.Original_NMR_STAR_version .
   _Entry.Experimental_method       NMR
   _Entry.Experimental_method_subtype    .
   _Entry.Details    .
   _Entry.BMRB_internal_directory_name   .
   loop_
      _Entry_author.Ordinal
      _Entry_author.Given_name
      _Entry_author.Family_name
      _Entry_author.First_initial
      _Entry_author.Middle_initials
      _Entry_author.Family_title
      _Entry_author.Entry_ID
      1 Regula Keller . M. . 336
      2 Kurt Wuthrich . . . 336
   stop_
save_
```

b

```
{
  "data": "336",
  "save_entry_information": [
    "Entry.Sf_category": "entry_information",
    "Entry.Sf_framecode": "entry_information",
    "Entry.ID": "336",
    "Entry.Title": "\n;\n1H-NMR studies of structural
homologies between the heme environments in horse
\ncytochrome c and in cytochrome c-552 from Euglena
gracilis\n;",
    "Entry.Type": "macromolecule",
    "Entry.Version_type": "update",
    "Entry.Submission_date": "1995-07-31",
    "Entry.Accession_date": "1996-04-12",
    "Entry.Last_release_date": ".",
    "Entry.Original_release_date": ".",
    "Entry.Origination": "BMRB",
    "Entry.NMR_STAR_version": "3.1.1.61",
    "Entry.Original_NMR_STAR_version": ".",
    "Entry.Experimental_method": "NMR",
    "Entry.Experimental_method_subtype": ".",
    "Entry.Details": ".",
    "Entry.BMRB_internal_directory_name": ".",
    "loop_0": [
      [
        "Entry_author.Ordinal",
        "Entry_author.Given_name",
        "Entry_author.Family_name",
        "Entry_author.First_initial",
        "Entry_author.Middle_initials",
        "Entry_author.Family_title",
        "Entry_author.Entry_ID"
      ],
      [
        {
          "Entry_author.Ordinal": "1",
          "Entry_author.Given_name": "Regula",
          "Entry_author.Family_name": "Keller",
          "Entry_author.First_initial": ".",
          "Entry_author.Middle_initials": "M.",
          "Entry_author.Family_title": ".",
          "Entry_author.Entry_ID": "336"
        },
        {
          "Entry_author.Ordinal": "2",
          "Entry_author.Given_name": "Kurt",
          "Entry_author.Family_name": "Wuthrich",
          "Entry_author.First_initial": ".",
          "Entry_author.Middle_initials": ".",
          "Entry_author.Family_title": ".",
          "Entry_author.Entry_ID": "336"
        }
      ]
    ]
  ]
}
```

Fig. 3 Internal StarFile object representation and correspondence to NMR-STAR format without comments: **a** An example of a NMR-STAR formatted file; **b** StarFile dictionary representation equivalent to the NMR-STAR formatted file and the JSONized version of the NMR-STAR file

OrderedDict class (Fig. 1c). This allows the user to access data in its original NMR-STAR organization using familiar Python dictionary syntax. The library provides facilities to read data from NMR-STAR formatted files into an internal StarFile object, to access and make modifications to this StarFile object, and to save the resulting StarFile object as a new NMR-STAR formatted file. It is also possible to create NMR-STAR files from scratch using this library; however, this requires the user to adhere to the recommended layout for NMR-STAR formatted files by adding keys and values to the StarFile object in the appropriate order.

The nmrstarlib module provides a memory-efficient read_files() generator function (Fig. 1c) that yields (emits) StarFile objects, one at a time for each file parsed. When reading an NMR-STAR formatted file (Fig. 2, Additional files 1 and 2), the read_files() generator function first opens the file and passes a file-handle to the StarFile.read() method that reads the text into Python as a string and passes that string into the bmrblex object that then splits the text into tokens. As the bmrblex lexical analyzer keeps emitting valid tokens, the StarFile object is constructed sequentially. The StarFile object decides what type of

Fig. 4 Example of output file: chemical shifts organized by amino acid residue type produced by csviewer module

Table 1 The nmrstarlib library performance test against NMR-STAR formatted files using pure Python and Python with C extension and against JSONized NMR-STAR files using the standard Python library json parser and the UltraJSON (ujson) 3rd party library

			NMR-STAR 2.1	NMR-STAR 3.1	JSONized NMR-STAR 2.1	JSONized NMR-STAR 3.1
Number of files			11,270	11,244	11,270	11,244
Total size of files, GB			1.1	1.8	4.6	22.0
Time, sec	Pure Python	json	326	1,100	30	130
	Python with C extension	[a]ujson	320	423	27	126
Average reading speed, KB/sec	Pure Python	json	3,290	1,700	158,549	176,479
	Python with C extension	[a]ujson	3,351	4,421	176,166	182,082

[a]We added support for the ujson library for versions of Python starting with Python 3.6, because the ujson library does not provide methods to keep the dict data structure in order when parsing from JSON files; however, starting with Python 3.6, the dict data structure is ordered by default

token it is dealing with and chooses which internal method to call in order to construct itself, i.e. calls to StarFile._build_starfile(), Starfile._build_saveframe(), or StarFile._build_loop(). For example, Fig. 2 shows the function call diagram during the StarFile object creation: the _build_saveframe() method is called 25 times and _build_loop() is called 37 times, meaning that the NMR-STAR file consists of 25 different saveframe categories and 37 loops. The total number of tokens processed is equal to 36,155 = 27 (from _build_starfile) + 786 (from _build_saveframe) + 35,342 (from _build_loop).

Each saveframe category is also an OrderedDict data structure that can be accessed by saveframe name as the key from the top-level StarFile object. Once a saveframe dictionary is constructed and populated with key-value pairs, it descends further into each loop and constructs a tuple of two lists: the first list corresponding to loop field keys (loop field names); the second list consists of OrderedDict objects corresponding to loop rows (loop records) in the original NMR-STAR file. By the end of parsing, a single nested dictionary/list structure in the form of a StarFile dictionary object (Fig. 3b) is constructed, emulating the structure of the original NMR-STAR formatted file (Fig. 3a). In addition, comments can be parsed and included as additional key-value pairs within the nested dictionary structure.

The nmrstarlib module provides a GenericFilePath (Figs. 1c and 2) object that is used by the read_files() generator function in order to open NMR-STAR formatted files from many different sources: a single file on a local machine; a URL address of a single file; a directory of files on a local machine; an archive of files on a local machine; a URL address of an archive of files; or the BMRB id of a single file.

To write from a StarFile object to an NMR-STAR formatted file, the library recursively crawls through the StarFile dictionary structure, formatting and printing each of the keys and corresponding values sequentially. This allows nmrstarlib to recall the sequential order of the original NMR-STAR formatted file, due to the stored ordering of key insertion from the underlying OrderedDict objects. Using Python's json library, the entire StarFile dictionary structure can be saved as JSON (JavaScript Object Notation), which is an open, human-readable, lightweight data exchange format that is readable by most programming languages via optimized parsing libraries. This JSON conversion of StarFile objects greatly facilitated the implementation of

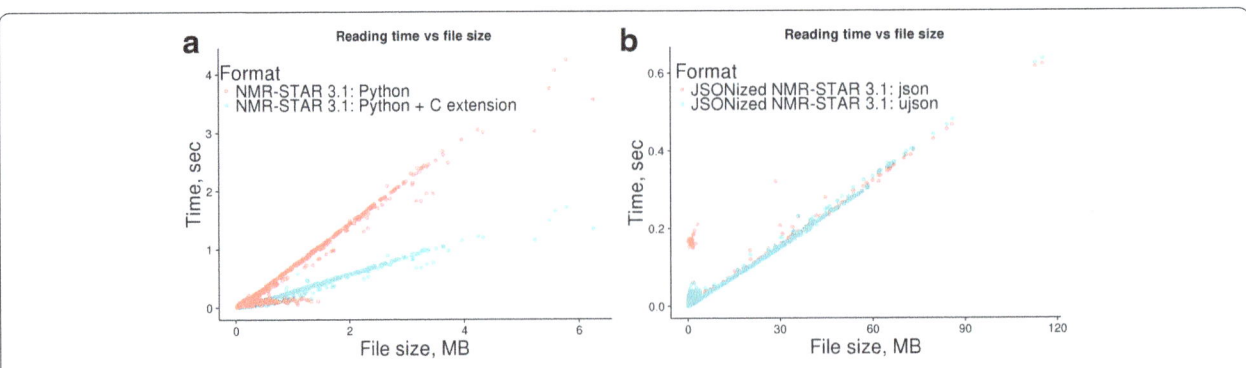

Fig. 5 Graph showing the dependency of loading time into StarFile object from the size of file: **a** Loading times for NMR-STAR 3.1 formatted files; **b** Loading times for JSONized NMR-STAR 3.1 files

Table 2 Converting NMR-STAR formatted files into their equivalent JSON format

	Directory		zip archive		tar.gz archive		tar.bz2 archive	
Format	NMR-STAR 2.1	NMR-STAR 3.1	NMR-STAR 2.1	NMR-STAR 3.1	NMR-STAR 2.1	NMR-STAR 3.1	NMR-STAR 2.1	NMR-STAR 3.1
Number of files	11,270	11,244	11,270	11,244	11,270	11,244	11,270	11,244
Time, min	8	20	9	22	12	27	15	68
Total size, MB	4,756	22,942	230	470	200	409	131	222

the `converter` module which converts original NMR-STAR formatted files into their equivalent JSONized NMR-STAR files and vice versa. The `converter` module (Fig. 1d) consists of a single `Converter` class which can convert in both one-to-one (single file) and many-to-many (directory or archive of files) modes. See "The `nmrstarlib` API Reference" documentation of the `converter` module for the full list of available conversion options (Additional file 3).

In order to simplify access to assigned chemical shift data, we created the `csviewer` module (Fig. 1e) that includes the `CSViewer` class that can access both the NMR-STAR version 2.1 and version 3.1 assigned chemical shifts loop and visualize (organize) chemical shift values by amino acid residue type, and save this visualization as an image file or a pdf document (Fig. 4). The `csviewer` module requires the `graphviz` Python library [12] in order to create an output file. In addition to visualizing chemical shift values, the `csviewer` module provide code example for utilizing the `nmrstarlib` library.

Overall, the `nmrstarlib` package can be used in two ways: 1) as a library for accessing and manipulating data stored in NMR-STAR formatted files, converting between NMR-STAR and its equivalent JSON format, and visualizing assigned chemical shift values; or 2) as a standalone command-line tool for converting files in bulk and visualizing assigned chemical shift values. We used the `docopt` Python library [13] to create the `nmrstarlib` package command-line interface.

Results
Performance on NMR-STAR formatted files

As part of `nmrstarlib`'s development process, we tested our library extensively against the entire BMRB (as of December 11, 2016) for both NMR-STAR version 2.1 and version 3.1 [14]. To measure the performance speed of the `nmrstarlib` library, we used a simple program that accesses NMR-STAR files from local directory one file at a time, which then creates a `StarFile` object and records how much time in seconds it took to create the object. Table 1 shows that our library was able to read the entire BMRB for both NMR-STAR version 2.1 and version 3.1 without any errors. With the pure Python implementation, it took 1,110 s (~18.3 min) and 326 s (~5.4 min) to read NMR-STAR version 3.1 and NMR-STAR version 2.1, respectively. With the more efficient Cython implementation, it took 423 s (~7 min) and 320 s (~5.3 min) to read NMR-STAR version 3.1 and NMR-STAR version 2.1, respectively. We used the metric kilobytes per second (KB/sec), because files/sec would be a misleading metric due to widely varying files sizes in the BMRB and because read times scale almost linearly (Fig. 5) with file size. As such, we found that `nmrstarlib`'s average reading speed is 1,700 KB/sec (NMR-STAR 3.1) and 3,290 KB/sec (NMR-STAR 2.1) for the Python implementation and 4,421 KB/sec (NMR-STAR 3.1) and 3,351 KB/sec (NMR-STAR 2.1) for the Cython implementation on the hardware used for testing. The NMR-STAR 3.1 is more comprehensive than

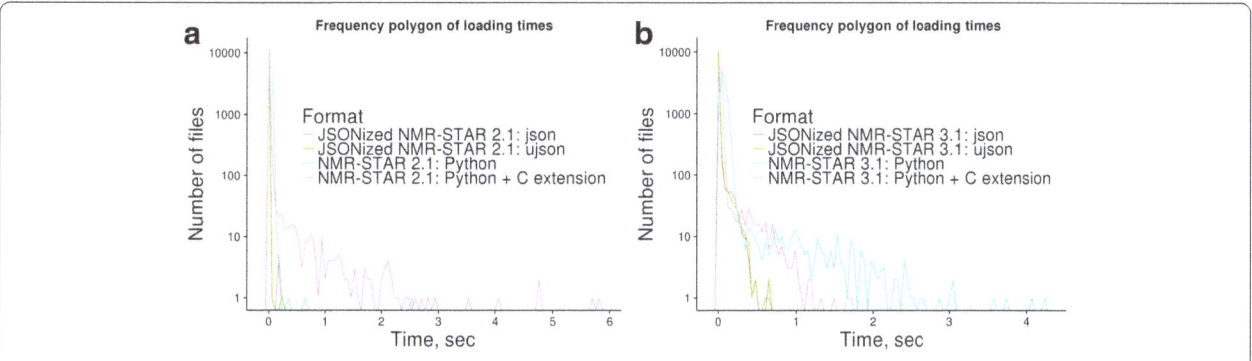

Fig. 6 Frequency polygon of loading times for NMR-STAR files: **a** Comparison of loading times between NMR-STAR 2.1 and JSONized NMR-STAR 2.1; **b** Comparison of loading times between NMR-STAR 3.1 and JSONized NMR-STAR 3.1

NMR-STAR 2.1 and usually represents more experimental information and details. This additional complexity is computationally harder to parse. However, for our Cython implementation average reading speed for NMR-STAR 3.1 was faster than for NMR-STAR 2.1 due to multiline text pre-processing discussed in more detail in the next section.

Performance on JSONized NMR-STAR files

Next, we converted both NMR-STAR version 2.1 and version 3.1 files into their equivalent JSON format and performed speed tests again (Table 1). We found that read times of both JSONized NMR-STAR version 2.1 and version 3.1 were significantly faster than read times of the original NMR-STAR formatted files: 130 s (~2.2 min) and 30 s (~0.5 min) for NMR-STAR version 3.1 and NMR-STAR version 2.1, respectively, for the entire BMRB data set. The average read speed was 176,479 KB/sec and 158,549 KB/sec for version 3.1 and version 2.1, respectively. Next, we tested performance using another compiled JSON parsing third-party library, UltraJSON (`ujson`) [15]. We found that reading times and average reading speeds of JSONized NMR-STAR files were slightly faster than using the built-in `json` parser: 127 s (182,082 KB/sec) and 27 s (176,166 KB/sec) for version 3.1 and version 2.1 respectively (Table 1). Table 2 shows how much time it took to convert the entire BMRB into its JSONized version and how much disk space it occupied as uncompressed directory and as compressed zip and tar archives. Compressed zip and tar formats represent the entire BMRB database in a single file and save disk space. In order to simplify access, our library provides facilities to directly read NMR-STAR files from zip and tar archives without the requirement to manually decompress and separate the archive into separate files first. Frequency polygons of loading times on Fig. 6 show that the majority of NMR-STAR and JSONized NMR-STAR files can be loaded

Table 4 Common usage patterns for the nmrstarlib module

Usage	Example
Reading:	`sf_gen = nmrstarlib.read_files('path')` `starfile = next(sf_gen)`
Access/Modification:	`starfile['saveframe']['key']` `starfile['saveframe']['key'] = new_value`
Writing:	`starfile.write(fileobj, fileformat='nmrstar')` `starfile.write(fileobj, fileformat='json')`

into `StarFile` object in less than 1 s per file and JSONized NMR-STAR files can be loaded much faster than the original NMR-STAR files. Figure 6a and b show that the fastest reading times were for parsing JSONized NMR-STAR files using the `ujson` and `json` parsers. However on Fig. 6a, it is clear that the pure Python implementation outperformed the Cython implementation for some of the NMR-STAR 2.1 files (e.g. BMRB ID: 17192, 16692). This is because those files contain saveframe categories deposited as very large multiline blocks of text and the majority of time is spent to pre-process them, equivalent NMR-STAR 3.1 files have those saveframes properly formatted and do not require extra time to pre-process multiline text blocks. For NMR-STAR 3.1 formatted files (Fig. 6b), the Cython implementation outperformed pure Python implementation in all cases.

Comparison to similar existing software

Using the entire BMRB, we performed and compared speed performance tests between our `nmrstarlib` package and the three other publically available Python libraries for reading NMR-STAR formatted files: PyStarLib [8], NMRPyStar [9], and PyNMRSTAR [10]. For each of these libraries, we wrote a simple Python

Table 3 Performance comparison of nmrstarlib to other Python libraries

		nmrstarlib	PyStarLib	NMRPyStar	PyNMRSTAR
Parsing NMR-STAR 2.1					
Number of files		11,270	11,270	11,270	11,270
Time, sec	Pure Python	326	239	N/A	547
	Python with C Extension	320	N/A	N/A	144
Success rate, %		100	99.57	0	100
Parsing NMR-STAR 3.1					
Number of files		11,244	11,244	11,244	11,244
Time, sec	Pure Python	1,100	796	56,569	2,354
	Python with C Extension	423	N/A	N/A	538
Success rate, %		100	95.92	100	100

Table 5 The nmrstarlib library command-line interface

Command	Description	Example

`convert`	Convert between NMR-STAR and JSON formats	`$ python3 -m nmrstarlib convert bmr18569.str 18569.json \`
		`-from_format=nmrstar -to_format=json`
		`$ python3 -m nmrstarlib convert 18569.json bmr18569.str \`
		`-from_format=json -to_format=nmrstar`

`csview`	View assigned chemical shifts	`$ python3 -m nmrstarlib csview 18569 \`
		`-csview_outfile=18569_cs_all`
		`-csview_format=png`
		`$ python3 -m nmrstarlib csview 18569 \`
		`-aminoacids=GLU,THR -atoms=CA,CB,CG,CG2 \`
		`-csview_outfile=18569_cs_GLU_THR_CA_CB_CG_CG2 \`
		`-csview_format=png`

program that loads a NMR-STAR formatted file from a directory, creates an object representation, and then reports how much time it took to process each file. Results of these comparisons are summarized in Table 3. For the pure Python implementation, PyStarLib showed the fastest reading time: 239 s (~4 min) and 796 s (~13.3 min) for NMR-STAR version 2.1 and version 3.1 respectively, but it was not able to parse 0.43% (48 files) NMR-STAR version 2.1 and 4.08% (459 files) NMR-STAR version 3.1. All errors occurred inside a function that is responsible for processing multiline quoted text, which uses regular expressions to collapse multiline quoted text into a single token. The most probable cause for these errors is a regular expression that is not capable of handling all edge cases. Examples of failures include files where: i) multiline quoted text included a semicolon character inside the text; ii) multiline quoted text that is not followed by the new line character; and iii) multiline quoted text followed by a loop (see Additional files 4, 5, 6, and 7 for list of failed files as of December 11, 2016 and particular fragments of files where the failure occurred for both NMR-STAR 2.1 and NMR-STAR 3.1 formatted files).

The pure Python implementation of the nmrstarlib package was the second fastest method 326 s (~5.4 min) and 1,110 s (~18.3 min) and, more importantly, parsed 100% of files for both NMR-STAR 2.1 and NMR-STAR 3.1, respectively. The NMRPyStar library showed the slowest results, taking 56,569 s (~15.7 h) to process NMR-STAR version 3.1 and was not able to read any of the NMR-STAR version 2.1 files (error status code was reported by the program during execution). Both the nmrstarlib and PyNMRSTAR provide Python + C extension implementations in order to speed up the tokenization process. The nmrstarlib performed faster than PyNMRSTAR on NMR-STAR 3.1 files: 423 s (~7 min) versus 538 s (~9 min). However, PyNMRSTAR was faster than nmrstarlib on NMR-STAR 2.1 files: 144 s (~2.4 min) versus 320 s (~5.3 min). Overall, the nmrstarlib (Python + C extension implementation) was the fastest method to read NMR-STAR 3.1 files, and PyNMRSTAR (Python + C extension implementation) was the fastest method to read NMR-STAR 2.1 files. However, when using the JSONized versions of NMR-STAR files with the nmrstarlib library, parsing speed can be further improved to 30 s for NMR-STAR 2.1 and 130 s for NMR-STAR 3.1 (see Table 1).

Table 6 Comparison of nmrstarlib to other Python libraries

Feature	nmrstarlib	PyStarLib	NMRPyStar	PyNMRSTAR
Read NMR-STAR 2.1	Yes	Yes	No	Yes
Read NMR-STAR 3.1	Yes	Yes	Yes	Yes
Supported Python version	2.7, 3.4+	2.7	2.7	2.6, 2.7, 3.3+
API Reference documentation	Yes	No	No	Yes
Tutorial documentation	Yes	No	No	Yes
PDF of documentation	Yes	No	No	Yes
User Guide documentation	Yes	No	Yes	No
Up to date online documentation	Yes	No	No	No
Open Source	Yes (GitHub)	Yes (SourceForge)	Yes (GitHub)	Yes (GitHub)

```
                         R Example using jsonlite library
> # install library
> install.packages("jsonlite")

> # load library
> library(jsonlite)

> # load data
> starfile <- fromJSON("bmr18569.str.json")

> # print saveframe names
> names(starfile)
 [1] "data"                      "save_entry_information"
 [3] "save_entry_citation"       "save_assembly"
 [5] "save_EVH1"                 "save_natural_source"
 [7] "save_experimental_source"  "save_sample_1"
 [9] "save_sample_2"             "save_sample_3"
[11] "save_sample_4"             "save_sample_conditions_1"
[13] "save_sample_conditions_2"  "save_sample_conditions_3"
[15] "save_sample_conditions_4"  "save_AZARA"
[17] "save_xwinnmr"              "save_ANSIG"
[19] "save_CNS"                  "save_spectrometer_1"
[21] "save_spectrometer_2"       "save_NMR_spectrometer_list"
[23] "save_experiment_list"      "save_chemical_shift_reference_1"
[25] "save_assigned_chem_shift_list_1" "save_combined_NOESY_peak_list"

> # access saveframe key-value data
> starfile$data
[1] "18569"
>
> starfile$save_entry_information$Entry.NMR_STAR_version
[1] "3.1.1.61"
>
> # access loop data
> starfile$save_entry_information$loop_1
[[1]]
[1] "Data_set.Type"   "Data_set.Count"   "Data_set.Entry_ID"

[[2]]
        Data_set.Type Data_set.Count Data_set.Entry_ID
1 assigned_chemical_shifts           1            18569
2       spectral_peak_list           1            18569
```

Fig. 7 Code example showing how to access data from JSONized NMR-STAR files using R programming language

All tests were performed on a single workstation desktop computer with Intel(R) Core(TM) i7-4930 K CPU @ 3.40GHz processor, 64 GB memory, and a solid-state drive. The latest stable version of Python (Python 3.6.0) was used to compare libraries. Python version 2.7 was used for libraries that do not support the latest version of Python.

Discussion
The nmrstarlib interface
To use nmrstarlib as a library, first import the library. Next, create a StarFile generator that will return StarFile instances one at a time from many different file sources: a local file, URL address of a file, directory, archive, BMRB id. Next, the Star-File object can be utilized like any built-in Python dict object. Table 4 shows common usage patterns for reading NMR-STAR files into StarFile objects, accessing and manipulating data using bracket accessors, and writing StarFile objects back to both NMR-STAR and JSONized NMR-STAR formats. For more detailed examples, see "The nmrstarlib Tutorial" documentation (Additional file 3).

The nmrstarlib command-line interface provides two commands: convert in order to convert between NMR-STAR format and its equivalent JSON format; the csview

```
                         JavaScript Example using jQuery
<!DOCTYPE html>
<html>
  <head>
    <title>Reading JSONized NMR-STAR with jQuery</title>
  </head>
  <body>
    <script src="https://ajax.googleapis.com/ajax/libs/jquery/3.1.0/jquery.min.js"></script>
    <script>
      $.getJSON("bmr18569.str.json", function(starfile) {
        console.log(starfile.data);               // prints data tag id
        console.log(starfile.save_entry_information); // prints entire saveframe data
        console.log(starfile.save_entry_information.loop_1); // prints loop_1 data
      });
    </script>
  </body>
</html>
```

Fig. 8 Code example showing how to access data from JSONized NMR-STAR files using JavaScript programming language

command for quick access to assigned chemical shift data of a single `StarFile`, organizing chemical shifts by amino acid residue type. Table 5 shows common usage examples for the `convert` and `csview` commands. For a full list of available conversion options and more detailed examples see "The `nmrstarlib` API Reference" and "The `nmrstarlib` Tutorial" documentation. Figure 4 shows example output of the `csview` command.

We also have developed the "User Guide", "The `nmrstarlib` Tutorial" and "The `nmrstarlib` API

C++ example using RapidJSON library

```cpp
#include <iostream>

// include rapidjson headers
#include "rapidjson/document.h"
#include "rapidjson/filereadstream.h"

using namespace std;

int main()
{
  // open file
  FILE* fp = fopen("bmr18569.str.json", "r"); // Windows use "rb"

  // read input stream via FILE pointer
  char readBuffer[65536];
  rapidjson::FileReadStream is(fp, readBuffer, sizeof(readBuffer));

  // create rapidjson::Document and parse input stream
  rapidjson::Document starfile;
  starfile.ParseStream(is);

  fclose(fp); // close file pointer

  // print saveframe names
  cout << "Accessing saveframe categories: \n";
  for (rapidjson::Value::ConstMemberIterator itr = starfile.MemberBegin();
    itr != starfile.MemberEnd(); ++itr)
  {
    cout << "  " << itr->name.GetString() << "\n";
  }
  // access saveframe key-value data
  cout << "Accessing saveframe data: \n";
  cout << "  " << "data: " << starfile["data"].GetString() << "\n";
  cout << "  " << "NMR-STAR version: " <<
      starfile["save_entry_information"]["Entry.NMR_STAR_version"].GetString() << "\n";

  // access loop data
  cout << "Accessing loop data:\n";
  const rapidjson::Value& loop_1_fields = starfile["save_entry_information"]["loop_1"][0];
  const rapidjson::Value& loop_1_values = starfile["save_entry_information"]["loop_1"][1];

  cout << "loop fields:\n";
  for (rapidjson::SizeType i = 0; i < loop_1_fields.Size(); i++)
  {
    cout << "  " << loop_1_fields[i].GetString() << "\n";
  }

  cout << "loop values:\n";
  for (rapidjson::SizeType i = 0; i < loop_1_values.Size(); i++)
  {
    for (rapidjson::Value::ConstMemberIterator itr = loop_1_values[i].MemberBegin();
      itr != loop_1_values[i].MemberEnd(); ++itr)
    {
      itr->name.GetString();
      cout << "  " << itr->name.GetString() << ": " << itr->value.GetString() << "\n";
    }
  }
}
```

Fig. 9 Code example showing how to access data from JSONized NMR-STAR files using C++ programming language

Reference" documentation that is available as a PDF file (Additional file 3) and up-to-date online documentation (Table 6).

Advantages of using nmrstarlib and JSONized NMR-STAR version

One of the main advantages of our library is that it provides a one-to-one mapping between each of the following representations of BMRB entries: NMR-STAR format, internal Python `OrderedDict`- and `list`-based objects, and JSONized NMR-STAR format. This makes the library more Python-idiomatic, providing a very intuitive programming interface for accessing and manipulating NMR data. Another benefit of our `nmrstarlib` package is that the `bmrblex` lexical analyser module is written in a generic fashion, making it easy to adapt for parsing data from other STAR-related formats, for example, the Crystallographic Information File (CIF) and its closely related macromolecular CIF (mmCIF) format.

JSON is an open, programming language independent, human-readable, data exchange standard that represents data objects in a nested dictionary/list ASCII format. JSON is one of the most common formats for asynchronous browser/server communication as an alternative to XML (Extensible Markup Language). We selected the JSON object representation, because it has a smaller overhead compared to common XML object representations, making it faster to parse and more human-readable when formatted for this purpose. But more importantly, it facilitates a one-to-one mapping with both nested Python data structures and BMRB's nested data representations of their entries. While XML is more flexible, it is not easily represented by a nesting of standard Python data structures that would produce an intuitive programming interface. Also, JSONization of the original NMR-STAR files provides several advantages: i) much faster reading times (see Table 1) and ii) makes the data stored in BMRB entries easily accessible to other programming languages that have JSON parsers, i.e. all modern programming languages, scripting as well as compiled, without requiring to write a specific parser for the specialized NMR-STAR format. Figures 7, 8, and 9 show code examples for accessing data from JSONized NMR-STAR files using R with the `jsonlite` library [16], JavaScript with the `jQuery` library [17], and C++ with the `RapidJSON` library [18] (Additional file 8 provides output of C++ example after compilation and execution), respectively.

But one disadvantage of using JSON format is that it is more verbose in comparison to the original NMR-STAR format. As a result, uncompressed JSONized NMR-STAR files occupy more disk space (Table 2). However, the `nmrstarlib` library offers the ability to read NMR-STAR files in both uncompressed (directory of files) and compressed (zip and tar archives) forms, making storage and access of JSONized NMR-STAR files very efficient.

Conclusions

The `nmrstarlib` package is a useful Python library, providing classes and other facilities for parsing, accessing, and manipulating data stored in NMR-STAR and JSONized NMR-STAR formats. Also, `nmrstarlib` provides a simple command-line interface that can convert from the NMR-STAR file format into its equivalent JSON file format and vice versa, as well as accessing and visualizing assigned chemical shift values. The library has an easy-to-use, idiomatic dictionary-based interface, usable in programs written in Python. The library also has extensive documentation including the "User Guide", "The `nmrstarlib` Tutorial", and "The `nmrstarlib` API Reference". Furthermore, the easy conversion into the JSONized NMR-STAR format facilitates utilization of BMRB entries by programs in any programming language with a JSON parser. This same basic approach can be used to quickly JSONize other older text-based scientific data formats, making the underlying scientific data easily accessible in a wide variety of programming languages. As demonstrated in this study, many available JSON parsers are highly optimized and typically much more efficient than specialized parsers for scientific data formats. Thus, JSONization of older scientific data formats provides easy steps for reaching Interoperability and Reusability goals of FAIR guiding principles [19].

Additional files

Additional file 1: Function call diagram of nmrstarlib.

Additional file 2: Profile of nmrstarlib execution.

Additional file 3: Documentation for nmrstarlib.

Additional file 4: List of failed NMR-STAR 2.1 files for PyStarLib.

Additional file 5: List of failed NMR-STAR 3.1 files for PyStarLib.

Additional file 6: Fragments of failed NMR-STAR 2.1 files for PyStarLib.

Additional file 7: Fragments of failed NMR-STAR 3.1 files for PyStarLib.

Additional file 8: Output of C++ example from Fig. 9.

Abbreviations
NMR: Nuclear magnetic resonance; BMRB: Biological Magnetic Resonance Data Bank; STAR: Self-defining text archival and retrieving; JSON: JavaScript Object Notation; XML: Extensible markup language; UML: Unified modeling language

Acknowledgements
We want to acknowledge the constant work and effect that the BMRB staff have done over the years to maintain and expand the BMRB public repository of NMR data.

Funding
This work was supported by National Science Foundation grant NSF 1252893 (Hunter N.B. Moseley); however, they played no role in the design or conclusions of this study.

Authors' contributions
AS, MA, and HNBM worked together on the design of the library and its API. AS and MA implemented the library. HNBM helped troubleshoot implementation issues. AS created library documentation. AS tested the library and compared performance to other libraries. AS and HNBM wrote the manuscript. All authors have read and approved the manuscript.

Competing interests
The authors declare that they have no competing interests.

Author details
[1]School of Interdisciplinary and Graduate Studies, University of Louisville, Louisville, KY 40292, USA. [2]Department of Computer Engineering and Computer Science, University of Louisville, Louisville, KY 40292, USA. [3]Department of Molecular and Cellular Biochemistry, University of Kentucky, Lexington, KY 40356, USA. [4]Markey Cancer Center, University of Kentucky, Lexington, KY 40356, USA. [5]Center for Environmental and Systems Biochemistry, University of Kentucky, Lexington, KY 40356, USA. [6]Institute for Biomedical Informatics, University of Kentucky, Lexington, KY 40356, USA.

References
1. Ulrich EL, Akutsu H, Doreleijers JF, Harano Y, Ioannidis YE, Lin J, Livny M, Mading S, Maziuk D, Miller Z, Nakatani E, Schulte CF, Tolmie DE, Kent Wenger R, Yao H, Markley JL. BioMagResBank. Nucleic Acids Res. 2008;36 Suppl 1:D402–8.
2. Berman H, Henrick K, Nakamura H, Markley JL. The worldwide Protein Data Bank (wwPDB): Ensuring a single, uniform archive of PDB data. Nucleic Acids Res. 2007;35 Suppl 1:D301–3.
3. Hall SR. The STAR file: a new format for electronic data transfer and archiving. J Chem Inf Model. 1991;31(2):326–33.
4. Van Rossum G, Drake FL Jr. The Python Language Reference. Technical report, Python Software Foundation; 2014.
5. Van Rossum G, Drake FL. The Python Library Reference, October. 2010. p. 1–1144.
6. Ronacher A, Hettinger R. PEP 372—Adding an ordered dictionary to collections. [Online]. Available: https://www.python.org/dev/peps/pep-0372/. Accessed June 2008.
7. Python 2.7 Countdown. [Online]. Available: https://pythonclock.org/.
8. Doreleijers J. PyStarLib. [Online]. Available: https://sourceforge.net/projects/pystarlib/. Accessed Oct 2014.
9. Fenwick M. NMRPyStar. [Online]. Available: https://github.com/mattfenwick/NMRPyStar. Accessed Dec 2014.
10. Wedell J. PyNMRSTAR. [Online]. Available: https://github.com/uwbmrb/PyNMRSTAR. Accessed Mar 2017.
11. Behnel S, Bradshaw R, Citro C, Dalcin L, Seljebotn DS, Smith K. Cython: The best of both worlds. Comput Sci Eng. 2011;13(2):31–9.
12. graphviz Python library. [Online]. Available: http://graphviz.readthedocs.io/en/latest/index.html. Accessed 2017.
13. docopt Python Library for creating command-line interfaces. [Online]. Available: http://docopt.readthedocs.io/en/latest/. Accessed Apr 2016.
14. Biological Magnetic Resonance Bank. [Online]. Available: http://www.bmrb.wisc.edu/. Accessed Mar 2017.
15. UltraJSON. UltraJSON is an ultra fast JSON encoder and decoder written in pure C with bindings for Python 2.5+ and 3. [Online]. Available: https://github.com/esnme/ultrajson. Accessed Feb 2017.
16. Ooms J, Lang TD, Lloyd H. jsonlite: A Robust, High Performance JSON Parser and Generator for R. [Online]. Available: https://cran.r-project.org/web/packages/jsonlite/index.html. Accessed Feb 2017.
17. jQuery is a cross-platform JavaScript library. [Online]. Available: http://jquery.com/. Accessed Jan 2017.
18. Yip M. RapidJSON - A fast JSON parser/generator for C++ with both SAX/DOM style API. [Online]. Available: http://rapidjson.org/. Accessed Mar 2017.
19. Wilkinson MD, Dumontier M, Aalbersberg IJ, Appleton G, Axton M, Baak A, Blomberg N, Boiten J-W, da Silva Santos LB, Bourne PE, Bouwman J, Brookes AJ, Clark T, Crosas M, Dillo I, Dumon O, Edmunds S, Evelo CT, Finkers R, Gonzalez-Beltran A, Gray AJG, Groth P, Goble C, Grethe JS, Heringa J, 't Hoen P a, Hooft R, Kuhn T, Kok R, Kok J, Lusher SJ, Martone ME, Mons A, Packer AL, Persson B, Rocca-Serra P, Roos M, van Schaik R, Sansone S-A, Schultes E, Sengstag T, Slater T, Strawn G, Swertz M a, Thompson M, van der Lei J, van Mulligen E, Velterop J, Waagmeester A, Wittenburg P, Wolstencroft K, Zhao J, Mons B. The FAIR Guiding Principles for scientific data management and stewardship. Sci Data. 2016;3:160018.

4

M-AMST: an automatic 3D neuron tracing method based on mean shift and adapted minimum spanning tree

Zhijiang Wan[1,2,3,4,5], Yishan He[2,3,4,5], Ming Hao[2,3,4,5], Jian Yang[2,3,4,5] and Ning Zhong[1,2,3,4,5]*

Abstract

Background: Understanding the working mechanism of the brain is one of the grandest challenges for modern science. Toward this end, the BigNeuron project was launched to gather a worldwide community to establish a big data resource and a set of the state-of-the-art of single neuron reconstruction algorithms. Many groups contributed their own algorithms for the project, including our mean shift and minimum spanning tree (M-MST). Although M-MST is intuitive and easy to implement, the MST just considers spatial information of single neuron and ignores the shape information, which might lead to less precise connections between some neuron segments. In this paper, we propose an improved algorithm, namely M-AMST, in which a rotating sphere model based on coordinate transformation is used to improve the weight calculation method in M-MST.

Results: Two experiments are designed to illustrate the effect of adapted minimum spanning tree algorithm and the adoptability of M-AMST in reconstructing variety of neuron image datasets respectively. In the experiment 1, taking the reconstruction of APP2 as reference, we produce the four difference scores (entire structure average (ESA), different structure average (DSA), percentage of different structure (PDS) and max distance of neurons' nodes (MDNN)) by comparing the neuron reconstruction of the APP2 and the other 5 competing algorithm. The result shows that M-AMST gets lower difference scores than M-MST in ESA, PDS and MDNN. Meanwhile, M-AMST is better than N-MST in ESA and MDNN. It indicates that utilizing the adapted minimum spanning tree algorithm which took the shape information of neuron into account can achieve better neuron reconstructions. In the experiment 2, 7 neuron image datasets are reconstructed and the four difference scores are calculated by comparing the gold standard reconstruction and the reconstructions produced by 6 competing algorithms. Comparing the four difference scores of M-AMST and the other 5 algorithm, we can conclude that M-AMST is able to achieve the best difference score in 3 datasets and get the second-best difference score in the other 2 datasets.

Conclusions: We develop a pathway extraction method using a rotating sphere model based on coordinate transformation to improve the weight calculation approach in MST. The experimental results show that M-AMST utilizes the adapted minimum spanning tree algorithm which takes the shape information of neuron into account can achieve better neuron reconstructions. Moreover, M-AMST is able to get good neuron reconstruction in variety of image datasets.

Keywords: M-AMST, Neuron reconstruction, Mean shift, Sphere model, Coordinate transformation

* Correspondence: zhong@maebashi-it.ac.jp
[1]Beijing Advanced Innovation Center for Future Internet Technology, Beijing University of Technology, Beijing, China
[2]Department of Life Science and Informatics, Maebashi Institute of Technology, Maebashi, Japan
Full list of author information is available at the end of the article

Background

Understanding how the brain works from the aspect of cognition and structure is one of the greatest challenges for modern science [1]. On one hand, it is meaningful to systematically investigate human information processing mechanisms from both macro and micro points of views by cooperatively using experimental, computational, cognitive neuroscience. On the other hand, acquiring knowledge of the neuron' morphological structure is also of particular importance to simulate the electrophysiological behavior which intricately links with cognitive function and promotes our understanding of brain. Based on the above views, several research journals (Brain Informatics [2], Bioimage Informatics [3] and Neuroinformatics [4]), worldwide neuron reconstruction contest (DIADEM [5]) and bench testing project (BigNeuron [6]) have been launched. One of their basic tasks is how to extract the neuronal morphology from the molecular and cellular microscopic images, namely neuron reconstruction or neuron tracing.

In order to get the neuron tracing algorithms with high performance as many as possible, the BigNeuron project aims at gathering a worldwide community to define and advance the state-of-the-art of single neuron reconstruction. The primary method to achieve that goal is to bench test as many varieties of automated neuron reconstruction methods as possible against as many neuron datasets as possible following standardized data protocols [6]. So far, varieties of neuron reconstruction methods based on image segmentation theories such as fuzzy set [7], level set [8, 9], active contour model [10–12], graph theory [13], and clustering [14, 15] have been contributed to the project. For example, APP2 algorithm based on level set theory can generate reliable tree morphology of neuron with the fastest tracing speed [9]. A neuron tracing algorithm named Micro-Optical Sectioning Tomography ray-shooting (MOST for short) achieves a good result in terabytes 3D datasets of the whole mouse brain [16]. Additionally, a neuron tracing algorithm named SIMPLE is a DT-based method and can produce better reconstruction in dragonfly thoracic ganglia neuron images than other methods [17]. A neuron tracing method based on graph theory, namely neuron tracing minimum spanning tree (N-MST for short), also gets reasonable reconstructions for several neuron image datasets. Due to the spatial nature of image, the methods mentioned above are all take the spatial information into account. However, in some segmentation scenarios, the objects of interest may be reasonably characterized by an intensity distribution. In the 3D image, the more voxels distributed in an image region, the region has a higher voxel density. For such a situation, it is important to integrate intensity information into a spatial algorithm. The neuron tracing method based on clustering is the algorithm which adopts

spatial information and intensity distribution of neuron simultaneously. Moreover, because the clustering algorithms are intuitive and, some of them, easy to implement, they are very popular and widely used in image segmentation. For instance, mean shift is a nonparametric density gradient estimation using a generalized kernel approach and is one of the most powerful clustering techniques. Cai et al. proposed a cross-sections of axons detection and connection method using nonlinear diffusion and mean shift [14]. The automatic method can shift the centroids of cross-section on slice A iteratively until the sample mean convergence on slice B. They concluded the centroid on slice A and the centroid on slice B correspond to the same axon. Comaniciu et al. proposed a robust approach for the analysis of a complex multimodal feature space and to delineate arbitrarily shaped clusters in it [18]. The basic computational module of the technique is the mean shift. They proved for discrete data the convergence of a recursive mean shift procedure to the nearest stationary point of the underlying density function and, thus, its utility in detecting the modes of the density. They also claimed that the mean shift algorithm is a density estimation-based non-parametric clustering approach that the data space can be regarded as the empirical probability density function (p.d.f) of the represented parameter. As we know, dense region in the data space corresponds to local maxima of the p.d.f, that is, to the modes of the unknown density. Once the location of a mode is determined, the cluster associated with it is delineated based on the local structure of the data space. As it happens, the neuron image generated by fluorescent probes has the characteristics of spatial distribution, intensity discretization and the portions around the neuron skeleton have a higher voxel density. Based on this, we developed a neuron tracing algorithm based on mean shift and minimum spanning tree as a contribution to the BigNeuron project in the beginning. Specifically, the algorithm can move each voxel to the local mean until automatically get the convergence region which has the local maxima of the p.d.f. Meanwhile, the voxels in the convergence regions can also be considered as the classification voxels which indicate the modes of unknown density, other voxels which shift toward and finally locate at the regions after several iterations can be marked as the same classification as the corresponding classification voxel. They also can be regarded as the voxels subordinated to the classification voxels. Based on the subordinate voxels belong to the different classification voxel, the local structure of the neuron can be delineated. In this method, not only the information of voxel density distribution of neuron image can be captured correctly, but also got the sufficient voxels belong to several modes to delineate the whole neuron topological structure. In the basis of the classification voxels and their own subordinate voxels, we can use the minimum spanning tree (MST) algorithm to reconstruct the neuron. It is worth noting that with MST, the information of spatial distribution of neuron is also adopted to get the neuron reconstruction.

However, the experimental result of M-MST shows that although the M-MST algorithm can reconstruct 120 images successfully, it generates less precise connection between some neuron segments since the MST just takes the spatial distance as the weight of edge to build the paternity between nodes. The node is defined as the voxel with the property of spatial coordinates, radius, node type and parent compartment. The situation about less precise connection between some neuron segments caused by MST can be illustrated as the following:

Node A and node B belong to the different neuron segments and have the nearest distance or the smallest edge weight in the neuron image. According to the topological structure of original neuron image, there is a gap between them. In this case, it is not suitable to use the minimum spanning tree algorithm to build paternity of the two nodes directly. If we can detect the gap between the two nodes and set their weight of edge in MST as a high value which is exceed a lot than the real spatial distance, the MST will choose other pair of nodes which have no gap between them to form the neuron segment. The pair of nodes which have no gap between them are more likely subordinate to the same neuron segment. Once the gap is detected, the shape information can be captured. Therefore, the weight calculation method considers the shape information of neuron will help for achieving more precise neuron reconstruction.

In this paper, we focus on introducing an improved algorithm, namely M-AMST, in which a rotating sphere model based on coordinate transformation is used to improve the weight calculation method in M-MST. Figure 1 gives an overview of the M-AMST and the related reconstruction result in four steps. The four steps can be described as follows. Firstly, input a single neuron image (Fig. 1(a)) into the Vaa3D [19]. Secondly, for each foreground voxel, the mean shift algorithm defines a window with certain spatial range and takes the voxel as the center of the window. Then it shifts the voxel to the local mean iteratively until getting the convergence region. It is worth noting that the voxel located at the convergence region cannot shift for one step. This kind of voxel could be considered as the classification voxel. By observing their position in the neuron image, the most of classification voxels are located around the neuron skeleton, in which the neuron segment with high intensity and voxel density located. Moreover, a radius calculation method is adopted to calculate the radius of every foreground voxel. The foreground voxel with radius can be called as foreground node. Due to the size of every foreground node is greater or equal than 1.0, some of them might be overlapped or even covered by others. Such case is deemed as the repeat expression or the over reconstruction of neuron. We need the nodes as few as possible to express the neuronal morphology as complete as possible. In response, a node pruning method based on the distance between the pair of nodes and their own radiuses is developed. After that, a slew of nodes can be retained and formed to be a node set. The node set will be considered as the seeds to be input into the adapted MST to build the tree structure of neuron. In Fig. 1(b), the node set in green color extracted by mean shift algorithm and pruned by the node pruning method are overlaid on top of original neuron image. Thirdly, a rotating sphere model based on coordinate transformation is implemented to extract a pathway between each pair of nodes. In Fig. 1(c), the initial reconstruction result is overlaid on top of original neuron image, the white line between green nodes pointed by the yellow arrow is the pathway extracted by the rotating sphere model based on coordinate transformation, the green line is not a pathway generated by the model since there is a gap between the two nodes. Fourthly, take the accumulating distance of the node list in the pathway between each pair of nodes as the weight and use the minimum spanning tree algorithm to reconstruct the

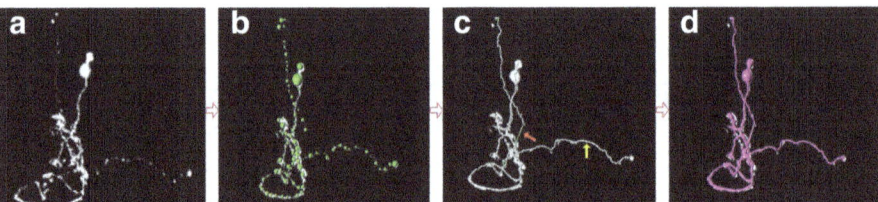

Fig. 1 Overview of the M-AMST and the related reconstruction result in four steps. **a** an original neuron image. **b** the node set in green color extracted by mean shift algorithm and pruned by the node pruning method are overlaid on top of original neuron image. **c** the initial reconstruction result is overlaid on top of original neuron image, the white line between green nodes pointed by the yellow arrow is the pathway extracted by the rotating sphere model based on coordinate transformation, the green line is not a pathway generated by the model since there is a gap between the two nodes. **d** final reconstruction result in red color is overlaid on top of original neuron image

neuron image. In Fig. 1(d), final reconstruction result in red color is overlaid on top of original neuron image.

Method

Topological structure segmentation

A. Voxel clustering using mean shift algorithm

On the whole, we follow the sequence of neuron reconstruction operations: binarization, skeletonization, rectification and graph representation [1]. In the binarization operation, we firstly define the voxel whose intensity is less than a threshold as the dark spot and otherwise the bright spot. For each voxel, the number of the dark spots and the bright spots among the 26 surrounding voxels are calculated respectively. And then, we calculate the ratio of the number of the dark spots and the bright spots, the ratio is compared with a threshold to determine whether the voxel is a foreground or not. In the skeletonization operation, we use mean shift algorithm to extract the neuron skeleton.

The implementation of mean shift in this paper is interpreted as the following steps:

(1) Mean shift involves shifting a kernel iteratively to a region with higher density until convergence. We shift the 3D coordinate of each voxel using a Gaussian kernel described as the following:

$$K(x) = \frac{1}{2\pi\delta^2} \exp\left(-C * \frac{|x-\bar{x}|^2}{2\delta^2}\right), \qquad (1)$$

C is a scaling coefficient, $\bar{x}$ is the average and δ is standard deviation. The calculation method of $\bar{x}$ and δ are illustrated as follow.

(2) Assume a sphere centered on voxel P and with radius r. Using X-axis as example, the x in the formula (1) can be calculated by

$$x = \left(x^r - x_i^r\right) * \left(x^r - x_i^r\right)/r * r, \qquad (2)$$

where x^r means the x-coordinate of P, x_i^r means the x-coordinate of a voxel in the sphere. The standard deviation δ is calculated by

$$\delta = \sqrt{\sum_{i=1}^{n}(x_i - \bar{x}) * (x_i - \bar{x})/n}, \qquad (3)$$

where x_i is a x-coordinate converting value which is obtained by formula (2), $\bar{x}$ is the average of the x-coordinate converting value of every voxel in the sphere. The average $\bar{x}$ is calculated by

$$\bar{x} = \sum_{i=1}^{n} x_i/n. \qquad (4)$$

(3) The new coordinate value of the sphere center in X-axis is calculated by

$$next_x = \frac{\sum_a K\left(\left(x^r - x_i^r\right) * \left(x^r - x_i^r\right)/r * r\right) * x_i^r}{\sum_a K\left(\left(x^r - x_i^r\right) * \left(x^r - x_i^r\right)/r * r\right)}. \qquad (5)$$

where $next_x$ is the coordinate values of the new center voxel in X-axis. The symbol a indicates the whole sphere region for the current foreground voxel. It is worth noting that the calculation method of the new coordinate value in Y-axis and Z-axis is as the same as the method mentioned above.

B. Covered node pruning

As mentioned above, we define the voxels which cannot shift for one iteration as the classification voxels which also can be called as marks, the other voxels that shift toward iteratively and finally located at the marks can be defined as the corresponding subordinate voxels. The two kinds of voxels can be used to reconstruct the whole neuronal topological structure. However, after calculating radius for the marks and the corresponding subordinate voxels, they might be overlapped or even covered by others. In order to get the nodes as few as possible to express as complete neuronal morphology as possible, we prune the marks and other nodes overlapped or covered by others using a node pruning method. The node pruning method adopts three steps listed as follows:

(1) For a pair of marks, prune the covered mark according to the distance between them and their own radiuses. Figure 2 gives three covering situations of mark. The red and purple dots represent two different marks and their own radiuses are r_1 and r_2 respectively. We define two marks should all be kept (Fig. 2a) if the difference between their Euclidean distance D and the sum of their radiuses is greater than a threshold. Conversely, prune one of marks (Fig. 2b and c) without defining a particular pruning priority. The threshold should be set greater than 3 according to the prior knowledge which claimed that the human eye can tell the detail variation above 3 voxels.

(2) Remark the subordinate nodes of the removing mark to the corresponding keeping mark. Specifically, due to some covered or overlapped marks are pruned, the nodes which are subordinate to the removing marks should be re-subordinate to the keeping marks. We deal with this using a two-

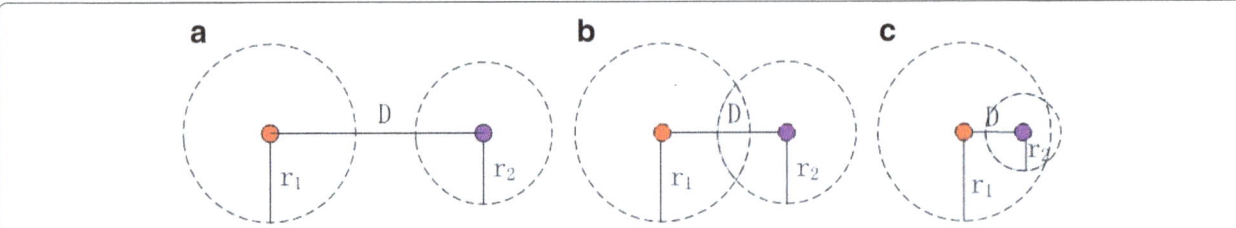

Fig. 2 Three covering situations of mark. Red and purple dots represent two different marks, their own radiuses are r_1 and r_2 respectively. D is their Euclidean distance. **a** Keep; **b** prune one mark and **c** prune one mark

fold QMap data structure, the keys of first and second QMap mean the order number of keeping mark and the removing mark in the original mark set respectively. The values of second QMap mean the subordinate node set belongs to the corresponding removing mark. After remarking, we get the keeping marks and their own subordinate node set.

(3) For each node set, prune the subordinate nodes overlapped or covered by others but always keep the marks. The pruning method is consistent with the first step.

Pathway extraction using a sphere model

Based on the marks and their own subordinate node set, we use MST algorithm to reconstruct the neuron and finish the graph representation step. The main principle of MST is to select a pathway with minimum total weights to connect all vertices in a connected and undirected graph. Taking the spatial distance between each pair of nodes as the weight is the most obvious weight calculation method. In M-MST, we build an undirected graph by connecting all extracted nodes firstly and then calculate the spatial distance between each pair of nodes as the weight of edge. However, building the paternity between each pair of nodes according to their spatial distance could easily lead to the less precise connection between some neuron segments. The weight calculation method combines the shape and spatial information of neuron will help for achieving more precise neuron reconstruction. Therefore, in order to get an accurate neuronal topological reconstruction, a rotating sphere model based on coordinate transformation is proposed to improve the weight calculation method of edge in M-MST. The core idea of the rotating sphere model is to move the sphere centered on each node in the node set to progressively approach other nodes. For each pair of nodes, we define the two nodes as starting node and terminal node respectively. A pathway could be extracted between the starting node and the terminal node if there is no gap between them or one node is not far away from the other. The main principle of the rotating sphere model based on coordinate transformation is described as the following.

For each pair of nodes, a sphere centered at the starting node can be constructed in the beginning. The sphere can be split up into several quadrants by the 3D coordinate axis, every quadrant contains several foreground voxels. We always define the line between the starting node and terminal node as the Y'-axis and the positive direction of Y'-axis starts from the starting node to the terminal node. Taking the position of starting node as reference, different foreground voxels in the sphere have different coordinate value of Y'-axis. Suppose that there are two vectors named A and B respectively. The vector A starts from the sphere center to the foreground voxel with positive coordinate value of Y'-axis, the vector B starts from the sphere center to the voxel with negative coordinate value of Y'-axis. The angle between vector A and the positive direction of Y'-axis will always less or equal than 90° and the cosine value of the angle is greater or equal than 0. The angle between vector B and the positive direction of Y'-axis will always greater than 90° and the cosine value of the angle is less than 0. We define the rotating direction of the sphere always follow the direction of the vector A. Due to the Vaa3D platform who possesses a self-defined 3D coordinate system (X, Y and Z), the coordinate value of every foreground voxel in the 3D coordinate system should be transformed to the new coordinate system (X', Y' and Z'). This operation aims at ensuring the sphere centered at the starting node could always be approaching to the terminal node. Toward this purpose, we use two rotation steps to recalculate the new coordinate value of every foreground voxel in the sphere.

(1) First rotation:

$$
\begin{aligned}
x_1 &= x \\
y_1 &= cos\theta * y - sin\theta * z \\
z_1 &= sin\theta * y + cos\theta * z
\end{aligned}
\tag{6}
$$

where θ means the angle between Y-axis and T1, T1 means the line between the projection point of terminal node in the YOZ plane and sphere center.

(2) Second rotation:

$$x_2 = cos\gamma * x_1 - sin\gamma * y_1$$
$$y_2 = sin\gamma * x_1 + cos\gamma * y_1 \quad (7)$$
$$z_2 = z_1$$

where γ means the angle between Y'-axis and T2, T2 means the line between the projection point of terminal node in the YOZ plane and sphere center.

It is worth noting that in every rotating step, we guarantee the new sphere center located at the voxel which intensity is greater than 0. Specifically, we select a foreground voxel which intensity is greater than 0 from one of the quadrants as the new center of sphere. Every foreground voxel in different quadrant can be assigned a weight which indicates the gravitation attracted by the terminal node. We accumulate the weight value of all voxels in every quadrant respectively and select the maximum as the candidate quadrant, the new center of sphere is the voxel which has the shortest distance to the terminal node in the candidate quadrant. Every step will iteratively repeat the method until the terminal node located within the radius range of the new sphere. The calculation formula of total gravitation F for each quadrant is shown as the following:

$$F = \sum_q cos\theta * (I_1 * I_2 / D^2) / n \quad (8)$$

where θ means the angle between the vector which is from the sphere center to the foreground voxel and the positive direction of Y'-axis. I_1 and I_2 indicate the intensity of the foreground voxel in the quadrant and terminal node respectively. D is the Euclidean distance between the foreground voxel and terminal node, n is the number of voxels included in the q-th quadrant.

Figure 3 gives the schematic map of the rotating sphere model. Part A indicates the coordinate system (X, Y and Z) in Vaa3D platform, the two red dots (S and T) imply the starting and terminal node respectively. Part B means the new coordinate system (X', Y' and Z'). Part C illustrates the rotating steps from the starting node (marked with red (a)) to the terminal node (marked with red (d)). Red (b) and red (c) are the schematic sphere center which are selected from the rotating procedures. Using the rotating sphere model based on coordinate transformation, a pathway can be extracted if there is no gap between the pair of nodes. It is worth noting that the pathway contains a node list since the sphere model continuously selects the foreground voxel as the new sphere center in every rotating step. Every node list represents the shape information between the corresponding pair of nodes.

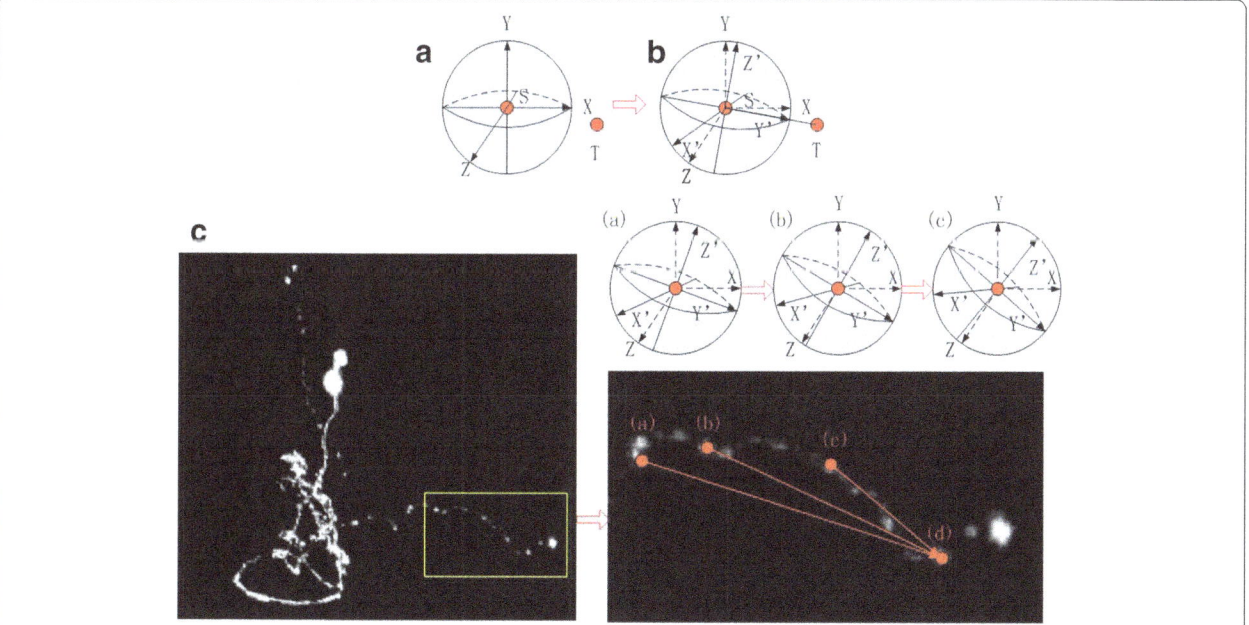

Fig. 3 The schematic map of the rotating sphere model. Part A indicates the coordinate system (X, Y and Z) in Vaa3D platform, the two red dots (S and T) imply the starting and terminal node respectively. Part B means the new coordinate system (X', Y' and Z'). Part C illustrates the rotating procedure from the starting voxel (marked with red (**a**)) to the terminal voxel (marked with red (**d**)). Red (**b**) and red (**c**) are the schematic sphere center which are selected from the rotating procedures

Neuron reconstruction adopting MST algorithm

For each rotating step, we take the sphere center chose in the previous rotating step as the parent node of the current sphere center. A node list can be got if the starting node approaches the terminal node successfully. The node list is also a list which is composed of voxels with paternity. We defined the node list as the pathway between the starting node and the terminal node. Comparing with using the Euclidean distance as the weight of edge of the MST, it is more reasonable to adopt the accumulating distance of the node list as the weight of edge. Specifically, the accumulating distance can be calculated by summing up the distance between each pair of the parent node and the child node in the list. And then, take the accumulating Euclidean distance as the weight of edge of the MST. Notable, the weight will be set to a numerical value which is far greater than the real spatial distance if the sphere rotating model fails to extract the pathway. The reason for the sphere rotating model failed can be concluded as two aspects, the one is a gap exists between the two nodes, the other is the rotating times beyond the pre-configured threshold since one node is far away from the other. The calculation method of the weight in M-AMST described as the following:

if pathway exist,

then $W_E = \sum_{i=1}^{n-1} Dis(p_i, p_{(i+1)})$;

else $W_E = M \quad Dis(S, T)$,

where W_E is the accumulating Euclidean distance of the pathway between the corresponding pair of nodes, p_i and p_{i+1} mean the child node and parent node in the node list respectively, Dis indicates the Euclidean distance between the two nodes, M means a positive integer value which is greater than 10.

After that, we use MST algorithm to build a graph representation in SWC format. Specifically, the weights between all pairs of nodes can form a weight diagonal matrix, which can be acted as the input of MST algorithm. In order to reconstruct the neuron image more elaborate, based on the neuron reconstruction by MST, we fulfill the node list into the pathway between the corresponding pair of nodes.

Results

Parameters in the implementation

We implemented the M-AMST algorithm as a plugin of the Vaa3D which is the common platform to implement algorithms for BigNeuron project (bigneuron.org) bench testing. On the whole, the implementation of the M-AMST algorithm can be split into four steps which are summarized as follows:

(1) Binarization. We define the voxel whose intensity which is less than a threshold as the dark spot and otherwise the bright spot. For each voxel, the number of the dark spots and the bright spots among the 26 surrounding voxels are calculated respectively. And then, we calculate the ratio of the number of the dark spots and the bright spots, the ratio is compared with a threshold to determine whether the voxel is a foreground or not. The intensity threshold and ratio threshold are set to be 30 and 0.3 respectively.

(2) Skeletonization. For each foreground voxel, the mean shift algorithm defines a sphere with certain spatial range and takes the voxel as the center of the sphere. Then it shifts the voxel to the local mean iteratively until getting the convergence region. In order to avoid the endless loop, the number of shifting times of the foreground voxel is set to 100. The radius of the sphere is set to 5.

(3) Rectification. We prune the nodes yielded from skeletonization step using a node pruning method. For a pair of nodes, we define two nodes should all be kept if the difference between their Euclidean distance D and the sum of their radiuses is greater than a certain voxel distance. The voxel distance is set to 3.

(4) Graph representation. In the implementation of the rotating sphere model, the radius of sphere is set to 2 and the sphere was split into 4 quadrants. We extract the pathway between each pair of nodes and calculate the length of it by accumulating the Euclidean distance between the parent node and child node in the node list. A weight diagonal matrix can be generated to act as the input of MST algorithm. Based on the neuron reconstruction by MST, we fulfill the node list into the pathway between the corresponding pair of nodes. After that, in order to remove the unnecessary or redundant spurs in the tree structure, we adopt a hierarchical pruning method utilized in APP2. The code can be downloaded from www.github.com/Vaa3D/vaa3d_tools/tree/master/bigneuron_ported/APP2_ported.

Experimental results

A. Neuron reconstruction efficiency comparison between M-AMST and M-MST

One hundred twenty confocal neuron images of the Drosophila were used to test the performance of M-AMST. Firstly, we selected two Drosophila neuron images from 120 images randomly to illustrate the difference between the reconstruction of the M-AMST and M-MST. The reconstruction results showed in Fig. 4 are overlaid on top of the original images for better visualization. For the first neuron image showed in the left part of Fig. 4, the yellow box contains the reconstruction results of the two methods and we can see them more

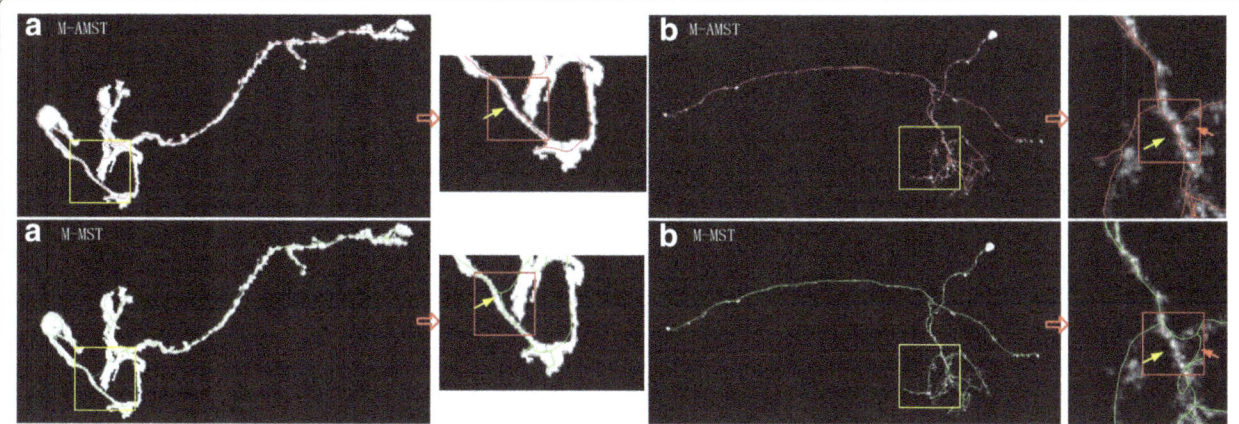

Fig. 4 The comparison of neuron reconstructions using M-AMST and M-MST for two Drosophila neuron images. The reconstructions are overlaid on top of the original images for better visualization. The yellow box points out the different reconstruction results and the red box with the arrow make them more clearly

clearly in the red box. For the result of M-MST displayed in the bottom left, the reconstruction is not accurate since it crosses the gap and connects two different segments (see green line and yellow arrow), but the result of the M-AMST shows in the up left is accurate. For the second neuron image showed in the right part of Fig. 4, there are two different parts which are showed in the red box with yellow and red arrow respectively. For the M-MST, a neuron portion is not reconstructed (see yellow arrow) successfully and the different neuron segments are connected by crossing the gap (see red arrow). The M-AMST method does not generate such bug and reconstructs the neuron image accurately. The result showed above proved that the reconstruction effect of M-MST is worse than M-AMST since the M-MST just considers the spatial distance between neuron segments and ignores the shape information of neuron morphology.

B. Running time comparison between M-AMST and M-MST

It is notable that all of the experiments were performed on a Panasonic laptop with 2.6 GHz Intel Core i5-4310U CPU and 4G RAM. Table 1 summarizes the running speed of M-AMST versus M-MST for several Drosophila neuron images. The running time of each step (binarization, skeletonization, rectification and graph representation) is represented by Tb, Ts, Tr and Tg respectively. M-AMST and M-MST indicates their own running time. Since the whole procedure of M-AMST is coherent with M-MST except the step of graph representation, we can envisage that the computational time of the adapted MST exceeds the original MST. The relative high computational complexity of M-AMST could be attributable to the efficiency of pathway extraction based on the sphere model. However, for some images, the running time of M-MST is measly outperformed M-AMST, which can be explained that the running time is affected

Table 1 Comparison of running time (seconds) of M-AMST and M-MST on few images

ID	Size	Memory (KB)	Running time (s)					
			Tb	Ts	Tr	Tg	M-AMST	M-MST
1	1024*1024*119	121857	35.147	0.046	0.016	4.384	39.593	38.438
2	1024*1024*109	111617	30.748	0.124	0.094	3.292	34.258	35.428
3	1024*1024*121	123905	33.494	0.156	0.078	5.647	39.375	33.431
4	1024*1024*111	113665	30.14	0.078	0.015	5.523	35.756	32.948
5	1024*1024*109	111617	31.138	0.031	0.031	5.835	37.035	33.119
6	1024*1024*125	128001	39.344	0.062	0.078	1.217	40.701	36.629
7	1024*1024*128	131073	36.115	0.062	0.031	3.838	40.046	40.155
8	1024*1024*115	117761	29.859	0.015	0.016	1.451	31.341	33.041
9	1024*1024*112	114689	31.527	0.032	0.031	3.245	34.835	30.577
10	1024*1024*117	119809	31.84	0.031	0.016	3.104	34.991	32.542

by the dynamic change of laptop performance. Comparing with Tb, the running time of pathway extraction based on sphere model is not the arch time consuming step.

C. Comparison with other reconstruction algorithms

We compared the 120 neuron reconstructions generated by M-AMST with another four tracing algorithms which listed as follows: M-MST, MOST, SIMPLE, N-MST. Notably, the four tracing algorithms are also developed by the member groups of BigNeuron and the corresponding code can be downloaded from www.github. com/Vaa3D/vaa3d_tools/tree/master/bigneuron_ported. Due to the APP2 tracing algorithm is the fastest tracing algorithm and is reliable in generating tree morphology of neuron among the existing methods, we select the reconstructions generated by APP2 as the reference to compare the effect of the five algorithms. Moreover, we calculate four difference scores (ESA, DSA, PDS and MDNN) of the reconstructions produced by APP2 and the five tracing algorithms. Correspondingly, the four difference scores measure the overall average spatial divergence between two reconstructions, the spatial distance between different structures in two reconstruction, and the percentage of the neuron structure that noticeably varies in independent reconstruction, as well as the maximum distance to the nearest reconstruction elements between two reconstructions. The smaller value the four difference scores get, the neuron reconstruction effect of the competing algorithm is closer to the reference algorithm. To make a fair comparison, the reported results of the competing algorithms correspond to the default parameters called by respective plugins.

The histogram and boxplot are adopted to compare the performance of the five algorithms. Figure 5 shows the average of the four difference scores of the five

algorithms compared with APP2 reconstructions. As we can see, MOST achieves the best reconstructions and the M-AMST gets the relative good results. Comparing with M-MST, M-AMST gets the lower average of difference score in ESA, PDS and MDNN. Although the average of DSA of M-AMST is lower than M-MST, the two values are very close. Moreover, comparing with N-MST, the average of three difference scores of M-AMST includes the ESA, DSA and MDNN are better. It indicates that utilizing the adapted MST proposed in this paper can achieve better neuron reconstructions. For both of the M-AMST and M-MST, the average of the percentage of different structure are lower than N-MST. This probably due to the node pruning method leads to the effect of over-deletion which means the method does not keep the sufficient nodes to delineate the whole topology. In order to illustrate the distribution of the four difference scores of the five competing algorithms, Fig. 6 shows the box plots of the four difference scores of the neuron reconstructions obtained by the algorithms. Due to the average of four difference scores of SIMPLE is far beyond other four competing algorithms, list each difference score of the four competing algorithm with the SIMPLE's together will decrease the observability of the figure. Therefore, Fig. 6 only shows the difference score of the four algorithms. From the distribution of the four difference scores show in the Fig. 6, we can get that compare with the M-MST and N-MST, the ESA and MDNN score of M-AMST is better, the number of the corresponding outlier is also less, which means the distance between the reconstruction of the M-AMST and APP2 is smaller than the distance between the reconstruction of the APP2 and the M-MST or N-MST, this fully proves the effective of the adapted MST method. However, from the DSA and PDS aspects, the M-AMST

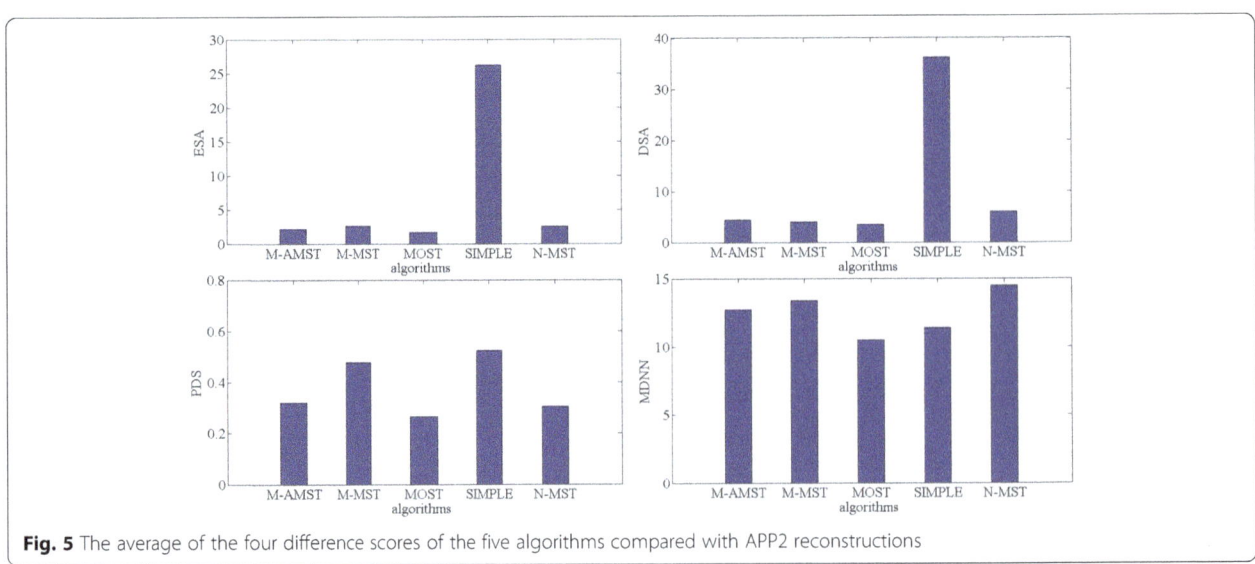

Fig. 5 The average of the four difference scores of the five algorithms compared with APP2 reconstructions

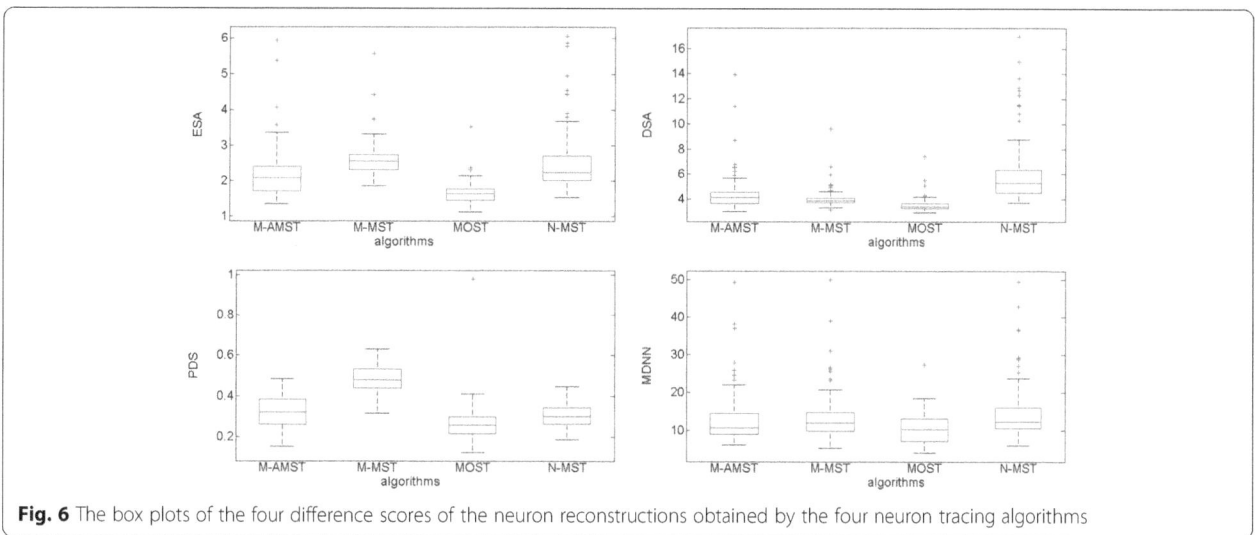

Fig. 6 The box plots of the four difference scores of the neuron reconstructions obtained by the four neuron tracing algorithms

does not get the best scores. Moreover, the four difference scores of the MOST are better than the M-AMST, which might be attributable to the MOST is better at reconstructing the confocal neuron images of the Drosophila. In order to further explain the effect of the M-AMST, we tested M-AMST on serval neuron datasets and compared the M-AMST reconstructions with the gold standard reconstructions.

D. Tracing different neuron images and comparing with the gold standard reconstruction

We test the 6 algorithms (M-AMST, M-MST, MOST, SIMPLE, N-MST and APP2) on the 7 neuron datasets respectively. The 7 datasets were one part of the data released in the first bench-testing of the BigNeuron which was started in the summer of 2015. The neuron images released in the first bench-testing phase are all with the gold standard reconstruction and aimed at fine-tuning the algorithms or training the classifiers of the BigNeuron contributor. Table 2 summarizes the basic information of the 7 neuron datasets and the reconstruction result of the 6 algorithms. (1)-(7) represents the dataset of checked6 frog scripts, checked6 human culturedcell Cambridge in vitro confocal GFP, checked6 janelia flylight part2, checked6 zebrafish horizontal cells UW, checked7 janelia flylight part1, checked7 taiwan flycirciut and checked7 utokyo fly respectively. Correspondingly, the number of the neuron images included in the each dataset is orderly indicated by N1-N7. N8 means the number of the neuron reconstructions successfully generated by each algorithm. The four difference scores (ESA, DSA,PDS and MDNN) are denoted by a-d in order, their values are obtained by comparing the SWC file of the reconstruction algorithm with the gold standard SWC file. According to the number of the neuron images included in each dataset, three

representations are adopted to illustrate the result of the four difference scores: (1) each difference score is illustrated by a float if the dataset includes one image; (2) each difference score is represented by the average and the standard variance if the dataset includes more than or equal to two images; (3) each difference score is denoted by "-" if the reconstruction algorithm fails to reconstruct the neuron or generate the SWC file. The four difference scores marked with the red bold font indicates that the related algorithm achieves the best score result compare with other algorithms. The four difference score marked with the black bold font means that the related algorithm gets the second-best score result. As is shown in the table, for the dataset of (4), (5) and (6), the M-AMST gets the best difference scores. For the dataset of (1) and (3), the M-AMST achieves the second-best difference scores. For the dataset of (1), the M-AMST is close to APP2 in the ESA, and the M-AMST is superior to APP2 in PDA. For the dataset of (2), the MOST obtains the best difference scores. However, 6 algorithms are all failed to get the good reconstructions, which means there is no significance to compare the reconstruction effect of the algorithms for this kind of dataset. For the dataset of (3), N-MST gets the best difference scores, after observing the raw image, we find the neurons in the image are all surround many white spots with low intensity and high density. As mentioned above, the M-AMST adopts a rigorous de-noise method to identify the foreground voxel in the binarization step, which causes many noisy voxels are deemed to be the foreground and makes M-AMST is inferior to APP2 in DSA and PDS. To sum up, using the gold standard SWC file as the reference and making a comparison between M-AMST and the other 5 algorithms, M-AMST is able to reconstruct variety of neuron datasets successfully and achieve the good difference scores.

Table 2 The basic information of the 7 neuron datasets and the reconstruction result of the 6 algorithms

		(1)	(2)	(3)	(4)	(5)	(6)	(7)
		N1　　1	N2　　4	N3　　3	N4　　1	N5　　20	N6　　12	Num7　　3
M-AMST	N8	1	4	2	1	19	12	2
	a	**5.0**	44.24 ± 15.21	**2.27 ± 0.01**	8.06	**2.26 ± 0.39**	**2.11 ± 0.22**	8.50 ± 3.71
	b	**8.18**	50.58 ± 14.40	**4.43 ± 0.48**	9.91	**3.97 ± 0.72**	**3.12 ± 0.40**	10.11 ± 4.31
	c	**0.45**	0.74 ± 0.05	**0.41 ± 0.02**	0.77	**0.40 ± 0.11**	**0.41 ± 0.06**	0.79 ± 0.01
	d	**22.5**	16.64 ± 6.62	**7.93 ± 2.34**	32.90	**9.45 ± 3.37**	**6.77 ± 2.12**	30.0 ± 9.03
M-MST	N8	1	4	2	1	20	12	3
	a	7.20	**38.92 ± 16.86**	4.46 ± 1.18	25.57	2.89 ± 0.65	3.19 ± 0.39	14.02 ± 4.9
	b	8.96	**46.46 ± 17.96**	6.71 ± 1.41	27.35	4.36 ± 0.84	4.04 ± 0.31	15.42 ± 4.87
	c	0.72	**0.74 ± 0.07**	0.53 ± 0.04	0.89	0.5 ± 0.08	0.65 ± 0.09	0.87 ± 0.05
	d	23.87	**28.58 ± 11.57**	8.07 ± 2.19	29.86	9.59 ± 3.82	7.67 ± 1.86	37.06 ± 0.77
MOST	N8	1	4	2	1	19	12	3
	a	61.82	**20.23 ± 12.31**	51.86 ± 9.18	74.62	45.28 ± 30.04	63.56 ± 20.49	33.10 ± 41.50
	b	123.67	**26.15 ± 13.20**	52.57 ± 9.89	75.87	54.27 ± 37.74	64.78 ± 20.02	35.72 ± 40.53
	c	0.49	**0.49 ± 0.11**	0.78 ± 0.21	0.60	0.52 ± 0.15	0.74 ± 0.10	0.61 ± 0.24
	d	1.08	**8.95 ± 3.21**	5.88 ± 2.54	4.29	4.49 ± 1.99	6.24 ± 3.17	26.65 ± 2.85
SIMPLE	N8	1	0	2	0	19	11	0
	a	15.32	-	10.04 ± 2.19	-	-	8.10 ± 8.06	21.79 ± 11.99
	b	21.09	-	13.77 ± 1.87	-	12.69 ± 10.10	27.74 ± 14.36	-
	c	0.60	-	0.66 ± 0.10	-	0.43 ± 0.22	-	0.62 ± 0.07
	d	18.22	-	22.84 ± 1.19	-	14.26 ± 9.39	7.67 ± 2.99	-
N-MST	N8	1	4	3	1	20	12	3
	a	88.29	163.6 ± 162.0	**2.28 ± 0.33**	**12.42**	5.96 ± 8.40	8.17 ± 18.23	**5.69 ± 2.32**
	b	176.36	166.2 ± 159.5	**4.46 ± 0.43**	**22.75**	10.08 ± 12.18	11.72 ± 24.06	**10.04 ± 3.07**
	c	0.49	0.63 ± 0.36	**0.28 ± 0.05**	**0.42**	0.30 ± 0.08	0.35 ± 0.08	**0.48 ± 0.12**
	d	1.66	158.5 ± 147.1	**3.44 ± 0.80**	**22.01**	9.18 ± 7.87	5.54 ± 1.92	**51.19 ± 26.49**
APP2	N8	1	4	3	1	20	12	3
	a	**4.47**	45.54 ± 16.35	6.10 ± 4.11	18.96	**3.15 ± 2.24**	7.46 ± 9.14	**6.12 ± 2.31**
	b	**11.89**	62.25 ± 14.31	8.07 ± 4.54	24.57	**4.63 ± 2.82**	12.36 ± 12.45	**12.32 ± 3.18**
	c	**0.27**	0.49 ± 0.08	0.38 ± 0.05	0.58	**0.34 ± 0.11**	0.31 ± 0.08	**0.41 ± 0.07**
	d	**20.65**	22.66 ± 4.86	3.0 ± 0.64	18.74	**5.60 ± 1.48**	4.83 ± 1.29	**34.85 ± 6.78**

The symbol "-" means the reconstruction algorithm fails to reconstruct the neuron or generate the SWC file. The four difference scores marked with the red bold font indicates that the related algorithm achieves the best score result compare with other algorithms. The four difference score marked with the black bold font means that the related algorithm gets the second-best score result

Discussion

In summary, the main steps of the M-AMST followed the sequence of neuron reconstruction operations: binarization, skeletonization, rectification and graph representation. In the skeletonization step, M-AMST adopts mean shift to extract nodes distributed around the centroid of the local neuron segments. The reason for using mean shift can be concluded as the two aspects. On one hand, mean shift is an automatic clustering algorithm which can move and cluster each foreground voxel to the local convergence region. The voxels located at the convergence region cannot shift for one step by mean shift. By observing their position in the neuron image, the most classification voxels are located around the neuron skeleton, in which the neuron portions with high intensity and voxel density located. On the other hand, other voxels which shift toward and finally locate at the regions after several iterations can be marked as the same classification as the corresponding classification voxels. That kind of voxels can be regarded as the subordinate voxels to the classification voxels. Based on the subordinate voxels belong to the different classification voxel, the local structure of the neuron can be delineated. Therefore, with mean shift, not only the information of voxel density distribution of neuron image can be captured correctly, but also got the sufficient voxels belong to several modes to delineate the whole neuron topological structure. Moreover, we develop a pathway extraction method using a rotating sphere model based on coordinate transformation to improve the weight calculation approach in MST. It is worth noting that the pathway contains a node list since selecting the voxel which intensity is greater than 0 as the new sphere center in every rotating step. Every node list represents the shape information between the corresponding pair of nodes. For a pathway between the pair of nodes, we calculate the length of it by sequentially accumulating the Euclidean distance between the parent node and child node in the list. And then, take the accumulating Euclidean distance as the weight of edge. Notable, the weight will be set to a numerical value which is far greater than the real spatial distance if the sphere rotating model fails to extract the pathway. In the experimental stage, two experiments are designed illustrate the effectiveness of the adapted MST method and the adoptability of M-AMST in reconstructing variety of neuron datasets. The result of experiment 1 shows that M-AMST gets lower difference scores than M-MST in ESA, PDS and MDNN. Meanwhile, M-AMST is better than N-MST in ESA and MDNN. It indicates that utilizing the adapted minimum spanning tree algorithm which took the shape information of neuron into account can achieve better neuron reconstructions. In order to testify the adoptability of M-AMST, 7 neuron datasets released in the first bench-testing stage of the BigNeuron are chose to be reconstructed by 6 algorithms in the experiment 2. The neuron reconstruction results generated by each algorithm are compared with the corresponding gold standard SWC file. The comparison result shows that M-AMST is able to achieve the best difference score in 3 datasets and get the second-best difference score in the other 2 datasets. This indicates that the M-AMST can reconstruct variety of neuron datasets successfully and achieve good difference scores. However, there are still several limitations in M-AMST which are listed as follows.

(1) Several neuron segments with low intensity are not reconstructed successfully. The reason can be concluded into two points. On one hand, the foreground identification method used in the binarization step might ignore the nodes with low intensity. On the other hand, the voxels with low intensity located around the neuron skeleton will be moved in several iterations and cannot be identified as the classification marks.

(2) After analyzing the reconstructions generated by M-AMST thoroughly, we find that M-AMST did not thoroughly solve the problem of connecting the different segments by crossing the gap. The main reason for this is the pathway extraction method failed when there is a gap between the pair of nodes. In this case, although a numerical value which is far greater than the real spatial distance is assigned to the weight of edge between them, MST might still select this pathway if there is no smaller weight can be chose.

(3) The node pruning method probably degrades the reconstruction accuracy of M-AMST. For each pair of nodes, we define one of the pair of nodes should be deleted if the difference between their Euclidean distance D and the sum of their radiuses is less than a threshold. A higher threshold could cause the method cannot keep the sufficient nodes to reconstruct the whole topology. Conversely, a lower threshold could cause the reconstruction with redundant neuron segments due to the method keeps the excessive nodes.

In response to limitations mentioned above, the implementation details of M-AMST algorithm will be further refined. In the near future, we will keep working on developing the neuron image de-noising and tracing algorithms based on the machine learning method so as to continue making contributions to the BigNeuron project.

Conclusions

We develop a pathway extraction method using a rotating sphere model based on coordinate transformation to improve the weight calculation approach in MST. The corresponding experiment shows that utilizing the adapted minimum spanning tree algorithm which took the shape information of neuron into account can achieve better neuron reconstructions. Moreover, the adoptability of M-AMST in reconstructing variety of neuron images is also testified. The result indicates that comparing with the gold standard reconstruction, the neuron reconstruction generated by the M-AMST is able to achieve good difference scores in variety of neuron datasets.

Abbreviations
APP2: Improved all-path pruning method; M-AMST: Mean shift and adapted minimum spanning tree; M-MST: Mean shift and minimum spanning tree; MOST: Micro-Optical Sectioning Tomography ray-shooting; N-MST: Neuron tracing minimum spanning tree; p.d.f: probability density function

Acknowledgements
The authors thank the BigNeuron project and Dr. Hanchuan Peng for providing the testing image data used in this article and many discussions.

Funding
This work is partially supported by the National Basic Research Program of China (No. 2014CB744600), National Natural Science Foundation of China (No. 61420106005), Beijing Natural Science Foundation (No. 4164080), and Beijing Outstanding Talent Training Foundation (No. 2014000020124G039).

The four projects are all funded the first workshop of BigNeuron hold in Beijing. The workshop aims at gathering Chinese neuron scientists to study and share the released neuron dataset, analyze and reconstruct the neuron image and give the interpretation for neuron reconstruction. The funding for writing and publishing the manuscript is provided by the National Natural Science Foundation of China.

Authors' contributions

WZ carried out the design and implementation of M-AMST algorithm and finished the work of writing the draft. HY and HM helped to generate the neuron reconstructions of 120 confocal images using 5 algorithms, they also compared the neuron reconstructions of 5 algorithms respectively and get the statistical results of four scores. WZ generated the neuron reconstruction of 7 datasets using the M-AMST, the reconstructions of other 5 algorithms were results of the bench-test finished by Dr. Hanchuan Peng's group. YJ provided the suggestion of manuscript revision. ZN agreed to be accountable for all aspects of the work in ensuring that questions related to the accuracy or integrity of any part of the work are appropriately investigated and resolved. All authors read and approved the final manuscript.

Authors' information

Not applicable.

Competing interests

The authors declare that they have no competing interests.

Author details

[1]Beijing Advanced Innovation Center for Future Internet Technology, Beijing University of Technology, Beijing, China. [2]Department of Life Science and Informatics, Maebashi Institute of Technology, Maebashi, Japan. [3]International WIC Institute, Beijing University of Technology, Beijing, China. [4]Beijing Key Laboratory of MRI and Brain Informatics, Beijing, China. [5]Beijing International Collaboration Base on Brain Informatics and Wisdom Services, Beijing, China.

References

1.   Meijering E. Neuron tracing in perspective. Cytometry A. 2010;77 A(7):693–704.
2.   Zhong N, Yau SS, Ma J, Shimojo S, Just M, Hu B, Wang G, Oiwa K, Anzai Y. Brain Informatics-Based Big Data and the Wisdom Web of Things. IEEE Intell Syst. 2015;30(5):2–7.
3.   Peng H. Bioimage informatics: a new area of engineering biology. Bioinformatics. 2008;24(17):1827–36.
4.   Ascoli GA. Neuroinformatics Grand Challenges. Neuroinformatics. 2008;6(1):1–3.
5.   Peng H, Meijering E, Ascoli GA. From DIADEM to BigNeuron. Neuroinformatics. 2015;13(3):259–60.
6.   Peng H, Hawrylycz M, Roskams J, Hill S, Spruston N, Meijering E, Ascoli Giorgio A. BigNeuron: large-scale 3d neuron reconstruction from optical microscopy images. Neuron. 2015;87(2):252–6.
7.   Pal SK. Fuzzy sets in image processing and recognition. In: Fuzzy Systems, 1992, IEEE International Conference on: 8-12 Mar 1992 1992. 119-126.
8.   Malladi R, Sethian JA. Level set and fast marching methods in image processing and computer vision. In: Image Processing, 1996 Proceedings, International Conference on: 16-19 Sep 1996 1996. 489-492 vol.481.
9.   Xiao H, Peng H. APP2: automatic tracing of 3D neuron morphology based on hierarchical pruning of a gray-weighted image distance-tree. Bioinformatics. 2013;29(11):1448–54.
10. Peng H, Long F, Myers G. Automatic 3D neuron tracing using all-path pruning. Bioinformatics. 2011;27(13):i239–47.
11. Wang Y, Narayanaswamy A, Tsai C-L, Roysam B. A broadly applicable 3-D neuron tracing method based on open-curve snake. Neuroinformatics. 2011;9(2):193–217.
12. Cai H, Xu X, Lu J, Lichtman JW, Yung SP, Wong ST. Repulsive force based snake model to segment and track neuronal axons in 3D microscopy image stacks. Neuroimage. 2006;32(4):1608–20.
13. Peng H, Ruan Z, Atasoy D, Sternson S. Automatic reconstruction of 3D neuron structures using a graph-augmented deformable model. Bioinformatics. 2010; 26(12):i38–46.
14. Cai H, Xu X, Lu J, Lichtman J, Yung SP, Wong ST. Using nonlinear diffusion and mean shift to detect and connect cross-sections of axons in 3D optical microscopy images. Med Image Anal. 2008;12(6):666–75.
15. Oliver A, Munoz X, Batlle J, Pacheco L, Freixenet J. Improving Clustering Algorithms for Image Segmentation using Contour and Region Information. In: 2006 IEEE International Conference on Automation, Quality and Testing, Robotics: 25-28 May 2006 2006. 315-320.
16. Wu J, He Y, Yang Z, Guo C, Luo Q, Zhou W, Chen S, Li A, Xiong B, Jiang T, et al. 3D BrainCV: Simultaneous visualization and analysis of cells and capillaries in a whole mouse brain with one-micron voxel resolution. NeuroImage. 2014;87: 199–208.
17. Yang J, Gonzalez-Bellido PT, Peng H. A distance-field based automatic neuron tracing method. BMC Bioinformatics. 2013;14(1):1–11.
18. Comaniciu D, Meer P. Mean shift: a robust approach toward feature space analysis. IEEE Trans Pattern Anal Mach Intell. 2002;24(5):603–19.
19. Peng H, Ruan Z, Long F, Simpson JH, Myers EW. V3D enables real-time 3D visualization and quantitative analysis of large-scale biological image data sets. Nat Biotech. 2010;28(4):348–53.

5

The JBEI quantitative metabolic modeling library (jQMM): a python library for modeling microbial metabolism

Garrett W. Birkel[1,2,8], Amit Ghosh[1,2,6], Vinay S. Kumar[1,2], Daniel Weaver[1,2], David Ando[1,2], Tyler W. H. Backman[1,2,8], Adam P. Arkin[1,4,5], Jay D. Keasling[1,2,3,4,7] and Héctor García Martín[1,2,8,9]* (iD)

Abstract

Background: Modeling of microbial metabolism is a topic of growing importance in biotechnology. Mathematical modeling helps provide a mechanistic understanding for the studied process, separating the main drivers from the circumstantial ones, bounding the outcomes of experiments and guiding engineering approaches. Among different modeling schemes, the quantification of intracellular metabolic fluxes (i.e. the rate of each reaction in cellular metabolism) is of particular interest for metabolic engineering because it describes how carbon and energy flow throughout the cell. In addition to flux analysis, new methods for the effective use of the ever more readily available and abundant -omics data (i.e. transcriptomics, proteomics and metabolomics) are urgently needed.

Results: The jQMM library presented here provides an open-source, Python-based framework for modeling internal metabolic fluxes and leveraging other -omics data for the scientific study of cellular metabolism and bioengineering purposes. Firstly, it presents a complete toolbox for simultaneously performing two different types of flux analysis that are typically disjoint: Flux Balance Analysis and ^{13}C Metabolic Flux Analysis. Moreover, it introduces the capability to use ^{13}C labeling experimental data to constrain comprehensive genome-scale models through a technique called two-scale ^{13}C Metabolic Flux Analysis (2S-^{13}C MFA). In addition, the library includes a demonstration of a method that uses proteomics data to produce actionable insights to increase biofuel production. Finally, the use of the jQMM library is illustrated through the addition of several Jupyter notebook demonstration files that enhance reproducibility and provide the capability to be adapted to the user's specific needs.

Conclusions: jQMM will facilitate the design and metabolic engineering of organisms for biofuels and other chemicals, as well as investigations of cellular metabolism and leveraging -omics data. As an open source software project, we hope it will attract additions from the community and grow with the rapidly changing field of metabolic engineering.

Keywords: Flux analysis, ^{13}C Metabolic Flux Analysis, -omics data, Predictive biology

Background

Metabolism is the set of physical and chemical processes through which living organisms extract energy and mass from their environment in order to sustain life [1]. Hence, an understanding of metabolism is fundamental in order to set the stage for, and understand the physical limitations of, every biological process. For example, metabolic disorders lie at the heart of a variety of human illnesses such as diabetes [2], obesity [3] and cancer [4].

Understanding of metabolism is also of central importance to the microbial production of drugs, bioproducts and biofuels through metabolic engineering [5]. This application of microbial bioengineering to produce biofuels and other commodity products has attracted significant attention due to its potential to reduce the CO_2 emissions which cause climate change, reduce society's reliance on fossil fuels for energy and chemicals,

*Correspondence: hgmartin@lbl.gov
[1]Biological Systems and Engineering Division, Lawrence Berkeley National Laboratory, Berkeley, CA, USA
[2]Joint BioEnergy Institute, Emeryville, CA, USA
Full list of author information is available at the end of the article

and provide for enhanced energy security. The production of commodity chemicals from lignocellulosic biomass is critical for achieving these goals, particularly in view of expected future population growth in emergent economies coupled with a marked improvement in their living standards [6].

Metabolic engineering rarely succeeds at the first attempt to introduce and activate a heterologous pathway in a host organism. Rather, most of the time in this process is spent troubleshooting the pathways for acceptable production [7]. Very often, the metabolic engineer diagnoses the problem in the bioengineered microbe by relying on a variety of -omics data (transcriptomics, proteomics, metabolomics) that are expected to provide an explanation of why the system does not behave as expected [8–10]. However, this diagnosing through -omics data is a non trivial process: often the amount of data is much larger than what metabolic engineers are typically trained to analyze, and it is non-trivial to extract actionable items (e.g. overexpress this specific gene or change that promoter) that will reliably improve the system's behavior [11, 12]. To complicate matters, the reason for the underperformance of a given pathway may not rely on the pathway itself but on other parts of the host metabolism that affect the pathway's performance in intricate and non-direct ways (e.g. cofactor availability or gene regulation) [13].

Mathematical modeling helps contextualize experimental data into a model that provides an explanation of the observed phenomena and provides predictions for changes in the system under study. Metabolic modeling, in particular, has been successfully used to e.g. increase product yield [14] or identify bottlenecks and competing pathways [15]. Moreover, data analysis and modeling is expected to increase in scope and importance as the creation of increasingly larger sets of -omics data sets are enabled by the rapid drop of DNA sequencing costs [16] and the availability of high-throughput mass spectrometry workflows [17, 18]. Hence, research institutes such as the Joint BioEnergy Institute (JBEI), tasked to produce the scientific and technological underpinnings that will enable lignocellulosic biofuels [19], have devoted research effort not only to develop drop-in biofuels [20] but also tools to speed up bioengineering through mathematical modeling.

Some of the mathematical tools created at JBEI to enable better prediction of bacterial and eukaryotic metabolism are presented here in the form of the JBEI Quantitative Metabolic Modeling library (jQMM). The jQMM library includes algorithms to measure and predict internal metabolic fluxes using three different techniques: ^{13}C Metabolic Flux Analysis (^{13}C MFA [21]), Flux Balance Analysis (FBA [22]) and two-scale ^{13}C Metabolic Flux Analysis (2S-^{13}C MFA [23]). It also includes methods to produce actionable insights from -omics data to improve pathway yield [24], and methodologies for the flux analysis of microbial communities [25]. We hope to add to the jQMM library in the future tools and algorithms currently being developed at JBEI, as well other institutions who may consider contributing to this library.

Methods
Key capabilities
The JBEI QMM library includes code to perform the following techniques: FBA [22], ^{13}C MFA [21], 2S-^{13}C MFA [23], Principal Component Analysis of Proteomics (PCAP [24]), and ^{13}C MFA for microbial communities [25]. Mathematical details for each method are provided in the supplementary material (Additional file 1). FBA, ^{13}C MFA and 2S-^{13}C MFA are techniques used to calculate internal metabolic fluxes; detailed explanations regarding the differences between each method can be found in Garcia Martin et al. [23]. This library contains sample demonstration scripts to, for example, replicate all the figures from Garcia Martin et al. in the form of Jupyter notebooks (see Table 1). The code in the jQMM library focuses on ^{13}C MFA, 2S-^{13}C MFA, since there is already an open source python-based library [26] for FBA. For ^{13}C MFA there exists some closed-source packages (e.g. METRAN [27], INCA [28], 13CFLUX2 [29]), a few open-source packages (OPENFLUX [30] and OPENFLUX2 [31]) but no open source python-based libraries. For 2S-^{13}C MFA, jQMM is the first public library.

Implementation
jQMM is implemented in Python 2.7, and we have aimed to provide a code base that is as modular as possible, in order to facilitate understanding and reusability. The

Table 1 Table of iPython Notebooks

iPython Notebook number	Topic
A0	module tests
A1	Core demo
A2	Enhanced lists demo
A3	ReactionNetworks demo
A4	FluxModels demo
A5	GAMSclasses demo
A6	Predictions demo
A7	Labeling and DB demo
B1	FBA demo
B2	TCA ^{13}C MFA demo
B3	Toya data ^{13}C MFA demo
B4	Toya data 2S-^{13}C MFA demo
B5	PCAP example
B6	^{13}C MFA for microbial communities

code is divided into a set of modules that incorporate classes and methods which will typically be used together. Figure 1 details the different modules and how they interact with each other for flux-based analysis. The following subsections provide a brief explanation of the purpose of each module and Figs. 2, 3 and 4 (and Additional file 1: Figures S1-S3) provide graphical depictions of each module together with their main classes and methods. Tutorials are provided for each module, along with unit tests.

In order to facilitate reproducibility, the lack of which represents one of the main problems afflicting biological research [32], all results and tutorials are provided in the form of Jupyter notebooks [33, 34] (see Table 1 and jupyter.org). This approach is meant to facilitate the reproduction of published results and the creation of new results based on these. The Jupyter notebooks are divided into two types: type A demonstrates the library's capabilities and type B reproduces published results using code from the jQMM library (Table 1). To further enhance reproducibility we have provided a docker container in the github repository that includes all dependencies. Docker (http://www.docker.com) containers wrap a piece of software in a complete filesystem containing everything needed to run: code, systems libraries, systems tools

and anything else that can be installed in a server. This guarantees that it will always run correctly and in the same way, regardless of the system environment it is running in. The jQMM docker container can be run on any computer and any cloud computing service such as Amazon Web Services (AWS), Google Cloud Platform or Microsoft Azure. The container does not include GAMS, CPLEX or CONOPT licences, which must be provided by the user.

Modules

This section contains a succint explanation for each module and some of its classes and capabilities. A more detailed description of each class containing all possible functionality can be found in the Jupyter notebooks (Table 1) and the jQMM code on the github website (see availability below).

Core

The core module contains the classes that constitute the base for all other functionality (Fig. 2). These basic classes include the *Reaction* and *Metabolite* classes, along with the derived *Product* and *Reactant* classes. These classes are defined as required to describe labeling experiments and, as such, are different from equivalently named classes in COBRApy [26] (defined for

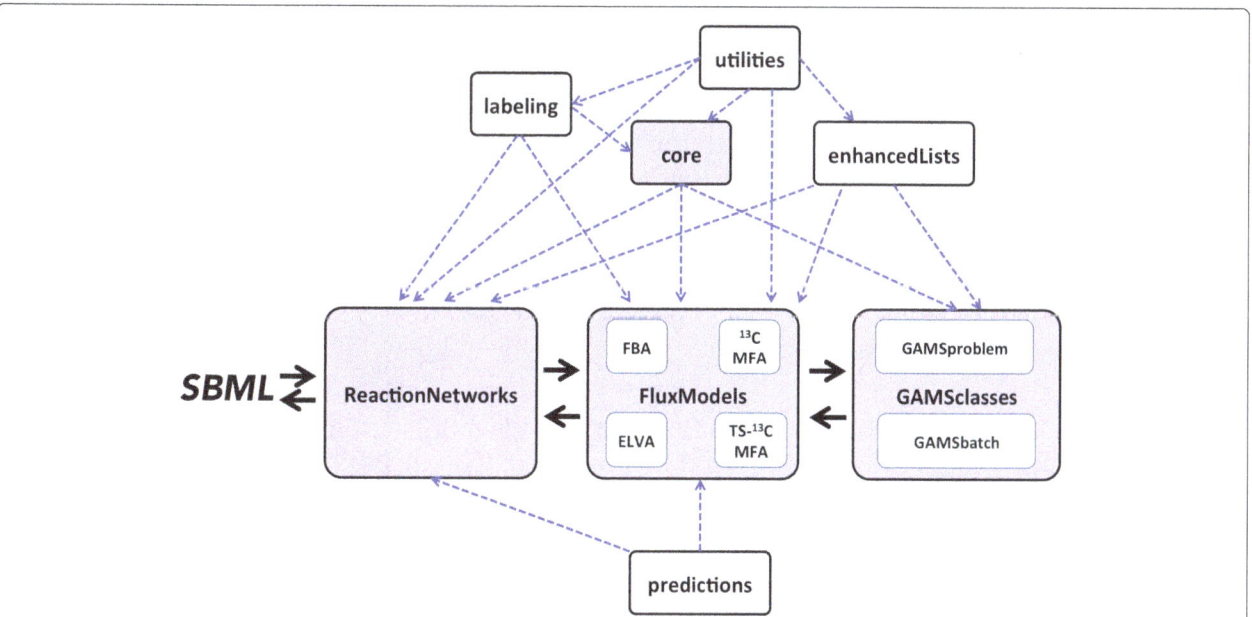

Fig. 1 Diagram of jQMM library module relationships and typical flow for flux analysis. Flux calculations typically start with information stored in an SBML (Systems Biology Markup Language) file and translated into a reaction network. That reaction network is enclosed in a flux model of the appropriate type for the desired method (FBA, 13C MFA, 2S-13C MFA, ELVA). The flux model instance uses the GAMSclasses module in order to solve the appropriate optimization problem, and turns it back to the flux model instance, which stores the information as a reaction network or an SBML file. The core, labeling, utilities and enhancedLists modules are used by these "main workflow" modules (in *darker shade*). Predictions use information from reaction networks and flux models to make flux predictions for genetic modifications

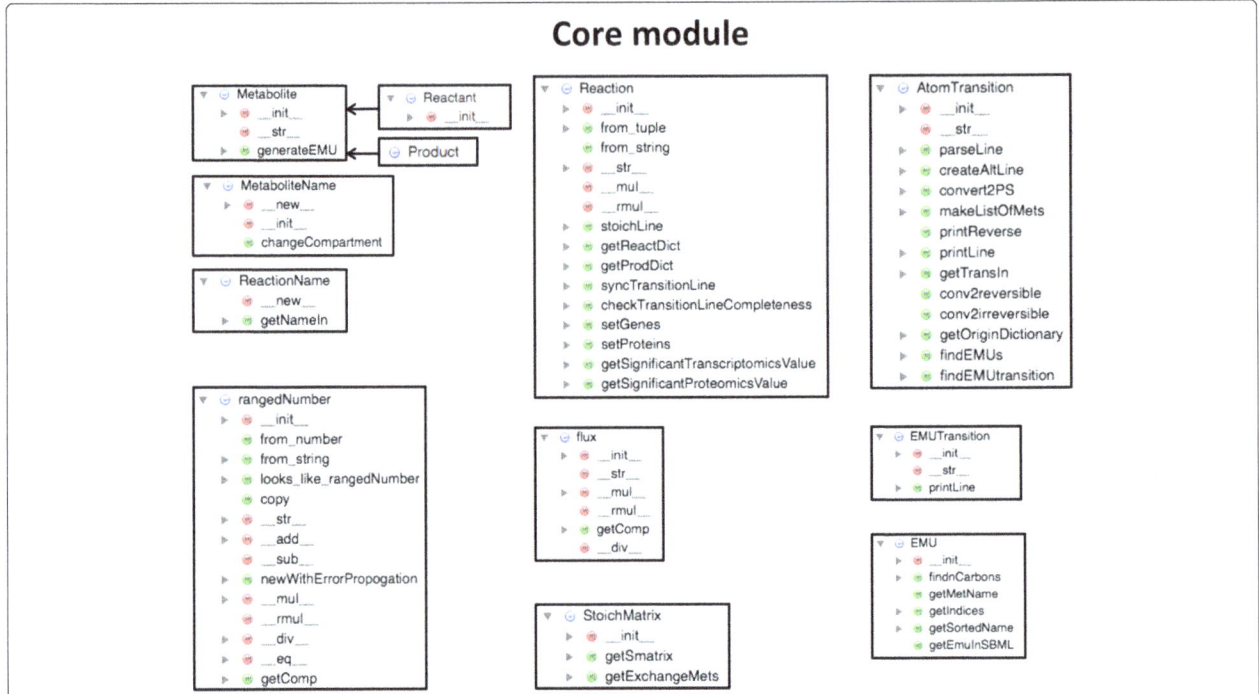

Fig. 2 Core module class diagram. The core module contains the core classes for the rest of the library: classes for metabolite, reaction, flux, elementary metabolite units (EMU), atom transitions, emu transitions and stoichiometry matrices

FBA purposes). However, we expect in the future to derive these classes from the corresponding COBRApy classes. Other basic classes include the *flux* class, which is used to describe fluxes for a given reaction (See Jupyter notebook A1).

Core classes that describe concepts exclusively used in ^{13}C labeling experiments include the *atomTransition*, *emu* and *emuTransition* classes. The *atomTransition* class is used to hold carbon transition rules (information on the fate for each carbon atom in a reaction [23]), e.g.:

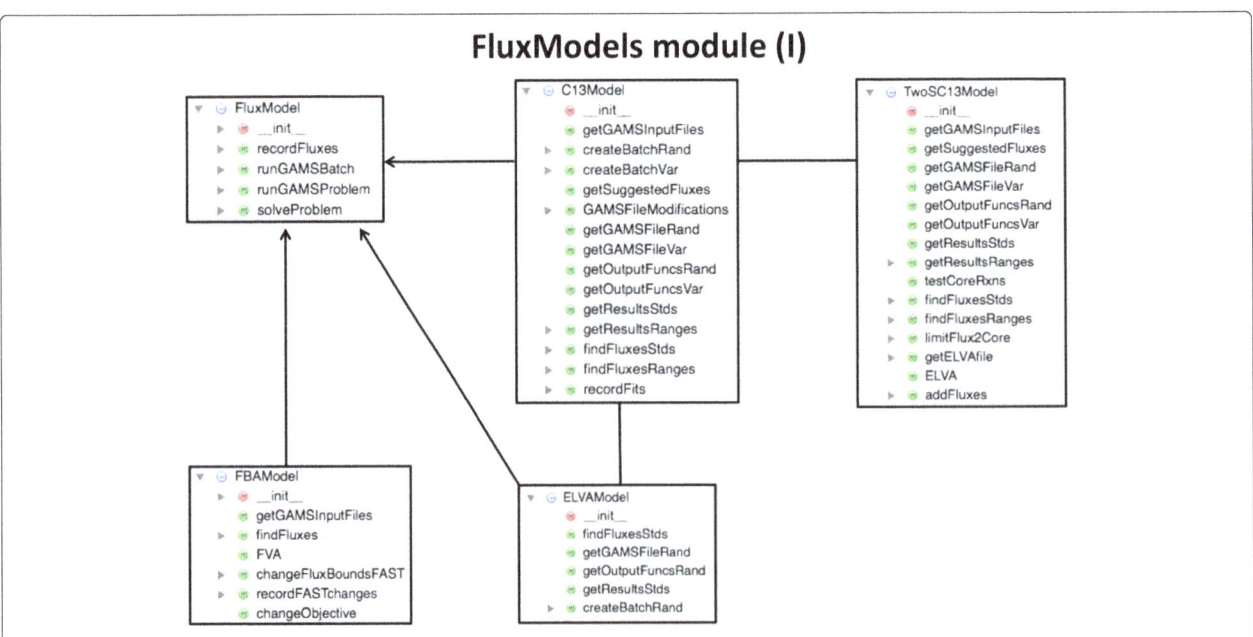

Fig. 3 FluxModels module class diagram (part I). The FluxModels module contains classes for the different types of models used for each flux analysis type: FBA, ^{13}C MFA, 2S-^{13}C MFA, ELVA, etc. *Arrows* indicate derived classes

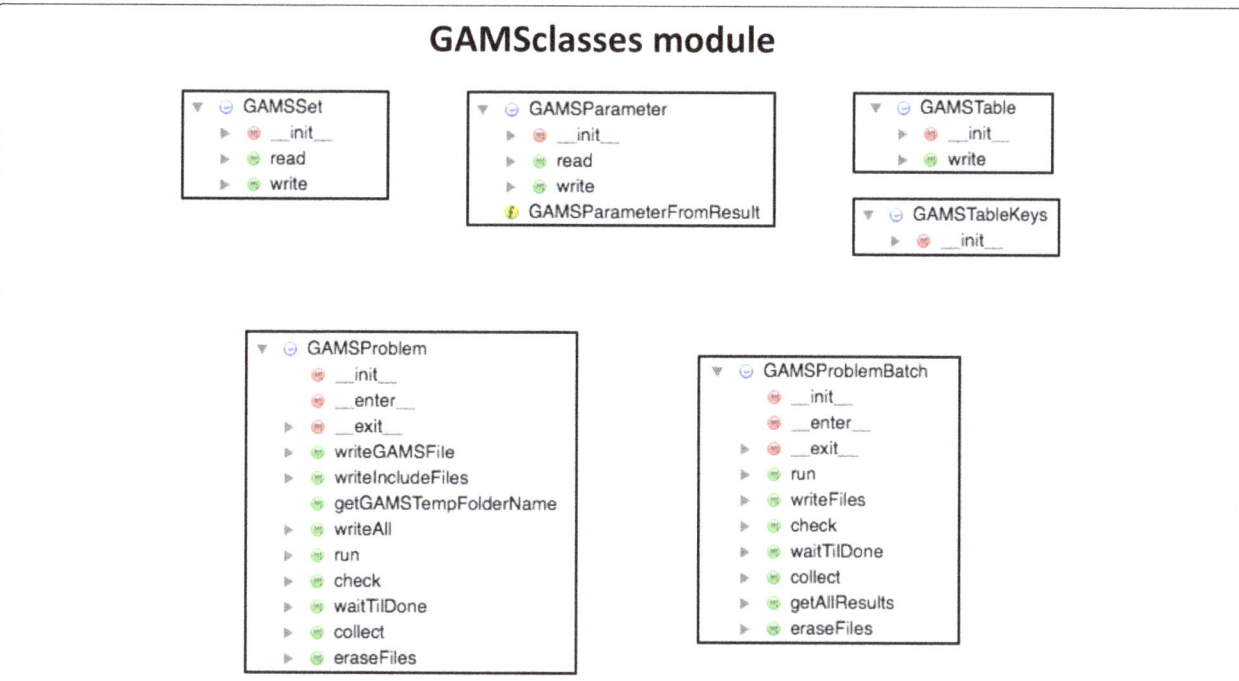

Fig. 4 GAMSclasses module class diagram. The GAMSclasses module contains the classes needed to solve the optimization problems through the GAMS algebraic modeling system. Each of the optimization problems is described through a GAMS file, GAMS sets, GAMS parameters or GAMS tables, which are used as input for the GAMSproblem instance (see Fig. 1). GAMS batches are sets of GAMS problems, to be solved in *series* or *parallel*

```
PDH: pyr -> co2 + accoa ; abc -> a +
bc
```

Note that this is different from the *Reaction* class, which would only hold information on the reaction name, products and reactants for a reaction:

```
PDH: coa[c] + nad[c] + pyr[c] ->
co2[c] + accoa[c] + nadh[c]
```

By using two classes to differentiate transitions and reactions, we minimize confusion between two related but different concepts. The emu class describes elementary metabolite units (emus, [35]), which are used to efficiently calculate labeling patterns corresponding to a flux profile. Emus represent moieties comprising any distinct subset of a compound's atoms: for example, if pyruvate consists of 3 carbons, an emu is a subset of any number of these 3 carbons: e.g. pyr_1_2, pyr_1_2_3 or pyr_2_3 are emus. The *EMUTransition* class describes how emus transition into each other for each reaction, for example for citrate synthase:

```
CS, accoa_c_1_2 + oac_c_2 ->
cit_c_3_4_5
```

indicates that emus accoa_c_1_2 and oac_c_2 condense into emu cit_c_3_4_5. This initial decomposition of atom transitions into emus and emu transitions

is fundamental to solving the forward problem (i.e. determine labeling from fluxes) in ^{13}C MFA, and is provided as part of the core module, as shown in Jupyter notebook A2.

A final core class is the *rangedNumber* class, which describes floating point numbers with a lower and upper limit, reflecting confidence intervals. These type of numbers are used, for example, to hold information for a flux obtained from 2S-^{13}C MFA, where we have a best fit value (0.5), plus upper (0.6) and lower limits (0.3) which represent the upper and lower flux values compatible with the experimental data (see Equation 23 in [23]):

```
PDH: [ 0.3 : 0.5 : 0.6 ]
```

ReactionNetworks

The ReactionNetworks module (see Additional file 1: Figure S1) contains the classes needed to store three types of reaction networks: stoichiometric (*reactionNetwork*, used for FBA), carbon transitions (*C13reactionNetwork*, for ^{13}C MFA) and stochiometric with carbon transitions only for a defined core (*TSreactionNetwork*, for 2S-^{13}C MFA). The first class contains the methods for manipulating purely stoichiometric reaction networks, such as those used in FBA, the second one contains the methods for dealing with carbon transition networks typically used in

^{13}C MFA and the final class contains the methods used to manipulate networks where genome-scale stoichiometric information is mixed with carbon transitions information (used in 2S-^{13}C MFA). Examples of these types of networks along with tutorials on how to use these classes to manipulate them can be found in Jupyter notebook A3.

FluxModels

The FluxModels module (Fig. 3, and Additional file 1: Figure S2) contains the classes used to solve each of the various types of flux problems solved in this library: FBA (*FBAModel*), ^{13}C MFA (*C13Model*), 2S-^{13}C MFA (*TwoSC13Model*) and ELVA (External Labeling Variability Analysis, *ELVAModel*, a subproblem of 2S-^{13}C MFA, [23]). Flux Variability Analysis (FVA [36]) and ^{13}C Flux Variability Analysis (^{13}C FVA [23]) can be solved from the same model as for FBA and ^{13}C MFA respectively, since they require the same information. Aditionally, classes for the results obtained from each of these types of models are included, which are used to plot, analyze and manipulate results (*FBAResults*, *FVAResults*, *C13Results*, *TSResults*, *ELVAResults*, *C13FVAResults* and *TSFVAResults*). More details can be found in Jupyter notebook A4.

EnhancedLists

The enhancedLists module (Additional file 1: Figure S3) holds lists that have been enhanced with methods useful for the objects contained. In this way, there is e.g. a *ReactionList* class that holds reactions and e.g. has a *carbonTransitionsOK()* method that tests if carbon transition and reaction information are compatible (same reactants, same products, etc). The types of lists included are *ReactionList*, *MetaboliteList*, *EMUList*, *EMUTransitionList* and *AtomTransitionList*. Their use is demonstrated in Jupyter notebook A2.

GAMSclasses

The GAMSclasses module contains the classes used to solve the optimization problems for each model (see below) through the GAMS modeling system (GAMS Development Corporation), either in series or in parallel (Fig. 4). As depicted in Fig. 1, all flux techniques require solving an optimization problem using a different type of solver (either CPLEX or CONOPT in GAMS), which are accessed through the GAMS framework. At this moment, this interaction is encoded in this class, but we expect to replace it soon by use of the GAMS python API. More information and tests can be found in Jupyter notebook A5.

Predictions

This module provides the classes needed to make flux predictions using the Minimization of Metabolic Adjustment (MoMA [36]) and Regulatory on/off minimization (ROOM [37]) methods that were used in the original 2S-^{13}C MFA paper [23]. A demonstration can be found in Jupyter notebook A6.

Labeling

The Labeling module contains all classes needed to analyze, manipulate and plot labeling patterns (a.k.a Mass Distribution Vector, MDV, the fraction of molecules with $n=1, 2, 3 ...$ labeled carbons) and fragments. The base class *labelingData* is used to derive *LCMSLabelData*, *CEMSLabelData* and *GCMSLabelData*, which are classes for each type of mass spectroscopy data. The *fragment* class contains all information for the MS fragment harboring the MDV and is used to derive classes *GCMSfragment*, *LCMSfragment* and *CEMSfragment*. More details can be found in Jupyter notebook A7.

DB

The DB module houses a database of all hard-coded information, conveniently located in a single module. This includes, for example, a dictionary of fragments with all their detailed information: name, abbreviation, formula, number of carbons, formula without carbon backbone, etc.

Utilities

The utilities module stores small functions that are useful in a variety of modules: e.g. a function to extract the atoms numbers from an elemental composition string ($'H6NO2Si' \Rightarrow [('H', 6), ('N', 1), ('O', 2), ('Si', 1)]$).

Results and discussion

We will now demonstrate the use of the jQMM library for each of the techniques included in the package using data from the literature.

^{13}C Metabolic Flux Analysis for TCA toy model

The first demonstration (Fig. 5) involves the toy model of the TCA cycle originally used by Antoniewicz et al. [35] to showcase the Elementary Metabolite Unit (EMU) method (Fig. 12 in [35]). The toy model is a simplification of the tricarboxylic acid (TCA) cycle that still retains its main features. Acetyl coenzyme A (AcCoA) and aspartate are the two substrates, and glutamate and carbon dioxide are the products. The labeling for glutamate is calculated assuming a mixture of 25% [2-^{13}C]AcCoA and 25% [1,2-^{13}C]AcCoA as the tracer input (with everything else unlabeled), while aspartate is considered unlabeled. Fluxes are fixed to predetermined values, since the goal in this case is to showcase the calculation of labeling patterns from fluxes. Jupyter Notebook B2 demonstrates how, using jQMM, EMU transitions are generated, and solved through GAMS and CONOPT without the need to decompose

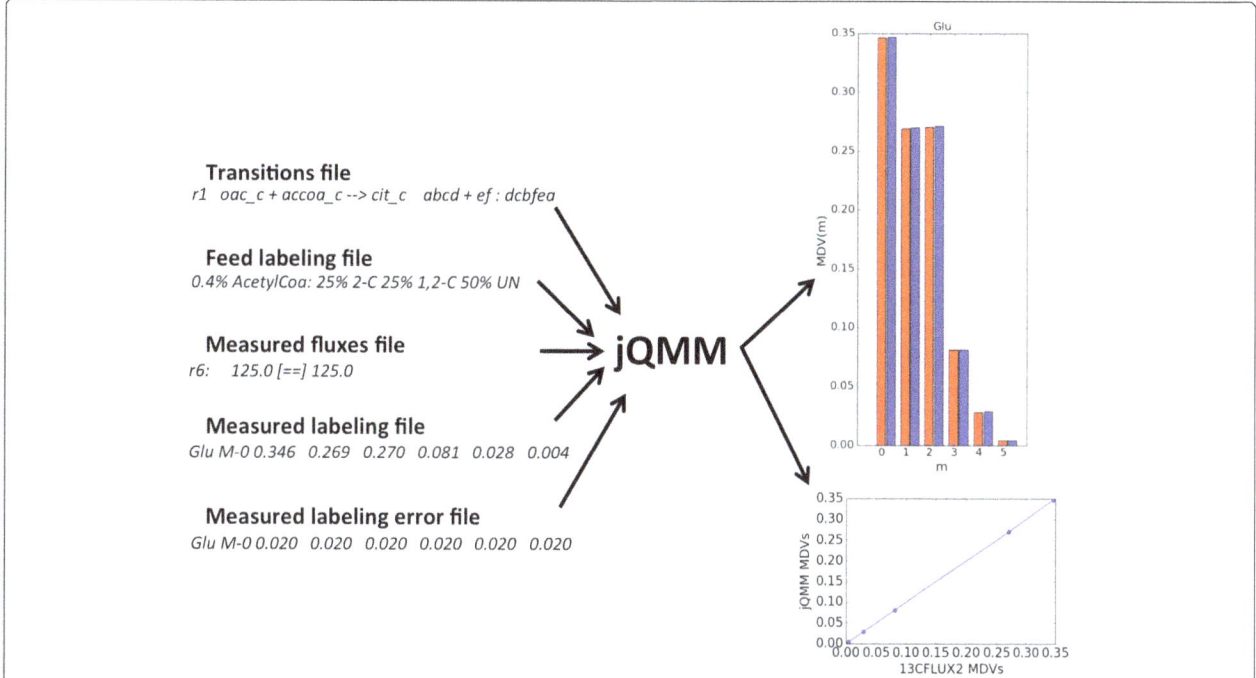

Fig. 5 Example jQMM inputs and outputs for TCA toy model. The TCA toy model represents a symplification of the TCA cycle that still retains its main features and is used to test ^{13}C MFA tools [35]. In this case, the input consists of the carbon transitions (the fate of each carbon for each the reaction), the feed labeling information (AcetylCoA 50% unlabeled, 25% labeled in the second carbon, 25% labeled in the first two carbons), the measured labeling patterns (glutamate MDVs), the error in these measurements, and the fluxes which are known (e.g. r6). File *line* examples are given in *cursive below*. In this case, all fluxes are assumed to be measured and the goal is to find the corresponding MDV for glutamate, which is given in the *top right part* of the figure (in *red* experimentally measured MDV, in *blue* theoretically calculated MDV). The *lower right part* of the figure provides a comparison with 13CFLUX2 [29], a well known ^{13}C MFA package (See Jupyter Notebook B2)

the network into decoupled EMU reaction networks. The computed labeling pattern is the same as obtained through 13CFLUX2 [29](see Fig. 5, and Jupyter Notebook B2 for more details).

^{13}C Metabolic Flux Analysis for *E. coli* data

The next demonstration of the jQMM library involves a traditional use of ^{13}C MFA to calculate fluxes using the data from Toya et al. [38], in order to reproduce supplementary figure S18 from Garcia Martin et al. [23]. Fluxes are calculated for wild type *E. coli* fitting labeling data for 9 different intracellular metabolites. This calculation uses the original small-scale model for central carbon metabolism that lumps several reactions together and displays only an approximate account of fluxes to biomass. For example, as discussed in Garcia Martin et al. [23], the large fluxes draining acetyl-CoA into biomass and to the exterior of the cell as acetate (each a third of the pyruvate dehydrogenase flux, PDH) imposed in the original publication have been overestimated. The typical biomass function used in these minimal ^{13}C MFA models with reactions lumped together involves a specific stoichiometry of central carbon intermediates converted to biomass. However, this represents an

approximation since some of the metabolites required for biomass growth are not present in the minimal network model and need to be substituted by their requirements in terms of intermediates considered in the minimal network (acetyl-CoA, in this case). These effects require significant effort to be accurately incorporated into a small-scale model, but they are properly handled by the genome-scale model, as demonstrated in the next example. Jupyter Notebook B3 shows how to calculate fluxes through ^{13}C MFA for this case.

2S-^{13}C Metabolic Flux Analysis for *E. coli* data

This example calculates fluxes for a full comprehensive genome-scale model [23] constrained by ^{13}C labeling data through the 2S-^{13}C MFA algorithm (Fig. 6). The input is the same as for the ^{13}C MFA case in the previous section, with the addition of the iJR904 genome-scale model [39]. As discussed in depth in Garcia Martin et al. [23], the solution displays nearly the same values as ^{13}C MFA for central metabolism (see e.g. Figure S4 and S18 in [23]). This similarity is expected since 2S-^{13}C MFA is designed to mimic ^{13}C MFA for this part of metabolism (see "Limiting flux to core reactions" section in [23]). The only difference for the

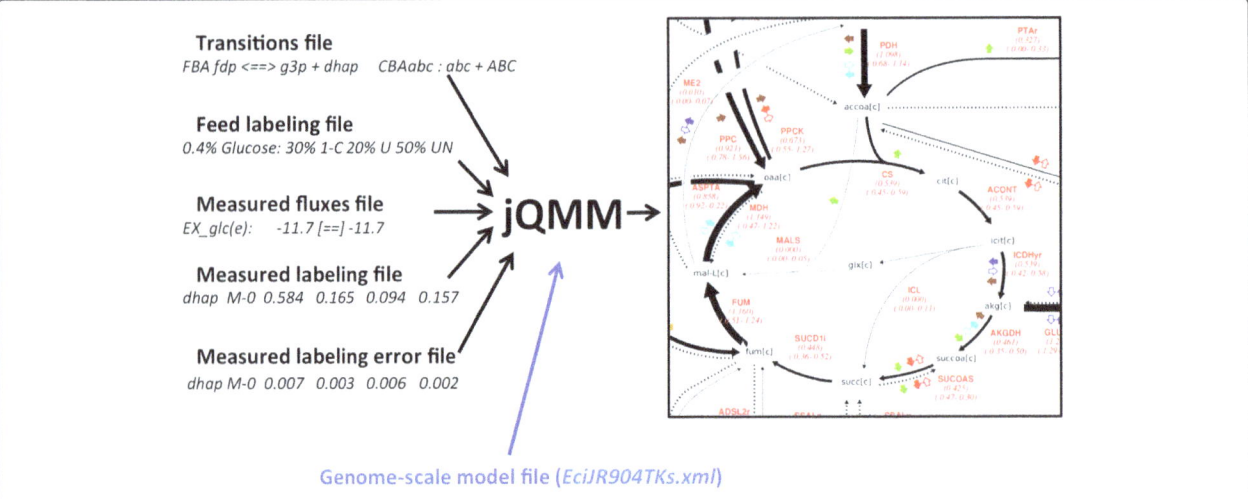

Fig. 6 Example jQMM inputs and outputs for E. coli 2S-^{13}C MFA. 2S-^{13}C MFA allows for the calculation of fluxes for genome-scale models constrained by ^{13}C labeling data [23]. The input is the same as for ^{13}C MFA(Fig.5), but with the addition of a genome-scale model. The output consists of fluxes for the genome-scale model, which are visualized in a *metabolic map* (see Jupyter Notebook B4)

Toya et al. data set can be found in the TCA cycle flux. These differences appear because genome-scale models account for fluxes to biomass in a more realistic manner and because they do not rule out unexpected metabolic routes compatible with the available data.

Furthermore, the use of genome-scale models allows for some information that can not be obtained from smaller, non-comprehensive models, such as an account of all reactions producing and consuming e.g. NADPH as constrained by ^{13}C labeling data (Fig. 6 in [23]). Other advantages of using genome-scale models include the prediction of non measured fluxes (Fig. S13 in [23]) and quantitative full predictions of directly measurable data (Fig. 10 in [23]). Jupyter Notebook B4 shows how to reproduce all figures from Garcia Martin et al. [23].

^{13}C Metabolic Flux Analysis for a soil microbial community

In order to show the general capabilities of the jQMM library for non-standard ^{13}C MFA we will show how one can use this library to calculate fluxes for microbial communities as shown in Hagerty et al. [25]. The problem presented and solved in this paper is related to ^{13}C MFA but, at the same time, the technical details differ notably. In this and other experiments [40, 41], Dijkstra et al. collected several soil samples and incubated them with two different carbon sources (glucose and pyruvate), where each source was chosen to have a different initial labeling (1-^{13}C or fully labeled for glucose and 1-^{13}C or 2,3-^{13}C for pyruvate). For each carbon source and labeling type, the labeling fraction of CO_2 emmited from the soil was measured. For each carbon source, the two relative CO_2 labeling profiles provided enough information to

fit a small model of carbon metabolism that represented the combination of reactions for all microbial entities in the soil. The good fits supported the assumption that all members of the community display a similar metabolic flux state, and the results are used to derive the carbon utilization efficiency (i.e. how much carbon is diverted into biomass and how much is lost as CO_2, a critical parameter for climate models [42]). Jupyter notebook B6 shows how to do these calculations using modular combinations of the jQMM library without recourse to complicated excel data sheets [25].

Flux balance analysis

The jQMM library provides the means to perform FBA (Jupyter notebook B1) and other COBRA methods for the sake of comparison with ^{13}C-based methods [23]. However, the main purpose of jQMM is not to provide a COBRA library, since actively developed open source python libraries such as COBRApy [26] and cameo (http://cameo.bio/) are available for this purpose. Methods for FBA, FVA, Minimization of Metabolic Adjustment (MoMA [36]) and Regulatory on/off minimization (ROOM [37]) are provided. In order to showcase these capabilities we replicated part of the tutorial from the COBRA v 1.0 package in Becker et al. [43], which can be found in Jupyter notebook B1. Future versions of the library will likely be changed to use cameo for this functionality.

Principal Component Analysis of Proteomics (PCAP)

While most of the jQMM library is focused on flux analysis, we have found it useful to statistically model -omics data without the recourse to mechanistic models [24], an approach which we expect to be increasingly popular in the coming years given the increasing availability of -omics data. In Jupyter notebook B5, we show how to use the jQMM library and quantitative proteomics data to increase biofuel yields by replicating the PCAP (Principal Component Analysis for Proteomics) analysis with a recreation of previously published figures [24] (Fig. 7).

This case represents a prototypical example of pathway engineering, in which a series of exogenous genes from a variety of organisms (*E. coli, S. cerevisiae, S. aureus*, and plants such as *M. spicata*) were assembled together in a pathway able to produce limonene [44], a jet-fuel [45]. However, when assembling this pathway together, there was no easy way to figure out the design parameters for optimal function: what promoter strength to use, when to induce? how much? In this case, a combinatorial approach was used, using promoters of different strength (weak, medium, strong), different induction strengths (low, medium, high) and induction times (low, medium, high). For each of these scenarios, limonene production and protein expression for all genes in the pathway were measured. The raw proteomics data did not provide clear trends but the use of a basic data analysis tool such as a Principal Component Analysis (PCA) produced actionable results which allowed for the improvement of limonene production by 40% and bisabolene production by 200% leading to the highest producing strain of this biodiesel replacement to that date [24]. Jupyter notebook B5 shows how to perform this analysis.

Conclusion

We have presented here the jQMM library, an open-source modular library for modeling metabolism and the first open-source library for ^{13}C MFA in Python. The jQMM library is also the first one to provide a method to calculate fluxes for genome-scale models constrained by ^{13}C labeling data through 2S-^{13}C MFA, and is the first to unify the FBA, ^{13}C MFA and 2S-^{13}C MFA in a single package. Unlike METRAN [27], INCA [28] and 13CFLUX2 [29]), it is open source and, unlike OPENFLUX [30] and OPENFLUX2 [31], it is written in Python. Furthermore, it includes an example on how to calculate fluxes for microbial communities through ^{13}C MFA, and the tools to leverage other types of -omics data (e.g. proteomics for biofuel production).

While the examples here are mainly focused on various types of flux analysis (FBA, ^{13}C MFA, 2S-^{13}C MFA, etc) and biofuels, the tools are of general aplicability to other fields (e.g. pharma, biomedicine, environmental microbiology) and include statistical methods as well (e.g. PCAP). jQMM presents not only methods for modeling and generating actionable items form -omics data, but also libraries for data manipulation and visualization (fluxes and labeling patterns see Figs. 6 and 5), including classes for most common concepts in metabolism (e.g. reactions, metabolites, GC-MS fragments, reaction networks) in order to facilitate their manipulation, and which are of utility beyond the examples presented here. Every module in jQMM is explained in detail in Jupyter notebooks that allow for immediate reproducibility. Python unit tests are

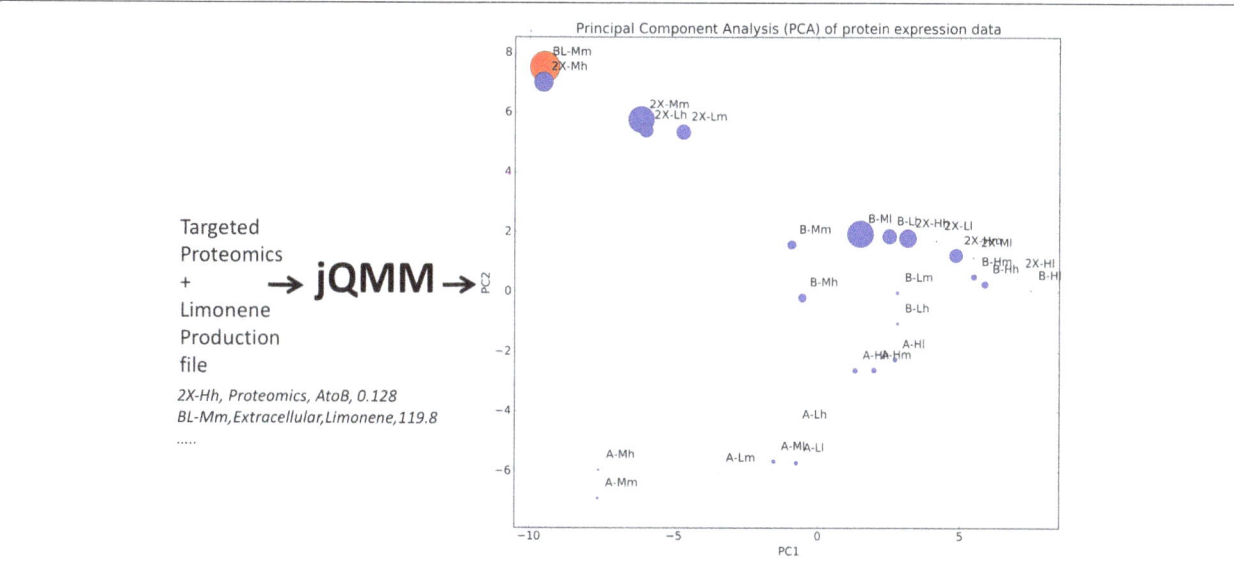

Fig. 7 Example jQMM inputs and outputs for Principal Component Analysis of Proteomics (PCAP). PCAP is a method developed to leverage proteomics data in order to improve bioproduct yield. The inputs are measured proteomic profiles and bioproduct production levels and the output consists of a *plot* that can be used to increase yields as shown in Alonso-Gutierrez et al. [24]. (see Jupyter Notebook B5). File *line* examples are provided in *cursive below*

included for all modules for ease of testing when adding functionality.

jQMM will continue growing as the methods developed and used at JBEI for biofuel production expand and mature, while other authors contribute to it. At the moment jQMM is heavily dependent on the closed-source commercial platform GAMS and the CPLEX and CONOPT solvers. Our first future improvement to the jQMM library is planned to involve providing an interface to open-source and free version of solvers equivalent to GAMS. Further future improvements will likely include integration with COBRApy [26], cameo (cameo.bio) or KBase (kbase.us).

Abbreviations
^{13}C MFA: ^{13}C Metabolic Flux Analysis; 2S-^{13}C MFA: Two-scale ^{13}C Metabolic Flux Analysis; AcCoa: Acetyl-CoA; API: Application Program Interface; COBRA: COnstraint Based Reconstruction and Analysis; ELVA: External Labeling Variability Analysis; EMU: Elementary Metabolite Unit; FBA: Flux Balance Analysis; FVA: Flux Variability Analysis; JBEI: Joint BioEnergy Institute; jQMM: JBEI Quantitative Metabolic Modeling; MDV: Mass Distribution Vector; MoMA: Minimum of Metabolic Adjustment; NADPH: Nicotinamide adenine dinucleotide phosphate; PDH: Pyruvate Dehydrogenase; PCAP: Principal Component Analysis of Proteomics; ROOM: Regulatory On/Off Minimization; TCA: trycarboxylic acid

Acknowledgements
Not applicable.

Funding
This work was part of the DOE Joint BioEnergy Institute (http://www.jbei.org) supported by the U. S. Department of Energy, Office of Science, Office of Biological and Environmental Research, and was part of the Agile BioFoundry (http://agilebiofoundry.org) supported by the U.S. Department of Energy, Energy Efficiency and Renewable Energy, Bioenergy Technologies Office, through contract DE-AC02-05CH11231 between Lawrence Berkeley National Laboratory and the U. S. Department of Energy. The United States Government retains and the publisher, by accepting the article for publication, acknowledges that the United States Government retains a non-exclusive, paid-up, irrevocable, world-wide license to publish or reproduce the published form of this manuscript, or allow others to do so, for United States Government purposes. This research is also supported by the Basque Government through the BERC 2014-2017 program and by Spanish Ministry of Economy and Competitiveness MINECO: BCAM Severo Ochoa excellence accreditation SEV-2013-0323.

Authors' contributions
HGM, GWB, AG, VSK, DW and DA wrote the code and analyzed the data. HGM, GWB, DA, AA and JDK wrote the paper. TWHB provided the docker containers. All authors have read and approve of the final version of the manuscript.

Competing interests
JDK has financial interests in Amyris and Lygos. All other authors have no competing interests to declare.

Author details
[1]Biological Systems and Engineering Division, Lawrence Berkeley National Laboratory, Berkeley, CA, USA. [2]Joint BioEnergy Institute, Emeryville, CA, USA. [3]Department of Chemical and Biomolecular Engineering, University of California, Berkeley, CA, USA. [4]Department of Bioengineering, University of California, Berkeley, CA, USA. [5]Environmental Genomics and Systems Biology Division, Lawrence Berkeley National Laboratory, Berkeley, CA, USA. [6]School of Energy Science and Engineering, Indian Institute of Technology (IIT), Kharagpur, India. [7]Novo Nordisk Foundation Center for Biosustainability, Technical University of Denmark, DK2970 Hørsholm, Denmark. [8]DOE Agile BioFoundry, Emeryville, CA, USA. [9]BCAM, Basque Center for Applied Mathematics, Bilbao, Spain.

References
1.   Nelson DL, Lehninger AL, Cox MM. Lehninger Principles of Biochemistry. New York: W. H. Freeman and Company; 2008.
2.   Xu F, Tavintharan S, Sum CF, Woon K, Lim SC, Ong CN. Metabolic signature shift in type 2 diabetes mellitus revealed by mass spectrometry-based metabolomics. J Clin Endocrinol Metab. 2013;98(6):. doi:10.1210/jc.2012-4132.
3.   Cummins TD, Holden CR, Sansbury BE, Gibb AA, Shah J, Zafar N, Tang Y, Hellmann J, Rai SN, Spite M, Bhatnagar A, Hill BG. Metabolic remodeling of white adipose tissue in obesity,. Am J Physiol Endocrinol Metab. 2014;307(3):262–77. doi:10.1152/ajpendo.00271.2013.
4.   Wu L, Gomes A, Sinclair D. Geroncogenesis: Metabolic changes during aging as a driver of tumorigenesis. 2014. NIHMS150003. doi:10.1016/j.ccr.2013.12.005.
5.   Chubukov V, Mukhopadhyay A, Petzold CJ, Keasling JD, Martín HG. Synthetic and systems biology for microbial production of commodity chemicals. npj Syst Biol Appl. 2016;2:16009. doi:10.1038/npjsba.2016.9.
6.   Lorek S, Spangenberg JH. Sustainable consumption within a sustainable economy – beyond green growth and green economies. J Clean Prod. 2014;63:33–44. doi:10.1016/j.jclepro.2013.08.045.
7.   Nielsen J, Keasling JD. Engineering cellular metabolism. Cell. 2016;164(6): 1185–97.
8.   Lee SY, Lee DY, Kim TY. Systems biotechnology for strain improvement. Trends Biotechnol. 2005;23(7):349–58.
9.   Han MJ, Yoon SS, Lee SY. Proteome analysis of metabolically engineeredescherichia coli producing poly (3-hydroxybutyrate). J Bacteriol. 2001;183(1):301–8.
10. George KW, Chen A, Jain A, Batth TS, Baidoo EE, Wang G, Adams PD, Petzold CJ, Keasling JD, Lee TS. Correlation analysis of targeted proteins and metabolites to assess and engineer microbial isopentenol production. Biotech Bioeng. 2014;111(8):1648–58.
11. Palsson B, Zengler K. The challenges of integrating multi-omic data sets. Nat Chem Biol. 2010;6(11):787–9.
12. Nielsen J, Fussenegger M, Keasling J, Lee SY, Liao JC, Prather K, Palsson B. Engineering synergy in biotechnology. Nat Chem Biol. 2014;10(5): 319–22.
13. Javidpour P, Pereira JH, Goh EB, McAndrew RP, Ma SM, Friedland GD, Keasling JD, Chhabra SR, Adams PD, Beller HR. Biochemical and structural studies of nadh-dependent fabg used to increase the bacterial production of fatty acids under anaerobic conditions. Appl Environ Microbiol. 2014;80(2):497–505.
14. Bhan N, Xu P, Khalidi O, Koffas MA. Redirecting carbon flux into malonyl-coa to improve resveratrol titers: proof of concept for genetic interventions predicted by optforce computational framework. Chem Eng Sci. 2013;103:109–14.
15. Lee KH, Park JH, Kim TY, Kim HU, Lee SY. Systems metabolic engineering of escherichia coli for l-threonine production. Mol Syst Biol. 2007;3(1):149.
16. Wetterstrand KA. DNA Sequencing Costs: Data from the NHGRI Genome Sequencing Program (GSP). 2016. http://www.genome.gov/ sequencingcosts. Accessed 05 Feb 2016.
17. Batth TS, Singh P, Ramakrishnan VR, Sousa MML, Chan LJG, Tran HM, Luning EG, Pan EHY, Vuu KM, Keasling JD, Adams PD, Petzold CJ. A targeted proteomics toolkit for high-throughput absolute quantification of Escherichia coli proteins. Metab Eng. 2014;26C:48–56.
18. Fuhrer T, Zamboni N. High-throughput discovery metabolomics. Curr Opin Biotechnol. 2015;31:73–8.
19. Blanch HW, Adams PD, Andrews-Cramer KM, Frommer WB, Simmons BA, Keasling JD. Addressing the need for alternative transportation fuels: the Joint BioEnergy Institute. ACS Chem Biol. 2008;3(1):17–20.
20. Beller HR, Lee TS, Katz L. Natural products as biofuels and bio-based chemicals: fatty acids and isoprenoids. Nat Prod Rep. 2015;32.10:1508–26.
21. Wiechert W. 13C metabolic flux analysis. Metab Eng. 2001;3(3):195–206.
22. Lewis NE, Nagarajan H, Palsson BO. Constraining the metabolic genotype-phenotype relationship using a phylogeny of in silico methods. Nat Rev Microbiol. 2012;10(4):291–305.
23. Garcia Martin H, Kumar VS, Weaver D, Ghosh A, Chubukov V, Mukhopadhyay A, et al. A Method to Constrain Genome-Scale Models with 13C Labeling Data. PLoS Comput Biol. 2015;11(9):e1004363. doi:10.1371/journal.pcbi.1004363.
24. Alonso-Gutierrez J, Kim EM, Batth TS, Cho N, Hu Q, Chan LJG, Petzold CJ, Hillson NJ, Adams PD, Keasling JD, Garcia-Martin H, Soon Lee T.

Principal component analysis of proteomics (PCAP) as a tool to direct metabolic engineering. Metab Eng. 2014;28:123–33.

25. Hagerty SB, van Groenigen KJ, Allison SD, Hungate BA, Schwartz E, Koch GW, Kolka RK, Dijkstra P. Accelerated microbial turnover but constant growth efficiency with warming in soil. Nat Clim Chang. 2014;4(10):903–6.

26. Ebrahim A, Lerman JA, Palsson BO, Hyduke DR. COBRApy: COnstraints-Based Reconstruction and Analysis for Python. BMC Syst Biol. 2013;7(1):74.

27. Crown SB, Ahn WS, Antoniewicz MR. Rational design of 13C-labeling experiments for metabolic flux analysis in mammalian cells. BMC Syst Biol. 2012;6(1):43.

28. Young JD. INCA: a computational platform for isotopically non-stationary metabolic flux analysis. Bioinforma (Oxford, England). 2014;30(9):1333–5.

29. Weitzel M, et al. 13CFLUX2Ûhigh-performance software suite for 13C-metabolic flux analysis. Bioinformatics. 2013;29.1:143–5.

30. Quek LE, Wittmann C, Nielsen LK, Krömer JO. OpenFLUX: efficient modelling software for 13C-based metabolic flux analysis. Microb Cell Factories. 2009;8(1):25.

31. Shupletsov MS, Golubeva LI, Rubina SS, Podvyaznikov DA, Iwatani S, Mashko SV. OpenFLUX2: 13 C-MFA modeling software package adjusted for the comprehensive analysis of single and parallel labeling experiments. Microb Cell Factories. 2014;13(1):152.

32. Plant AL, Locascio LE, May WE, Gallagher PD. Improved reproducibility by assuring confidence in measurements in biomedical research. Nat Methods. 2014;11(9):895–8.

33. Wilson G, Aruliah DA, Brown CT, Chue Hong NP, Davis M, Guy RT, Haddock SHD, Huff KD, Mitchell IM, Plumbley MD, Waugh B, White EP, Wilson P. Best Practices for Scientific Computing. PLoS Bio. 2014;12(1):. arXiv:1210.0530v3.

34. Pérez F, Granger BE. IPython: a system for interactive scientific computing. Comput Sci Eng. 2007;9(3):21–9.

35. Antoniewicz MR, Kelleher JK, Stephanopoulos G. Elementary metabolite units (EMU): A novel framework for modeling isotopic distributions. Metab Eng. 2007;9(1):68–86. NIHMS150003.

36. Segrè D, Vitkup D, Church GM. Analysis of optimality in natural and perturbed metabolic networks. Proc Natl Acad Sci U S A. 2002;99(23):15112–7.

37. Shlomi T, Berkman O, Ruppin E. Regulatory on/off minimization of metabolic flux changes after genetic perturbations,. Proc Natl Acad Sci U S A. 2005;102(21):7695–00.

38. Toya Y, Ishii N, Nakahigashi K, Hirasawa T, Soga T, Tomita M, Shimizu K. 13C-Metabolic flux analysis for batch culture of Escherichia coli and its pyk and pgi gene knockout mutants based on mass isotopomer distribution of intracellular metabolites. Biotechnol Prog. 2010;26(4):975–92.

39. Reed JL, Vo TD, Schilling CH, Palsson BO. An expanded genome-scale model of Escherichia coli K-12 (iJR904 GSM/GPR). Genome Biol. 2003;4(9):54.

40. Dijkstra P, Dalder JJ, Selmants PC, Hart SC, Koch GW, Schwartz E, Hungate BA. Modeling soil metabolic processes using isotopologue pairs of position-specific 13C-labeled glucose and pyruvate. Soil Biol Biochem. 2011;43(9):1848–57.

41. Dijkstra P, Thomas SC, Heinrich PL, Koch GW, Schwartz E, Hungate BA. Effect of temperature on metabolic activity of intact microbial communities: Evidence for altered metabolic pathway activity but not for increased maintenance respiration and reduced carbon use efficiency. Soil Biol Biochem. 2011;43(10):2023–31.

42. Cox PM, Betts RA, Jones CD, Spall SA, Totterdell IJ. Acceleration of global warming due to carbon-cycle feedbacks in a coupled climate model. Nature. 2000;408(6809):184–7.

43. Becker SA, Feist AM, Mo ML, Hannum G, Palsson BO, Herrgard MJ. Quantitative prediction of cellular metabolism with constraint-based models: the COBRA Toolbox. Nat Protoc. 2007;2(3):727–38.

44. Alonso-Gutierrez J, Chan R, Batth TS, Adams PD, Keasling JD, Petzold CJ, Lee TS. Metabolic engineering of Escherichia coli for limonene and perillyl alcohol production. Metab Eng. 2013;19:33–41.

45. Chuck CJ, Donnelly J. The compatibility of potential bioderived fuels with Jet A-1 aviation kerosene. Appl Energy. 2014;118:83–91.

6

Improving the accuracy of high-throughput protein-protein affinity prediction may require better training data

Raquel Dias and Bryan Kolaczkowski[*]

Abstract

Background: One goal of structural biology is to understand how a protein's 3-dimensional conformation determines its capacity to interact with potential ligands. In the case of small chemical ligands, deconstructing a static protein-ligand complex into its constituent atom-atom interactions is typically sufficient to rapidly predict ligand affinity with high accuracy (>70% correlation between predicted and experimentally-determined affinity), a fact that is exploited to support structure-based drug design. We recently found that protein-DNA/RNA affinity can also be predicted with high accuracy using extensions of existing techniques, but protein-protein affinity could not be predicted with >60% correlation, even when the protein-protein complex was available.

Methods: X-ray and NMR structures of protein-protein complexes, their associated binding affinities and experimental conditions were obtained from different binding affinity and structural databases. Statistical models were implemented using a generalized linear model framework, including the experimental conditions as new model features. We evaluated the potential for new features to improve affinity prediction models by calculating the Pearson correlation between predicted and experimental binding affinities on the training and test data after model fitting and after cross-validation. Differences in accuracy were assessed using two-sample t test and nonparametric Mann–Whitney U test.

Results: Here we evaluate a range of potential factors that may interfere with accurate protein-protein affinity prediction. We find that X-ray crystal resolution has the strongest single effect on protein-protein affinity prediction. Limiting our analyses to only high-resolution complexes (≤2.5 Å) increased the correlation between predicted and experimental affinity from 54 to 68% ($p = 4.32 \times 10^{-3}$). In addition, incorporating information on the experimental conditions under which affinities were measured (pH, temperature and binding assay) had significant effects on prediction accuracy. We also highlight a number of potential errors in large structure-affinity databases, which could affect both model training and accuracy assessment.

Conclusions: The results suggest that the accuracy of statistical models for protein-protein affinity prediction may be limited by the information present in databases used to train new models. Improving our capacity to integrate large-scale structural and functional information may be required to substantively advance our understanding of the general principles by which a protein's structure determines its function.

Keywords: Protein-protein, Binding affinity, Machine learning, Intermolecular interactions, Scoring functions

* Correspondence: bryank@ufl.edu
Department of Microbiology and Cell Science, University of Florida,
Gainesville, FL, USA

Background

Proteins are involved in the majority of chemical reactions that take place within living cells, making them essential for all aspects of cellular function. Proteins never work in isolation; their functional repertoire is determined by how they interact with various small-molecule, DNA/RNA, protein or other ligands. Ligand affinity is largely determined by a protein's 3-dimensional structure, which determines the spatial conformation of attractive and repulsive forces between the protein and a potential ligand [1–3]. The affinity with which a protein interacts with various ligands–typically expressed as the dissociation constant (Kd or pKd = −log Kd)—provides critical information about protein function and biochemistry, and has been used for the discovery and optimization of novel pharmaceuticals [4–6].

High-throughput prediction of protein-ligand affinity is typically conducted using a fast statistical "scoring function" that decomposes binding affinity into component atom-atom interaction terms representing the attractive and repulsive forces acting across the protein-ligand complex [7, 8]. Although scoring functions can be derived directly from physical chemistry principles [9], the most effective approaches are usually "trained" using large databases of structural complexes with associated experimentally-determined binding affinities [10–12]. After training, a model's expected predictive accuracy can be gauged by correlating its predicted affinities with experimentally-determined values across a novel dataset not included in training [13, 14].

Many scoring functions are capable of using only the atomic interactions extracted from crystal structures to rapidly predict protein-small molecule affinities with >70% correlation, which is commonly considered adequate to support structure-based drug design [11, 15–21]. Recently, we developed efficient statistical models capable of predicting protein-DNA/RNA affinities with similar accuracy [22]. Our structure-based prediction models also revealed that different combinations of atom-atom interactions are important for predicting different types of protein-ligand complexes. However, no statistical models we examined were capable of predicting protein-protein affinity with >60% correlation, even under the 'best case' scenario in which the protein-protein complex was known experimentally.

Accurate prediction of protein-protein interactions is a major goal of computational structural biology, and many approaches have been examined to improve the accuracy of protein-protein affinity prediction [23]. The structural basis of protein-protein interactions is typically more complex and flexible than other protein-ligand interactions, suggesting that entropic forces may be more important in protein-protein interactions [24, 25]. Physics-based approaches such as molecular dynamics can model entropic factors and produce highly-accurate affinity predictions but are too computationally complex to support high-throughput analyses [10–12, 17, 26, 27]. As an alternative approach, smaller manually-curated affinity benchmarks have been proposed to improve the accuracy of high-throughput statistical affinity prediction [28]. However, predictive accuracy on manually-curated datasets rarely exceeds ~60% correlation [29], and accuracy achieved using carefully curated datasets may not generalize well to new data. Importantly, the specific factors that may influence statistical prediction of protein-protein affinity have not been identified, making it difficult to devise reasonable strategies to improve current methods.

Methods
Structural dataset curation
X-ray and NMR structures of protein-protein complexes and their associated binding affinities (−$\log_{10}$-transformed dissociation constants, pKds) were obtained from PDBbind [30] and from the protein-protein affinity benchmark database [28]. Complexes with ambiguous ligand information were excluded, as were complexes with multiple ligands or mulitimeric proteins, similar to previous approaches applied for building refined protein-ligand data sets [11, 30, 31]. From each protein – protein complex, we extracted a suite of non-redundant atom-atom interactions thought to potentially correlate with ligand affinity. Details on how each atom-atom interaction is defined and calculated are presented in our previous work [22]. We included only those atomic interactions that could be determined entirely from atomic coordinates and atom types in a standard PDB file.

For each protein-protein complex, we extracted additional information on structure acquisition method, temperature, pH, and crystal resolution from the Protein Data Bank [32]. We constructed data sets of 569 protein-protein complexes with assigned temperature data, 545 complexes with pH information and 622 complexes with acquisition method and resolution information. When several temperature values were available for the same structure, we used the mean temperature. We constructed filtered data sets based on structural resolution and acquisition method, with 205 high-resolution structures (≤2.5 Å) and 165 NMR structures.

For each complex, we extracted additional information on binding assay pH, temperature, and methodology from the protein-protein affinity benchmark database, which is a nonredundant set of 144 protein-protein complexes with detailed information on the experimental methods used for measuring binding affinities [28]. We extracted pH data for 127 complexes, temperature data for 103 complexes and binding assay technology for 136 complexes. Information available for each protein-protein complex is provided in Additional file 1: Table S1.

Statistical modeling, model selection and cross-validation

We updated the original version of our protein-protein affinity prediction model [22] by adding parameters for estimating hydrophobic surface tension and hydrophobicity. The hydrophobicity algorithm used is adapted from [33]. Each amino acid in the surface has a pre-defined hydrophobicity score, modulated by peptide endings and varying between approximately −1 (most hydrophilic) and +2 (most hydrophobic). The surface tension parameter was calculated by summing the atomic contributions of each amino acid to the protein surface tension. These atomic contribution scores were adapted from [34]. Other atom-atom interaction terms in the present statistical model are identical to those defined and evaluated in our previous work [22].

Statistical models were implemented using a generalized linear model framework (GLM, implemented in the GLMULTI package in R), assuming a Gaussian error distribution with logarithmic link function. We used the GLMULTI genetic algorithm to generate 500 candidate models (default parameters, except population size = 500, level = 2, and marginality enabled) and selected the best-fit model for each training dataset using the Akaike information criterion (AIC).

We evaluated the potential for new features to improve affinity prediction models by calculating the Pearson correlation (r^2) between predicted and experimental binding affinities on the training data after model fitting, which represents the 'best-case' possible accuracy.

We then used cross-validation to estimate the expected accuracy of each trained statistical model when applied to new data and to evaluate possible model over-fitting to training data [13, 14]. For each model, we performed 100 replicates of leave-one-out cross-validation. For each replicate, we randomly partitioned the structural data into a testing data set of size $n = 1$, with the remaining complexes used to train the regression model. We calculated the Pearson correlation (r^2) and root mean squared deviation (RMSD) between predicted and experimentally-determined binding affinities on the unseen testing data and report the average r^2 and RMSD of each model over the 100 replicates.

Differences in accuracy were assessed using the parametric two-tailed, two-sample t test, assuming unequal variances, and the nonparametric Mann–Whitney U test. For evaluating the effects of dataset subsampling on predictive accuracy, we used Fisher's z-transformation, which incorporates a correction for comparing results obtained on a subsample to results from the full dataset (33). In addition, we performed 1000 replicates of random subsampling to evaluate the expected effect of subsampling on predictive accuracy.

Results

Statistical prediction of protein-protein binding affinity relies on information extracted from large structure-affinity databases [29, 35]. Accuracy and generalizability of predictive models is therefore expected to depend on the quantity and quality of information in the training database as well as the particular types of information available [36]. To evaluate how various aspects of structure-affinity databases affect the accuracy of protein-protein affinity prediction, we examined 1577 protein-protein complexes from the PDBbind database, a comprehensive collection of experimentally-determined affinity measurements assigned to 3-dimensional structural complexes, commonly used to evaluate affinity prediction algorithms [30].

We found that nearly 2/3 of the protein-protein complexes in PDBbind had ambiguous affinity measurements or multiple ligands, making it difficult to confidently assign affinity information to specific components of the structural complex (see Additional file 1: Text S1). We identified 955 ambiguous complexes, with an additional 20 complexes removed due to missing coordinates and/or steric clashes [37, 38]. Removing these complexes resulted in a filtered training database of 622 protein-protein dimers.

Consistent with results from previous studies [11, 19, 35, 39, 40], we found that removing complexes with ambiguous, missing or unreliable data was required to support robust training of affinity prediction models (see Additional file 1: Figure S1 and associated text). Models trained using either the complete PDBbind (1557 complexes) or the filtered database of 622 dimers performed very poorly when applied to the complete PDBbind dataset. For the remainder of this study, we therefore focus our analyses on the filtered PDBbind database of 622 dimers.

Incorporating additional structural features improves protein-protein affinity prediction

We have previously developed statistical approaches for predicting protein-protein affinity incorporating a wide range of atom-atom interaction terms expected to impact macromolecular interactions [22]. However, in those analyses, protein-protein affinities could not be predicted with >0.49 correlation, after cross-validation. Applying these models to our filtered PDBbind dataset resulted in a correlation between predicted and experimentally-determined binding affinities of 0.44 in cross-validation analyses (Fig. 1a).

We did find that including additional atom-atom interaction terms can improve the accuracy of affinity-prediction models. For example, hydrophobicity [33] and surface tension parameters [34] are weakly correlated with binding affinity across the filtered PDBbind dataset (Spearman correlation >0.10, $p < 7.48 \times 10^{-3}$; Additional file 1:

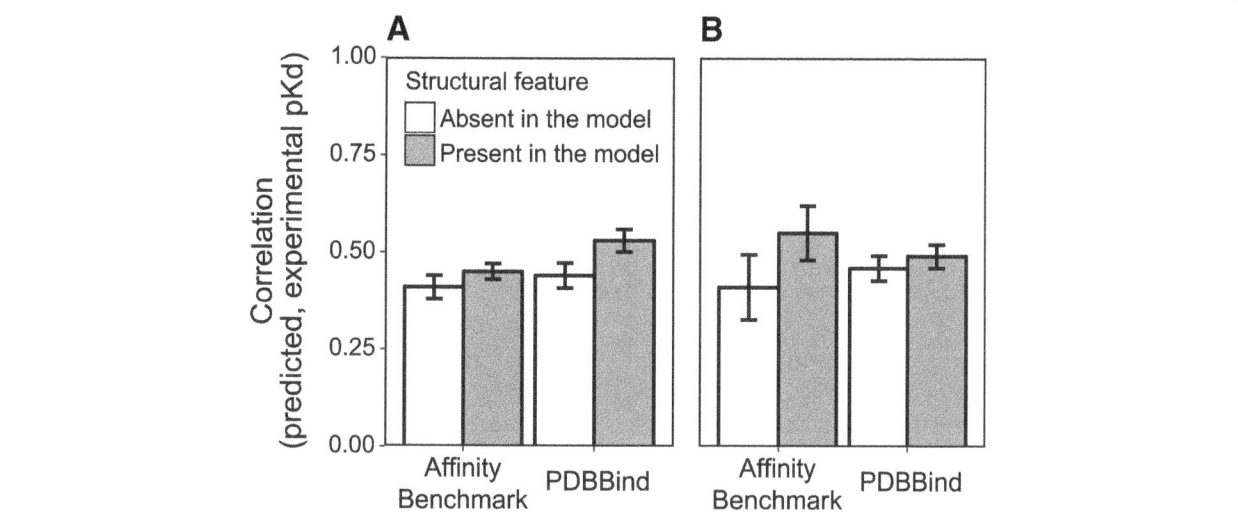

Fig. 1 Including additional structural features improves prediction of protein-protein binding affinity. In addition to the atom-atom interaction terms evaluated in our previous study [22] we extracted additional features from protein-protein complexes in our filtered training datasets from PDBbind and the Binding Affinity Benchmark and performed cross-validation to evaluate the expected accuracy of affinity-prediction models trained using these features, when applied to new data (see Methods). We plot the Pearson correlation between predicted and experimentally-determined binding affinities for the original model (*white*) and the model including additional features (*gray*). Bars indicate standard errors. **a** Hydrophobicity and surface tension parameters were extracted from structural data and incorporated into the prediction model. **b** We calculated the root mean squared deviation (RMSD) between unbound and bound forms of the components of each protein-protein complex as well as differences in the area of each protein accessible to solvent (see Methods). These features were incorporated into prediction models. For complexes in the PDBbind database, we simulated the unbound forms by using homology modeling

Figure S2A). Incorporating these parameters into the predictive model improved the correlation between predicted and experimentally-determined affinities in cross-validation analyses from 0.44 to 0.54 (one-tailed Fisher's z = 2.04, p = 0.0207; Fig. 1a).

We also evaluated the relationship between binding affinity and structural changes caused by protein-protein binding by examining the change in conformational entropy upon complex formation. This can be roughly calculated by comparing the structure of the bound complex (*holo*) to the unbound (*apo*) structures and has been successfully applied for predicting binding affinity in a small dataset of 17 protein-protein complexes [41].

We extracted 143 *holo* complexes with corresponding *apo* structures from the protein-protein affinity benchmark database (Additional file 1: Table S2). Differences between *holo* and the *apo* forms were characterized by calculating root mean squared deviations (RMSDs) and changes in the accessible-to-solvent surface area upon complex formation. Although RMSD was not correlated with experimental binding affinity (Spearman correlation = 0.02, p = 0.73), we did observe a significant correlation between binding affinity and the change in accessible-to-solvent area caused by formation of the protein-protein binding interface, suggesting that this parameter may be useful for improving affinity prediction (Spearman correlation = −0.28, p = 8.63x10^{-4}, Additional file 1: Figure S2B).

Cross-validation analysis confirmed that including changes in the accessible-to-solvent area as an explanatory variable improved affinity prediction accuracy, both on the Affinity Benchmark database (r^2 = 0.41 vs. 0.55; William's test p = 2.3x10^{-3}) and the filtered PDBbind database (r^2 = 0.46 vs. 0.49; William's test p = 2.6x10^{-3}; Fig. 1b).

Scoring functions that exhibit a Pearson correlation >0.72 and an RMSD <2 Å between predicted and experimental binding affinity in cross-validation analyses are commonly characterized as providing robust affinity inferences [11, 40, 42–44]. While our results do suggest that incorporating additional structural information can improve protein-protein affinity prediction, the improvements in accuracy we observed were generally incremental, and even best-case accuracy currently remains too low to support robust affinity inferences.

Existing studies of protein-protein affinity prediction have occasionally identified models capable of accurately predicting affinities on carefully curated small datasets, but accurate prediction of protein-protein affinity across large structure-affinity databases has remained unobtainable [23]. This could be due to lack of generalizability, possibly because curation of small datasets may inadvertently select for a subset of the structural features present in large-scale databases. Alternatively, the quality of both structural and affinity data in large databases could be highly variable, making some complexes more 'difficult' to accurately predict than others and potentially misleading model training procedures. Currently, almost nothing is known about how variation in characteristics of the structural and experimental data in structure-affinity databases might impact protein-protein affinity prediction. We examine this potential issue for the remainder of this study.

Crystal resolution affects protein-protein affinity prediction accuracy

Crystallographic resolution is proportional to the precision of the 3-dimensional coordinates of the atoms in the structure. Typically, high-resolution structures (<2.5 Å) exhibit correct folding, have very small numbers of incorrect rotamers and present accurate surface loops. In contrast, low-resolution structures (>3.5 Å) are more likely to result in folding errors or incorrectly-modeled surface loops [45, 46]. We hypothesized that high-resolution structures would produce more reliable atom-atom interaction calculations and result in more accurate binding affinity predictions.

We did observe a weak but significant correlation between crystal resolution and the difference between predicted and experimental binding affinities ($r^2 = 0.11$, $p = 5.26 \times 10^{-3}$; Additional file 1: Figure S3A). However, including crystal resolution as a parameter in the affinity prediction model did not improve the correlation between predicted and experimental affinities, even across complexes included in the training dataset ($r^2 = 0.64$ vs. 0.65; Fisher one-tailed z = 0.16, $p = 0.87$; Fig. 2a). These results suggest that using crystal resolution as an explanatory variable in the model is unlikely to improve affinity prediction accuracy, even in the 'best case' scenario in which new data 'look' exactly like the data used for training.

However, constraining our training dataset to only include high-resolution structures (<2.5 Å, resulting in 205 protein-protein complexes) improved the correlation between predicted and experimental affinities on training data from 0.64 to 0.85 (Fisher one-tailed z = 6.14, $p = 7.97 \times 10^{-10}$; Fig. 2b). Furthermore, cross-validation analysis using only high-resolution structures resulted in increased predictive accuracy when applied to new data not included in training ($r^2 = 0.54$ vs. 0.68; Fisher two-tailed z = 2.85, $p = 4.32 \times 10^{-3}$; Fig. 2c,d).

Although Fisher's z-transformation incorporates a correction for comparing results obtained on a subsample to results from the full dataset [47], we were concerned that selecting a subsample of the original testing data could lead to a spurious improvement in predictive accuracy, irrespective of the effects of higher crystal resolution. However, when we randomly selected subsets of equivalent size, accuracy never

improved to the extent observed for the high-resolution dataset ($p < 4 \times 10^{-3}$; Fig. 2b).

Although these results suggest that training protein-protein affinity prediction using high-resolution structures may improve predictive accuracy, different types of complexes are likely to crystalize at different resolutions. If complexes whose affinities are more difficult to predict for inherent reasons also tend to crystalize at lower resolution, the effects of resolution on predictive accuracy may be indirect.

To address this issue, we grouped protein-protein complexes into clusters based on 90% sequence identity. Within each cluster of similar complexes, we calculated the correlation between crystal resolution and affinity prediction accuracy (Additional file 1: Table S3). Although there were only 21 clusters with >3 similar complexes in our dataset, we did find that all the clusters exhibiting a significant correlation between resolution and affinity prediction were consistent with the expectation that higher-resolution structures produced more accurate affinity predictions. While it is not possible to generalize from such limited data, these results do suggest that higher resolution structures may improve affinity prediction accuracy across at least some groups of similar protein-protein complexes.

Restricting training and testing data to either high-resolution or NMR data also resulted in a reduction of root-mean squared deviation (RMSD) between predicted and experimental affinities, compared to the original dataset (high-resolution and NMR RMSDs = 1.79 and 1.56, respectively, vs. 1.90 for the original dataset, t-test $p < 0.03$, Mann–Whitney $p < 0.01$; Fig. 2d). Together, these results suggest that training statistical models using high-resolution crystal structures or NMR data may improve affinity prediction accuracy when trained models are applied to new data.

It is interesting that restricting training data to NMR structures also improved predictive accuracy (see Fig. 2), as the resolution of NMR structures is typically lower than X-ray crystal structures. However, NMR structures are determined from proteins in solution, which may more accurately reflect the native functional environment of the protein [48, 49]. The capacity to capture protein-protein interactions in solution may contribute to the improved predictive accuracy of models trained using NMR data, particularly for cases in which the crystallographic process might introduce structural artifacts.

Experimental conditions such as temperature and pH are critical for the formation and stability of a protein crystal [50]. However, the optimal conditions for crystallization may differ from those used for measuring binding affinity, potentially creating a mismatch between a crystalized protein-protein complex and that same complex in experimental solution. To examine the

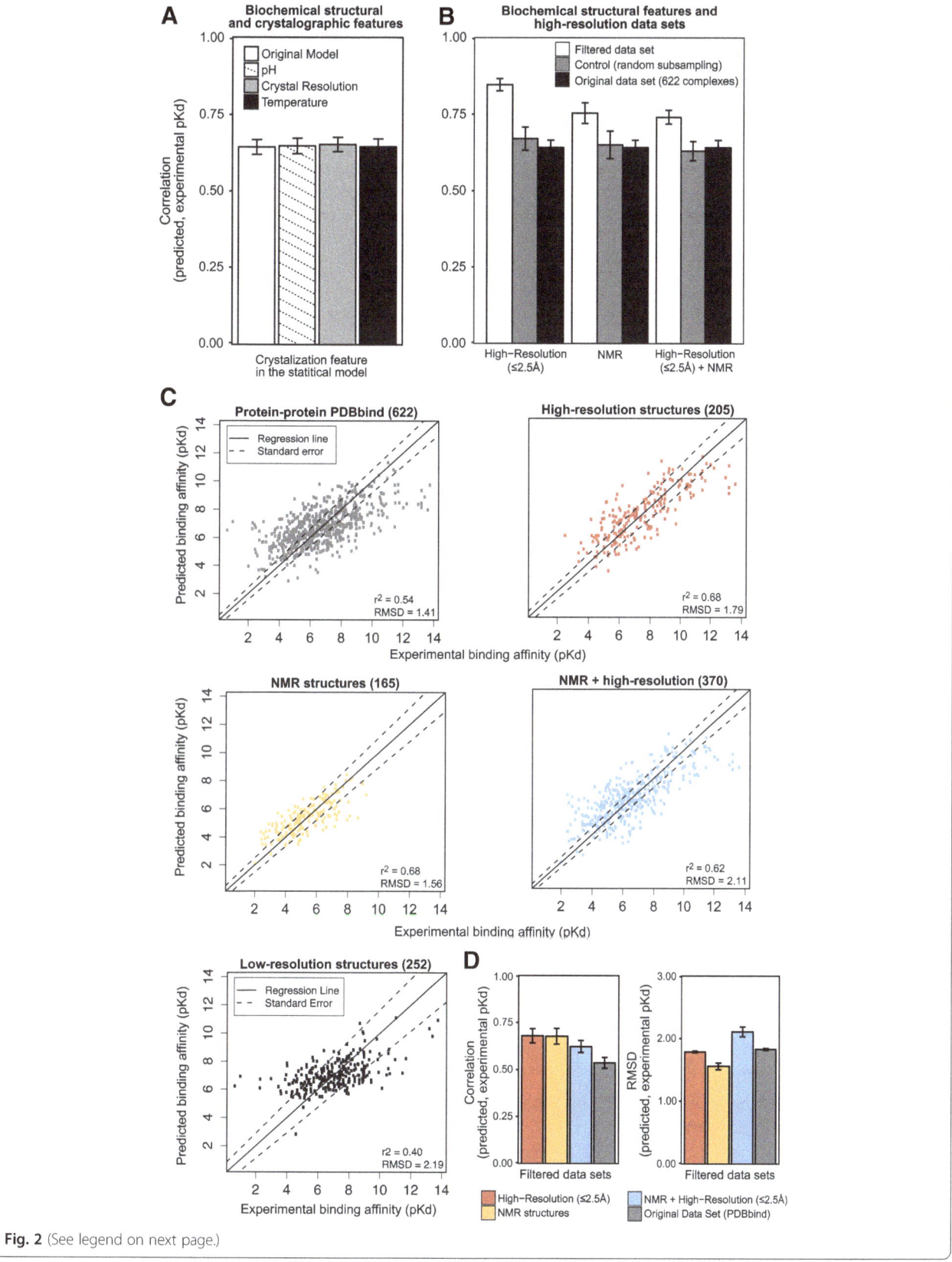

Fig. 2 (See legend on next page.)

(See figure on previous page.)
Fig. 2 High-resolution structural information improves protein-protein binding affinity prediction. **a** We plot the Pearson correlation between predicted and experimentally-determined binding affinities (pKds) for the original affinity-prediction model incorporating only biochemical structural data (see Methods) and three models incorporating crystallographic features (temperature, pH and resolution) as additional parameters. See Methods for model training details. Bars indicate standard errors. **b** We trained affinity prediction models using high-resolution crystallographic data (≤2.5 Å), NMR structures or both high-resolution and NMR data. We plot the correlation between predicted and experimentally-determined affinities (pKds) for models trained using each type of filtered data set (*white series*) and compare results to models trained using the complete database of 622 protein-protein dimers (*black*) and models trained using randomly-selected subsets of the original data set of equal size to the high-resolution training data (*gray*). Bars indicate standard errors. **c** We performed leave-one-out cross-validation to evaluate the expected accuracy of affinity-prediction models applied to new data (see Methods). We plot the predicted vs. experimentally-determined binding affinities (pKds) of each cross-validated structural complex for models trained using the complete data set of 622 protein-protein dimers (*gray*), high-resolution crystallographic data (205 complexes with resolution ≤2.5 Å, red), 165 NMR complexes (*orange*) and the combined high-resolution + NMR data (370 complexes, *blue*). We report the best-fit regression line and its standard error as well as the Pearson correlation between predicted and experimentally-determined affinities (r^2) and the RMSD between predicted and experimental affinities. **d** plots the Pearson correlation and RMSD, respectively, for models trained using each type of filtered data set, with bars indicating standard errors

potential effects of crystallization conditions on protein-protein affinity prediction, we extracted temperature and pH information from the Protein Data Bank [51] for the complexes in our training dataset and evaluated the effect of including this information on predictive accuracy. Although both crystallization temperature and pH were weakly negatively correlated with experimental binding affinity ($r^2 = -0.26$, $p = 4.91 \times 10^{-10}$ and $r^2 = -0.16$, $p = 1.96 \times 10^{-4}$ for temperature and pH, respectively. Additional file 1: Figure S3B), we observed no improvement in predictive accuracy when these parameters were incorporated into the statistical model (Fisher one-tailed $z < 0.1$, $p > 0.92$; Fig. 2a).

Overall, these results suggest that the quality of structural data can affect the accuracy of statistical affinity prediction, and that training models using high-quality structures may be one avenue available to improve protein-protein and other affinity predictions. While limiting training data to high-resolution structures was not required for accurate prediction of protein-small molecule or protein-DNA/RNA affinities in our previous analysis [22], protein-protein complexes typically have larger numbers of atoms at the protein-ligand interface and may be more sensitive to potential errors induced by lower crystal resolution. Differences in crystal resolution across protein-small molecule, protein-DNA/RNA and protein-protein training datasets may also contribute to differences in affinity prediction accuracy.

Lack of information on binding assay conditions impairs protein-protein affinity prediction

In addition to crystallographic conditions or resolution, variation in the experimental conditions and assays used to measure binding affinity could affect prediction accuracy. Different proteins can have dramatically different activities across temperature, pH and concentration of ions or cofactors [52, 53], and assay conditions have been shown to strongly affect reaction rates [54–56].

Even though experimental conditions can be critical for evaluating affinity measurements, this information is not available in the major structure-affinity databases used for training statistical predictors [30, 57]. Detailed experimental information is available for a small protein-protein affinity benchmark database [28]. After excluding complexes with missing data, we obtained 127 protein-protein complexes with data indicating the pH of the binding affinity experiment and 103 complexes with temperature information (Additional file 1: Table S2).

We found no significant increase in the correlation between predicted and experimental affinities when binding experiment pH was included as an explanatory variable, even across data used to train the model ($r^2 = 0.67$ vs. 0.69, William's t = 0.43, $p = 0.34$; Fig. 3a). Although pH has been shown to affect binding affinity measurements in some systems [58–60], the effect of pH on pKd may depend on particular properties of the specific interacting proteins. We did observe a small, marginally-significant increase in correlation when including temperature in the model ($r^2 = 0.70$ vs. 0.76, William's t = −1.81, $p = 0.04$; Fig. 3a). Cross-validation analysis confirmed that binding assay temperature has the capacity to improve affinity prediction accuracy when applied to unseen testing data ($r^2 = 0.46$ vs. 0.52; William's t = −29.84, $p = 9.03 \times 10^{-188}$; Fig. 3b).

Although the dataset examined in this analysis was small, compared to the training data available in the filtered PDBbind data set (~100 vs. ~600 complexes), these results suggest that incorporating parameters describing the experimental conditions used to measure protein-protein binding affinity may be important for training affinity-prediction models. The lack of information describing binding assay conditions in large structure-affinity databases may impose a limit on the accuracy of statistical models constructed from these databases.

Protein-protein affinity can be measured by a variety of approaches, some of which may more strongly impact

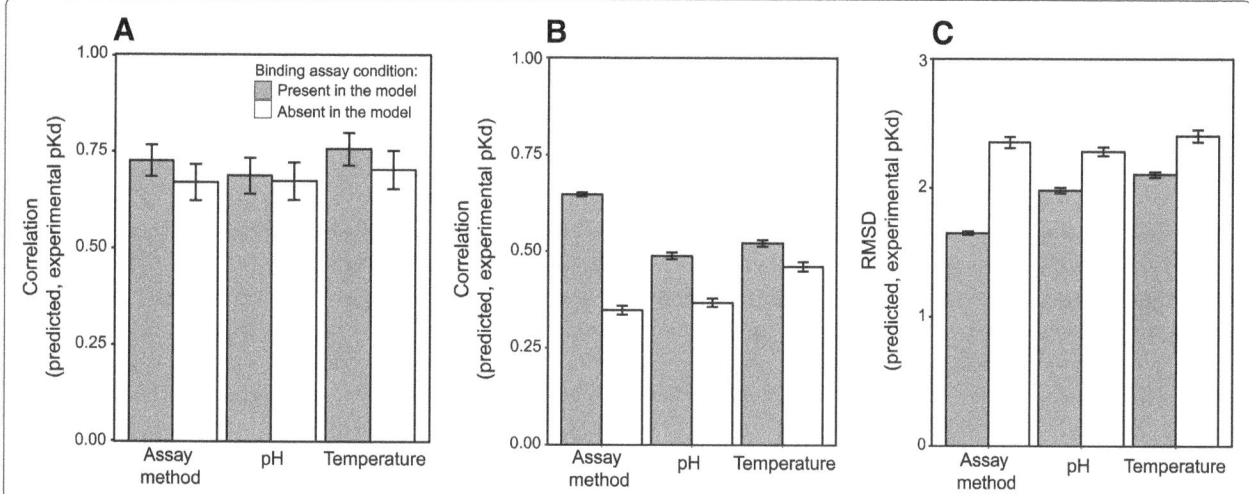

Fig. 3 Incorporating information about binding assay conditions improves protein-protein affinity prediction. **a** We plot the Pearson correlation between predicted and experimentally-determined binding affinities (pKd) using models with (*white*) and without (*gray*) three features describing the conditions under which binding affinities were measured experimentally (temperature, pH and the assay method). **b** We plot the correlation between predicted and experimentally-determined binding affinities for the same models examined in (**a**), using unseen testing data generated by leave-one-out cross validation (see Methods). **c** We plot the RMSD between predicted and experimentally-determined binding affinities for the same models in (**a** and **b**), using leave-one-out cross-validation. In each panel, bars indicate standard errors

affinity prediction than others. Common technologies used in the protein-protein affinity benchmark database [28] are Isothermal Titration Calorimetry (ITC [61]), surface plasmon resonance (SPR [62]), and inhibition assays [63–66].

We observed a weak but significant correlation between the use of inhibition assays and experimentally-determined affinity values (Spearman correlation = 0.33, $p = 1.16 \times 10^{-4}$), whereas ITC was weakly negatively correlated with affinity (Spearman correlation = −0.36, $p = 1.84 \times 10^{-5}$; Additional file 1: Figure S3C). These results suggest that reported affinity measurements are somewhat dependent on the type of assay used: inhibition assays typically result in higher affinities, whereas ITC tends to produce lower affinity values. It is not clear whether this "assay effect" represents a general bias in one or more of the methodologies used to assess binding affinity, or if different methodologies tend to be applied to complexes with higher vs. lower biological affinities.

When we included experimental assay method as an explanatory variable in the statistical model, we observed a strong increase in predictive accuracy, assessed by cross-validation ($r^2 = 0.35$ vs. 0.65; William's t = −29.84, $p = 9.03 \times 10^{-188}$; Fig. 3b). In addition, there was no significant difference between training and cross-validation correlation results (Fisher's z = 1.22, $p = 0.11$; Fig. 3a,b), suggesting that the optimized statistical model—including assay method—exhibits minimum over-fitting.

Incorporating binding assay conditions in our statistical model also resulted in a significant decrease in RMSD between predicted and experimentally-determined affinities (Fig. 3c). Adding binding assay pH to the statistical model reduced RMSD from 2.28 to 1.98 (*t*-test t = −7.61, $p = 2.09 \times 10^{-12}$; Mann–Whitney w = 2272, $p = 2.66 \times 10^{-11}$). Similarly, RMSD decreased from 2.40 to 2.10 when temperature was incorporated (*t*-test t = −5.80, $p = 4.41 \times 10^{-8}$; Mann–Whitney w = 3221, $p = 1.39 \times 10^{-5}$). Finally, RMSD decreased from 2.36 to 1.65 when binding assay method was included as a model parameter (*t*-test t = −15.58, $p = 2.80 \times 10^{-30}$; Mann–Whitney w = 383, $p = 1.65 \times 10^{-29}$).

Overall, these results suggest that incorporating information about the experimental conditions used to measure protein-protein affinity can have a strong effect on the predictive accuracy of statistical models. Detailed experimental conditions are generally not incorporated into large-scale structure-affinity databases, which may place a practical upper bound on the accuracy of statistical affinity prediction.

Could database errors limit predictive accuracy?

Manual examination of specific examples in the protein-protein affinity benchmark database [28] revealed that, in some cases, the conditions of the crystalized complex are so different from the conditions of the binding assay that it is not clear they are biochemically comparable. For example, the crystal structure of Nuclease A (NucA) in complex with intracellular inhibitor NuiA is a D121A mutant (PDB ID 2O3B), whereas the affinity assay was performed using the wild-type NuiA [67]. This particular complex was the 3rd worst prediction made by our statistical model, with a predicted pKd based on the mutant structure of 7.06 vs. an experimental pKd of the wild-type protein of 11.49. It is not clear whether this misprediction is due to a poor statistical fit to this complex or to differences between the binding affinities of the mutant vs. wild-type proteins.

To evaluate the potential effects of this mismatch on predicted binding affinity accuracy, we generated the wild-type structure of NuiA using homology modeling [68] and re-estimated the binding affinity using the trained statistical model. Modeling the wild-type NuiA in complex with NucA increased the predicted pKd from 7.06 (mutant NuiA) to 7.49, decreasing the difference between predicted and experimentally-determined affinity somewhat, possibly because the D121A mutation disrupted a hydrogen bond between wild-type NuiA's D121 and NucA's E24 (Fig. 4).

Although the significance of the improvement in this single 'case study' cannot be evaluated statistically—and is likely to depend on the specific scientific question being considered—this result does suggest that small differences between the crystalized protein-protein complex and the complex whose binding affinity is measured—in this case a single amino acid mutation—may have a measurably negative affect on the accuracy of affinity prediction. The extent of similar errors in large-scale structure-affinity databases is unknown.

Missing information about key affinity-determining factors from either the crystallization or affinity experiments could also affect affinity prediction accuracy. In one example, the crystal structure of GTP-Bound Rab4Q67L GTPase in complex with the central Rab binding domain of Rabenosyn-5 (PDB ID 1Z0K) has an additional cofactor, 2-(N-morpholino)-ethanesulfonic

acid or MES, which was not present in the binding assay [69]. This complex was the 4th worst prediction made by the statistical model (predicted pKd = 9.14; experimental pKd = 5.11). Although the extent to which the presence/absence of the MES cofactor may have affected affinity prediction is unclear, that the crystalized complex does not correspond to the complex assayed in the experimental affinity measurement raises concerns about the accuracy of this database entry.

Other examples of missing information likely to affect affinity prediction could be manually identified from the affinity benchmark database [28], many of which appear to have had a negative impact on affinity predictions made by our statistical model (see Additional file 1: Text S2). Overall, we found 4 of the 10 worst predictions made by the statistical model were for complexes with obvious mismatches between crystallization and binding assay conditions or cases in which information potentially impacting affinity measurement or crystallization was missing from the database (Additional file 1: Table S4).

These potential database errors were identifiable due to the amount of detail provided in small curated databases like the protein-protein affinity benchmark [28]. However, we expect similar potential issues exist in large-scale databases like PDBbind and BindingDB, which together contain >10,000 protein-ligand structures [30, 57]. Many of the entries in these large structure-affinity databases lack information concerning binding assay and/or crystallization conditions. Our results suggest that this information may be critical for supporting accurate, high-throughput affinity

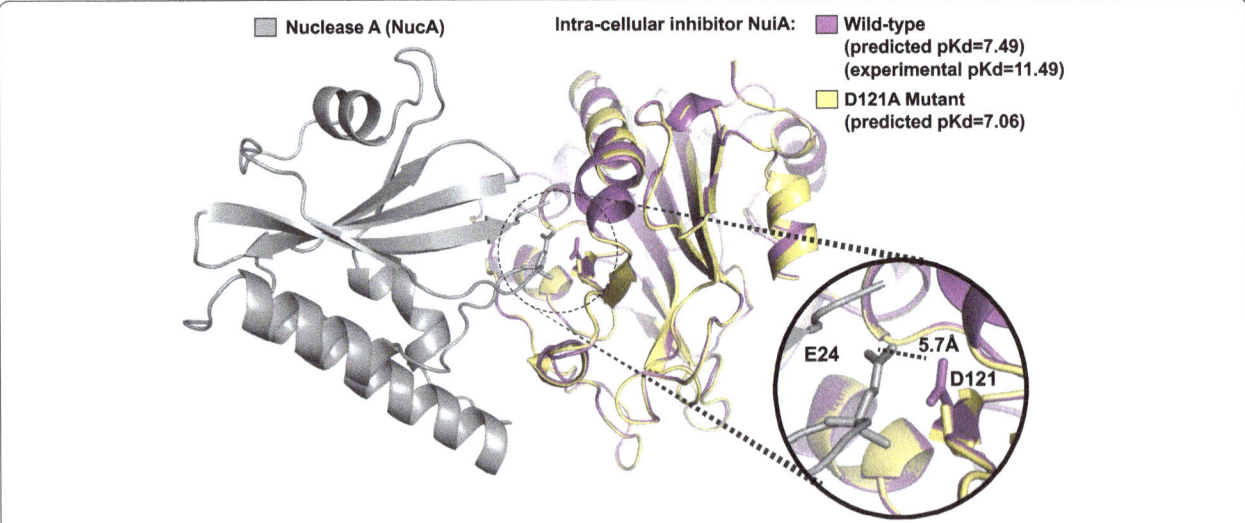

Fig. 4 Database errors may interfere with affinity prediction accuracy. We predicted the binding affinities of mutant D121A intracellular inhibitor NuiA (NuiA; *yellow*) and wild-type NuiA (*purple*) in complex with Nuclease A (nucA; *gray*) using the trained statistical affinity-prediction model. The wild-type NuiA structure was inferred by homology modeling, using the mutant structure as a template. We plot the structure of each complex and report predicted and experimentally-determined binding affinities (pKds). Inset displays a close-up of the D121A mutation, showing an inferred hydrogen bond between D121 and E24 of NucA (*dashed line*)

prediction and also important for identifying potential database errors.

We also identified a handful of cases in which binding affinity values were incorrectly entered in the PDBbind database. For example, the binding affinity (pKd) assigned to Human prolactin (hPRL) in complex with its receptor is 0.67 in PDBbind, whereas the experimentally-determined binding affinity from the literature is >5.65, depending on pH [70]. Similarly, the prolactin and prolactin receptor mutant complex has an assigned pKd of 1.03 in PDBbind, whereas the affinity from the literature is 6.14 [70]. Although these particular cases were manually corrected in the filtered dataset used for this study, the extent to which various errors are present in large structure-affinity databases remains unknown, making it difficult to characterize the potential effects of database errors on affinity prediction.

Discussion

The accuracy of machine learning and other statistical prediction methods depends on having a large quantity of high-quality training data. Errors in the training data can impair the inferred model's predictive performance [71], whereas a too-small training dataset can interfere with generalizability to new data [72]. Our results suggest that curation errors, lack of information about experimental conditions and low-quality data present in large structure-affinity databases could reduce the maximum achievable accuracy of protein-protein affinity prediction models developed from these databases.

We have shown that limiting training data to high-resolution crystal structures—easily extracted from structural information—can dramatically improve affinity prediction. However, we are cautious that the resulting reduction in breadth of training data may limit the generalizability of inferred models to new problems, particularly complex structural interactions that may not crystalize at high resolution due to inherent flexibility.

We have also shown that incorporating information about the experimental conditions used to measure binding affinity may be important for producing accurate affinity predictions from structural data, probably due to their effects on resulting affinity measurements. Unfortunately, most large structure-affinity databases do not include detailed experimental information, and databases that do include this information appear to have at least some examples of dramatic mismatches between crystallographic and affinity-measurement conditions. The extent to which these types of potential errors are present in large-scale databases is not known, making it difficult to assess the general impact of these potential problems on affinity prediction.

Conclusion

Although careful manual curation can be used to develop high-quality structure-affinity databases, this approach is unlikely to scale up to the number of structures required for training robust, generalizable predictive models. A possible computational approach to building high-quality, large-scale structure-affinity databases would be to extract detailed information about crystallographic and affinity-measurement conditions directly from scientific literature using text-mining approaches [73–75], although errors in text-mining could then potentially propagate to training databases. Alternatively, authors could be encouraged to directly supply the required information as part of a database submission policy associated with scientific publication. This approach has been successfully used to develop the Protein Data Bank [32], Genbank [76] and similar community resources. Ultimately, it may be up to the community of researchers to develop the standards and practices necessary to support large-scale investigations of the general structural basis for protein-protein interactions.

Acknowledgements
This work was supported by the National Science Foundation (Molecular and Cellular Biology, grant number 1412442) and by University of Florida. We also gratefully acknowledge the anonymous reviewers that helped to improve this manuscript.

Funding
Publication of this article was funded by the National Science Foundation (Molecular and Cellular Biology, grant number 1412442) and by the University of Florida.

Authors' contributions
All authors contributed equally to this work. RD and BK participated at the design of the present project. RD performed the implementation and statistical analysis, and worked on the writing of the present manuscript. BK supervised the implementation and statistical analysis of the present statistical, reviewed and edited the present manuscript. Both authors read and approved the final manuscript.

Competing interests
The authors declare that they have no competing interests.

About this supplement
This article has been published as part of BMC Bioinformatics Volume 18 Supplement 5, 2017: Selected works from the Joint 15th Network Tools and Applications in Biology International Workshop and 11th Integrative Bioinformatics International Symposium (NETTAB / IB 2015). The full contents of the supplement are available online at https://bmcbioinformatics.biomedcentral.com/articles/supplements/volume-18-supplement-5.

References
1. Mazza C, Ohno M, Segref A, Mattaj IW, Cusack S. Crystal structure of the human nuclear cap binding complex. Mol Cell. 2001;8(2):383–96.
2. Liu S, Song X, Chrunyk BA, Shanker S, Hoth LR, Marr ES, Griffor MC. Crystal structures of interleukin 17A and its complex with IL-17 receptor A. Nat Commun. 2013;4:1888.
3. Duss O, Michel E, Diarra dit Konte N, Schubert M, Allain FH. Molecular basis for the wide range of affinity found in Csr/Rsm protein-RNA recognition. Nucleic Acids Res. 2014;42(8):5332–46.

4. Zhang B, Zhang T, Sromek AW, Scrimale T, Bidlack JM, Neumeyer JL. Synthesis and binding affinity of novel mono- and bivalent morphinan ligands for kappa, mu, and delta opioid receptors. Bioorg Med Chem. 2011;19(9):2808–16.

5. Trapani G, Franco M, Latrofa A, Ricciardi L, Carotti A, Serra M, Sanna E, Biggio G, Liso G. Novel 2-phenylimidazo[1,2-a]pyridine derivatives as potent and selective ligands for peripheral benzodiazepine receptors: synthesis, binding affinity, and in vivo studies. J Med Chem. 1999;42(19):3934–41.

6. Hog S, Wellendorph P, Nielsen B, Frydenvang K, Dahl IF, Brauner-Osborne H, Brehm L, Frolund B, Clausen RP. Novel high-affinity and selective biaromatic 4-substituted gamma-hydroxybutyric acid (GHB) analogues as GHB ligands: design, synthesis, and binding studies. J Med Chem. 2008;51(24):8088–95.

7. Bren U, Martinek V, Florian J. Decomposition of the solvation free energies of deoxyribonucleoside triphosphates using the free energy perturbation method. J Phys Chem B. 2006;110(25):12782–8.

8. Bren M, Florian J, Mavri J, Bren U. Do all pieces make a whole? Thiele cumulants and the free energy decomposition. Theor Chem Acc. 2007;117(4):535–40.

9. Lee MC, Duan Y. Distinguish protein decoys by using a scoring function based on a new AMBER force field, short molecular dynamics simulations, and the generalized born solvent model. Proteins. 2004;55(3):620–34.

10. Dias R, Timmers LFSM, Caceres RA, de Azevedo WF. Evaluation of molecular docking using polynomial empirical scoring functions. Curr Drug Targets. 2008;9(12):1062–70.

11. Wang R, Lai L, Wang S. Further development and validation of empirical scoring functions for structure-based binding affinity prediction. J Comput Aided Mol Des. 2002;16(1):11–26.

12. Bohm HJ, Stahl M. Rapid empirical scoring functions in virtual screening applications. Med Chem Res. 1999;9(7–8):445–62.

13. Shao J. Linear-model selection by cross-validation. J Am Stat Assoc. 1993; 88(422):486–94.

14. Efron B. Estimating the error rate of a prediction rule - improvement on cross-validation. J Am Stat Assoc. 1983;78(382):316–31.

15. Kruger DM, Ignacio Garzon J, Chacon P, Gohlke H. DrugScore(PPI) knowledge-based potentials used as scoring and objective function in protein-protein docking. PLoS One. 2014;9(2):e89466.

16. Hsieh JH, Yin S, Liu S, Sedykh A, Dokholyan NV, Tropsha A. Combined application of cheminformatics- and physical force field-based scoring functions improves binding affinity prediction for CSAR data sets. J Chem Inf Model. 2011;51(9):2027–35.

17. Eldridge MD, Murray CW, Auton TR, Paolini GV, Mee RP. Empirical scoring functions: I. The development of a fast empirical scoring function to estimate the binding affinity of ligands in receptor complexes. J Comput Aided Mol Des. 1997;11(5):425–45.

18. De Azevedo Jr WF, Dias R. Evaluation of ligand-binding affinity using polynomial empirical scoring functions. Bioorg medchem. 2008;16(20):9378–82.

19. Cheng T, Liu Z, Wang R. A knowledge-guided strategy for improving the accuracy of scoring functions in binding affinity prediction. BMC Bioinformatics. 2010;11:193.

20. Brylinski M. Nonlinear scoring functions for similarity-based ligand docking and binding affinity prediction. J Chem Inf Model. 2013;53(11):3097–112.

21. Ashtawy HM, Mahapatra NR. BgN-Score and BsN-Score: bagging and boosting based ensemble neural networks scoring functions for accurate binding affinity prediction of protein-ligand complexes. BMC Bioinformatics. 2015;16 Suppl 4:S8.

22. Dias R, Kolazckowski B. Different combinations of atomic interactions predict protein-small molecule and protein-DNA/RNA affinities with similar accuracy. Proteins. 2015;83(11):2100–14.

23. Kastritis PL, Bonvin AM. Are scoring functions in protein-protein docking ready to predict interactomes? Clues from a novel binding affinity benchmark. J Proteome Res. 2010;9(5):2216–25.

24. Kastritis PL, Bonvin AM. On the binding affinity of macromolecular interactions: daring to ask why proteins interact. J R Soc Interface. 2013; 10(79):20120835.

25. Keskin O, Gursoy A, Ma B, Nussinov R. Principles of protein-protein interactions: what are the preferred ways for proteins to interact? Chem Rev. 2008;108(4):1225–44.

26. De Paris R, Quevedo CV, Ruiz DD, Norberto de Souza O, Barros RC. Clustering molecular dynamics trajectories for optimizing docking experiments. Comput Intell Neurosci. 2015;2015:916240.

27. de Vries SJ, van Dijk M, Bonvin AM. The HADDOCK web server for data-driven biomolecular docking. Nat Protoc. 2010;5(5):883–97.

28. Kastritis PL, Moal IH, Hwang H, Weng Z, Bates PA, Bonvin AM, Janin J. A structure-based benchmark for protein-protein binding affinity. Protein Sci. 2011;20(3):482–91.

29. Yan Z, Guo L, Hu L, Wang J. Specificity and affinity quantification of protein-protein interactions. Bioinformatics. 2013;29(9):1127–33.

30. Wang R, Fang X, Lu Y, Yang CY, Wang S. The PDBbind database: methodologies and updates. J Med Chem. 2005;48(12):4111–9.

31. Cheng T, Li X, Li Y, Liu Z, Wang R. Comparative assessment of scoring functions on a diverse test set. J Chem Inf Model. 2009;49(4):1079–93.

32. Sussman JL, Lin D, Jiang J, Manning NO, Prilusky J, Ritter O, Abola EE. Protein data bank (PDB): database of three-dimensional structural information of biological macromolecules. Acta Crystallogr D Biol Crystallogr. 1998;54(Pt 6 Pt 1):1078–84.

33. Fauchere JL, Pliska V. Hydrophobic parameters-Pi of amino-acid side-chains from the partitioning of N-acetyl-amino-acid amides. Eur J Med Chem. 1983; 18(4):369–75.

34. Vasina EN, Paszek E, Nicolau Jr DV, Nicolau DV. The BAD project: data mining, database and prediction of protein adsorption on surfaces. Lab Chip. 2009;9(7):891–900.

35. Li X, Zhu M, Li X, Wang H-Q, Wang S. Protein-Protein Binding Affinity Prediction Based on an SVR Ensemble. In: Intelligent Computing Technology. Edited by Huang D-S, Jiang C, Bevilacqua V, Figueroa J, vol. 7389. Heidelberg: Springer Berlin Heidelberg; 2012. p. 145–51.

36. Beyene J, Atenafu EG, Hamid JS, To T, Sung L. Determining relative importance of variables in developing and validating predictive models. BMC Med Res Methodol. 2009;9:64.

37. Hooft RW, Vriend G, Sander C, Abola EE. Errors in protein structures. Nature. 1996;381(6580):272.

38. Vriend G, Sander C. Quality-control of protein models - directional atomic contact analysis. J Appl Crystallogr. 1993;26:47–60.

39. Camacho CJ, Zhang C. FastContact: rapid estimate of contact and binding free energies. Bioinformatics. 2005;21(10):2534–6.

40. Krammer A, Kirchhoff PD, Jiang X, Venkatachalam CM, Waldman M. LigScore: a novel scoring function for predicting binding affinities. J Mol Graph Model. 2005;23(5):395–407.

41. Grunberg R, Nilges M, Leckner J. Flexibility and conformational entropy in protein-protein binding. Structure. 2006;14(4):683–93.

42. Sotriffer CA, Sanschagrin P, Matter H, Klebe G. SFCscore: scoring functions for affinity prediction of protein-ligand complexes. Proteins. 2008;73(2):395–419.

43. Wang JC, Lin JH, Chen CM, Perryman AL, Olson AJ. Robust scoring functions for protein-ligand interactions with quantum chemical charge models. J Chem Inf Model. 2011;51(10):2528–37.

44. Ouyang X, Handoko SD, Kwoh CK. CScore: a simple yet effective scoring function for protein-ligand binding affinity prediction using modified CMAC learning architecture. J Bioinforma Comput Biol. 2011;9 Suppl 1:1–14.

45. Sweet RM. Outline of Crystallography for Biologists. By David Blow. Oxford University Press, 2002. Price GBP 25 (paperback). ISBN-0-19-851051-9. Acta Crystallographica Section D Volume 59, Issue 5. Acta Crystallographica Section D. 2003;59(5):958.

46. Warren GL, Do TD, Kelley BP, Nicholls A, Warren SD. Essential considerations for using protein-ligand structures in drug discovery. Drug Discov Today. 2012;17(23–24):1270–81.

47. Gayen AK. The frequency distribution of the product–moment correlation coefficient in random samples of any size drawn from non-normal universes. Biometrika. 1951;38(1–2):219–47.

48. Silverstein RM, Webster FX, Kiemle DJ. Spectrometric identification of organic compounds. 7th ed. Hoboken: John Wiley & Sons; 2005.

49. Cavalli A, Salvatella X, Dobson CM, Vendruscolo M. Protein structure determination from NMR chemical shifts. Proc Natl Acad Sci U S A. 2007; 104(23):9615–20.

50. Patrick MH. Crystallography made crystal clear A guide for users of macromolecular models (3rd Ed.), biochemistry and molecular biology education. Biochem Mol Biol Educ. 2007;35(5):387–8.

51. Rose PW, Prlic A, Bi C, Bluhm WF, Christie CH, Dutta S, Green RK, Goodsell DS, Westbrook JD, Woo J, et al. The RCSB protein data bank: views of structural biology for basic and applied research and education. Nucleic Acids Res. 2015;43(Database issue):D345–56.

52. Maun HR, Wen XH, Lingel A, de Sauvage FJ, Lazarus RA, Scales SJ, Hymowitz SG. Hedgehog pathway antagonist 5E1 binds hedgehog at the pseudo-active site. J Biol Chem. 2010;285(34):26570–80.

53. Arac D, Boucard AA, Ozkan E, Strop P, Newell E, Sudhof TC, Brunger AT. Structures of neuroligin-1 and the neuroligin-1/neurexin-1 beta complex

reveal specific protein-protein and protein-Ca2+ interactions. Neuron. 2007; 56(6):992–1003.

54. Svec F, Yeakley J, Harrison 3rd RW. The effect of temperature and binding kinetics on the competitive binding assay of steroid potency in intact AtT-20 cells and cytosol. J Biol Chem. 1980;255(18):8573–8.

55. Reverberi R, Reverberi L. Factors affecting the antigen-antibody reaction. Blood Transfus. 2007;5(4):227–40.

56. Voet D, Voet JG, Pratt CW. Fundamentals of Biochemistry, 3rd edn. Hoboken: Wiley; 2008.

57. Liu T, Lin Y, Wen X, Jorissen RN, Gilson MK. BindingDB: a web-accessible database of experimentally determined protein-ligand binding affinities. Nucleic Acids Res. 2007;35(Database issue):D198–201.

58. Hianik T, Ostatna V, Sonlajtnerova M, Grman I. Influence of ionic strength, pH and aptamer configuration for binding affinity to thrombin. Bioelectrochemistry. 2007;70(1):127–33.

59. Watanabe H, Matsumaru H, Ooishi A, Feng Y, Odahara T, Suto K, Honda S. Optimizing pH response of affinity between protein G and IgG Fc: how electrostatic modulations affect protein-protein interactions. J Biol Chem. 2009;284(18):12373–83.

60. Gillard M, Chatelain P. Changes in pH differently affect the binding properties of histamine H1 receptor antagonists. Eur J Pharmacol. 2006; 530(3):205–14.

61. Pierce MM, Raman CS, Nall BT. Isothermal titration calorimetry of protein-protein interactions. Methods. 1999;19(2):213–21.

62. Rich RL, Myszka DG. Higher-throughput, label-free, real-time molecular interaction analysis. Anal Biochem. 2007;361(1):1–6.

63. Barrett S, Mohr PG, Schmidt PM, McKimm-Breschkin JL. Real time enzyme inhibition assays provide insights into differences in binding of neuraminidase inhibitors to wild type and mutant influenza viruses. PLoS One. 2011;6(8):e23627.

64. Alexander PW, Rechnitz GA. Enzyme inhibition assays with an amperometric glucose biosensor based on a thiolate self-assembled monolayer. Electroanal. 2000;12(5):343–50.

65. Meyer-Almes FJ, Auer M. Enzyme inhibition assays using fluorescence correlation spectroscopy: a new algorithm for the derivation of k(cat)/K-M and K-i values at substrate concentrations much lower than the Michaelis constant. Biochemistry-Us. 2000;39(43):13261–8.

66. Widemann BC, Balis FM, Adamson PC. Dihydrofolate reductase enzyme inhibition assay for plasma methotrexate determination using a 96-well microplate reader. Clin Chem. 1999;45(2):223–8.

67. Ghosh M, Meiss G, Pingoud AM, London RE, Pedersen LC. The nuclease a-inhibitor complex is characterized by a novel metal ion bridge. J Biol Chem. 2007;282(8):5682–90.

68. Kelley LA, Mezulis S, Yates CM, Wass MN, Sternberg MJE. The Phyre2 web portal for protein modeling, prediction and analysis. Nat Protoc. 2015;10(6): 845–58.

69. Eathiraj S, Pan X, Ritacco C, Lambright DG. Structural basis of family-wide Rab GTPase recognition by rabenosyn-5. Nature. 2005;436(7049):415–9.

70. Kulkarni MV, Tettamanzi MC, Murphy JW, Keeler C, Myszka DG, Chayen NE, Lolis EJ, Hodsdon ME. Two independent histidines, one in human prolactin

and one in its receptor, are critical for pH-dependent receptor recognition and activation. J Biol Chem. 2010;285(49):38524–33.

71. Domingos P. A Few useful things to know about machine learning. Commun ACM. 2012;55(10):78–87.

72. Chapelle O, Vapnik V, Bengio Y. Model selection for small sample regression. Mach Learn. 2002;48(1–3):9–23.

73. Shah PK, Perez-Iratxeta C, Bork P, Andrade MA. Information extraction from full text scientific articles: where are the keywords? BMC Bioinformatics. 2003;4:20.

74. Peng FC, McCallum A. Information extraction from research papers using conditional random fields. Inform Process Manag. 2006;42(4):963–79.

75. Wang HC, Kooi TK, Kao HY, Lin SC, Tsai SJ. Using positive and negative patterns to extract information from journal articles regarding the regulation of a target gene by a transcription factor. Comput Biol Med. 2013;43(12):2214–21.

76. Benson DA, Karsch-Mizrachi I, Lipman DJ, Ostell J, Sayers EW. GenBank. Nucleic Acids Res. 2009;37(Database issue):D26–31.

7

Geminivirus data warehouse: a database enriched with machine learning approaches

Jose Cleydson F. Silva[1,2], Thales F. M. Carvalho[1], Marcos F. Basso[2], Michihito Deguchi[2], Welison A. Pereira[2], Roberto R. Sobrinho[2], Pedro M. P. Vidigal[3], Otávio J. B. Brustolini[2], Fabyano F. Silva[5], Maximiller Dal-Bianco[2], Renildes L. F. Fontes[6], Anésia A. Santos[2,7], Francisco Murilo Zerbini[2,8], Fabio R. Cerqueira[1,9†] and Elizabeth P. B. Fontes[2,4*†]

Abstract

Background: The *Geminiviridae* family encompasses a group of single-stranded DNA viruses with twinned and quasi-isometric virions, which infect a wide range of dicotyledonous and monocotyledonous plants and are responsible for significant economic losses worldwide. Geminiviruses are divided into nine genera, according to their insect vector, host range, genome organization, and phylogeny reconstruction. Using rolling-circle amplification approaches along with high-throughput sequencing technologies, thousands of full-length geminivirus and satellite genome sequences were amplified and have become available in public databases. As a consequence, many important challenges have emerged, namely, how to classify, store, and analyze massive datasets as well as how to extract information or new knowledge. Data mining approaches, mainly supported by machine learning (ML) techniques, are a natural means for high-throughput data analysis in the context of genomics, transcriptomics, proteomics, and metabolomics.

Results: Here, we describe the development of a data warehouse enriched with ML approaches, designated geminivirus.org. We implemented search modules, bioinformatics tools, and ML methods to retrieve high precision information, demarcate species, and create classifiers for genera and open reading frames (ORFs) of geminivirus genomes.

Conclusions: The use of data mining techniques such as ETL (Extract, Transform, Load) to feed our database, as well as algorithms based on machine learning for knowledge extraction, allowed us to obtain a database with quality data and suitable tools for bioinformatics analysis. The Geminivirus Data Warehouse (geminivirus.org) offers a simple and user-friendly environment for information retrieval and knowledge discovery related to geminiviruses.

Keywords: Machine learning, Random Forest, Knowledge discovery, Data mining, Data Warehouse, Geminivirus

Background

The advancement of high-throughput sequencing technologies has enabled the rapid increase of genomic data in public databases and introduced genomics into the era of massive data generation. The biggest challenges, thus, turned out to be how to acquire, classify, store, and analyze huge datasets and extract knowledge from them [1]. Furthermore, the processing of massive data analysis has additional challenges, such as how to feasibly address bulky data, how to speed up the processing, and how to maintain the data veracity.

To extract and process data of interest, it is recommended to use the process known as Knowledge Discovery in Databases (KDD) process by which the data are selected, preprocessed, transformed, mined, and evaluated [2, 3]. The data mining step includes the application

* Correspondence: bbfontes@ufv.br
†Equal contributors
2National Institute of Science and Technology in Plant-Pest Interactions/BIOAGRO, Universidade Federal de Viçosa, Viçosa, Brazil
4Departamento de Bioquímica e Biologia Molecular, Universidade Federal de Viçosa, Viçosa, Brazil
Full list of author information is available at the end of the article

of unsupervised and supervised methods such as clustering analysis, classification, and rule learning techniques [4]. Machine learning (ML) techniques and data mining applications have been suggested for high-throughput data analysis in plants as well for all levels of studies, i.e., in genomics, transcriptomics, proteomics, and metabolomics [5], including taxonomic classification in metagenomic data [6]. The current high-throughput sequencing methods, metagenomics analysis approaches, and powerful bioinformatics tools accelerated knowledge acquisition of a number of viromes, allowing the identification of several viral agents in a wide range of cultivated and uncultivated plants. Furthermore, using rolling-circle amplification approaches, thousands of full-length geminivirus and satellite genome sequences have been amplified and have become available in public databases [7–9] (www.ictvonline.org).

The *Geminiviridae* family is a group of single-stranded DNA (ssDNA) viruses, with twinned and quasi-isometric virions, which infects a wide range of dicotyledonous and monocotyledonous plants and is responsible for important economic losses in tropical and subtropical regions worldwide. The *Geminiviridae* family is composed of nine genera: *Becurtovirus*, *Begomovirus*, *Curtovirus*, *Eragrovirus*, *Mastrevirus*, *Topocuvirus*, *Turncurtovirus*, *Capulavirus* and *Grablovirus* [10, 11]. The current classification is based on their insect vector, host range, genome organization, and phylogeny reconstruction [7, 8]. Except for viruses in the genus *Begomovirus*, which can be monopartite (single genomic DNA) or bipartite (two DNA components, referred to as DNA-A and DNA-B), all geminiviruses from the other genera have a single genomic component. The DNA-A of begomoviruses contains genes required for DNA replication (*Rep*, *REn*), gene expression control (*TrAP*), suppression of host defenses (*TrAP* and *AC4*), and viral genome encapsidation (*CP*), whereas the DNA-B encodes two proteins involved in intra- and intercellular movement (NSP and MP) [9, 12]. The single genomic component of mastreviruses encodes four proteins: a movement protein (pre-coat), a coat protein (CP), and two splicing variants of the replication-associated protein (Rep) [13]. The genomic structure of becurtoviruses contains five genes: the pre-coat gene, a *CP*, two *Reps*, and possibly a regulatory gene (*Reg*) [14, 15]. Viruses from the genera *Eragrovirus* and *Turncurtovirus* encode a pre-coat protein, CP, Rep, and transactivator protein (TrAP). However, turncurtoviruses encodes two additional proteins, the replication enhancer (Ren) and Symptom determinant/possible symptom determinant proteins (Sd/p.sd) [7, 16–18]. The genomic structure of curtoviruses is composed of seven genes, including the pre-coat gene, *Reg*, *CP*, *Ren*, *TrAP*, *Rep*, and *Sd/p.sd*

[19, 20]. The genus *Topocuvirus* has only one genome sequence deposited in public databases, which is organized into six genes, a pre-coat gene, *CP*, *Ren*, *TrAP*, *Rep*, and *Sd/p.sd* [21]. The recently discovered *Capulavirus* and *Grablovirus* genera encompass viruses that share similar genomic organization with becurtoviruses and eragroviruses [11].

Typically, the "Old World Geminiviruses" (from Europe, Asia, and Africa) are predominantly monopartite and commonly associated with alpha- or betasatellite DNAs, whereas "New World Geminiviruses" (from the Americas) are predominantly bipartite and may be associated with alphasatellites [22]. The betasatellite genome is approximately 1.35-kb long and harbors a single Open Reading Frame (ORF), βC1 [23]. The alphasatellite genome is approximately 1.37-kb long and contains a single ORF, which encodes a rolling-circle replication-initiator protein (*Rep*) [24, 25].

Unlike other important viral pathogens, such as Hepatitis C (hcv.lanl.gov) and HIV (hiv.lanl.gov), no database has been developed, which integrates all relevant information and provides user-friendly tools enriched by ML approaches for the easy manipulation of geminivirus data. Large amounts of information are distributed in a wide range of databases and in different file formats (for example GenBank, UniProt, VIPR, and ViralZone). Acquiring access to this information is usually a complex and time-consuming task. Additionally, a high level of computational expertise is required. To overcome these limitations, we developed a new data warehouse, designated Geminivirus Warehouse (geminivirus.org), using the concepts of the KDD process. The data warehouse geminivirus.org uses the ETL (Extract, Transform, Load) process, commonly applied for data warehouses, to choose, curate sequences, and standardize data. The geminivirus.org data warehouse is enriched with ML methods for the classification of the viral genus using the genomic sequence and the identification of gene coding sequences. The computational tools also comprise species demarcation, and include advanced bioinformatics tools for basic local alignment search, pairwise sequence comparison, including construction of the respective identity matrix, and phylogenetic analysis. Furthermore, we developed an algorithm for the ORF prediction from the genomes of each genus with high accuracy and which is capable of identifying possible intron regions.

Construction and content
Implementation
The geminivirus.org data warehouse was implemented using an MVC (Model-View-Controller) software

architecture pattern with modules programmed in the Java programming language and with MYSQL server relational databases. The data warehouse structure was organized in SQL tables in the star format [26] and geminivirus.org was developed using the KDD concepts, in which the ETL process was applied and ML algorithms from the Weka library v3.7.11 [27] were used. The design and workflow are summarized in Fig. 1 and detailed in the following sections.

Data source

Initially, geminivirus-related data were obtained in a document format from the GenBank nucleotide database. This file format contains information related to the complete genome, country of origin, geographic coordinates of the region or province, collection date, host, author responsible for the collection, among

others. Notably, all of these data are not always present, and the document structure is often out of the standard. Next, the article abstracts relating to the retrieved sequences were retrieved from the PubMed database. Finally, the geographic information such as the country name, geographic region or geographical coordinates obtained from GenBank were used to retrieve the geographical coordinates in the UTM (Universal Transverse Mercator) format of Google Maps.

Raw data extraction

Preliminarily, information was extracted from the GenBank file, as described in Table 1. Then, each file record was inspected using the following criteria concerning the full-length genome sequences: i) the length must be greater than or equal to the

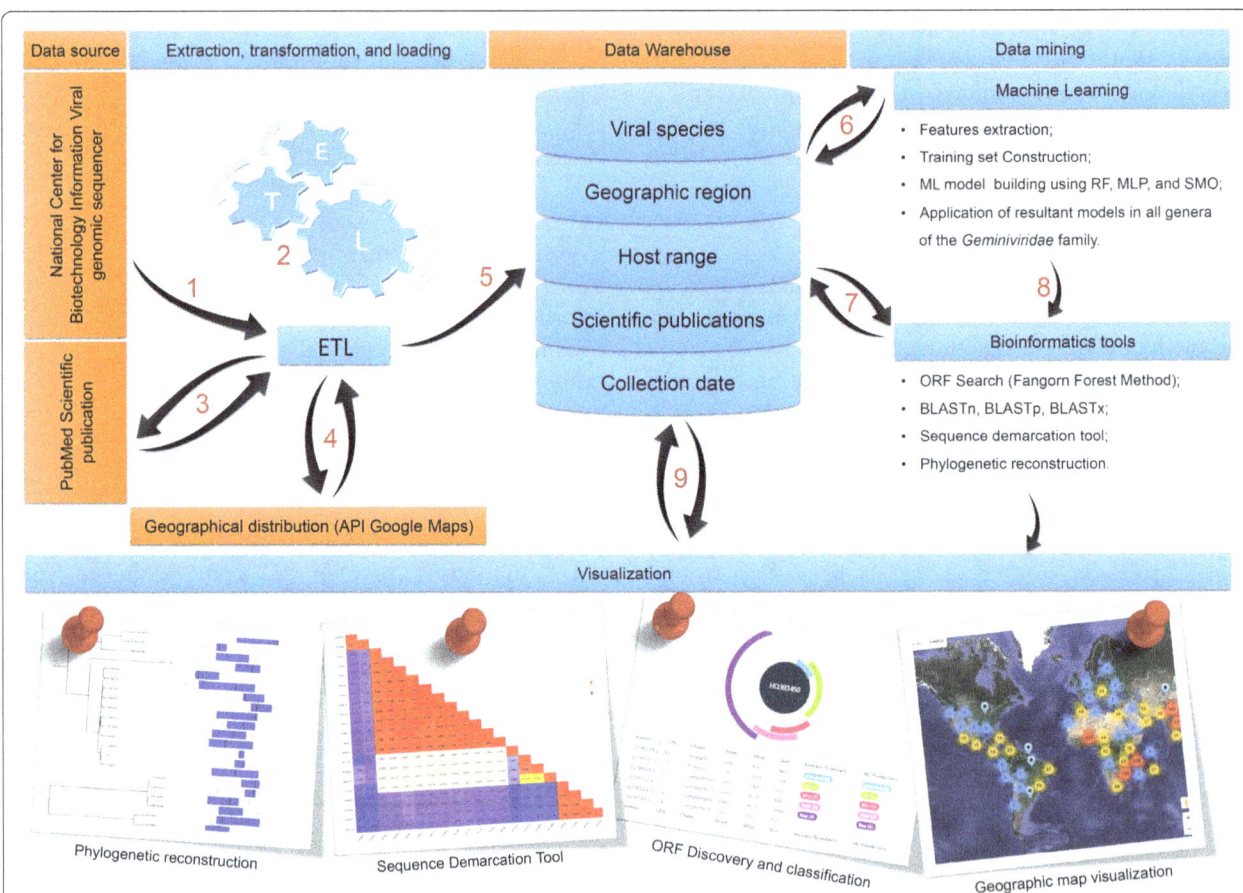

Fig. 1 Overview of the geminivirus.org framework. Initially, the geminivirus data were recovered from GenBank in the GenBank file format (1). The data were extracted, transformed, and standardized using algorithms based on rules and machine learning (ML) approaches (2). Next, the abstracts of the scientific publications were recovered from PubMed (https://www.ncbi.nlm.nih.gov/pubmed) (3) and the geographic coordinates of the isolates were retrieved from Google Maps (4). Data were merged and loaded into the relational database (5) in different dimensions such as the collection date, host range, geographic region, genomic data, associated publications, and organism data. The data were used to define the training set for building ML models to classify genera using Random Forest (RF), Multilayer Perceptron (MLP), and Sequential Minimal Optimization (SMO) learning algorithms (6). Information and analytical tools, such as basic local alignment tools (BLAST), sequence demarcation tools, and phylogenetic reconstruction, were embedded in the system (7) and an ORF Search tool for classification of ORFs based on ML procedures (8) was implemented. All analysis results are visible and freely available (9)

Table 1 Example of information extracted from the GenBank file and stored in geminivirus.org

TAGs	Value
LOCUS	KJ939916
DEFINITION	Soybean chlorotic spot virus isolate BR:Flt14:11 segment DNA-A, complete sequence.
ORGANISM	Soybean chlorotic spot virus
PUBMED	25028472
AUTHORS	Sobrinho,R.R., Xavier,C.A.D., Pereira,H.M.B., Lima,G.S.A.,,Assuncao,I.P., Mizubuti,E.S.G., Duffy,S. and Zerbini,F.M.
„JOURNAL Submitted	Departamento de Fitopatologia, BIOAGRO, Universidade Federal de Vicosa, Av. Peter Henry Rolfs s/n, Vicosa, Minas Gerais 36570–900, Brazil
Assembly Method	CodonCode Aligner v. 4.1.1 DEMO
Sequencing Technology	Sanger dideoxy sequencing
host	Macroptilium lathyroides
taxon	1221206
country	Brazil
segment	DNA-A
lat_lon	
collection_date	18-Mar-2012
collected_by	
CDS	199..954
gene	
note	coat protein
product	CP
protein_id	AIN36521.1
translation	MVKRDAPWRHMAGTSKVSRSSNFSPRGGGGPKNNRTSEWVNRPM …
ORIGIN	ACCGGATGGCCGCGCGATTTTTTATGGGCCTTATCTTTTGGCT CGTTCTTTTGGACCGAGTGTATTTGAATTAAAGTAAAGTTATTC CCTGTCCAA……………

minimum predefined size, ii) the length cannot exceed the maximum length, and iii) they must fit one of the nine genera of the *Geminiviridae* family (Additional file 1: Table S1). Genomes of alpha- and betasatellites were also included, as well as unassigned genomes that have not yet been classified by ICTV (https://talk.ictvonline.org/taxonomy). Thereafter, each pre-selected record was stored as a candidate to join the data warehouse sequences.

Data transformation

After the extraction step, it is necessary to transform and standardize the data, as well as to correct errors and relate different information or data from heterogeneous sources to improve the quality and consolidate the data [28]. In addition, it is necessary to associate metadata to data of interest entered into the

database [29]. To perform these steps, the pre-selected records were processed using the following criteria incorporated in Algorithm 1:

(i) **Origin of replication.** Firstly, corrections were performed in genome sequences that do not start at the expected genomic coordinates. These genome regions were adjusted to start in the first nucleotide after the cleavage site (dash) within the conserved nonanucleotide at the geminivirus replication origin (TAATATT-AC) and geminivirus-associated alpha- and betasatellite DNAs;

(ii) **Repairing the Open Read Frame coordinates in the genome.** The start and stop codon coordinates of each gene belonging to the genome sequence adjusted in step i were redefined according to the adjustment performed;

(iii) **Genus classifier in the *Geminiviridae* family.** The genera were confirmed using ML approaches;

(iv) **Checking the consistency of ORFs.** We verified whether all coding DNA sequences had start and stop codons, and whether the amino acid sequences were not truncated;

(v) **Standardization of gene acronyms.** The standardization of acronyms for gene identification was conducted following the genomic organization of the nine genera of the *Geminiviridae* family (Table 2). The following acronyms were used: CP, capsid protein; Rep, replication-associated protein; TrAP, transactivator protein; Ren, replication enhancer; MP, movement protein; NSP, nuclear shuttle protein; Reg, regulatory gene; Sd, symptom determinant; Ss, silencing suppressor; Tgs, transcriptional gene silencing. Note that the DNA-A component of old-world bipartite begomoviruses contains a V2 ORF, defined as a pre-coat in our standardization [8];

(vi) **ORF classifier in each genus of the *Geminiviridae* family.** In this step, we confirmed whether the ORFs were correctly standardized using ML approaches;

(vii) **Standardization of country abbreviation.** Country and continent abbreviations for all genera were standardized;

(viii) **Standardization of species name.** The species names were replaced following a list of begomovirus species, as of January 2015 [9], available at the ICTV website (https://talk.ictvonline.org/ictv_wikis/geminiviridae/m/files_gemini/5120/download);

(ix) **Recovering geographic coordinates.** We recovered geographical coordinates with exact (deposited coordinates) or approximate positions through secondary information, such as the informed country;

(x) **Recovering scientific publications.** All scientific publications related to a deposited sequence were recovered.

Algorithm 1: Data transformation step.

```
Input: Organisms selected in the extraction stage
1:  SELECTED ← set of organisms selected in the extraction stage
2:  while SELECTED ≠ ∅ do
3:      select ← SELECTED
4:      sequence ← select['genome']
5:      if sequence does not contain replication origin then
6:        sequence ← corrects replication origin of sequence
7:        select['genome'] ← sequence
8:        select['ALL ORFs'] ← corrects all ORFs start and stop codon from select['ALL ORFs']
9:      end if
10:     select['ML classification'] ← classifies the virus genera using ML
11:     ORFs ← select['ALL ORFs']
12:     size ← quantity of ORFs
13:     while i ≤ size do
14:         if ORFs[i]['sequence'] does not contain start codon then
15:            ORFs[i]['erro'] ← no start codon
16:         end if
17:         if ORFs[i]['sequence'] does not contain stop codon then
18:            ORFs[i]['erro'] ← no stop codon
19:         end if
20:         if ORFs[i]['sequence'] is truncated then
21:            ORFs[i]['erro'] ← truncated sequence
22:          end if
23:         ORFs[i]['Varsani Standard'] ← standardize ORFs[i]['cds']          ▸ see Table 2
24:         ORFs[i]['ML classification'] ← classifies the type of ORF using ML
25:         i ← i + 1
26:     end while
27:     select['ALL ORFs'] ← ORFs
28:     if select['country'] /= ∅ then
29:        select['acronym'] ← recover the acronym using select['country']
30:        select['continent'] ← recover the continent using select['country']
31:     end if
32:     select['coordinate'] ← search coordinates in Google Maps
33:     select['publication'] ← search publication in PubMed
34:     save select in Geminivirus DW
35: end while
```

Our programs were developed using object-oriented programming concepts. We implemented a collection of classes designated as Object Geminivirus (OG). The instances of the OG classes have the purpose of storing information, performing tasks (e.g., create, read, update, delete data in database), and communicating between the application (user interface) and the database. OG objects are instantiated with the data just after their transformation, preparing them to be loaded into our data warehouse.

Data load

The storage structure of our data warehouse was modeled in SQL tables with an adapted star scheme (Additional file 1: Figure S1). The star scheme is composed of one or more fact tables that represent data as facts. For instance, each isolate or organism can have these facts: (i) genome sequence and open reading frames, (ii) geographical localization, (iii) collection date, (iv) host range, (v) authors and related institutions and (vi) scientific publication reference. To insert the data into SQL tables, the transformed data were loaded to the OG object, by which each full-length genome and its associated metadata were inserted into the database, maintaining the integrity of data in different star scheme tables. It is worth mentioning that the OG object allows for control of all changes and transformations performed in a sequence and their associated metadata to record the history of changes and additional information. Other information regarding genome sequences which was not

Table 2 Terms used to name CDS in NCBI

Genera	CDS term NCBI	Varsani standard
Betasatellite	"beta" or "c1"	betaC1
Alphasatellite	"alpha" or "rep"	alphaRep
Begomovirus	"bv1" or "nsp" or "nuclear shuttle"	NSP
Begomovirus	"bc1" or "bc2" or "mp"	MP
All genera	"c1" or "ac1" or "rep" or al1	Rep
All genera	"c2" or "ac2" or "trap" or "al2" or "transcription activator protein"	TrAP
All genera	"c3" or "ac3" or "ren" or al3	REn
All genera	"c4" or "ac4" or al4	sd/p.sd
All genera	"c5" or "ac5"	AC5
All genera	"v1" or "av1" or "cp" or "ar1" or "capsid protein" or "coat protein"	CP
All genera	"v2" or "av2" or "pre-coat" or "precoat" or ar2	pre-coat
All genera	"v3" or "av3"	Reg

available on the GenBank Database was manually updated by our team, who inspected several scientific articles and afterwards inserted the information into the data warehouse.

Data mining
Machine learning

Datasets As mentioned above in steps iii and vi of data transformation, the genome sequences and complete ORFs are classified using ML approaches. For genus classification, complete genomes of species from the nine genera as well as satellite genome sequences were used to create the training set instances. As a result, the possible class labels are *Begomovirus, Mastrevirus, Curtovirus, Becurtovirus, Eragrovirus, Turncurtovirus, Topocuvirus, Capulavirus, Grablovirus*, alphasatellite, and betasatellite. The genus genomes were selected according to taxonomic reviews [7, 9, 11, 12, 19]. The betasatellite sequences, in turn, were chosen using the study of Briddon et al. [30], while alphasatellites were randomly selected from our curated repository. In addition, a test set was created using sequences contained in geminivirus.org, which were not present in the training set. For ORF classification, the training set was built using the ORFs pertaining to the genome sequences with which the genus training set was constructed. In this case, the instance labels are betaC1, alphaRep, Rep, TrAP, REn, Sd/p.sd, AC5, CP, pre-coat, Reg, MP, and NSP. A test set was also created with ORFs contained in the genomic sequences used for the genus test set mentioned above. The number of sequences, in each class, used to compose the training/test sets for genus and ORF classification, is shown in Additional file 1: Table S2.

Classification attributes In the case of genus classification, each genome sequence, selected to produce an instance in the training or test set, is split into four pieces of same (or nearly same) size, and the following attributes are then collected: proportions of A, T, C, and G of the whole sequence; and proportion of A, T, C, and G as well as GC content of each of the four pieces, totaling 24 attributes.

For ORF classification, attributes were extracted from every coding DNA sequence (CDS) and respective amino acid sequence to produce each instance. In this case, the attributes are proportions of A, T, C, and G of the CDS; and the proportion of each one of the 20 amino acids in the translated CDS.

Machine learning algorithm selection The ML algorithms Sequential Minimal Optimization (SMO), Random Forest (RF), and Multilayer Perceptron (MLP) are easily adaptable to handle multiclass classification problems [31–33], and are largely applied in several recent solutions for bioinformatics problems [34–37]. For this reason, we decided to perform some experiments with these ML approaches to select the best one to be incorporated in geminivirus.org. These algorithms were trained with the training sets presented previously, by means of the WEKA API (v3.7) [27], using the default parameters. The generalization of the resultant models was evaluated using three different techniques:

(i). The use of a completely independent test set;
(ii). 10-fold cross-validation [38]; and
(iii). leave-one-out, which is an *n*-fold cross validation, where *n* is the number of training instances.

To evaluate the performance of classification models, we used the statistical metrics accuracy, precision, recall, and F-measure (equations presented in Additional file 1: Equations S1) as well as the area under the ROC curve (AUC) [39].

Accessing the predictive power of the three models built from the SMO, RP, and MLP approaches for genus classification, MLP and RF presented similar results, with a slight superiority of MLP. Both performed better than SMO. MLP presented the following mean values for accuracy, precision, recall, and AUC, respectively: 0.966, 0.974, 0.967, and 0.986. See detailed results of all tests in Additional file 1:Table S3.

In the same way, the three above-mentioned ML algorithms were tested for ORF classification using the same three evaluation methods. This time, the three resultant models presented a similar predictive power, with a slight superiority of RF over MLP and SMO. RF could achieve mean values for accuracy, precision, recall, and AUC of 0.975, 0.976, 0.976, 0.991, respectively (detailed results in Additional file 1: Table S4). Consequently, geminivirus.org applies the MLP model for genus classification, and the RF model for ORF classification.

Bioinformatics tools

The data warehouse geminivirus.org provides a user-friendly web interface for the easy usage of advanced bioinformatics tools to search for viral information and to perform basic local alignment search, species demarcation, optimized phylogenetic analysis, ORF discovery and classification, as well as geographical visualization of geminiviruses and satellite-related data:

(i). **User-friendly search modules.** The web interface contains user-friendly search modules for viral sequences and scientific publications. The user can perform a search using keywords, such as viral name, host plant, GenBank Database accession number, country of origin, genome segment (DNA-A, DNA-B, monopartite genome or alpha- and betasatellite), collection year and sequence submission year. The search for scientific publications can also be performed using keywords such as PubMed ID, author name, virus name, scientific journal, and sequence publication year;

(ii). **Basic local alignment search.** To perform a basic local alignment search with sequences of genomes, amino acids, or CDS, we embedded the BLAST software [40] (BLASTn, BLASTp, and BLASTx algorithms) in our platform with pre-adjusted p-value parameters.

(iii). **Species demarcation.** We also incorporated the SDT v1.0 software [41] into geminivirus.org, which enables pairwise-sequence comparison analyses. Query sequences are used for pairwise alignments using MAFFT [42], MUSCLE [43], or ClustalW [44] algorithms. Based on the percentage of sequence identities, desired sequences can be selected to generate a comparative identity matrix. Thus, this matrix can be viewed in geminivirus.org or downloaded to the user's computer, opened with the original SDT software, and can be edited using any image editing software.

(iv). **Phylogenetic reconstruction analysis.** An automated phylogenetic analysis may be performed in geminivirus.org. The user initially enters at least one query sequence and then performs a search for sequence homology using BLAST algorithms. Query sequences are used to perform pairwise alignments using MAFFT, MUSCLE, or ClustalW algorithms, and the alignment output is automatically loaded into the FastTree software [45]. The phylogenetic analysis is performed using the maximum-likelihood method with 1000 bootstrap replications and other default parameters. The FastTree 2 software uses minimum-evolution subtree-pruning-regrafting and maximum-likelihood NNIs (nearest-neighbor interchange) to search for better trees. We also embedded the Phytools R package into our platform for visualization and additional analysis, for which the fastBM simulation function is used [46]. In addition, the phylogenetic tree output can also be downloaded in the Newick format to the user's computer, then opened and edited using, for example, the FigTree v1.4.2 software (http://tree.-bio.ed.ac.uk/software/figtree).

(v). **Data Visualization.** All information related to geminiviruses and geminivirus-associated satellites, such as viral species and geographical distribution, can be visualized using a graphic interface developed in the Google Maps API (https://developers.-google.com/maps/?hl=en) and Google MarkerClusterer (https://github.com/googlemaps/js-marker-clusterer). Additionally, statistical information about the amount of full-length genome sequences per country, viral species, year, and related scientific publications are also shown in charts using the Google Charts API (https://developers.-google.com/chart/?hl=en).

(vi). **Discovery and classification of ORFs**. We have developed an algorithm for prediction and classification of genes. Moreover, the algorithm allows the classification of the viral genus based on the genomic sequence using ML approaches.

Utility and discussion

Geminiviruses infect a wide range of dicotyledonous and monocotyledonous plants causing expensive losses worldwide. A wide range of studies have been published in the literature using genomic data and different bioinformatics, such as studies of molecular interaction mechanisms among viral and host plants [47, 48], population biology [49], species taxonomy [8, 9, 50], and discovery of new viral species by analysis of genetic diversity [51]. In spite of the geminivirus relevance, inflicting serious threat to agriculture in tropical and subtropical areas, there are no databases integrating all relevant related information and providing user-friendly tools for easily manipulating the data. The lack of comprehensive bioinformatics tools for geminivirus analyses motivated the development of a specific database for geminivirus, including automated pipelines to boost findings and the exchange of information among researchers.

The high diversity and amount of viral species complicate the recovery and interpretation of viral genomic and proteomic data. After the advent of the rolling-circle amplification

(RCA), using the phi-29 DNA polymerase along with current high-throughput sequencing methods, thousands of full-length sequences have become available from public databases in the last 10 years. This large amount of data is available in a wide range of databases or as supplemental material in scientific publications, such as the full-length genome, coding DNA sequence, geographical localization, host range, data collection, species names, and species identifiers (by acronym). All of these data have great potential to result in new knowledge when unified. Approximately 274 full-length genomic sequences of geminiviruses became available in GenBank databases from 1990 to 2003. Nonetheless, this number has increased exponentially (approximately 34 times) up to the current date (9255 full-length sequences). In parallel, a significant number of scientific papers involving geminiviruses have been reported during the same period.

The number of full-length genomic sequences is distributed among the nine genera and other quantitative information can be found in the data warehouse (http://geminivirus.org:8080/geminivirusdw/statistics.jsp). Furthermore, recently discovered geminiviruses showed that the genetic diversity among genera reaches high levels and, in some cases, presents specific genome architectures [8]. Considering this highly divergent genomic content, we have built a web platform that includes associated metadata, search modules, bioinformatics tools, and ML methods, which retrieve information of interest, demarcate species, and classify genera and ORFs. The following sections provide detailed information on the use of geminivirus.org to retrieve or discover information about geminiviruses.

Sequence search and data visualization

The search for geminivírus, DNA satellite and gene sequences into Geminivirus Data Warehouse is available through the menu designated Virus. The search and analysis tools provide various searching criteria on both nucleotide sequences (full-length genomes or genes) and amino acid sequences (proteins). The metadata provides the users with the ability to perform searches based on parameters (or combinations thereof), such as viral name, host plant, access number into GenBank Database, country of origin, or genome segment (DNA-A, DNA-B begomoviruses components, alpha- or betasatellite DNAs). After the search, the results are shown in a table format, in which the columns refer to the accession number, sequence description, collection date, submission date in GenBank, host, country, and sequence length. The resultant accession number links all information related to the complete sequence (http://geminivirus.org:8080/geminivirusdw/viewOrgServlet?id=KC706589). In addition to the metadata, the information about the sequence authorship, the funding institutions and those responsible for the data collection (when available in GenBank) was preserved.

Information concerning authorship is accessible from the abstract of publications linked to the complete sequences. Furthermore, other detailed information about the consistency and quality of data is presented:

(i). **About ORFs.** Relevant information, such as gene names, virion-sense or complementary strain of the genome, protein sequence and coding DNA sequence are also presented. In addition, the quality of these sequences (presence of start codon, stop codon, and truncation) is inspected and accepted with the appearance of the light blue star and a notice indicating the status of the algorithm verification. However, neglected annotation of submitted sequences is quite common. To overcome this problem, ORFs were classified using learning models constructed with the Random Forest algorithm, as previously mentioned. Thus, the result of the classification is presented to the user, indicating the gene name and the resulting likelihood of classification. This way, the consistency of the ORF and its annotation is reinforced.

(ii). **View of the genome architecture.** The genomic architecture can be viewed in an interactive circular diagram. Furthermore, the genes are shown in a table comprising summary information.

(iii). **Revisions.** During the process of cleaning and processing the data, some changes are performed in the complete genome, CDS, or protein, whereas metadata is included in other data sources or manually entered. The added and changed information is stored in a database, to register the change history. In addition, the history is visible to the users in a change timeline.

The treatments and aggregated information of the submitted sequences are important and positively assist in conducting several studies, such as migration studies, phylogeography analysis, recombination analysis, genetic diversity, and species demarcation, among others. The associated metadata is rated by intuitive icons that represent the sequence quality and reliability. For example, the viral sequences approved in the initial filter receive a yellow medal. On the other hand, sequences associated with at least one publication receive a green medal, and sequences that are inspected and corrected manually receive a red medal. In addition, those sequences confirmed by the Random Forest learning model receive a blue medal. Stars are also used and refer to the existence of a particular metadata. An empty star denotes that the associated metadata is in the process of manual inspection.

Searching publications

The search for scientific publications can also be performed using keywords, such as PubMed ID, author name, virus name, scientific journal name, and sequence publication year. The search results are presented in a table containing titles, authors, publication year, and PMID number. Clicking on the provided link to each publication enables the access to its abstract, along with other information such as the scientific journal and accession number of the associated virus sequence.

Basic local alignment search tools

A basic local alignment search can be performed against a query sequence (nucleotide or amino acid) using the BLASTn, BLASTx, BLASTp algorithms embedded in our system. BLAST serves as a tool for searching sequences with higher similarity. The alignment results can be chosen to be automatically used as input in the SDT v1.0 software for species demarcation and also in FastTree for phylogenetic analysis, both embedded in the data warehouse. The BLAST results are merged with other associated metadata, including sequence match, collection date, host, and geographic region. Thus, the tabulated results may help researchers in making decision based on sequence comparisons, host range, and geographical location.

Species demarcation

A SDT (Sequence Demarcation Tool) was recently implemented for viral species demarcation which provides standardization for all parameters, such as alignments and processing gaps, to calculate the percentage of sequence identity between genomes or gene sequences [41]. We incorporated an adapted parallel version of the SDT software into geminivirus.org (http://geminivirus.org:8080/geminivirusdw/SDT_demarcation.jsp). This enables genome sequences of geminiviruses and associated satellite DNAs to be directly compared and eliminates the need for a local installation of the SDT desktop version in the user's computer. Briefly, the analysis performs a preliminary comparison of a query sequence to other available sequences in geminivirus.org using BLAST algorithms, which enables a pre-selection of closely related sequences. Then, SDT performs all of the comparisons between the query sequences provided by the user and those sequences that were preselected in the previous step of the BLAST results.

Another advantage of using SDT from geminivirus.org is that the algorithm only performs comparisons involving query sequences provided by the user against those available in geminivirus.org, which are the subject sequences of interest. It reduces the analysis complexity and duration needed to generate results. It is important to highlight that the implementation of the SDT program into geminivirus.org in our data warehouse enables

the usage of this software from various platforms. Finally, a color array can be obtained, representing the identity percentage values, and can be downloaded as a list containing the results of all pairwise comparisons.

Phylogenetic reconstruction

The phylogenetic analysis from geminivirus.org enables a rapid visualization of phylogenetic relationships and groupings from the input sequence dataset (http://geminivirus.org:8080/geminivirusdw/phylogeny.jsp). Initially, the sequence of interest is submitted against the geminivirus sequences using the BLASTn algorithm. The selected sequences from the BLAST results are then automatically given as inputs to the MUSCLE algorithm to perform multiple sequence alignments. Next the MUSCLE output is automatically loaded into FastTree. The FastTree is a tool that enables phylogeny inference for alignments with up to hundreds of sequences. It is slightly more accurate than its former version and 100–1000-fold faster than other tools.

Prediction and classification of ORFs in full-length genome sequences

Collectively, geminiviruses contain ten different known genes. In addition, the alphasatellites encode alphaRep, while betasatellites encode betaC1. The most common way to identify such genes is through the ORF finder tool (www.ncbi.nlm.nih.gov/projects/gorf/). However, prediction and in silico annotation of these ORFs require computational expertise and time to process and analyze the data. To address this restriction, we developed a method of prediction and classification of ORFs designated the Fangorn Forest method. In addition, a complete pipeline can optionally be used to classify the viral genus. The Fangorn Forest tool is freely available at http://geminivirus.org:8080/geminivirusdw/discoveryGeminivirus.jsp.

The geminivirus.org warehouse is structured to accommodate information about geminiviruses and related DNA satellites that become available regularly. Our platform will be frequently updated with new information extracted from GenBank, scientific publications, meetings, and abstracts. The inclusion of new data sources will enhance the wealth of data contained in our data warehouse and will promote an expansion of our system to accommodate further information that can assist in the interpretation of bioinformatics analysis results. Future improvements will permit further development of meta-analysis tools and natural language processing to extract knowledge from published studies and standardize sequences to be deposited directly into the data warehouse. We plan to develop a mobile application to assist data collection and information exchange among researchers and geminivirus.org.

Conclusions

The geminivirus.org database is an integrated and open-access data warehouse that optimizes complicated and comprehensive searches that are difficult to perform using currently existing tools. Therefore, it is efficient in assisting targeted searches and provides accurate and concise information on all geminiviruses and geminivirus-associated satellites to the scientific community. It provides a user-friendly environment to retrieve information about (i) the geographic distribution of geminiviruses throughout the globe through an interactive map; (ii) the circular genomic structure through interactive visualization; (iii) advanced graphs with statistical information and results provided by species demarcation, phylogenetic analysis, and ORF search. Its flexibility enables the addition or analysis of various taxonomy types, genome, sampling, or biological data to facilitate and update information sources. Furthermore, the implementation of algorithms based on ML approaches allows the prediction and classification of viral genes as well as the identification of the genus based on viral genomic sequences. The data sources and additional analytical tools will greatly facilitate searches in the geminivirus.org information management system. The geminivirus.org data warehouse is freely available and will represent a valuable resource for the research community.

Availability and requirements

geminivirus.org is available from http://geminivirus.org. The application was built in Debian Linux, GlassFish Server Open Source Edition 4.1.1, Java v8, JavaServer Pages (JSP), and MySQL5 environment. The geminivirus.org front-end layer uses HTML5, Bootstrap CSS library, JavaScript and jQuery. It is compatible with Chrome v57, Firefox v52, and Safari v10. The free tools used by geminivirus.org are: R v3.1.1, blastp/blastn v2.2.29, CLUSTAL v2.1, MAFFT v7.205, MUSCLE v3.8.31 and WEKA v3.7.11. geminivirus.org is free for academic use.

Abbreviations

CDS: Coding DNA sequence; CP: Capsid protein; ETL: Extract, Transform, Load; KDD: Knowledge Discovery in Databases; ML: Machine learning; MLP: Multilayer Perceptron; MP: Movement protein; MVC: Model View Controller; NNIs: Nearest-neighbor interchange; NSP: Nuclear shuttle protein; OG: Object Geminivirus; ORF: Opening reading frame; RCA: Rolling-circle amplification; Reg: Regulatory gene; REn: Replication enhancer; Rep: Replication-associated protein; RF: Random Forest; Sd/p.sd: Symptom determinant/possible symptom determinant proteins; SDT: Sequence Demarcation Tool; SMO: Sequential Minimal Optimization; ssDNA: Single-stranded DNA; TrAP: Transactivator protein; UTM: Universal Transverse Mercator

Acknowledgements

We acknowledge the authors of geminivirus papers, sequences and characterization, which were not included in the reference list, but are listed and stored in the Geminivirus Data Warehouse (geminivírus.org).

Funding

The authors are grateful to the National Institute of Science and Technology in Plant-Pest Interactions (INCT-IPP), Fundação de Amparo à Pesquisa do estado de Minas Gerais (FAPEMIG), Coordenação de Aperfeiçoamento de Pessoal de Nível Superior (CAPES) and Conselho Nacional de Desenvolvimento Científico e Tecnológico (CNPq) for financial support. The funding bodies did not play any role in the design of the study, in the analysis and interpretation of data.

Authors' contributions

JCFS designed, developed the geminivírus.org warehouse and wrote the first draft of the paper; TFMC, developed the geminivírus.org warehouse; MFB, MD, WAP, RRS, PMPV, AAS, FMZ .MD-B manually curated the data; OJBB and FFS validated ML methods; RLFF edited the final draft of the paper and wrote the webpage; FRC supervised the development of the ML algorithms and wrote the paper; EPBF directed the project and wrote the paper. All authors read and approved the final manuscript.

Competing interests

The authors declare that they have no competing interests.

Author details

[1]Departamento de Informática, Universidade Federal de Viçosa, Viçosa, Brazil. [2]National Institute of Science and Technology in Plant-Pest Interactions/ BIOAGRO, Universidade Federal de Viçosa, Viçosa, Brazil. [3]Núcleo de Biomoléculas, Universidade Federal de Viçosa, Viçosa, MG, Brazil. [4]Departamento de Bioquímica e Biologia Molecular, Universidade Federal de Viçosa, Viçosa, Brazil. [5]Departamento de Zootecnia, Universidade Federal de Viçosa, Viçosa, Brazil. [6]Departamento de Solos, Universidade Federal de Viçosa, Viçosa, Brazil. [7]Departamento de Biologia Geral, Universidade Federal de Viçosa, Viçosa, Brazil. [8]Departamento de Fitopatologia, Universidade Federal de Viçosa, Viçosa, MG, Brazil. [9]Departamento de Engenharia de Produção, Universidade Federal Fluminense, Petrópolis, Rio de Janeiro, Brazil.

References

1. Stephens ZD, Lee SY, Faghri F, Campbell RH, Zhai C, Efron MJ, Iyer R, Schatz MC, Sinha S, Robinson GE. Big data: astronomical or genomical? PLoS Biol. 2015;13: e1002195.
2. Tsai CW, Lai CF, Chao HC, Vasilakos AV. Big Data analytics: a survey. J Big Data. 2015;2(1):1–32.
3. Dunkel B, Soparkar N, Szaro J, Uthurusamy R. Systems for KDD: from concepts to practice. Futur Gener Comput Syst. 1997;13(2):231–42.
4. Olshannikova E, Ometov A, Koucheryavy Y, Olsson T. Visualizing Big Data with augmented and virtual reality: challenges and research agenda. J Big Data. 2015;2(1):1–27.
5. Ma C, Zhang HH, Wang X. Machine learning for Big Data analytics in plants. Trends Plant Sci. 2014;19(12):798–808.
6. Rasheed Z, Rangwala H. Metagenomic taxonomic classification using extreme learning machines. J Bioinform Comput Biol. 2012;10(05):1250015.
7. Varsani A, Navas-Castillo J, Moriones E, Hernández-Zepeda C, Idris A, Brown JK, Zerbini FM, Martin DP. Establishment of three new genera in the family Geminiviridae: Becurtovirus, Eragrovirus and Turncurtovirus. Arch Virol. 2014; 159(8):2193–203.
8. Brown JK, Fauquet CM, Briddon RW, Zerbini FM, Moriones E, Navas-Castillo J. Family Geminiviridae. In: King AMQ, Lefkowitz E, Adams MJ, Carstens EB, editors. Virus Taxonomy: Ninth Report of the International Committee on Taxonomy of Viruses. New York: ELSEVIER Academic Press; 2012. p. 351–73.
9. Brown JK, Zerbini FM, Navas-Castillo J, Moriones E, Ramos-Sobrinho R, Silva JC, Fiallo-Olivé E, Briddon RW, Hernández-Zepeda C, Idris A, Malathi VG. Revision of Begomovirus taxonomy based on pairwise sequence comparisons. Arch Virol. 2015;160(6):1593–619.
10. Hanley-Bowdoin L, Bejarano ER, Robertson D, Mansoor S. Geminiviruses: masters at redirecting and reprogramming plant processes. Nat Rev Microbiol. 2013;11(11):777–88.
11. Varsani A, Roumagnac P, Fuchs M, JNavas-Castillo J, Moriones E, Idris A, Briddon RW, Rivera-Bustamante R, Zerbini, FM, Martin DP. Capulavirus and Grablovirus: two new genera in the family Geminiviridae. Arch Virol. 2017; doi:10.1007/s00705-017-3268-6.

12. Krenz B, Jeske H, Kleinow T. The induction of stromule formation by a plant DNA-virus in epidermal leaf tissues suggests a novel intra-and intercellular macromolecular trafficking route. Front Plant Sci. 2012;3:291.

13. Muhire B, Martin DP, Brown JK, Navas-Castillo J, Moriones E, Zerbini FM, Rivera-Bustamante R, Malathi V, Briddon RW, Varsani A. A genome-wide pairwise-identity-based proposal for the classification of viruses in the genus *Mastrevirus* (family *Geminiviridae*). Arch Virol. 2013;158(6):1411–24.

14. Yazdi HB, Heydarnejad J, Massumi H. Genome characterization and genetic diversity of beet curly top Iran virus: a geminivirus with a novel nonanucleotide. Virus Genes. 2008;36(3):539–45.

15. Heydarnejad J, Keyvani N, Razavinejad S, Massumi H, Varsani A. Fulfilling Koch's postulates for beet curly top Iran virus and proposal for consideration of new genus in the family Geminiviridae. Arch Virol. 2013;158(2):435–43.

16. Briddon RW, Heydarnejad J, Khosrowfar F, Massumi H, Martin DP, Varsani A. Turnip curly top virus, a highly divergent geminivirus infecting turnip in Iran. Virus Res. 2010;152(1):169–75.

17. Razavinejad S, Heydarnejad J, Kamali M, Massumi H, Kraberger S, Varsani A. Genetic diversity and host range studies of turnip curly top virus. Virus Genes. 2013;46(2):345–53.

18. Varsani A, Shepherd DN, Dent K, Monjane AL, Rybicki EP, Martin DP. A highly divergent South African geminivirus species illuminates the ancient evolutionary history of this family. Virol J. 2009;6(1):1.

19. Stanley J, Markham PG, Callis RJ, Pinner MS. The nucleotide sequence of an infectious clone of the geminivirus beet curly top virus. EMBO J. 1986;5(8):1761–7.

20. Varsani A, Martin DP, Navas-Castillo J, Moriones E, Hernández-Zepeda C, Idris A, Zerbini FM, Brown JK. Revisiting the classification of curtoviruses based on genome-wide pairwise identity. Arch Virol. 2014;159(7):1873–82.

21. Briddon RW, Bedford ID, Tsai JH, Markham PG. Analysis of the nucleotide sequence of the treehopper-transmitted geminivirus, tomato pseudo-curly top virus, suggests a recombinant origin. Virology. 1996;219(2):387–94.

22. Briddon RW, Patil BL, Bagewadi B, Nawaz-ul-Rehman MS, Fauquet CM. Distinct evolutionary histories of the DNA-A and DNA-B components of bipartite begomoviruses. BMC Evol Biol. 2010;10(1):1.

23. Cheng X, Wang X, Wu J, Briddon RW, Zhou X. βc1 encoded by tomato yellow leaf curl china betasatellite forms multimeric complexes in vitro and in vivo. Virology. 2011;409(2):156–62.

24. Briddon R, Stanley J. Subviral agents associated with plant single-stranded DNA viruses. Virology. 2006;344(1):198–210.

25. Fiallo-Olivé E, Martínez-Zubiaur Y, Moriones E, Navas-Castillo J. A novel class of DNA satellites associated with New World begomoviruses. Virology. 2012;426(1):1–6.

26. Boehnlein M, Ulbrich-vom FA. Deriving initial data warehouse structures from the conceptual data models of the underlying operational information systems. Proceedings of the 2nd ACM International Workshop on Data Warehousing and OLAP. 1999. p. 15–21.

27. Hall M, Frank E, Holmes G, Pfahringer B, Reutemann P, Witten IH. The WEKA data mining software: an update. SIGKDD Explor. 2009;11(1):10–8.

28. Kumar V, Thareja R. A simplified approach for quality management in data warehouse. IJDKP. 2013;3(5):61–9.

29. Bala M, Boussaid O, Alimazighi Z. Big-ETL: extracting-transforming-loading approach for Big Data. In: Int'l Conf Par and Dist Proc Tech and Appl. 2015. p. 462.

30. Briddon RW, Brown JK, Moriones E, Stanley J, Zerbini M, Zhou X, Fauquet CM. Recommendations for the classification and nomenclature of the DNA-β satellites of begomoviruses. Arch Virol. 2008;153(4):763–81.

31. Platt JC. Fast training of support vector machines using sequential minimal optimization. In: Schölkopf B, Burges CJC, Smola AJ, editors. Advances in Kernel Methods. Cambridge: MIT Press; 1999. p.185–208.

32. Breiman L. Random Forests. Mach Learn. 2001;45(1):5–32.

33. Gardner MW, Dorling S. Artificial neural networks (the multilayer perceptron)—a review of applications in the atmospheric sciences. Atmos Environ. 1998;32(14):2627–36.

34. Cai Y, Liao Z, Ju Y, Liu J, Mao Y, Liu X. Resistance gene identification from Larimichthys crocea with machine learning techniques. Sci Rep. 6;6 doi:10. 1038/srep38367.

35. Kushwaha SK, Chauhan P, Hedlund K, Ahrén D. NBSPred: a support vector machine-based high-throughput pipeline for plant resistance protein NBSLRR prediction. Bioinformatics. 2016;32(8):1223–5.

36. Sunseri J, Ragoza M, Collins J, Koes DR. A D3R prospective evaluation of machine learning for protein-ligand scoring. J Comput Aided Mol Des. 2016; 30(9):761–71.

37. Liao Z, Wang X, Zeng Y, Zou Q. Identification of DEP domain-containing proteins by a machine learning method and experimental analysis of their expression in human HCC tissues. Sci Rep. 2016;6. doi:10.1038/srep39655.

38. Sylvain A, Celisse A. A survey of cross-validation procedures for model selection. Stat Surv. 2010;4:40–79.

39. Bradley AP. The use of the area under the ROC curve in the evaluation of machine learning algorithms. Pattern Recogn. 1997;30(7):1145–59.

40. Altschul SF, Gish W, Miller W, Myers EW, Lipman DJ. Basic local alignment search tool. J Mol Biol. 1990;215(3):403–10.

41. Muhire BM, Varsani A, Martin DP. SDT: a virus classification tool based on pairwise sequence alignment and identity calculation. PLoS One. 2014;9(9):108277.

42. Katoh K, Misawa K, Kuma Kei-ichi, Miyata T. Mafft: a novel method for rapid multiple sequence alignment based on fast Fourier transform. Nucleic Acids Res. 2012;30(14):3059–66.

43. Edgar RC. MUSCLE: multiple sequence alignment with high accuracy and high throughput. Nucleic Acids Res. 2004;32(5):1792–7.

44. Li K-B. Clustalw-mpi: ClustalW-MPI: ClustalW analysis using distributed and parallel computing. Bioinformatics. 2003;19(12):1585–6.

45. Price MN, Dehal PS, Arkin AP. FastTree 2–approximately maximum-likelihood trees for large alignments. PLoS One. 2010;5(3):9490.

46. Revell LJ. phytools: an R package for phylogenetic comparative biology (and other things). Methods Ecol Evol. 2012;3(2):217–23.

47. Zorzatto C, Machado JPB, Lopes KV, Nascimento KJ, Pereira WA, Brustolini OJ, Reis PA, Calil IP, Deguchi M, Sachetto-Martins G, et al. NIK1-mediated translation suppression functions as a plant antiviral immunity mechanism. Nature. 2015;520(7549):679–82.

48. Brustolini OJ, Machado JPB, Condori-Apfata JA, Coco D, Deguchi M, Loriato VA, Pereira WA, Alfenas-Zerbini P, Zerbini FM, Inoue-Nagata AK, et al. Sustained NIK-mediated antiviral signalling confers broad-spectrum tolerance to begomoviruses in cultivated plants. Plant Biotechnol J. 2015;13(9):1300–11.

49. Rocha CS, Castillo-Urquiza GP, Lima AT, Silva FN, Xavier CA, Hora-Júnior BT, Beserra-Júnior JE, Malta AW, Martin DP, Varsani A, et al. Brazilian begomovirus populations are highly recombinant, rapidly evolving, and segregated based on geographical location. J Virol. 2013;87(10):5784–99.

50. Briddon RW, Martin DP, Owor BE, Donaldson L, Markham PG, Greber RS, Varsani A. A novel species of Mastrevirus (family *Geminiviridae*) isolated from *Digitaria Didactyla* grass from Australia. Arch Virol. 2010;155(9):1529–34.

51. Rosario K, Marr C, Varsani A, Kraberger S, Stainton D, Moriones E, Polston JE, Breitbart M. Begomovirus-associated satellite DNA diversity captured through Vector-Enabled Metagenomic (VEM) surveys using whiteflies (Aleyrodidae). Viruses. 2016;8(2):36.

8

Inclusion of the fitness sharing technique in an evolutionary algorithm to analyze the fitness landscape of the genetic code adaptability

José Santos[*] and Ángel Monteagudo

Abstract

Background: The canonical code, although prevailing in complex genomes, is not universal. It was shown the canonical genetic code superior robustness compared to random codes, but it is not clearly determined how it evolved towards its current form.

The error minimization theory considers the minimization of point mutation adverse effect as the main selection factor in the evolution of the code. We have used simulated evolution in a computer to search for optimized codes, which helps to obtain information about the optimization level of the canonical code in its evolution.

A genetic algorithm searches for efficient codes in a fitness landscape that corresponds with the adaptability of possible hypothetical genetic codes. The lower the effects of errors or mutations in the codon bases of a hypothetical code, the more efficient or optimal is that code.

The inclusion of the fitness sharing technique in the evolutionary algorithm allows the extent to which the canonical genetic code is in an area corresponding to a deep local minimum to be easily determined, even in the high dimensional spaces considered.

Results: The analyses show that the canonical code is not in a deep local minimum and that the fitness landscape is not a multimodal fitness landscape with deep and separated peaks. Moreover, the canonical code is clearly far away from the areas of higher fitness in the landscape.

Conclusions: Given the non-presence of deep local minima in the landscape, although the code could evolve and different forces could shape its structure, the fitness landscape nature considered in the error minimization theory does not explain why the canonical code ended its evolution in a location which is not an area of a localized deep minimum of the huge fitness landscape.

Keywords: Genetic code adaptability, Evolutionary computing, Genetic algorithm, Fitness sharing

Background

The canonical or standard genetic code is redundant since the 64 possible codons encode only 21 labels, the 20 amino acids present in proteins and the "stop" signal that defines the end of the protein translation process. The canonical code, although prevailing in complex genomes, is not universal. The existence of other different codes, like the one of mitochondrial DNA, altered the "frozen accident",

as coined by Crick [1], so extensive research has been performed in order to analyze the reasons behind the establishment of the canonical genetic code.

There are three main theories about the genetic code organization and development, which could have influenced the canonical code organization. The stereochemical theory states that the stereochemical interactions between bases and amino acids influenced the primordial code, probably in the RNA World or earlier [2, 3]. Thus, the physicochemical affinity between amino acids and the cognate codons determined the codon assignments [4].

---
*Correspondence: jose.santos@udc.es
Department of Computer Science, University of A Coruña, Campus de Elviña s/n, 15071 A Coruña, Spain

The second one is the coevolution theory [5], which maintains that at an early stage of the genetic code development few precursor amino acids were encoded. The other amino acids (product amino acids) developed biosynthetically from such precursor amino acids and these initial amino acids passed part or their whole codon domain to their biosynthetically produced amino acids. Finally, the error minimization theory or physicochemical theory considers the minimization of point mutation adverse effects as the main selection factor in the evolution of the code [4, 6, 7].

According to this last theory, the genetic code structure evolved to maximize its robustness, that is, to minimize the consequences of code mutations on the function of the encoded proteins [4]. In favor of this theory is the fact that similar codons encode amino acids with similar chemical properties: the codons that share two bases tend to code amino acids with similar hydrophobicity.

In this last alternative, a huge number of alternative "genetic codes" are possible. The number of possible alternative codes is $1.4 \cdot 10^{70}$ [8], when each amino acid is coded by 6 codons at maximun (like the canonical code case). If only permutations of the amino acids encoded in the 20 codon sets of the canonical code are allowed, there are 20! ($2.43 \cdot 10^{18}$) possible codes. This alternative, which maintains the canonical codon set structure, is the most used in the studies regarding the error minimization theory. Finally, more than $1.51 \cdot 10^{84}$ codes can be defined, when no restrictions in the associations between the 64 codons and the 21 meanings are considered [9].

In this error minimization theory, the efficiency or optimality of a code is defined taking into consideration the possible errors or mutations in the codon letters. Typically, all the point mutations in the codons are applied to quantify the change between the encoded amino acids before and after each point mutation. That change is measured taking into account an amino acid property, the polar requirement property being the one most used. Once such changes are averaged over all the possible mutations, the lower that error value of a code the more efficient or optimal is the corresponding code, since it means smaller phenotypic changes in the encoded proteins when mutations happen.

Moreover, two different analyses were considered to assess the optimality or adaptability of the canonical genetic code. In the first one, named "statistical analysis", many randomly generated codes are defined and then the probability of more efficient random codes with respect to the canonical one is quantified. The lower this probability the more optimized is the canonical code. According the authors that follow this analysis [2, 10–12], the pattern of codon assignments of the canonical code appears nearly optimal.

For example, in this statistical approach, Freeland and Hurst [10] used alternative codes with the same codon set block structure of the canonical code. In an ample sample of 1,000,000 possible alternative codes only 114 codes were more efficient than the canonical code (the criteria for defining such hypothetical codes are summarized in the "Methods" section, together with the measure to quantify code optimality). This low number of random better codes allows the authors to state that the canonical genetic code evolved under the selection of the error minimization. Moreover, when the authors weighted transition mutations differently from transversion mutations, only 1 in every million randomly alternative codes was better than the canonical genetic code, as the article title states [10].

Extensions of these initial works include the analyses by Gilis et al. [13], where the authors considered the role of amino-acid frequencies in the efficiency of the canonical genetic code, and the work by Torabi et al. [14] who have experimented with the role of aminoacyl-tRNA synthetases in decreasing the effects of mistranslations in the code evolution. Along this line, Zhu et al. [15] also took into consideration the codon usage of individual species in the code optimization for error minimization and Marquez et al. [16] checked whether organisms optimize the genetic code at the same time that the codon usage.

The second approach to assess code efficiency is the "engineering approach" [17, 18]. First, the most efficient code is found, typically by a computational search method. Afterwards, the error value of the canonical code is compared with the average error value of randomly generated codes and the error value of the best obtained code. The relative position of the canonical code error value with respect to the others provides a measure of the optimization level of the canonical code. The results with this approach show that the canonical code is not so close to the optimal as the statistical approach claims. The authors in [17, 19] have discussed and debated these two alternatives.

For example, in this engineering approach, according to Di Giulio [20], the canonical genetic code achieved 72.7% minimization of polarity distance when comparing the error value of the canonical code with the error values of random codes defined with the same codon set structure of the canonical one and with the value of the best possible code (The same methodology explained in the "Methods" section was used to assess the adaptability level of a possible code). The best code (necessary in the comparison) was obtained by Di Giulio [20] using simulated annealing. Novozhilov et al. [21], in order to investigate the optimization of codes for the maximum attainable robustness, used also a greedy minimization algorithm for searching better alternative codes. Their search method employed swaps of four-codon or two-codon series, in alternative codes defined with the same block structure and the same degeneracy degree of the canonical code. According to the authors' results, the canonical code is much closer to its local fitness minimum than the majority of the random codes with similar robustness to the canonical one.

Following other alternatives, Gardini et al. [22] focused their analysis on searching clues for the canonical code robustness. Given the dependence of protein function on the tridimensional structure conformation, they took into consideration the function of amino acids in their specific structural environment, analyzing the role of each amino acid in inner or outer regions of the protein structure. In their work [22], the Protein Data Bank ample information was used through a structural bioinformatics approach to search for unambiguous clues of the rationale of the canonical genetic code in assigning from one to six different codons for the different amino acids. For example, Leu and Arg offer a clear clue, since both appear in the canonical code with six assigned codons, therefore with a high protection from translational errors, and those also appear as the most abundant amino acids in protein-protein and protein-nucleic acid interactions.

Our work is focused on the error minimization theory. Previously, we have used a Genetic Algorithm (GA), as a search method for finding better adapted codes than the canonical one [23]. The GA provides a global search in the fitness landscape associated with the adaptability of possible hypothetical codes, allowing to obtain clues about the difficulty to obtain better optimized codes. Moreover, we also employed a model of alternative codes which reflects the known codon reassignments [24]. In line with the engineering approach, our results with simulated evolution revealed that the canonical code is far from the best possible optimized codes. Extending our previous work [23], Oliveira et al. [25] also proposed a multiobjective approach since two or more objectives were simultaneously optimized. They used as objectives code robustness against mutations, considering the changes in the polar requirement of amino acids (objective 1), and code robustness with respect to the hydropathy index or molecular volume changes under mutations of possible hypothetical codes (objective 2). The comparison between the evolution with only one objective and the use of a multiobjective evolutionary algorithm shown that the multiobjective alternative obtains optimized solutions closer to the canonical genetic code. Moreover, Oliveira and Tinós [26] proposed a function which uses entropy with the aim to increase the variability in the number of codons assigned to amino acids. That is, the effects against mutations of a code should be minimized while its entropy should be maximized. With this consideration, their results also indicate that the canonical genetic code is slightly better optimized with respect to not using

the entropy term. Also, Błażej et al. [27], inspired by our work with the adapted GA [23], analyzed the effectiveness of using various combinations of mutation and crossover probabilities under three models of the genetic code, assuming different restrictions on its structure.

In evolutionary computing, the so-called "fitness sharing technique" [28] is a "niching" method that allows the evolutionary algorithm search to be simultaneous performed in different areas (niches) corresponding to different local (or global) optima, that is, the technique permits the identification and localization of multiple optima in the search space. It should be noted that the optimization of the code adaptability turns into a minimization problem, where the "code fitness" or adaptability is inversely related to the error cost of a code (the more robust a code is against base mutations, the larger the fitness and the lower the error cost, "Methods" section). In the present work, the fitness sharing technique is introduced into the evolutionary algorithm, which allows the extent to which the canonical genetic code is in an area corresponding to a deep local minimum to be easily determined, even in the high dimensional spaces considered.

It is clearly not difficult to discern whether the canonical code is in a local minimum regarding the fitness landscape associated with the code adaptability. It is only necessary to consider the adaptability level in its neighborhood, that is, inspecting the fitness landscape in its close neighborhood. For example, Novozhilov et al. found that "The standard genetic code appears to be a point on an evolutionary trajectory from a random point (code) about half the way to the summit of the local peak" [21]. The fitness landscape is clearly rugged [21], but the question to answer in this paper is the following: is the standard genetic code in an area corresponding to a deep and separated local peak in the vast fitness landscape? Moreover, another related aspect is about the possible multimodal nature of the fitness landscape. That is, does the landscape, even with its rugged nature with many non-deep peaks, present localized areas of high fitness (adaptability) separated by low fitness barriers?

The answers to the questions are relevant since many previous works and authors discussed the possibility of the location of the standard genetic code in a local minimum (or close to it), regarding error cost, to explain its non-optimum adaptability. That information about the general surface of the fitness landscape could provide clues about the difficulty of the possible evolution of the canonical genetic code. Since an exhaustive search of the landscape is not possible, evolutionary computing was used to search in the promising areas of the fitness landscape, incorporating the aforementioned useful technique in evolutionary algorithms (fitness sharing) in order to obtain clues about the multimodal nature of the fitness landscape and the relative depth of its peaks.

In the rest of the paper the "Methods" section details the definitions of the alternative codes, their encoding in the GA population, the GA operators, the fitness definition in the landscape of such alternative codes, the fitness sharing technique as well as the measures used to quantify the canonical genetic code adaptability level. The "Results" section expounds the experiment results when the fitness sharing technique is introduced in the GA. Finally, the last sections present a discussion of the results and final conclusions.

Methods
Generation of variant genetic codes
Two possibilities of hypothetical codes were considered. The first possibility reflects the current genetic code translation table and is the most used in previous works [10, 12, 13]. When hypothetical codes were generated, two restrictions were considered:

1. The codon space (64 codons) was divided into 21 nonoverlapping sets of codons observed in the standard genetic code, each set comprising all codons specifying a particular amino acid in the standard code. Twenty sets correspond to the amino acids and one set to the 3 stop codons.
2. Each alternative code is formed by randomly assigning each of the 20 amino acids to one of these sets. The three stop codons remain invariant in position for all the alternative codes. Moreover, these three codons are the same stop codons of the standard genetic code (UAA, UAG and UGA).

This conservative restriction, which maintains the pattern of synonymous coding found with the standard genetic code, controls, as indicated by Freeland [29], possible biochemical restrictions on code variation and the level of redundancy inherent in the canonical code [11]; or, as stated by Novozhilov et al. [21], "The premise behind this choice is that the block structure of the code is a direct, mechanistic consequence of the mode of interaction between the ribosome, mRNA, and the cognate tRNA".

Although these authors (Novozhilov et al. [21]) indicate that codes with different block structures are not unviable or impossible "but they are likely to be substantially less fit than those with the canonical block structure", we used a second possibility with the definition of hypothetical codes with only one restriction: three codons for the stop signal are only imposed. The aim of the introduction of this last possibility, also used by Di Giulio et al. [18], is a comparison between the restrictive and non-restrictive hypothetical codes in terms of optimal values that can be obtained and in terms of location of the canonical genetic code in the fitness landscape associated with these two genetic code models.

Genetic algorithm adapted to the problem
Evolutionary computing was used for searching for optimal codes. A classical GA [30] with ad hoc operators for our problem was implemented [23]. The genetic population encodes possible hypothetical codes, whereas the fitness function is associated with the robustness against base mutations in each code. These aspects are detailed in the following subsections.

Encoding
Each individual of the genetic population must encode a hypothetical code. In our solution, in the case of non-restrictive codes, each individual has 64 positions, which correspond to the 64 codons, and each position encodes the particular amino acid associated with the codon (or the stop signal). As in [18], the stop signal is defined by three codons in each possible code.

In the case of restrictive codes, each individual has 20 positions, which correspond to the 20 codon sets, and each position encodes the particular amino acid associated with the codon set. In the encoding of a possible code, there is not a genotype position for the stop signal, since, as mentioned previously, the same codon set of the standard genetic code was used in all the individuals to define the stop signal.

With the non-restrictive codes, the individuals of the initial population correspond with random assignments of amino acids and the stop signal to the 64 codons, ensuring that all individuals encode, at least in one position, the 20 amino acids, in addition that three codons encode the stop signal. In the case of restrictive codes, the initial individuals are defined by random assignments between the 20 amino acids and the 20 codon sets.

Genetic operators
In the case of non-restrictive codes, a mutation operator and a swap operator were used. A mutation changes the amino acid encoded in each of the 64 positions, with a mutation probability, to a different one. The mutation does not operate if the amino acid to mutate is the only one in the whole code. These mutations simulate the possible errors in the transcription process from DNA to RNA and in the translation process when incorrect transfer RNAs join a given codon of the messenger RNA. From our application point of view, it is the operator that varies the number of codons associated with a particular amino acid.

The other genetic operator is a swap operator which interchanges the contents of two genotypic positions, that is, once two positions are randomly selected, the amino acids (or stop signal) codified by the two respective codons are swapped. The bottom part of Fig. 1 shows the basic functioning of these operators.

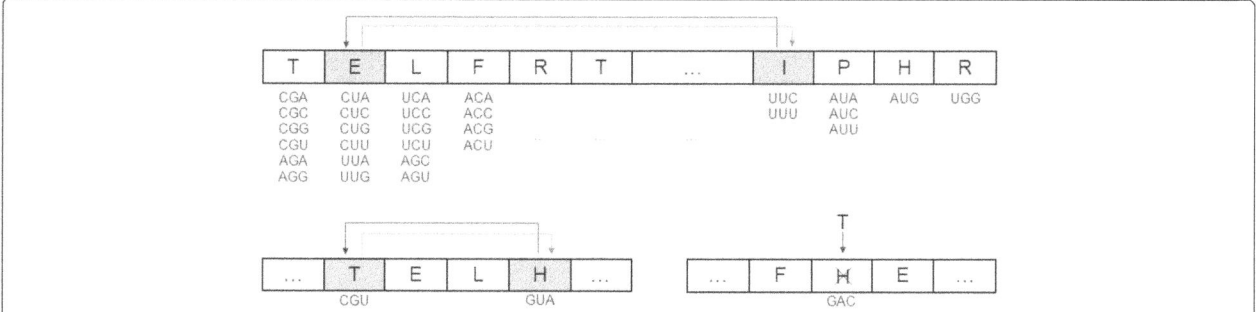

Fig. 1 Genetic operators. *Upper part*: Encoding of a genetic code and functioning of the swap operator with the restrictive codes. *Bottom part*: Swap operator (*left*) and mutation operator (*right*) used with the unrestrictive codes

The two operators guarantee that the 20 amino acids are always represented in the individuals. Other operators, such as the classical crossover operator, do not guarantee this important restriction.

In the restrictive codes case, as commented previously, each individual has 20 positions encoding the particular amino acid associated with a codon set present in the canonical code. As also mentioned, a fixed number of three codons are used for the stop label, which are the same used in the canonical code.

The GA only uses a swap operator with the restrictive hypothetical codes. As in the previous case, the operator interchanges the contents of two randomly selected positions (codon sets). The upper part of Fig. 1 shows the encoding of a given code and how this operator works.

Finally, as the selection operator, the tournament selection was used with both hypothetical codes. The operator selects the best individual in a window of randomly selected individuals from the population. Hence, the size of the window determines the required selective pressure. Moreover, elitism of the best individual was used; that is, this individual is retained in the next generation without changes.

Fitness function and optimality measure
The measure applied, for example, by Haig and Hurst [12] and Freeland and Hurst [10] to quantify the relative efficiency of any given code was used as fitness function. The measure calculates the mean squared (MS) change in an amino acid property resulting from all possible changes to each base of all the codons within a given code. Any one change is calculated as the squared difference between the property value of the amino acid coded by the original codon and the value of the amino acid coded by the new (mutated) codon. The final error is an average of the effects of all the substitutions over the whole code. Therefore, the error $\triangle$ (MS) is defined as:

$$\triangle^2 = \frac{\sum_{i,j} w_{i,j} \left(X_i - X_j\right)^2}{\sum_{i,j} N_{ij}} \qquad (1)$$

where N_{ij} is the number of times the $i - th$ amino acid changes into the $j - th$ amino acid, and X_i is the value of the amino acid property of the $i - th$ amino acid. $w_{i,j}$ is a weight associated with each possible mutation in each letter, which is set as 1 in the simple MS measure when an equal transition/transversion bias is considered. The changes from and to "stop" codons are ignored, while synonymous changes (the mutated codon encodes the same amino acid) are included in the calculation. Thus, the GA works with a minimization problem, where the MS measure is the objective to minimize by the GA operators. Therefore, as commented in the introduction, the adaptability or code fitness is inversely related to this error cost measure, since the lower the MS value the better the adaptability.

Like most authors, we have used the polar requirement as the amino acid property. The property can be considered as a measure of hydrophobicity and it was introduced by Carl Woese as a measure for the polarity of an amino acid, which is defined as a partitioning coefficient of an amino acid in a water/pyrimidine system [31].

Moreover, the previous equation can take into account the relative frequencies of transition/transversions mutations as well as mistranslations in the different bases. As stated by Freeland [29], the unequal chemical similarity of the 4 nucleotides to one another means that transition errors (substitution of a purine base into another purine, or a pyrimidine into another pyrimidine, i.e., $C \leftrightarrow T$ and $A \leftrightarrow G$) occur more frequently than transversions (interchange of pyrimidines and purines, i.e. $C, T \leftrightarrow A, G$).

To quantify the relative frequencies of mutations, we employed the rules from [10] used to consider the empirical data, which are summarized as:

1. Mistranslation of the second base is much less frequent than the other two positions, and mistranslation of the first base is less frequent than the third base position.
2. The mistranslations at the second base appear to be almost-exclusively transitional in nature.

3. At the first base, mistranslations appear to be fairly heavily biased toward transitional errors.
4. At the third codon position, there is very little transition bias.

Table 1 summarizes the quantification of mistranslation used to weigh the relative efficiency of the three bases. The weights $w_{i,j}$ in Eq. 1 correspond with the particular weights in Table 1. Therefore, the MS calculation takes into account those rules and, following the same terminology of Freeland and Hurst [10], we term the MS variant as tMS. For example, the MS value of the canonical code is 5.19 whereas its tMS value is 2.63.

As commented previously, the "engineering approach" compares the standard genetic code with the best possible alternative. The approach uses a "percentage distance minimization" (p.d.m.), which determines code optimality on a linear scale, as it is calculated as the percentage in which the canonical genetic code is in relation to the randomized mean code and the most optimized code. It is therefore defined as:

$$p.d.m. = \frac{\triangle_{mean} - \triangle_{code}}{\triangle_{mean} - \triangle_{low}} \cdot 100 \qquad (2)$$

where $\triangle_{mean}$ is the average error value, obtained by averaging over many random codes, and $\triangle_{low}$ is the best (or approximated) $\triangle$ value.

The measure can be interpreted as the optimization level reached during genetic code evolution [20]. For example, as previously indicated, Di Giulio et al. [18] reported a p.d.m. value of 72.7% in the case of codes with only amino acid permutations in the 20 sets of codons (restrictive codes), using a simulated annealing technique for obtaining the value of the best possible code, whereas we reported a p.d.m. value of 71% [23], using a GA for searching for the best possible code.

Fitness sharing
Fitness sharing is a classical technique in evolutionary computing for dividing the population into different subgroups according to the similarity of the individuals. In this present work this technique is incorporated into the GA. This fitness sharing concept was introduced by Holland [32] and extended, for example, by Goldberg and Richarson [28]. The shared fitness for the ith individual is defined as:

Table 1 Quantification of translational errors to measure the relative efficiency of a code (tMS)

Combined weighting	First base	Second base	Third base
For transitions	1	0.5	1
For transversions	0.5	0.1	1

$$f_{shared}(i) = \frac{f_{original}(i)}{\sum_{j=1}^{N} sh(d_{ij})} \qquad (3)$$

where the sharing function is calculated as:

$$sh(d_{ij}) = \begin{cases} 1 - \left(\frac{d_{ij}}{\sigma_{share}}\right)^{\alpha} & if\ d_{ij} < \sigma_{share} \\ 0 & otherwise \end{cases} \qquad (4)$$

being d_{ij} the distance between individuals i and j, σ_{share} the sharing radius, N the population size and α a constant called the sharing level.

In this application, the distance d_{ij} is measured by taking into account the difference in polar requirement between the amino acids encoded in the same positions by code i and code j of the population. It is defined as the root squared deviation between both codes:

$$d_{ij} = \frac{\sqrt{\sum_{k=1}^{L} (X_{ik} - X_{jk})^2}}{Max\_RSD} \qquad (5)$$

where Max_RSD is the maximum root squared deviation between two possible codes, taking into account the largest and lowest polar requirement values of amino acids (13 in Asp and 4.8 in Cys). The index k refers to a genotype position and L stands for the length of the genotypes in the individuals (20 for restrictive codes and 64 for unrestrictive codes). This ensures that distances are always in the range [0,1]. This procedure is the same in both hypothetical codes, except that in the case of unrestrictive codes the genotype positions where one of the codes encodes a stop signal are ignored (as in Eq. 1). Note that this definition of distance gives more information about the closeness of two codes than a simple calculation of how many different amino acids are encoded in the same position in the two codes. For example, in this last case, a simple swap of two amino acids in a code (distance 2 regarding different encoded amino acids) can correspond to different distances regarding Eq. 5, depending on the polar requirement values of the swapped amino acids.

Finally, if the application requires the minimization of the fitness (as in our case with MS), instead of its maximization, the formula in Eq. 3 turns to be a multiplication between the two terms ($f_{shared}(i) = f_{original}(i) \cdot \sum_{j=1}^{N} sh(d_{ij})$). Therefore, the fitness sharing technique increases the objective to minimize (MS value) in densely populated regions.

This way, fitness sharing modifies the search landscape by reducing the payoff in densely populated regions. The main drawback of the technique is its complexity, $O(N^2)$, because of the calculation of inter-distances. On the contrary, the important property is that fitness sharing tends to encourage searches in unexplored regions of the space and favors the formation of stable subpopulations [33].

Figure 2 shows an example with a multimodal function commonly used as benchmark in evolutionary computing. This function [33], defined on [0,1], consists of five unequally spaced peaks of nonuniform height. Maxima are located at approximate x values of 0.080, 0.247, 0.451, 0.681 and 0.934. A classical genetic algorithm was run to obtain the value that maximizes that function. A simple mutation operator (the encoded parameter x changes to a random close value) and a tournament selection were used. The population size was 100 and the probability of

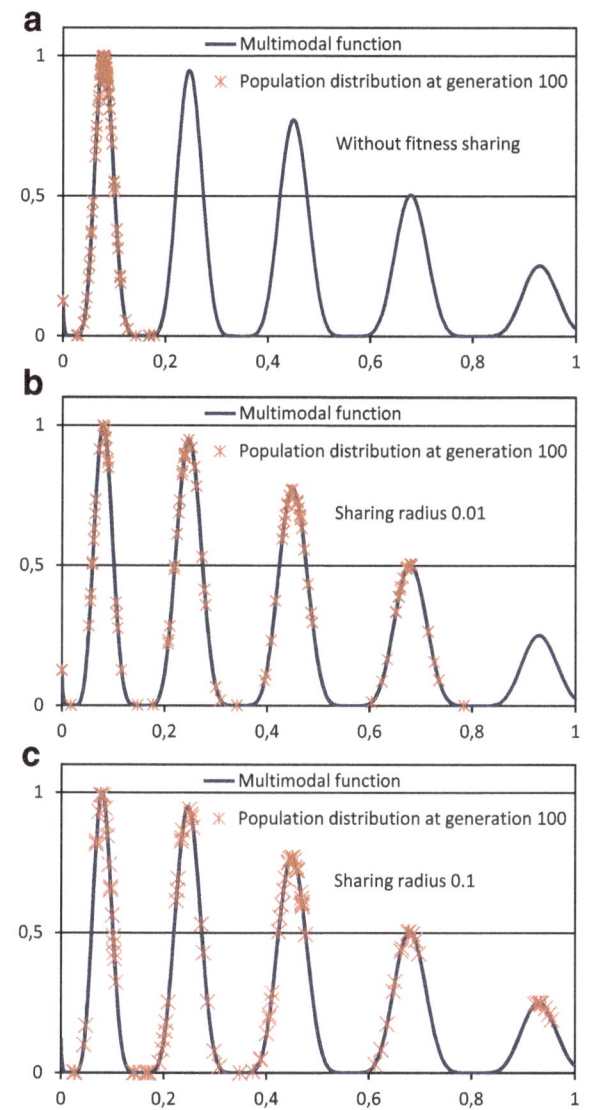

Fig. 2 GA final population distribution to maximize a multimodal function. The function has unequally spaced peaks of nonuniform height. Final population, at generation 100, in 3 cases: **a** Without fitness sharing. **b** Using fitness sharing with $\sigma_{share} = 0.01$, where the population is clustered around most of the peaks of the multimodal function. **c** Using fitness sharing with $\sigma_{share} = 0.1$, where the population is clustered around the different peaks of the multimodal function

mutation 0.25, running the GA across 100 generations. Without fitness sharing (Fig. 2a), the population tends to progressively move to the highest peak, even with the low selective pressure applied (a tournament size of 3%). However, with the introduction of fitness sharing ($\alpha=1$ in Eq. 4), now the population tends to converge at the niches of high fitness, as shown in Fig. 2b and Fig. 2c. Now the individuals are distributed in the peaks with a number that depends on the relative fitness of each peak. Using a higher σ_{share} (Fig. 2c, $\sigma_{share}=0.1$) the population tends to be more expanded, so it is more difficult to leave a local maximum with respect to the use of a lower σ_{share} (Fig. 2b, $\sigma_{share}=0.01$).

Figure 3 shows another example with a simple function and the same GA setup. The objective is to obtain the only minimum of the parabola. This is obviously a toy example, but the interest is to show in this function the effect of the introduction of fitness sharing that will be useful for our application. The upper part (Fig. 3a) shows the straightforward convergence of the population towards the global minimum without the use of fitness sharing. However, the introduction of fitness sharing ($\alpha=1$) tends to uniformly distribute the population around the minimum (Fig. 3b), with a distribution that depends on the sharing radius (σ_{share}). For example, with the largest sharing radius ($\sigma_{share} = 0.1$) in Fig. 3c, the population tends to be distributed in the whole range considered for the encoded parameter x ([-1,1]).

That is, the interest of the fitness sharing technique is to search in all the promising areas with high fitness found in the landscape, performing a better exploration of the search space. However, the introduction of the technique also has other consequence, because it provides clues about the fitness landscape. In the first chosen example (multimodal function), the distribution of the population reveals the multimodal nature of the fitness landscape, indicating the existence of several local maxima. In the second chosen example, the population is not clustered around local minima, and therefore the population is uniformly distributed around the global minimum and according to the sharing radius. Note that this information would not be obtained without the introduction of fitness sharing. These considerations are therefore going to be taken into account in the analysis of the (high dimensional) fitness landscape when searching for hypothetical codes with optimized adaptability.

Results

Evolutionary algorithm setup

The implemented GA, with the incorporation of fitness sharing, was tested by searching for optimized codes, using the two code models explained in the previous section. The GA parameters for the different experiments are: population size of 1000 individuals, mutation

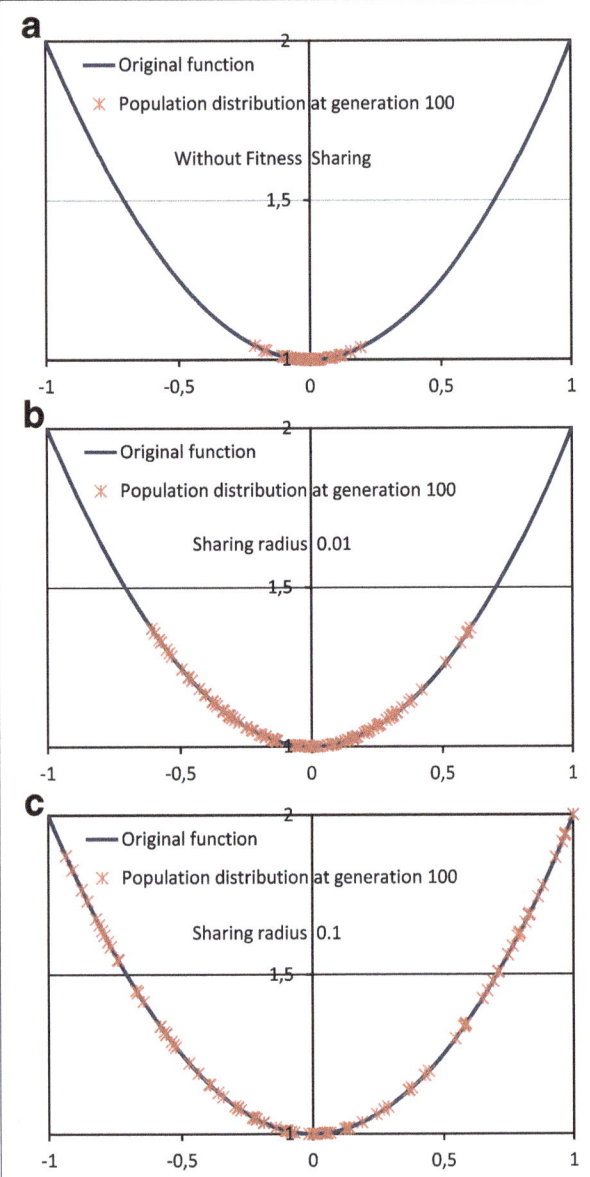

Fig. 3 GA final population distribution to minimize the function $1+x^2$. Final population, at generation 100, in 3 cases: **a** Without fitness sharing. Most of the individuals are close to the optimal value. **b** Using fitness sharing with $\sigma_{share} = 0.01$, with the population expanded around the minimum. **c** Using fitness sharing with $\sigma_{share} = 0.1$, with the population more expanded around the minimum

probability of 0.01 and a swap probability of 0.5. As explained, the restrictive model only uses the swap operator to interchange the 20 amino acids among the 20 codon sets. On the contrary, both operators are used with the non-restrictive model, where the mutation allows changing the number of codons assigned to an amino acid. Tournament selection with a tournament size of 3% of the population was used, which provides a low selective pressure. These genetic parameters are the same as those used

in our previous work [23], selected to provide an appropriate balance between exploration and exploitation in the GA search.

Regarding fitness sharing, the value of parameter α (Eq. 4) was set to 1 as it is sufficient to generate the possible clustering of the population into different niches. The sharing radius (σ_{share}) was varied to observe its effect on the evolution of possible codes, whereas d_{ij} was calculated using Eq. 5 ("Methods" section), which considers the root squared deviation between code i and code j of the population, taking into account the polar requirement of the amino acids encoded in each genotype position.

Evolution of restrictive codes

In a first experiment, the evolutionary algorithm was used to search for optimized codes, using the model of restrictive codes. Figure 4 shows the fitness (MS) evolution through 100 generations of the genetic algorithm. Figure 4 includes the evolution without fitness sharing and with two values for the parameter sharing radius ($\sigma_{share} = 0.01$ and $\sigma_{share} = 0.1$). The graphs in Fig. 4 are an average of 10 independent runs of the GA, beginning with different random initial populations. It is not easy to determine the appropriate values to use for the parameter σ_{share}, since it requires a previous knowledge about the landscape surface to easily obtain the possible clustering of the population into niches in a multimodal fitness landscape. Since we have no such a priori knowledge, we experimented with different values for the parameter σ_{share}, starting with a low value ($\sigma_{share} = 0.01$) and also using a larger value ($\sigma_{share} = 0.1$) of an order of magnitude.

The evolution without fitness sharing shows that it is easy to discover better adapted codes than the canonical one. In about 50 generations, an average best value of $MS = 3.50$ is obtained (by averaging the best values of the

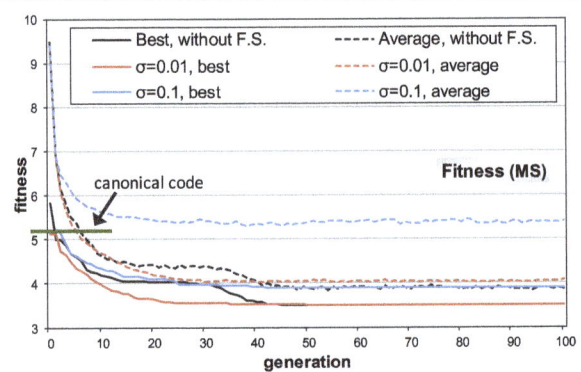

Fig. 4 MS evolution through generations with the restrictive codes. All graphs are an average of 10 independent GA runs, without fitness sharing and fitness sharing with two values for the parameter σ_{share} (sharing radius). The graph includes the canonical code MS value for comparison (*horizontal line*)

10 independent runs). The p.d.m. value is 71%, taking into account the MS value of the canonical code ($MS = 5.19$), the average value of the random codes at the initial generation of the GA and the best value in all the GA runs [23]. Even the average fitness of the population is lower with respect to the MS value of the canonical code. This shows that it is very easy for the GA to discover better adapted codes with respect to the canonical code, which denotes that the canonical code can be adapted but it is clearly far from the best possible adapted code.

When fitness sharing is considered in the GA evolution with a low sharing radius ($\sigma_{share} = 0.01$), the fitness progression is similar but more continuous. The reason is that fitness sharing tends to maintain the individuals at least with a distance of 0.01 between them, so it is more difficult for many individuals to correspond to the same solution, as can occur without the use of fitness sharing, even with the low selective pressure applied. With a larger sharing radius ($\sigma_{share} = 0.1$), the evolution of better codes is logically more difficult, although best values than the canonical code are obtained. The average fitness is now greater with respect to the canonical code and it is slightly variable through the evolutionary generations. This is because when several individuals fall in the same vicinity (easier with large values of σ_{share}), their fitness is penalized, and therefore in the next generation other individuals can be selected, which generates the variability in the average fitness. Table 2 summarizes the basic statistic information about the evolutions of Fig. 4, which shows the greater variability of the final results with larger values of the parameter σ_{share}.

Nevertheless, these previous graphs of evolution of the fitness do not provide information regarding the fitness landscape in the surroundings of the canonical genetic code. We want to discern whether the canonical code is in an area corresponding to a deep local optimum in the huge search space. Figure 5 helps to visualize this.

The columns in Fig. 5 correspond with a run of the GA with three cases: In column (a) the GA was run without fitness sharing whereas in columns (b) and (c) the

GA was run with fitness sharing with $\sigma_{share} = 0.01$ and $\sigma_{share} = 0.1$ respectively. In Fig. 5 the x-axis represents the distance of each encoded code in the population to the canonical code, whereas the y-axis corresponds to the MS value of each code. Therefore, Fig. 5 shows this correspondence between the MS value of each code and its corresponding distance to the canonical code in different generations of the GA.

It should be noted that if many individuals fell in the same local minimum, their distances to the canonical code would be similar as well as their MS values, that is, a cluster should be appreciated in the graph. The graphs show that the distances of the different codes of the population with respect to the canonical code vary, in most cases and after the initial generation, in the whole range between 0.4 and 0.55, which denotes that the hypothetical codes, more optimized than the canonical code in progressive generations, are far from the canonical code. This fact also indicates that the canonical code is not in an area corresponding to a deep local minimum, since all the individuals are far from it, without any individual close to it, as would occur if the canonical code was close to a deep local minimum (as explained in the "Methods" section, "Fitness sharing" subsection). The analyses with different values of σ_{share}, with a sweep of σ_{share} in an ample range, between 0.001 and 0.5 (not shown in the Figures), indicate the same conclusions.

It would be useful that an individual of the GA fell close to the canonical code, since it would also provide knowledge regarding whether the canonical code is in an area corresponding to a deep local peak. Since this is very difficult with the limited number of individuals of the population, in these GA runs, the canonical code was introduced in the initial population. This allows to test whether the canonical code can "survive" in the evolutionary progress towards optimized codes. Thus, in the initial generation (with random codes except the canonical one), there is a point that corresponds to the canonical code with distance 0 and $MS = 5.19$ (with larger size in Fig. 5). The other random codes present different MS values, most of them with larger values than the canonical MS value. Nevertheless, at generation 25, the canonical code has disappeared from the population (in all GA runs), indicating that it is not competitive with other hypothetical codes with better MS values at that generation. This can be obvious without fitness sharing. However, with fitness sharing, if the canonical code was in an area close to a local minimum, it would be difficult for this code to disappear from the population, as explained previously ("Methods" section. Fitness sharing), since some individuals would remain in a niche that would correspond to that area where the canonical genetic code is located. The fact that the canonical code disappears, even with a large sharing radius, is a second piece of evidence that indicates that

Table 2 Summary of statistics regarding Fig. 4. The values are an average of 10 independent runs of the GA

		Average final value	Standard deviation
Without fitness sharing	Best fitness	3.50	0.01
	Average fitness	3.88	0.04
Fitness sharing, $\sigma_{share} = 0.01$	Best fitness	3.50	0.01
	Average fitness	4.07	0.04
Fitness sharing, $\sigma_{share} = 0.1$	Best fitness	3.89	0.12
	Average fitness	5.41	0.07

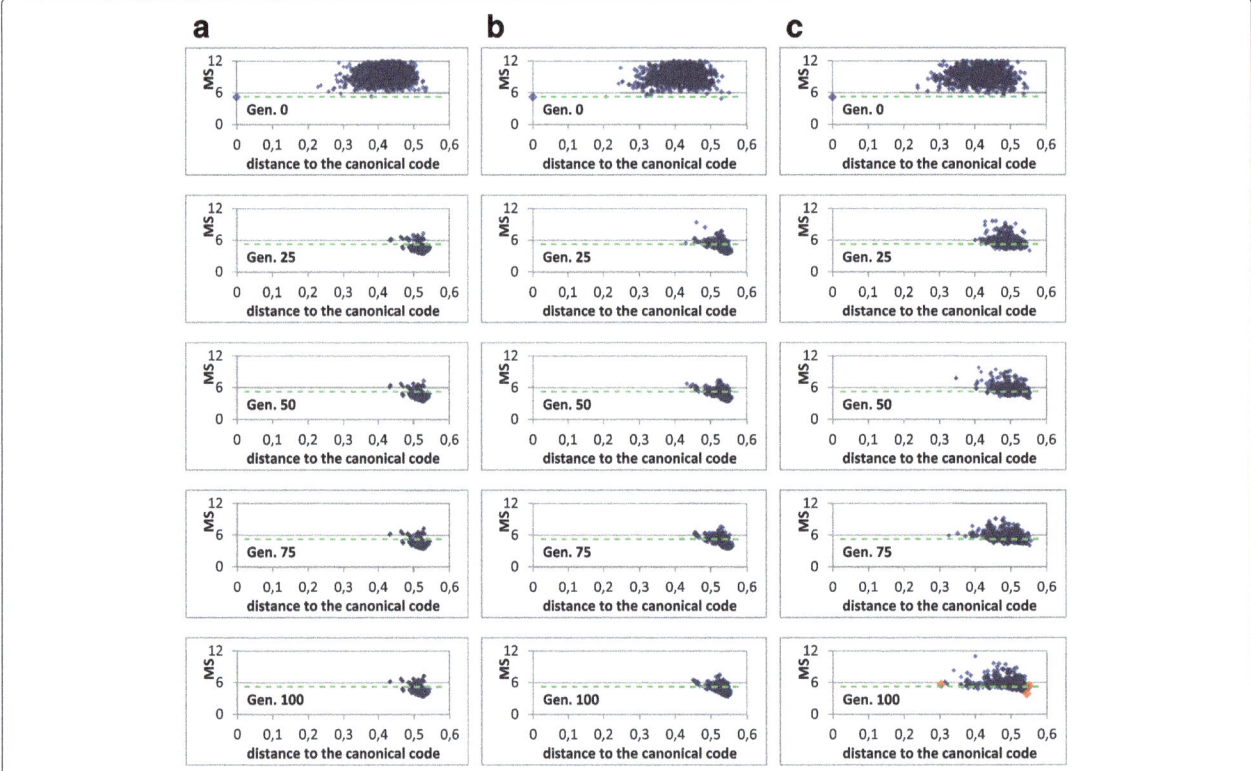

Fig. 5 MS value of the genetic population codes vs. their distances to the canonical code. The same graph is shown in different generations of a GA run, in 3 cases with the restrictive codes: **a** without fitness sharing, **b** fitness sharing with $\sigma_{share} = 0.01$ and **c** fitness sharing with $\sigma_{share} = 0.1$. The canonical code was introduced in the initial population in the 3 cases (point with larger size). The green dashed line represents the MS value of the canonical genetic code

the canonical code is not in an area corresponding to a deep local minimum.

To illustrate possible evolved codes and their distances, three alternative codes were selected from the final evolved ones in Fig. 5c (red points in the final population). Two of those codes correspond with the nearest code (*code 1*) and the furthest code (*code 2*) with respect to the canonical code (distances 0.31 and 0.55). The third one (*code 3*) is the code that has the best (minimum) MS value in that GA run of Fig. 5c. Figure 6 shows the codes with the assignments of amino acids to the different codons. In addition, each amino acid is represented with a gray level corresponding to its polar requirement: the brighter the gray level the higher the polar requirement. The stop signal is represented in white. Moreover, Fig. 6 includes the canonical code, which helps to visualize the distance between those codes and the canonical one. For example, the differences in polar requirement (gray level) between the amino acids in the same codon positions between the canonical code and *code 1* are lower with respect to the other cases, and *code 2* and *code 3* have assignments of amino acids with similar polar requirement in the same positions. Therefore, the distance (Eq. 5) between *code 2* and *code 3* is low. However, *code 3* has the minimum

MS value since it has amino acid assignments with similar polar requirement where possible mutations in the codon letters change the amino acids. For instance, *code 3* locates amino acids with very similar polar requirement in its third column (Tyr-5.4, Leu-4.9, Phe-5.0, Trp-5.2, Cys-4.8, Ile-4.9), since a mutation in the third codon letter can imply a change between such encoded amino acids.

Finally, the previous graphs show the distances of the encoded codes with respect to a "reference point" (the canonical code), but do not show how the different codes are far away from each other. It would also be interesting to know whether the final population is clustered around a small neighborhood of the search space or it is spread in the high dimensional landscape. For this, the inter-distances of individuals of the final population were calculated, and a histogram of these distances is plotted in Fig. 7. Inter-distances close to 0 mean the possibility of a cluster (niche) of the population around a local optimum. In Fig. 7 the *x*-axis is sampled in intervals of 0.01, that is, each of the 100 intervals in the *x*-axis specifies the number of inter-distances of the population in that interval. Although the distances are normalized in the interval [0,1] (Eq. 5), it is difficult to obtain inter-distances larger than 0.6, because it would imply large changes in

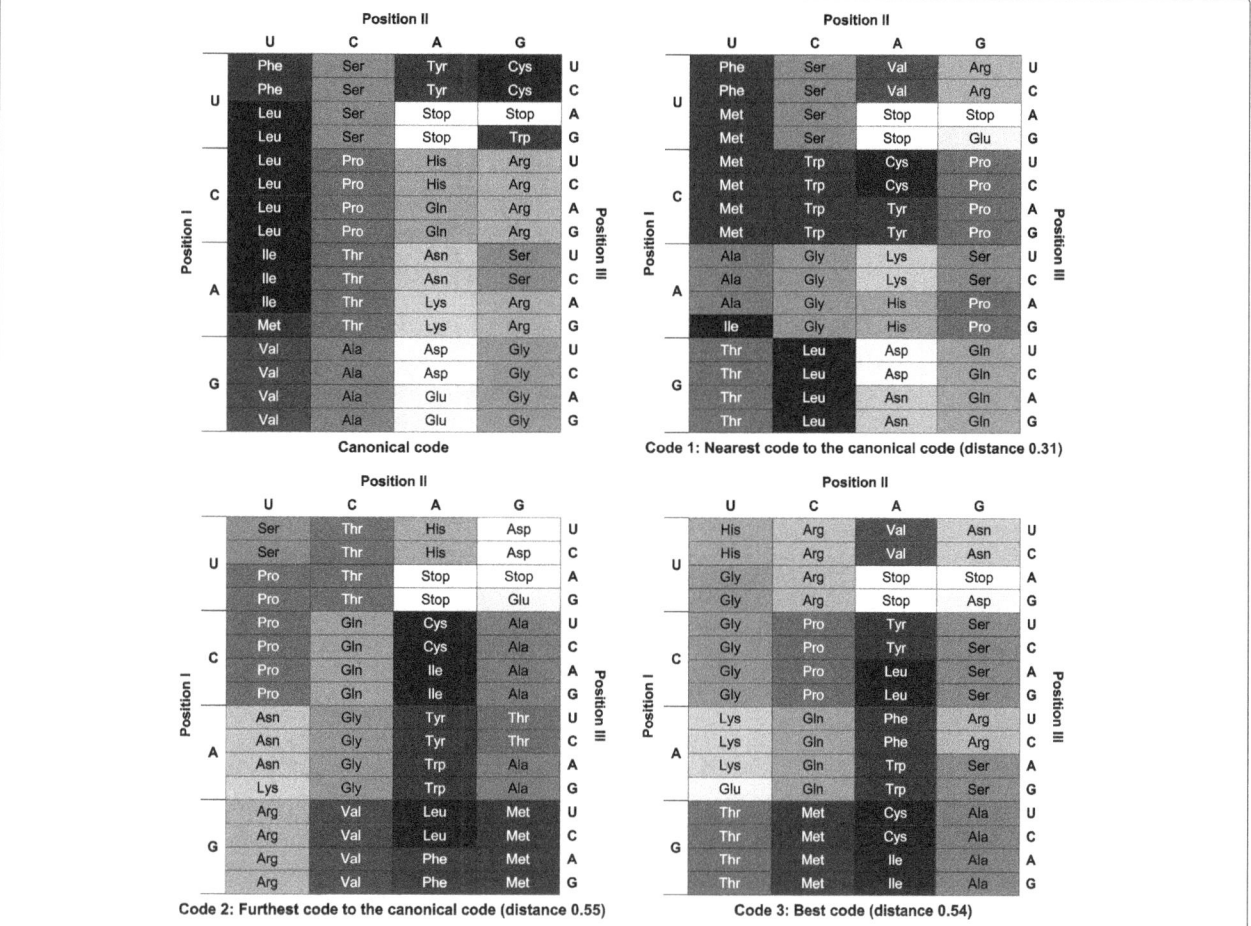

Fig. 6 Selected evolved codes. Three codes were selected from the final population in Fig. 5.c. *Code 1*: nearest code to the canonical code. *Code 2*: furthest code to the canonical code. *Code 3*: best code. The canonical code is included for comparison. $d_{Code1,Code2} = 0.40$, $d_{Code1,Code3} = 0.41$, $d_{Code2,Code3} = 0.18$, $MScode1 = 5.66$, $MScode2 = 5.49$, $MScode3 = 3.93$

the polar requirement of the encoded amino acids of the codes, and many amino acids have a similar value of polar requirement.

Figure 7 shows that the codes are further away from each other as the sharing radius becomes larger. Without fitness sharing (Fig. 7a), the individuals are closer to the best solution. In fact, at the final generation, many individuals correspond to the same solution (in this case the individuals with inter-distance 0). With the incorporation of fitness sharing and with the lower sharing radius (Fig. 7b, $\sigma_{share} = 0.01$) the population is slightly more expanded through the fitness landscape, with most of the inter-distances between 0.02 and 0.32. The main difference with respect to the case without fitness sharing is that now it is more difficult that two individuals correspond to the best obtained solution. Note that the histogram shows a low height in the first interval [0,0.01] of the x-axis; however, this does not imply that the inter-distances correspond to equal individuals. Moreover, if it is taken into account that using fitness sharing with

$\sigma_{share} = 0.01$, the average quality of the population (4.07, Table 2 when several GA runs are averaged) is close to the value without fitness sharing (3.88, Table 2), and even when the individuals are far away from each other, it means that there are many distant areas of the fitness landscape with better adapted codes than the canonical one. With a larger sharing radius (Fig. 7c, $\sigma_{share} = 0.1$), the expansion of the population in different and distant areas is more pronounced, as indicated now by the greater inter-distances, where most of them are in a continuous range between 0.14 and 0.46. It should also be noted that these inter-distances between the evolved codes are lower than the distances of those codes to the canonical one. It means that the canonical code is far from the area where most of the evolved codes are located. Therefore, the fitness landscape seems more like a vast space with a broad transition area to the optimum, in the sense that the results resemble the search of the global minimum in the parabola example of Fig. 3. The difference is that now the space surface can be rugged with

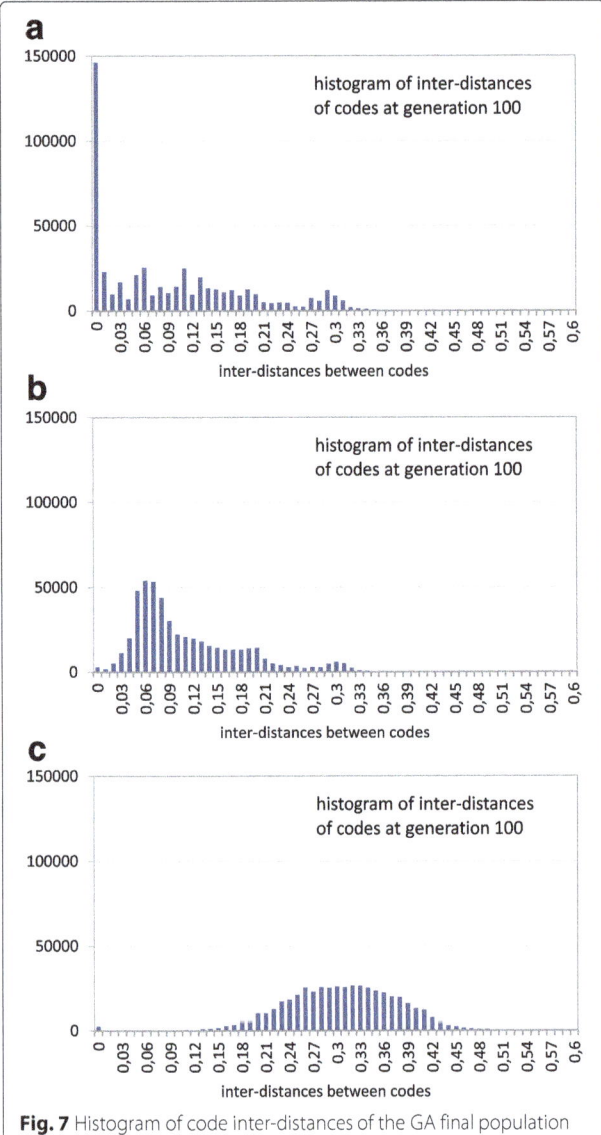

Fig. 7 Histogram of code inter-distances of the GA final population using the restrictive codes. The histogram corresponds to the final population (generation 100) in a GA run with three cases: **a** without fitness sharing, **b** fitness sharing with $\sigma_{share} = 0.01$ and **c** fitness sharing with $\sigma_{share} = 0.1$

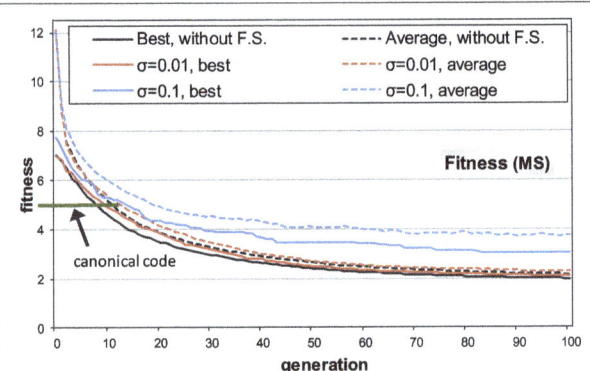

Fig. 8 MS evolution through generations with the unrestrictive codes. All graphs are an average of 10 independent GA runs, without fitness sharing and fitness sharing with two values for the parameter σ_{share} (sharing radius). The horizontal line shows the canonical code MS value

used. The evolution graphs correspond again to an average of 10 independent runs of the GA with different initial populations.

The same conclusions can be obtained with respect to the restrictive codes case. Now all the evolutions are more continuous, given the larger possibilities of codes and landscape areas to explore. The evolution without fitness sharing obtains a possible best code that is more optimized with respect to the use of restrictive codes. This is because the GA has more possibilities to search for adaptive codes against base mutations. Taking into account the average MS value of random codes and the best value of the GA runs, the p.d.m. value is 67% [23], meaning that the canonical code is less optimized (considering unrestrictive codes) with respect to the previous case (p.d.m. value 71% with restrictive codes). This is logical, since the restrictive codes impose constraints to obtain optimized codes. Table 3 summarizes the statistic information regarding the evolutions of Fig. 8, which shows a slight increase in the variability of the final results with larger values of the parameter σ_{share}.

many non-deep (in comparison with the depth of the best solutions) local peaks.

Evolution of non-restrictive codes

We repeated the analysis with the model of unrestrictive codes. The number of possible codes is now close to 10^{84} [9] with respect to the previous case ($2.43 \cdot 10^{18}$ possible restrictive codes). The huge increase in the possible codes will allow to check whether the same conclusions are obtained with respect to the previous case. Once more, Fig. 8 shows the evolution across 100 generations of the GA with the same setup as in the previous case, except that now both genetic operators (swap and mutation) are

Table 3 Summary of statistics regarding Fig. 8. The values are an average of 10 independent runs of the GA

		Average final value	Standard deviation
Without fitness sharing	Best fitness	1.97	0.02
	Average fitness	2.16	0.03
Fitness sharing, $\sigma_{share} = 0.01$	Best fitness	2.07	0.05
	Average fitness	2.25	0.05
Fitness sharing, $\sigma_{share} = 0.1$	Best fitness	3.11	0.25
	Average fitness	3.80	0.25

In the analysis of a variety of the best non-restrictive codes, the amino acids that appear in most of the codons have an intermediate value of polar requirement. This helps to minimize the MS error, when most of the mutation changes are among the intermediate values. It is the same idea expressed by Di Giulio [18], when the author used a simulating annealing algorithm to find optimized non-restrictive codes. As the author stated "by maximizing the number of synonymous changes in the code, it is reasonable to suppose that the objective function value is taken towards the absolute minimum" [18]. For example, the author obtained a best code with 42 synonymous codons to one amino acid, one codon to the remaining nineteen amino acids, in addition to the three codons for the termination meaning. However, in our case, the global search of the genetic algorithm finds optimized codes with a more balanced number of codes per amino acid, where only the amino acids with extreme values of polar requirement are associated with only one codon.

Figure 9 illustrates, in different generations of the evolutionary algorithm, the MS value of each code (y-axis) vs. the distance of each encoded code in the population to the canonical code (x-axis). The canonical genetic code was inserted in the initial population. Throughout the evolutionary process, the whole population is moving towards better values of adaptability (lower MS values). In fact, in the three cases considered, without fitness sharing (Fig. 9a) and with fitness sharing with two sharing radius (Fig. 9b and c), the average value of the population has a better value than the MS value of the canonical code (5.19), as previously seen also in Fig. 8. As in the previous case with restrictive codes, the canonical code does not survive in the first generations of the evolutionary process, even with the use of fitness sharing. This fact, together with the large distances of the optimized solutions at final generations with respect to the canonical code, are evidences for supporting the finding that, using the unrestrictive codes, the canonical code is clearly not located in a broad and deep peak; that is, the canonical code is not in a niche or promising area where some of the solutions of the genetic population could be refining their search. Again, the analyses with different values of σ_{share}, in an ample range between 0.001 and 0.5, present the same evidence, so these are not shown in the Figures.

The same previous analysis considering the inter-distances of the final population was performed in order to detect how far such final optimized solutions are, that is, how many areas are being simultaneously searched

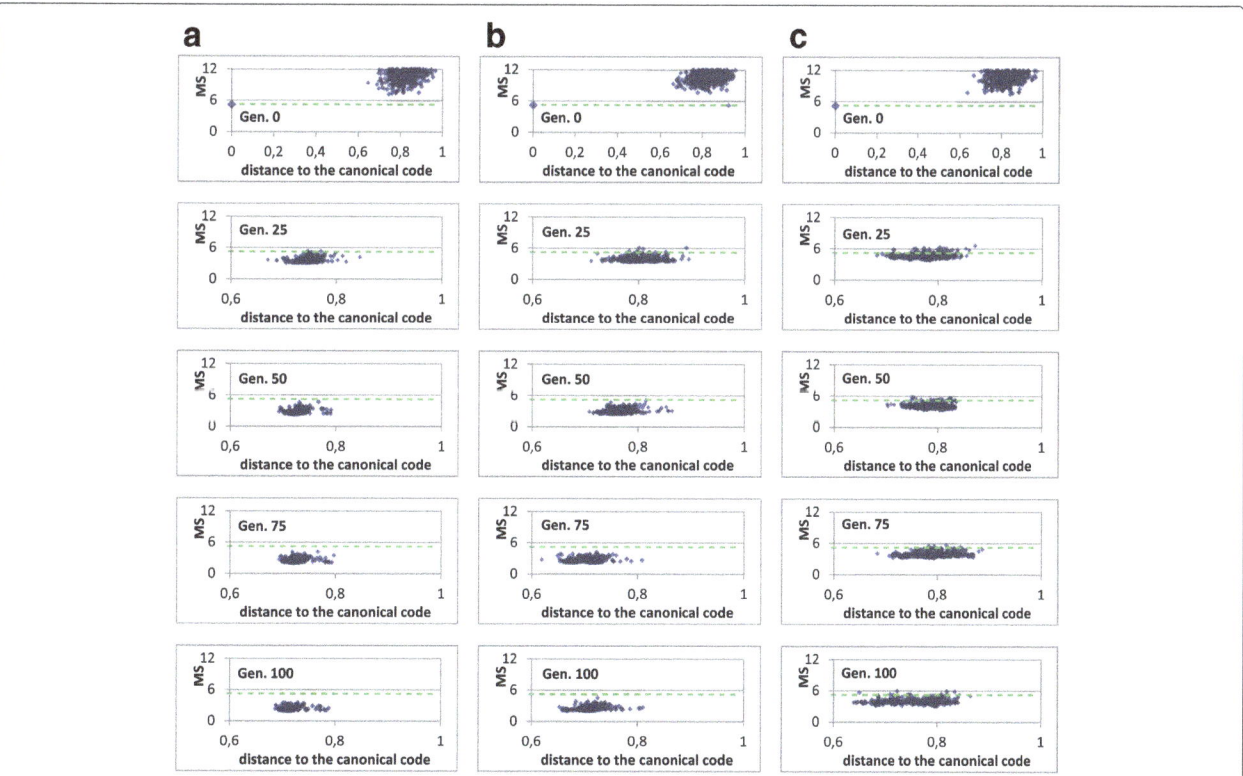

Fig. 9 MS value of the genetic population codes vs. their distances to the canonical code. The same graph is shown in different generations of a GA run, in 3 cases with the unrestrictive codes: **a** without fitness sharing, **b** fitness sharing with $\sigma_{share} = 0.01$ and **c** fitness sharing with $\sigma_{share} = 0.1$. The canonical code was introduced in the initial population in the 3 cases (point with larger size). The green dashed line represents the MS value of the canonical genetic code

even when the evolutionary process is ended. Note that the inter-distances can have greater values with respect to the previous case with restrictive codes, since the differences between the extreme values of the polar requirement can occur more times with the unrestrictive codes (the same amino acids with those extreme values can be encoded by many codons). Figure 10 shows that the codes are further away from each other as the sharing radius becomes larger. In Fig. 10a, without fitness sharing, some individuals correspond with the best solution (inter-distance 0). When fitness sharing is used, there are very few solutions corresponding with the best code. Figure 10b, using a low sharing radius ($\sigma_{share} = 0.01$), shows that the inter-distances are slightly larger, therefore the individuals are further away from each other with respect to the case without fitness sharing, that is, the final population is more expanded through the fitness landscape. With a larger sharing radius (Fig. 10c, $\sigma_{share} = 0.1$), the expansion of the population through the fitness landscape is clearly more pronounced, as can be seen with the high increase of the inter-distances, most of them being between 0.27 and 0.69. Even with the expansion of the population in distant areas of the fitness landscape, the average quality is better than that of the canonical code (Fig. 8). Consequently, the conclusion is again that the fitness landscape does not present a multimodal nature, since there is no niching or clustering effect of the population in promising areas.

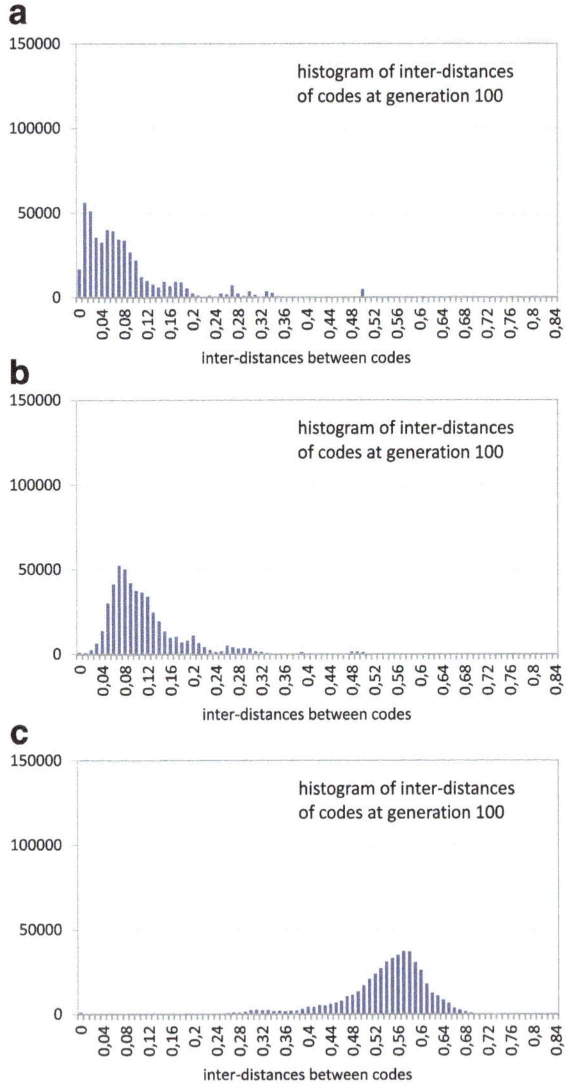

Fig. 10 Histogram of code inter-distances of the GA final population using the unrestrictive codes. The histogram corresponds to the final population (generation 100) in a GA run with three cases: **a** without fitness sharing, **b** fitness sharing with $\sigma_{share} = 0.01$ and **c** fitness sharing with $\sigma_{share} = 0.1$

Introduction of transition/transversion and translational biases

Previous works [10, 23] have demonstrated that the canonical code has better adaptability levels when the code fitness takes into account the different probabilities of transition and transversion mutations, together with the mistranslation of mRNA, which implies different probabilities of mutations in the three codon bases. For example, using tMS ("Methods" section), in [23], with the restrictive codes case, the p.d.m is 84%. The comparison of this value with respect to the use of MS as fitness (p.d.m=71%) means that the canonical genetic code is better adapted when the tMS calculation is taken into account. Consequently, a test was conducted to check whether these considerations in the tMS error measure imply a change in the fitness surface nature.

Figure 11 summarizes the results with the restrictive codes. It includes the analysis of tMS versus distances to the canonical genetic code, only at the final generation of the GA runs, in addition to the histograms of inter-distances of the final population from the evolutionary algorithm. As in the previous cases, a GA run without fitness sharing (Fig. 11a) and with fitness sharing (Figures 11b and 11c) were considered, and the canonical code was introduced in the initial population in the different GA runs.

The comparison with the previous cases using MS as objective to minimize (Figs. 5 and 7) shows that the inter-distances are quite similar when tMS is used. For example, the comparison between the histograms in Figs. 11 and 7 shows how the inter-distances are similar using tMS in the three cases (with and without fitness sharing) without significant variations. However, the distances of each code with respect to the canonical code are now lower. Using tMS these distances to the canonical code are between 0.16 and 0.5, with a larger interval with greater values of σ_{share}. In the case of using MS, these distances were between 0.3 and 0.55 (Fig. 5), also with greater intervals

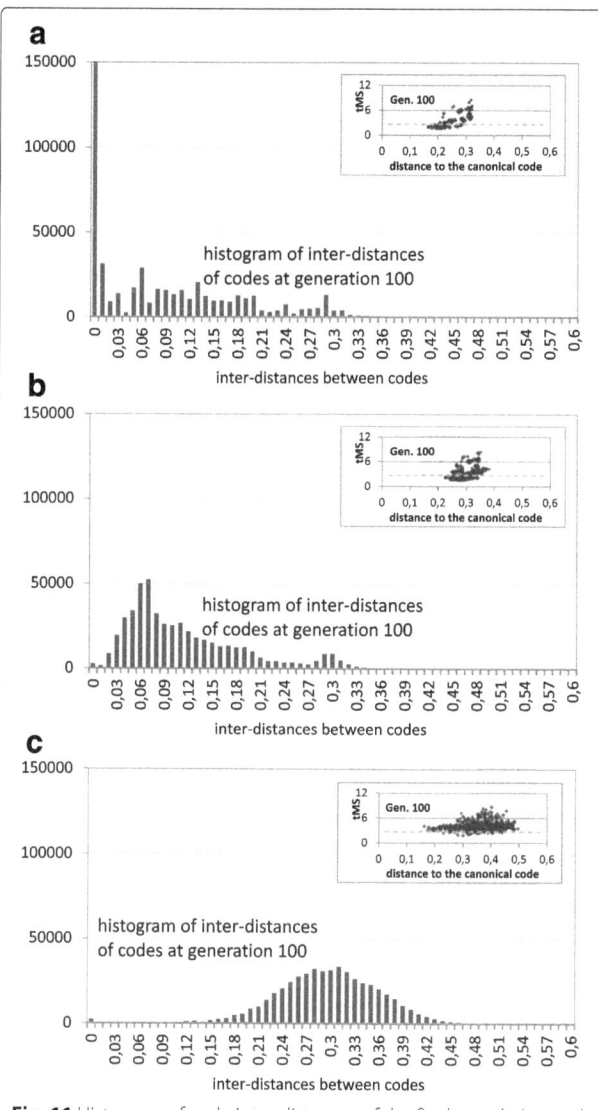

Fig. 11 Histogram of code inter-distances of the final population and using the restrictive codes with tMS. The insets show the tMS value of the hypothetical codes of the genetic population vs. their distances to the canonical code. The *green dashed line* represents the tMS value of the canonical genetic code. Each subfigure corresponds with the histograms of inter-distances of the final populations. The graphs correspond to the final population (generation 100) of a GA run in 3 cases: **a** without fitness sharing, **b** fitness sharing with $\sigma_{share} = 0.01$ and **c** fitness sharing with $\sigma_{share} = 0.1$

Figure 12 illustrates the same analysis with the unrestrictive codes. The histograms of Figs. 12 (tMS) and 10 (MS) are again very similar. However, there is a clear difference in the comparison of the distances of the individuals of the final population with respect to the canonical code (insets of Figs. 12 and 9). With tMS, the distances are between 0.4 and 0.8, whereas with MS those distances are approximately between 0.65 and 0.85, in both cases (tMS and MS) with a tendency of larger intervals with larger

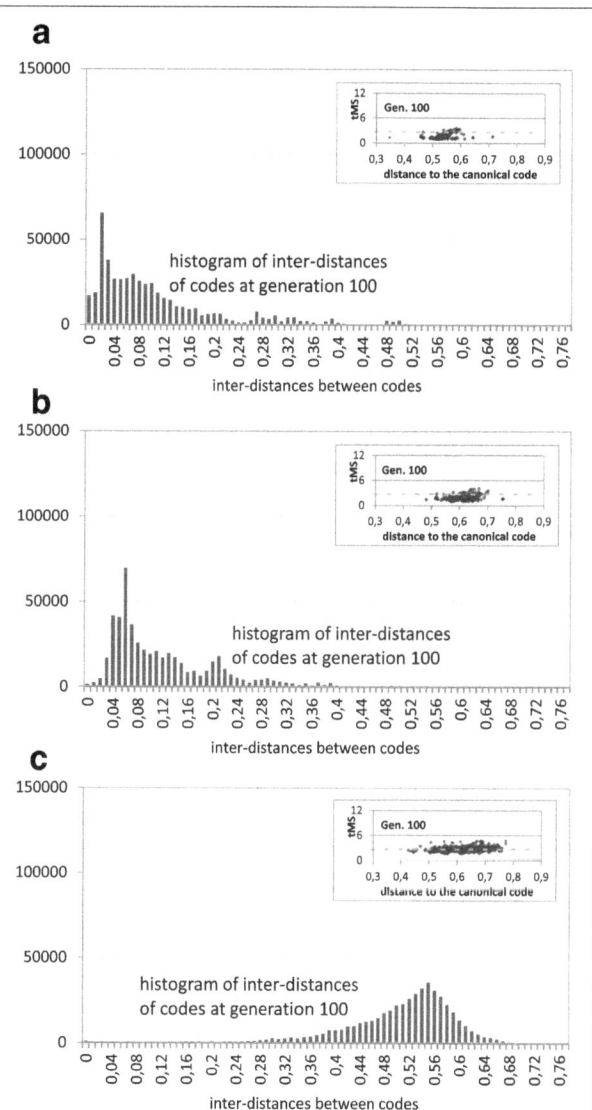

Fig. 12 Histogram of code inter-distances of the final population and using the unrestrictive codes with tMS. The insets show the tMS value of the hypothetical codes of the genetic population vs. their distances to the canonical code. The green dashed line represents the tMS value of the canonical genetic code. Each subfigure corresponds with the histograms of inter-distances of the final populations. The graphs correspond to the final population (generation 100) of a GA run in 3 cases: **a** without fitness sharing, **b** fitness sharing with $\sigma_{share} = 0.01$ and **c** fitness sharing with $\sigma_{share} = 0.1$

with larger values of σ_{share}. This means that the incorporation of tMS implies that the optimized codes are closer to the canonical code, showing again better adaptability of the canonical code considering tMS. However, even with the improvement of adaptability when considering tMS, the final solutions are far from the canonical code while the population is extended across an ample area of the fitness landscape, and without any presence of clear niches in the landscape.

values of σ_{share}. Thus, although the final codes are similarly extended in the search space using tMS or MS, the codes of the final population are closer to the canonical code using tMS, which is a further evidence of the better optimization of the code considering tMS.

Discussion

An adapted genetic algorithm for searching for possible alternative genetic codes better adapted than the canonical code was used in this work. The results of the GA clearly indicate that there are many alternative codes with a better adaptability than the canonical one, being the canonical code relatively far from the best possible adapted one. However, another question to explore is whether the canonical code is in an area corresponding or not to a deep local minimum in relation to the vast fitness landscape, which has been continuously discussed in previous works. This is the issue that has been explored in this work.

As indicated in the Introduction, this is not the same as discerning whether the canonical code is in a local peak (or close to it) regarding its immediate neighborhood. As stated by Novozhilov et al. "The fitness landscape of code evolution appears to be extremely rugged, containing numerous peaks with a broad distribution of heights, and the standard code is relatively unremarkable, being located on the slope of a moderate height peak" [21] (in their case, the authors weighted differently transition and transversion biases in the codon bases). Instead, our aim was to discern about the nature of the vast fitness landscape, inspecting whether it is a multimodal landscape with clear niches, with separated and deep broad MS peaks, together with the location of the canonical code in such a huge landscape.

As explained, due to the huge dimensionality of the search landscape, the incorporation of the fitness sharing technique in the evolutionary algorithm helps to determine and visualize whether the alternative codes, as well as the canonical code, are in an area corresponding to a deep local minimum.

The limitations of the fitness sharing technique are well-known, and basically entail that: i) Setting the dissimilarity threshold (sharing radius) requires a priori knowledge of how far apart the optima are and ii) The complexity per generation is $O(N^2)$ as a consequence of the distance calculations. Even with the complexity drawback, the technique was used because it helps to visualize the fitness landscape (the possible multimodal nature of the landscape). In order to overcome the first limitation, an analysis with different sharing radii was performed.

In the discussion about the location of the canonical code, the general idea is to consider that the canonical code was trapped in a local minimum. For example, according to Crick "there is no reason to believe, however,

that the present code is the best possible [...]. Instead, it may be frozen at a local minimum which it has reached by a rather random path" [1]. Or, as Knight et al. state "although search algorithms can sample billions of different codes, evolution is unlikely to have had similar opportunity given the extreme cost of changing an already functional code, and so we might either expect the code to be trapped at a local, rather than global, optimum" [34].

The engineering approach for obtaining better adapted codes was used in our work, since a computational search method was employed for obtaining possible better adapted codes than the canonical one. On the other hand, in the statistical approach, its main idea is that the standard code minimizes hydrophobicity errors far more than can be explained by chance. We are not in disagreement with this idea since the code is more optimized with respect to random codes, but at the same time our analysis indicates that the code is clearly far from possible optimal codes. Some authors within the statistical approach, such as Freeland [29], have argued against the engineering approach bases because the criticisms about the fact that the canonical code is far from optimal "overlook the fact that we know very little about the connectivity or accessibility of local and global optima for patterns of codon assignment". Nevertheless, the present study helps to visualize the possible formation of local minima in the fitness landscape, indicating that the canonical genetic code is not in an area of a broad and deep local minimum, since it is not captured in a niche with the fitness sharing technique.

Note that our objective was not to explain the possible evolutionary paths to the canonical code. Our objective was only to locate the canonical genetic code in the fitness landscape when possible hypothetical codes are considered, from an scenario of hypothetical codes without restrictions in the assignments to the consideration of restrictive hypothetical codes with the same codon set structure as the canonical genetic code. Moreover, the fitness landscape depends also on the definition of the fitness function, that is, how the adaptability level of a code is measured. We considered the basic property, the polar requirement of amino acids, the most used one in previous studies, in order to define the fitness landscape, since it is the main property for the folding of a protein and consequently to define its protein function. In this sense, Freeland is correct in the criticism "The evolutionary similarity of amino acids (meaning their substitutability within proteins) is unlikely to be perfectly represented by a single physiochemical measure (e.g. polar requirement) or indeed by any simple combination of two or three such indices" [29]. This analysis with other amino acid properties has been performed in several previous works [12, 23, 25, 35]. For example, when different amino acid properties (hydropathy index, isoelectric point and molecular volume, the same as those used by Haig and Hurst [12]) were employed, the results in [23] indicated that polar requirement is the property that provides the most significant evidence of error minimization. Thus, we have included only the most meaningful analysis with the polar requirement property.

Conclusions

The conclusions obtained in this study can be briefly summarized as follows:

1. The canonical code is better optimized with respect to random codes, when the effects of mutations are taken into account within the error minimization theory. However, the GA search indicates that the canonical code is far from the best possible codes.

2. When the fitness sharing technique is introduced in the GA search, it indicates that there are no clear niches in the vast fitness landscape, that is, localized areas of high fitness (low MS) separated by barriers of low fitness (high MS values). This is not in contradiction with the rugged nature of the fitness landscape. For example, Novozhilov et al. stated that a huge number of taller fitness peaks (with respect to the peak in which the canonical code is situated) exist in the landscape [21]. On the contrary, the results in this study denote that there are many connected areas (not clearly separated) with higher fitness than the canonical code, as inferred from the results of the distances of the optimized codes with respect to the canonical code and the inter-distances between the optimized codes of the GA final populations. That is, the fitness landscape is rugged but does not have a multimodal nature with clear and deep niches separated by fitness barriers. This explains why any search algorithm easily discovers better adapted codes than the canonical one.

Even when the canonical code shows better adaptability levels when the code fitness takes into account different weights for transition and transversion errors in the different codon bases, together with the mistranslational error weights in the three codon bases [10, 23], the same conclusion about the multimodal nature of the fitness landscape is obtained when such biases are considered.

3. The canonical code is clearly far away from those areas of higher fitness in the landscape. Given the non-presence of deep local minima in the landscape, although the code could evolve and different forces could shape its structure, the fitness landscape nature considered in the minimization theory does not explain why the canonical code ended its evolution in a location which is not an area of a localized deep MS minimum of the huge fitness landscape.

Abbreviations

GA: Genetic algorithm; MS: Mean squared; P.D.M: Percentage distance minimization; tMS: Mean squared considering transition/transversion and translational biases

Acknowledgements
None.

Funding
Ministry of Economy and Competitiveness of Spain (project TIN2013-40981-R) and Xunta de Galicia (project GPC ED431B 2016/035).

Authors' contributions
Both authors contributed equally in the writing of the manuscript and in the experimental design. Both authors read and approved the final manuscript.

Competing interests
The authors declare that they have no competing interests.

References
1. Crick F. The origin of the genetic code. J Theor Biol. 1968;38:367–79.
2. Knight RD, Freeland SJ, Landweber LF. Selection, history and chemistry: the three faces of the genetic code. Trends Biochem Sci. 1999;24:241–7.
3. Knight RD, Landweber LF. The early evolution of the genetic code. Cell. 2000;101:569–72.
4. Koonin EV, Novozhilov AS. Origin and evolution of the genetic code: the universal enigma. Life. 2009;61:99–111.
5. Wong JT. A co-evolution theory of the genetic code. Proc Nat Acad Sci USA. 1975;72:1909–12.
6. Di Giulio M. The origin of the genetic code: theories and their relationship, a review. Biosystems. 2005;80:175–84.
7. Chechetkin VR, Lobzin VV. Stability of the genetic code and optimal parameters of amino acids. J Theor Biol. 2011;2691(1):57–63.
8. Yockey HP. Information Theory, Evolution, and the Origin of Life. NY: Cambridge University Press; 2005.
9. Schönauer S, Clote P. How optimal is the genetic code? In: Frishman D, Mewes H, editors. Computer Science and Biology, German Conference on Bioinformatics (GCB 97). 1997. p. 65–67. http://www2.tcs.ifi.lmu.de/~clote/.
10. Freeland SJ, Hurst LD. The genetic code is one in a million. J Mol Evol. 1998;47(3):238–48.
11. Freeland SJ, Hurst LD. Load minimization of the genetic code: history does not explain the pattern. Proc R Soc. 1998;265:2111–9.
12. Haig D, Hurst LD. A quantitative measure of error minimization in the genetic code. J Mol Evol. 1991;33:412–7.
13. Gilis D, Massar S, Cerf NJ, Rooman M. Optimality of the genetic code with respect to protein stability and amino-acid frequencies. Genome Biol. 2001;2(11). https://www.ncbi.nlm.nih.gov/pmc/articles/PMC60310/.
14. Torabi N, Goodarzi H, Najafabadi HS. The case for an error minimizing set of coding amino acids. J Theor Biol. 2007;244(4):737–44.
15. Zhu CT, Zeng XB, Huang WD. Codon usage decreases the error minimization within the genetic code. J Mol Evol. 2003;57:533–7.
16. Marquez R, Smit S, Knight R. Do universal codon-usage patterns minimize the effects of mutation and translation error? Genome Biol. 2005;6(11):R91.
17. Di Giulio M. The origin of the genetic code. Trends Biochem Sci. 2000;25(2):44.
18. Di Giulio M, Capobianco MR, Medugno M. On the optimization of the physicochemical distances between amino acids in the evolution of the genetic code. J Theor Biol. 1994;168:43–51.
19. Freeland SJ, Knight RD, Landweber LF. Measuring adaptation within the genetic code. Trends Biochem Sci. 2000;25(2):44–5.
20. Di Giulio M. The extension reached by the minimization of the polarity distances during the evolution of the genetic code. J Mol Evol. 1989;29:288–93.
21. Novozhilov AS, Wolf YI, Koonin EV. Evolution of the genetic code: partial optimization of a random code for robustness to translation error in a rugged fitness landscape. Biol Direct. 2007;2:24. http://biologydirect.biomedcentral.com/articles/10.1186/1745-6150-2-24.
22. Gardini S, Cheli S, Baroni S, Di Lascio G, Mangiavacchi G, Micheletti N, Monaco CL, Savini L, Alocci D, Mangani S, Niccolai L. On nature's strategy for assigning genetic code multiplicity. PLoS ONE. 2016;11(2). http://journals.plos.org/plosone/article?id=10.1371/journal.pone.0148174.
23. Santos J, Monteagudo A. Study of the genetic code adaptability by means of a genetic algorithm. J Theor Biol. 2010;264(3):854–65.

24. Santos J, Monteagudo A. Simulated evolution applied to study the genetic code optimality using a model of codon reassignment. BMC Bioinforma. 2011;12:56.

25. de Oliveira LL, de Oliveira PS, Tinós R. A multiobjective approach to the genetic code adaptability problem. BMC Bioinforma. 2015;16:52.

26. de Oliveira LL, Tinós R. Entropy-based evaluation function for the investigation of genetic code adaptability. In: Proceedings of the ACM Conference on Bioinformatics, Computational Biology and Biomedicine BCB 2012. New York: ACM; 2012. p. 558–560.

27. Blażej P, Wnętrzak M, Mackiewicz P. The role of crossover operator in evolutionary-based approach to the problem of genetic code optimization. Biosystems. 2016;150:61–72.

28. Goldberg DE, Richarson J. Genetic algorithms with sharing for multimodal function optimization. In: Proceedings 2nd International Conference on Genetic Algorithms. Cambridge; 1987. p. 41–49. http://dl.acm.org/citation.cfm?id=42519.

29. Freeland SJ. The Darwinian genetic code: an adaptation for adapting? Genet Programm Evol Mach Kluwer Acad Publ. 2002;3:113–27.

30. Goldberg DE. Genetic Algorithms in Search, Optimization and Machine Learning. Boston, MA, USA: Addison-Wesley Longman Publishing Co., Inc.; 1989.

31. Woese CR. On the evolution of the genetic code. Proc Natl Acad Sci USA. 1965;54:1546–52.

32. Holland J. Adaptation in Natural and Artificial Systems. Cambridge, MA, USA: An Arbor MI: University of Michigan Press; 1975.

33. Sareni B, Krähenbühl L. Fitness sharing and niching methods revisited. IEEE Trans Evol Comput. 1998;2(3):97–106.

34. Knight RD, Freeland SJ, Landweber LF. Adaptive evolution of the genetic code. Genet Code Origin Life. 2004;80:175–84.

35. Higgs PG. A four-column theory for the origin of the genetic code: tracing the evolutionary pathways that gave rise to an optimized code. Biol Direct. 2009;4:16. http://biologydirect.biomedcentral.com/articles/10.1186/1745-6150-4-16.

9

Power analysis for RNA-Seq differential expression studies

Lianbo Yu[*] (iD), Soledad Fernandez[†] and Guy Brock[†]

Abstract

Background: Sample size calculation and power estimation are essential components of experimental designs in biomedical research. It is very challenging to estimate power for RNA-Seq differential expression under complex experimental designs. Moreover, the dependency among genes should be taken into account in order to obtain accurate results.

Results: In this paper, we propose a simulation based procedure for power estimation using the negative binomial distribution and assuming a generalized linear model (at the gene level) that considers the dependence between gene expression level and its variance (dispersion) and also allows equal or unequal dispersion across conditions. We compared the performance of both Wald test and likelihood ratio test under different scenarios. The null distribution of the test statistics was simulated for the desired false positive control to avoid excess false positives with the usage of an asymptotic chi-square distribution. We applied this method to the TCGA breast cancer data set.

Conclusions: We provide a framework for power estimation of RNA-Seq data. The proposed procedure is able to properly control the false positive error rate at the nominal level.

Keywords: RNA-Seq, Power, Wald test, Likelihood ratio test

Background

Discovering differential expression has been the main focus of many biological experiments and many patient cohort studies for several decades. Since the invention of microarray chips twenty years ago, a huge amount of data has been generated by profiling thousands of genes in various cell lines, model organisms, and human samples. Recently, RNA-Seq technology became the replacement of array technology because of its ability to not only quantify the transcriptome but also detect gene isoforms, novel transcripts, and gene fusion [1–3]. Similar to microarray studies, sample size calculations and power estimation are still some of the key issues in designing RNA-Seq experiments, but face some new challenges given the nature of RNA-Seq data.

RNA-Seq studies generate count based data. Several earlier published papers used Poisson distribution to model the count data [4–6]. Due to the restraint of the mean equal to the variance under the Poisson distribution,

the negative binomial (NB) distribution is a natural choice to provide a better fit for RNA-Seq data by allowing an over-dispersion parameter to capture extra variability over the mean. Thus, several specialized software packages have been developed to model RNA-Seq data based on the negative binominal distribution. Robinson et al. [7] developed the R package *edgeR*, which provides an exact test for two group comparisons initially and then was expanded to allow multifactor designs by a generalized linear model. Additionally, Love et al. [8] developed the R package *DESeq2* for differential expression analysis, which provides shrinkage estimators for both log fold change and dispersion by imposing a hierarchical model on them.

For testing differential expression with RNA-Seq experiments, several studies have attempted to provide sample size calculation and power estimation at a single gene level in the recent literature. Fang and Cui [6] introduced a simulation based power estimation approach using Wald test and likelihood ratio test (LRT). Li et al. [9] proposed an exact test method for calculating sample size at a single gene level or the marginal level, which is implemented in a web tool called *RNAseqPS* [10] and an R package called

*Correspondence: Lianbo.Yu@osumc.edu
†Equal contributors
Center for Biostatistics, Department of Biomedical Informatics, The Ohio State University, 1800 Cannon Dr., 43210 Columbus, OH, USA

RnaSeqSampleSize. Other studies have been published for sample size calculation and power estimation at a data set level by evaluating the proportion of true discoveries. Shyr and Li [11] proposed a sample size calculation method using TCGA data. Ching et al. [12] simulated data from six public data sets and compared power in the paired and unpaired designs. The *PROPER* method by Wu et al. [13] is a prospective power assessment approach, which simulated data based on an actual RNA-Seq data set, assessed several empirical error rates and empirical power levels, and stratified them by mean expression and dispersion. However these methods require simulations of all genes based on pilot data or data with similar biological context, and the specification of effect sizes of all genes simultaneously is a big challenge.

The above-mentioned literature on RNA-Seq sample size calculation and power estimation employed common analysis approaches, such as edgeR or DESeq2, that assume the negative binomial distribution. However, all these NB-based approaches have resulted in an inflated type I error rate as reported in several papers [14–17]. Accurate sample size calculation and power estimation should rely on an appropriate control of false positive error rate. Thus the major contribution of our study is that we addressed this issue by using a simulation based empirical approach. This approach properly controls the false positive error rate at the desired level. The idea of using the simulation based approach was originally proposed for modeling brain lesion counts in a multiple slerosis clinical trial by Rettiganti and Nagaraja [18]. But in their study, a simulation based method, called an exact parametric test, was developed for determining the critical values for testing treatment effect. The authors showed that the chi-square test used for Wald, Score, and LRT fails to maintain the nominal significance level, especially for small sample size studies. To overcome this deficiency in sample size calculation and power estimation approaches and to accommodate designs with multiple groups or multiple factors, we provide a framework that can be implemented for power estimation. The proposed simulation based procedure at a single gene level or marginal level uses the exact parametric test for power estimation to ensure the false positive error rate is properly controlled at the nominal level. In addition, we extended this procedure to unequal dispersion parameter cases for RNA-Seq sample size calculation and power estimation, which has not been proposed before. Simulations were conducted for the proposed procedure under both scenarios.

Methods
Negative binomial model
A negative binomial random variable X with mean μ and dispersion ϕ is denoted as $NB(\mu, \phi)$. It has variance $\mu + \mu^2 \phi$ and probability mass function as follows:

$$P(X = x) = \frac{\Gamma(x + \phi^{-1})}{\Gamma(\phi^{-1})\Gamma(x + 1)} \left(\frac{1}{1 + \mu\phi}\right)^{\phi^{-1}} \left(\frac{\mu}{\phi^{-1} + \mu}\right)^x, \quad (1)$$

$x = 0, 1, 2, \cdots; \mu > 0; \phi > 0.$

Dispersion as a function of mean expression
Love et al. [8] assume that the dispersion ϕ follows a log-Normal prior distribution with mean as a function of μ. The dispersion's functional trend is modeled as

$$\phi_{tr}(\mu) = \frac{a_1}{\mu} + a_0. \quad (2)$$

To estimate this functional form, gene-wise dispersion estimators were regressed against the means of the normalized counts. This approach provides gene-wise shrinkage estimators of the dispersion parameter by assuming the mean-dispersion dependence for all genes and shows adequate power for detecting differential expression especially in small sample size experiments.

Likelihood ratio test
Without loss of generality, we use γ to denote the fold ratio of a gene between two biological conditions. We are interested in testing the hypothesis $H_0 : \gamma = 1$ vs. hypothesis $H_1 : \gamma \neq 1$. Let $x_1, x_2, \cdots, x_{n_1}$ and $y_1, y_2, \cdots, y_{n_2}$ represent the gene expression counts from each condition. The LRT statistic is given by

$$L = -2log\left(\frac{sup_{\Theta_0}L(\mu, \gamma, \phi)}{sup_{\Theta}L(\mu, \gamma, \phi)}\right). \quad (3)$$

According to Rettiganti and Nagaraja [18], the maximum likelihood estimate (MLE) of μ under Θ_0 is

$$\tilde{\mu} = \frac{n_1 \bar{x} + n_2 \bar{y}}{n_1 + n_2}. \quad (4)$$

While under Θ, the MLE of μ is $\bar{x}_1$, and the MLE of γ is $\bar{y}/\bar{x}$. Dispersion ϕ is estimated by numerically maximizing the likelihood.

Wald test
A Wald test for testing log transformed γ with $H_0 : log(\gamma) = 0$ vs. $H_1 : log(\gamma) \neq 0$ is given by

$$W(log(\gamma)) = \left(\frac{log(\hat{\gamma})}{\hat{\sigma}_\gamma/\hat{\gamma}}\right)^2. \quad (5)$$

False positive error rate control
With thousands of genes tested in an RNA-Seq study, multiple comparison adjustment is necessary. While the Bonferroni method for controlling the family-wise error rate (probability of one or more false rejections among all comparisons) is very conservative, a less conservative procedure, named the extended interpretation of the Bonferroni method, for controlling the mean number of false positives can be used for multiplicity adjustment [19]. In

other words, the procedure controls the per family error rate (PFER) or per comparison error rate (PCER). It can be made as powerful as the Benjamimi-Hochberg FDR control procedure, and shows greater stability than the FDR. In our simulations, the nominal false positive error rate α will be the PCER for a single gene or at the marginal level.

Empirical parametric test

Inferences on the Wald test and the LRT typically rely on the chi-square distribution by asymptotic theory for large sample sizes. But this may lead to liberal results for small sample sizes since asymptotic theory may not work as expected. To address this issue for small sample sizes, the simulation based test by Rettiganti and Nagaraja [18] is used to provide a proper false positive error rate control. In summary, the empirical null distribution of the test statistics (Wald or LRT) is obtained from simulated experimental data under the null hypothesis for a large number of iterations (e.g., 100,000). The $100(1 - \alpha)th$ percentile from the null test statistics will be used as a significance cutoff for testing under the alternative hypothesis by comparing the test statistics with this percentile cutoff value.

Power estimation procedure

1. Specify all input parameters: sample size per condition n; mean expression μ; fold ratio between conditions γ, nominal false positive error rate α, number of simulations T.
2. Estimate the mean-dispersion functional form using pilot data set or specify an assumed functional form.
3. Calculate dispersion ϕ using Eq. 2 with mean expression μ.
4. Simulate count data from $NB(\mu, \phi)$ T times under both null and alternative hypotheses using the input parameters.
5. Fit NB model and obtain the test statistics (Wald or LRT) under the null hypothesis.
6. Calculate $100(1 - \alpha)th$ percentile as the significance cutoff.
7. Fit NB model and obtain test statistics (Wald or LRT) under the alternative hypothesis.
8. Calculate power for the specified input parameters.

Results

Simulations

Parameter settings

Count data were simulated from a negative binomial distribution under two experimental conditions (e.g. control vs. treatment) with equal dispersion parameters or unequal dispersion parameters (ratio of 1.5) between conditions. The parameters needed to calculate power of a single gene or marginal level are

sample size per condition n, mean expression μ of control group, treatment-to-control fold ratio γ, and nominal false positive error rate α. Simulation settings were $n = 5, 10, 20, 30, 40; \mu = 4, 20, 100, 500; \gamma = \frac{1}{3}, \frac{1}{2}, \frac{1}{1.5}, 1, 1.5, 2, 3; \alpha = 0.01, 0.005, 0.001, 0.0005$. The dispersion parameter ϕ was calculated for each μ from the mean-dispersion functional form $\phi = 0.26 + 3.65/\mu$ estimated from an unpublished canine thyroid RNA-Seq data set (Fig. 1). To explore the effect of the mean-dispersion functional form, a low-dependency ($\phi = 1.56 + 3.65/\mu$) and a high-dependency ($\phi = 0.032 + 3.65/\mu$) were also considered in the simulations as shown in Fig. 1. At each setting, 100,000 simulations were run under the null hypothesis and 10,000 simulations were run under the alternative hypothesis. The critical value was estimated by the empirical $100(1 - \alpha)th$ percentile from the null Wald and LRT statistics.

Equal dispersion

Figures 2 and 3 show the QQ plots for the Wald statistics and the LRT statistics under the null hypothesis at sample size $n = 5, 10, 20, 40$ with mean expression $\mu = 20$. When sample size increases, the distribution of either Wald or LRT statistics converges toward the chi-square distribution with 1 degree of freedom with a faster convergence for the LRT. The discrepancy is quite large when the sample size is small. Figure 4 shows the critical values using

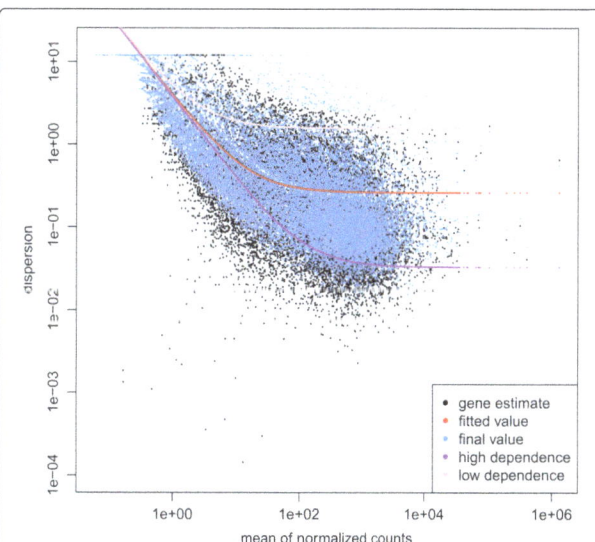

Fig. 1 Mean-dispersion functional form for simulations. *DESeq2* method was applied on a pilot data of unpublished canine thyroid RNA-Seq data set for setting up simulation parameters. The plot shows the estimated mean-dispersion function form (*red dots*) relative to the mean of the normalized counts. *Black dots* represent per-gene estimates of the dispersion while *blue dots* represent moderated estimates calculated by *DESeq2*. The fitted functional form and a lower and higher dependency functional forms were used in the simulation studies

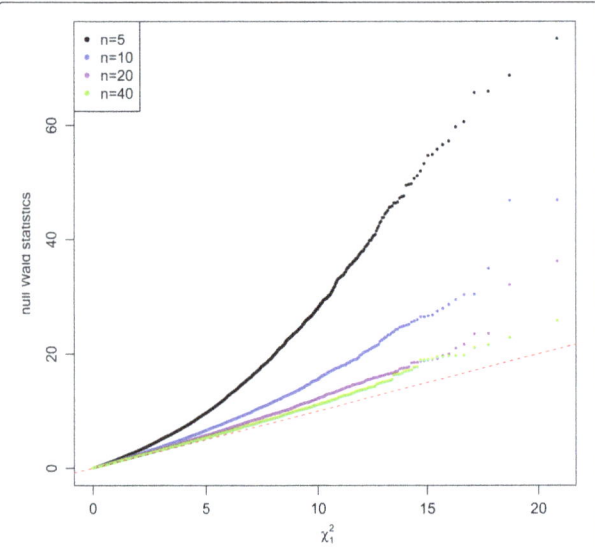

Fig. 2 QQ plot of null Wald statistics with equal dispersion parameters. Data were simulated 100,000 times with $\mu = 20$ under the null hypothesis. Sample sizes were set at $n = 5, 10, 20,$ and 40. Wald test was used for testing mean difference between two conditions. The discrepancy of the null Wald statistics from chi-square distribution with 1 degree of freedom gets smaller when sample size increases

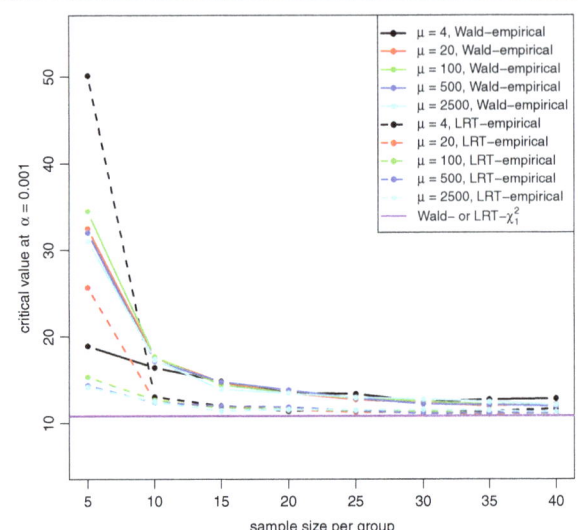

Fig. 4 Critival values plot for both Wald test and LRT with equal dispersion parameters. Critical values were calculated at the nominal false positive error rate of 0.001 from empirical percentile of null statisitics at 5 different mean expression levels for both Wald test (*solid line*) and LRT (*dashed line*), and for the chi-square distribution with 1 degree of freedom (*purple line*). Both Wald test and LRT with the empirical distribution have larger critical values than both Wald test and LRT with the chi-square distribution, and the Wald test has much larger values than the LRT with the empirical distribution

the empirical parametric test and the chi-square distribution at 8 different sample sizes and 5 different mean expression levels for both Wald test and LRT. The critical values get smaller with larger sample sizes. The empirical parametric approach for both Wald test and LRT has

much higher critical values than the chi-square distribution at smaller sample sizes, and the differences decrease when the sample size gets larger. The Wald test has larger critical values than the LRT in general. At each sample size and each mean expression level, the false positive error rate is controlled at the nominal level by the empirical parametric test. However, the estimated false positive error rate of either Wald test or LRT (Fig. 5) following the asymptotic chi-square distribution with 1 degree of freedom is much larger than the nominal false positive error rate, especially for small sample sizes.

Figure 6 and Additional file 1: Figure S1 show power at 8 different sample sizes and 6 different fold changes with mean expression $\mu = 100$ for the Wald test and the LRT at $\alpha = 0.001$. In both plots, power increases with larger sample sizes and larger absolute fold changes. Figure 7 and Additional file 1: Figure S2 show power at 8 different sample sizes and 5 different mean expression level with fold change $\gamma = 2$ under the alternative hypothesis and $\alpha = 0.001$. In both plots, power increases with larger sample sizes and larger mean expression levels. The Wald test and the LRT have similar power at different parameter values. Compared to the results for the medium-dependency functional form, both low-dependency and high-dependency functional forms have similar critical values and false positive error rates, but power estimation is lower(higher) for low(high)-dependency. (See Additional file 1: Figures S3–S8 for results using $\alpha = 0.01$

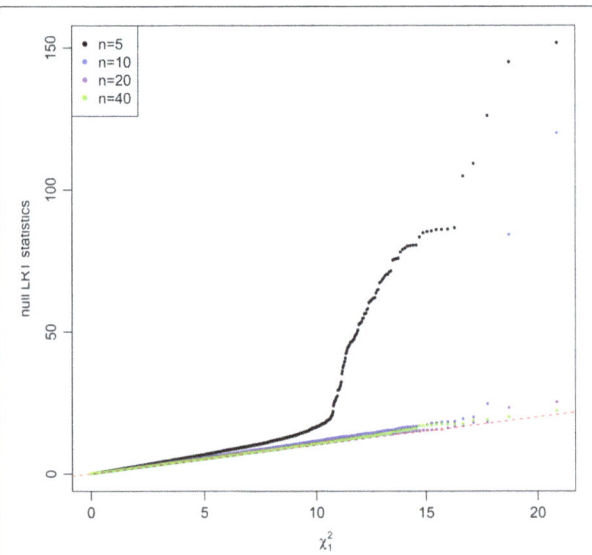

Fig. 3 QQ plot of null LRT statistics with equal dispersion parameters. Data were simulated 100,000 times with $\mu = 20$ under the null hypothesis. Sample sizes were set at $n = 5, 10, 20,$ and 40. The LRT was used for testing mean difference between two conditions. The discrepancy of the null LRT statistics from chi-square distribution with 1 degree of freedom gets smaller when sample size increases

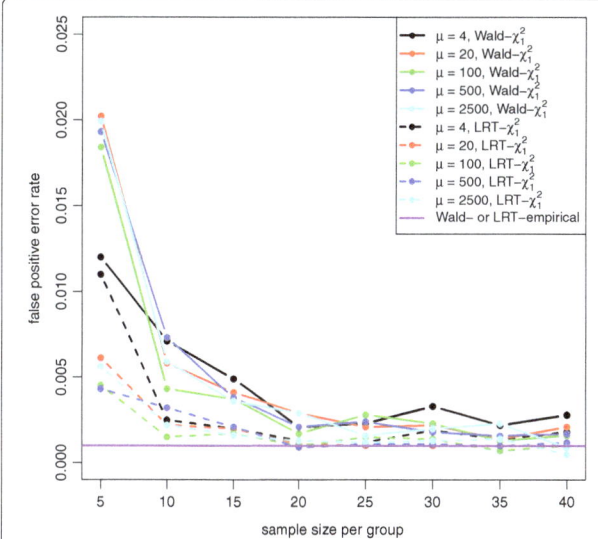

Fig. 5 False positive error rate plot for both Wald test and LRT with equal dispersion parameters. False positive error rate was calculated for both Wald test (*solid line*) and LRT (*dashed line*) following a chi-square distribution with 1 degree of freedom at 5 different mean expression levels. The nominal false positive error rate for both Wald and LRT with the empirical distribution is shown in *purple line* ($\alpha = 0.001$). Both tests with the chi-square distribution have the inflated false positive error rates

and Additional file 1: Figures S9–S20 for results with low(high)-dependency functional form).

Unequal dispersion
The QQ plot of the Wald test under the unequal dispersion setting (Additional file 1: Figure S21) is similar

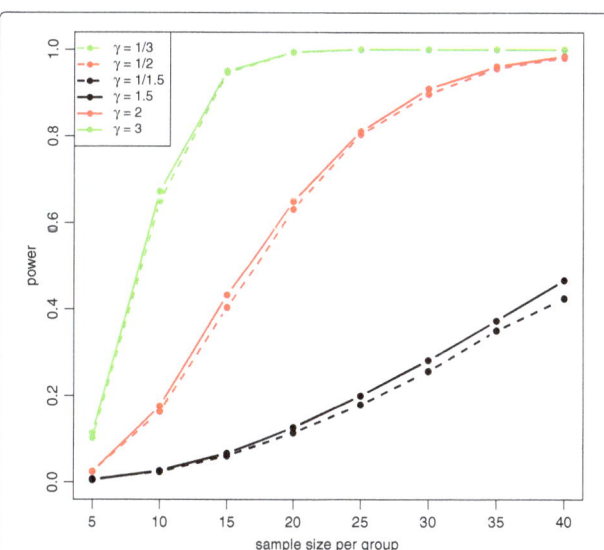

Fig. 6 Power plot at $\mu = 100$ for the Wald test with equal dispersion parameters. Power was calculated at 8 different sample sizes and 6 different fold changes under the alternative hypothesis with $\mu = 100$ and $\alpha = 0.001$. Power is higher for larger sample sizes and higher absolute fold changes

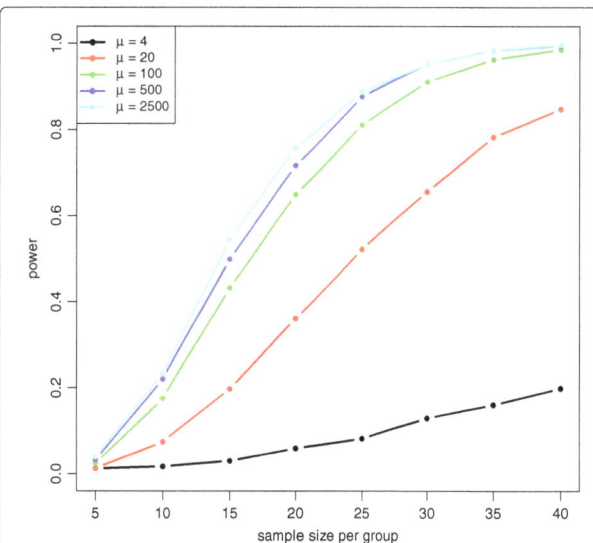

Fig. 7 Power plot at $\gamma = 2$ for the Wald test with equal dispersion parameters. Power was calculated at 8 different sample sizes and 5 different expression levels with $\gamma = 2$ under the alternative hypothesis and $\alpha = 0.001$. Power is higher for larger sample sizes and higher expression levels

to the QQ plot of the equal dispersion setting, but the LRT (Additional file 1: Figure S22) has minor differences between QQ plots of different sample sizes. Figure 8 shows the critical values of the empirical parametric distribution and the chi-square distribution at 4 different

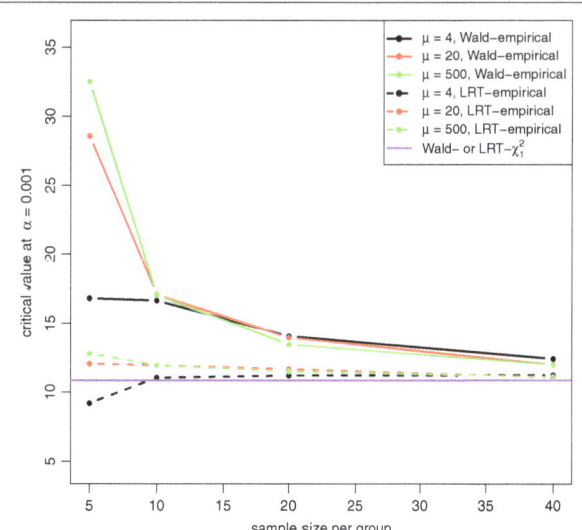

Fig. 8 Critival values plot for both Wald test and LRT with unequal dispersion parameters. Critical values were calculated at the nominal false positive error control level of 0.001 from empirical percentile of null statisitics at 3 different mean expression levels for both Wald test (*solid line*) and LRT (*dashed line*), and for a chi-square distribution with 1 degree of freedom (*purple line*). Both Wald test and LRT have larger critical values than the chi-square distribution, and the Wald test has much larger values than the LRT

sample sizes and 3 different mean expression levels for both Wald test and LRT. Similar to the equal dispersion setting, the empirical parametric test of the Wald test has much higher critical values than the chi-square distribution at small sample sizes, and the differences get smaller when sample size gets larger. However, the LRT has slightly higher critical values than the chi-square distribution. For false positive error rate (Fig. 9), the Wald test with the chi-square distribution has much higher values than the nominal level, while the LRT with the chi-square distribution has slightly higher values. Similar to the equal dispersion setting, the power of the Wald test (Figs. 10 and 11) and the LRT (Additional file 1: Figures S23 and S24) at $\alpha = 0.001$ is increased with larger sample sizes, larger mean expression levels, and larger absolute fold changes. (See Additional file 1: Figures S25–S30 for results using $\alpha = 0.01$ and Additional file 1: Figures S31–S42 for results with low(high)-dependency functional form).

Applications
TCGA data set
To study and demonstrate the proposed power estimation procedure in a real data application, we used the TCGA breast cancer data set as a pilot data for designing a new study for detecting differential expression. The TCGA breast cancer data set, acquired in Sep. 2015 from cBioPortal for Cancer Genomics, contains 1003 tumor samples with clinical information and 17866 gene features with non-zero counts. We chose the comparison between

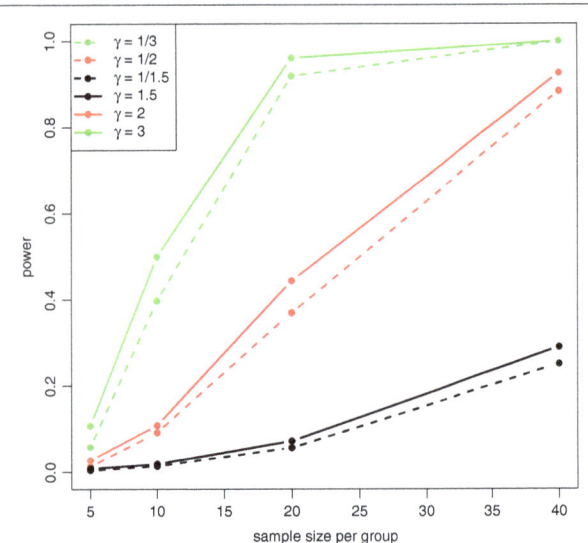

Fig. 10 Power plot at $\mu = 20$ for the Wald test with unequal dispersion parameters. Power was calculated at 4 different sample sizes and 6 different fold changes under the alternative hypothesis with $\mu = 20$ and $\alpha = 0.001$. Power is higher for larger sample sizes and higher absolute fold changes

two tumor stage categories I-II (746 samples) vs. III-IV (238 samples) when fitting the *DESeq2* package for estimating the mean-dispersion functional form (Additional file 1: Figure S43). The estimated mean expression levels were 27, 496, 2501 at the 10*th*, 50*th*, 90*th* percentiles for all gene features, respectively. Figure 12 shows the power as a function of sample sizes (range 3-100) at these mean

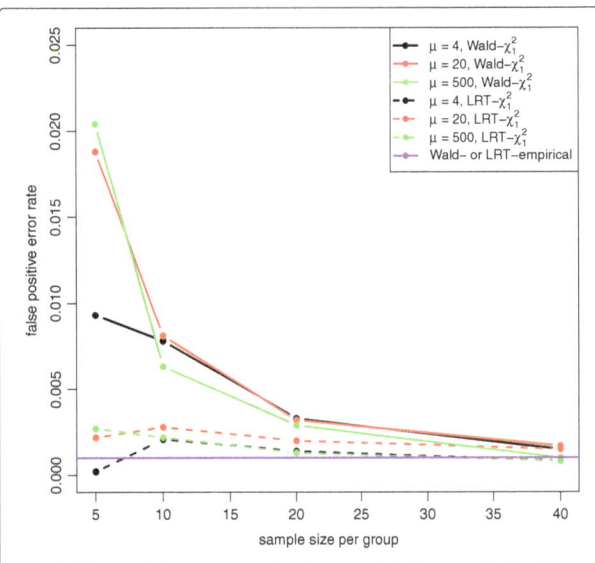

Fig. 9 False positive error rate plot for both Wald test and LRT with unequal dispersion parameters. False positive error rate was calculated for both Wald test (*solid line*) and LRT (*dashed line*) following a chi-square distribution with 1 degree of freedom at 3 different mean expression levels. The nominal false positive error rate control level for the empirical parametric test is shown in *purple line* ($\alpha = 0.001$). Both tests have inflated false positive error rates

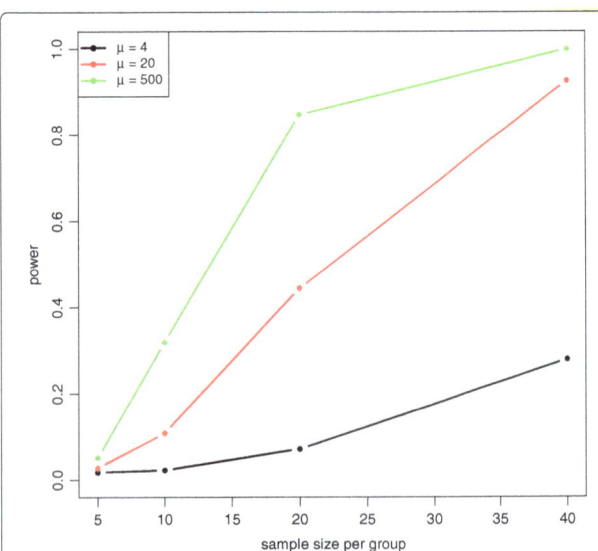

Fig. 11 Power plot at $\gamma = 2$ for the Wald test with unequal dispersion parameters. Power was calculated at 4 different sample sizes and 3 different expression levels with $\gamma = 2$ under the alternative hypothesis and $\alpha = 0.001$. Power is higher for larger sample sizes and higher expression levels

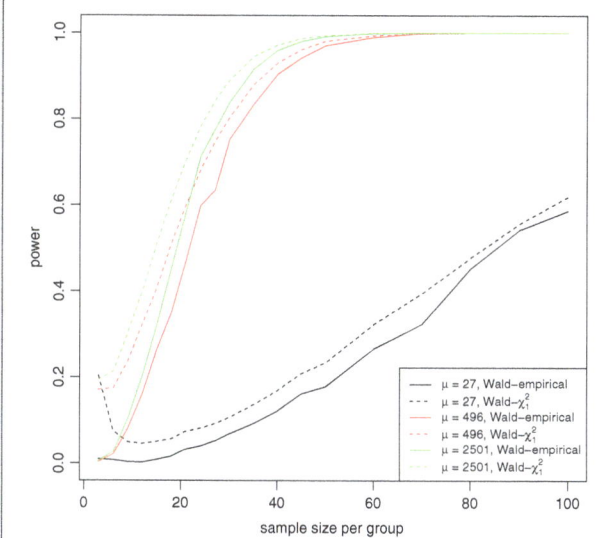

Fig. 12 Power plot for the Wald test with equal dispersion parameters for TCGA breast cancer data set. Power was calculated for the Wald test with the empirical distribution (*solid line*) or with a chi-square distribution with 1 degree of freedom (*dotted line*) at 3 different mean expression levels, 19 different sample sizes (range 3-100), and a fold change of 2 under the alternative hypothesis with $\alpha = 0.001$

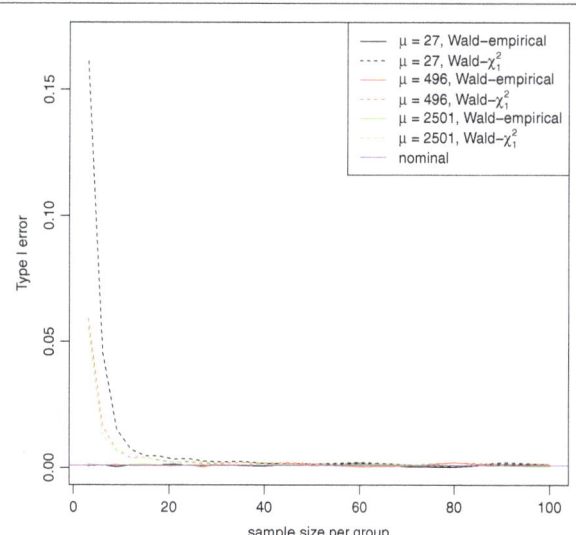

Fig. 13 False positive error rate plot for the Wald test with equal dispersion parameters for TCGA breast cancer data set. False positive error rate was calculated for the Wald test with the empirical distribution (*solid line*) or with a chi-square distribution with 1 degree of freedom (*dotted line*) at 3 different mean expression levels and 19 different sample sizes (range 3-100). The nominal false positive error rate for the Wald test with the empirical distribution is shown in *purple line* ($\alpha = 0.001$). Wald test with the chi-square distribution has the inflated false positive error rates

expression percentiles for a 2-fold difference between two patient subgroups. Wald test for the proposed empirical distribution and for the chi-square distribution were used at $\alpha = 0.001$. Figure 13 shows the false positive error rate at $\alpha = 0.001$. Even though the Wald test for the chi-square distribution has a little higher power at smaller sample sizes, this is mainly due to the failure to properly control false positive rate. To design a new study with 80% power, we will need $n = 33$ samples per group to detect a 2-fold difference for genes at the mean expression level of 496. The computation time for this power estimation is about 8 hours on a standard windows laptop with Intel Core i7-6820HQ CPU at 2.70GHz and 32GB RAM.

Discussion

Many published methods on identifying differentially expressed genes are based on the negative binomial distribution, and the inference mainly relies on asymptotic theory which is biased for small sample sizes. Several studies by Leng et al. [14], Lund et al. [15], Reeb and Steibel [16], and Rocke et al. [17] have reported the excess false positives by using these methods for differential expression detection with RNA-Seq data. The main reason is that the use of the significance cutoff from biased asymptotic distribution leads to the inflated false positive error rate especially for small sample sizes. Our simulation results confirm the great downward bias in the significance cutoff values when an asymptotic chi-square distribution is applied for both Wald test and LRT. Using the empirical

parametric test for estimating the critical values, we are able to control the false positive error rate at the desired nominal level for both tests (Additional file 1: Figures S44–S45).

In all current published methods on differential expression detection and power estimation, the dispersion parameter is assumed equal across conditions. Under this assumption, the power will be misestimated if dispersion values are very different across conditions. Therefore in the simulations we allowed the dispersion parameter to be equal or unequal across conditions to achieve accurate power. When the dispersion parameter is assumed equal, the exact test method by Li et al. [9] can be used for power estimation. However this exact test method only works for two group comparisons and it can not be adapted to allow for unequal dispersion across conditions.

The proposed work not only can be applied to multiple groups and multiple factor designs through generalized linear models, it can also be extended to the data set level. In this case, the null distribution of the test statistics could be simulated for each gene or for a group of genes with similar expression profile for a proper control of the false positive error rate.

Conclusions

With the emergence of RNA-Seq technology in recent years, RNA-Seq experiments have been widely used as an alternative to microarrays in biomedical research. Due to different data types, data analysis and power estimation are also different. New methods on sample size calculations and power estimation using the negative binomial distribution have already been proposed for this new technology. To overcome some of the limitations in current methods, we provide a framework for power estimation of RNA-Seq experiments by proposing a simulation based procedure, which provides a proper false positive control and can be applied in generalized linear model settings.

Additional file

Additional file 1: This file provides all supplementary figures referenced in results section. **Figure S1.** Power plot at $\mu = 100$ for the LRT with equal dispersion parameters. **Figure S2.** Power plot at $\gamma = 2$ for the LRT with equal dispersion parameters. **Figure S3.** Critival values plot for both Wald test and LRT with equal dispersion parameters at $\alpha = 0.01$. **Figure S4.** False positive error rate plot for both Wald test and LRT with equal dispersion parameters at $\alpha = 0.01$. **Figure S5.** Power plot at $\mu = 100$ for the Wald test with equal dispersion parameters at $\alpha = 0.01$. **Figure S6.** Power plot at $\mu = 100$ for the LRT with equal dispersion parameters at $\alpha = 0.01$. **Figure S7.** Power plot at $\gamma = 2$ for the Wald test with equal dispersion parameters at $\alpha = 0.01$. **Figure S8.** Power plot at $\gamma = 2$ for the LRT with equal dispersion parameters at $\alpha = 0.01$. **Figure S9.** Critival values plot for both Wald test and LRT with equal dispersion parameters assuming the high dependency functional form. **Figure S10.** False positive error rate plot for both Wald test and LRT with equal dispersion parameters assuming the high dependency functional form. **Figure S11.** Power plot at $\mu = 100$ for the Wald test with equal dispersion parameters assuming the high dependency functional form. **Figure S12.** Power plot at $\mu = 100$ for the LRT with equal dispersion parameters assuming the high dependency functional form. **Figure S13.** Power plot at $\gamma = 2$ for the Wald test with equal dispersion parameters assuming the high dependency functional form. **Figure S14.** Power plot at $\gamma = 2$ for the LRT with equal dispersion parameters assuming the high dependency functional form. **Figure S15.** Critival values plot for both Wald test and LRT with equal dispersion parameters assuming the low dependency functional form. **Figure S16.** False positive error rate plot for both Wald test and LRT with equal dispersion parameters assuming the low dependency functional form. **Figure S17.** Power plot at $\mu = 100$ for the Wald test with equal dispersion parameters assuming the low dependency functional form. **Figure S18.** Power plot at $\mu = 100$ for the LRT with equal dispersion parameters assuming the low dependency functional form. **Figure S19.** Power plot at $\gamma = 2$ for the Wald test with equal dispersion parameters assuming the low dependency functional form. **Figure S20.** Power plot at $\gamma = 2$ for the LRT with equal dispersion parameters assuming the low dependency functional form. **Figure S21.** QQ plot of null Wald statistics with unequal dispersion parameters. **Figure S22.** QQ plot of null LRT statistics with unequal dispersion parameters. **Figure S23.** Power plot at $\mu = 20$ for the LRT with unequal dispersion parameters. **Figure S24.** Power plot at $\gamma = 2$ for the LRT with unequal dispersion parameters. **Figure S25.** Critival values plot for both Wald test and LRT with unequal dispersion parameters at $\alpha = 0.01$. **Figure S26.** False positive error rate plot for both Wald test and LRT with unequal dispersion parameters at $\alpha = 0.01$. **Figure S27.** Power plot at $\mu = 20$ for the Wald test with unequal dispersion parameters at $\alpha = 0.01$. **Figure S28.** Power plot at $\mu = 20$ for the LRT with unequal dispersion parameters at $\alpha = 0.01$. **Figure S29.** Power plot at $\gamma = 2$ for the Wald test with unequal dispersion parameters at $\alpha = 0.01$. **Figure S30.** Power plot at $\gamma = 2$ for the LRT with unequal dispersion parameters at $\alpha = 0.01$. **Figure S31.** Critival values plot for both Wald test and LRT with unequal dispersion parameters assuming the high dependency functional form. **Figure S32.** False positive error rate plot for both Wald test and LRT with unequal dispersion parameters assuming the high dependency functional form.

Figure S33. Power plot at $\mu = 20$ for the Wald test with unequal dispersion parameters assuming the high dependency functional form. **Figure S34.** Power plot at $\mu = 20$ for the LRT with unequal dispersion parameters assuming the high dependency functional form. **Figure S35.** Power plot at $\gamma = 2$ for the Wald test with unequal dispersion parameters assuming the high dependency functional form. **Figure S36.** Power plot at $\gamma = 2$ for the LRT with unequal dispersion parameters assuming the high dependency functional form. **Figure S37.** Critival values plot for both Wald test and LRT with unequal dispersion parameters assuming the low dependency functional form. **Figure S38:** False positive error rate plot for both Wald test and LRT with unequal dispersion parameters assuming the low dependency functional form. **Figure S39.** Power plot at $\mu = 20$ for the Wald test with unequal dispersion parameters assuming the low dependency functional form. **Figure S40.** Power plot at $\mu = 20$ for the LRT with unequal dispersion parameters assuming the low dependency functional form. **Figure S41.** Power plot at $\gamma = 2$ for the Wald test with unequal dispersion parameters assuming the low dependency functional form. **Figure S42.** Power plot at $\gamma = 2$ for the LRT with unequal dispersion parameters assuming the low dependency functional form. **Figure S43.** Mean-dispersion functional form of TCGA breast cancer data set. **Figure S44.** False positive error rate plot for both Wald test and LRT with equal dispersion parameters. **Figure S45.** False positive error rate plot for both Wald test and LRT with unequal dispersion parameters.

Abbreviations
LRT: Likelihood ratio test; MLE: Maximum likelihood estimation; NB: Negative binomial; PCER: Per comparison error rate; PFER: Per family error rate; TCGA: The cancer genome atlas

Acknowledgements
The authors wish to acknowledge the anonymous reviewers for their comments and suggestions which helped improve the manuscript.

Funding
This research was partially supported by NIH grants 2P30CA016058-40 and UL1TR001070. The funding body played no role in the design or conclusions of this study.

Authors' contributions
All authors were involved in method development. LY and SF generated the original idea. LY performed the simulations and wrote the manuscript. GB and SF guided the research and revised the manuscript. All authors read and approved the final version of this manuscript.

Competing interests
The authors declare that they have no competing interests.

References
1. Shendure J. The beginning of the end for microarrays?. Nat Methods. 2008;5(7):585–7.
2. Wang Z, Gerstein M, Snyder M. RNA-Seq: a revolutionary tool for transcriptomics. Nat Rev Genet. 2009;10(1):57–63.
3. Griffith M, Griffith OL, Mwenifumbo J, Goya R, Morrissy AS, Morin RD, Corbett R, Tang MJ, Hou YC, Pugh TJ, Robertson G, Chittaranjan S, Ally A, Asano JK, Chan SY, Li HI, McDonald H, Teague K, Zhao Y, Zeng T, Delaney A, Hirst M, Morin GB, Jones SJ, Tai IT, Marra MA. Alternative expression analysis by RNA sequencing. Nat Methods. 2010;7(10):843–7.
4. Marioni JC, Mason CE, Mane SM, Stephens M, Gilad Y. RNA-seq: an assessment of technical reproducibility and comparison with gene expression arrays. Genome Res. 2008;18(9):1509–17.
5. Wang L, Feng Z, Wang X, Wang X, Zhang X. DEGseq: an R package for identifying differentially expressed genes from RNA-seq data. Bioinformatics. 2010;26(1):136–8.
6. Fang Z, Cui X. Design and validation issues in RNA-seq experiments. Brief Bioinform. 2011;12(3):280–7.
7. Robinson MD, McCarthy DJ, Smyth GK. edgeR: a Bioconductor package for differential expression analysis of digital gene expression data. Bioinformatics. 2010;26(1):139–40.
8. Love MI, Huber W, Anders S. Moderated estimation of fold change and dispersion for RNA-seq data with DESeq2. Genome Biol. 2014;15(12):550.

9. Li CI, Su PF, Shyr Y. Sample size calculation based on exact test for assessing differential expression analysis in RNA-seq data. BMC Bioinforma. 2013;14:357.

10. Guo Y, Zhao S, Li CI, Sheng Q, Shyr Y. RNAseqPS: A Web Tool for Estimating Sample Size and Power for RNAseq Experiment. Cancer Inform. 2014;13(Suppl 6):1–5.

11. Shyr D, Liu Q. Next generation sequencing in cancer research and clinical application. Biol Proced Online. 2013;15(1):4.

12. Ching T, Huang S, Garmire LX. Power analysis and sample size estimation for RNA-Seq differential expression. RNA. 2014;20(11):1684–96.

13. Wu H, Wang C, Wu Z. PROPER: comprehensive power evaluation for differential expression using RNA-seq. Bioinformatics. 2015;31(2):233–41.

14. Leng N, Dawson J, Thomson J, Ruotti V, Rissman A, Smits B, Haag J, Gould M, Stewart R, Kendziorski C. EBSeq: An empirical Bayes hierarchical model for inference in RNA-Seq experiments. Technical report, Department of Biostatistics and medical informatics: University of Wisconsin; 2012.

15. Lund S, Nettleton D, McCarthy DJ, Smyth GK. Detecting differential expression in RNA-sequencing data using quasi-likelihood with shrunken dispersion estimates. Stat Appl Genet Mol Biol. 2012;11(5):article 8.

16. Reeb PD, Steibel JP. Evaluating statistical analysis models for RNA sequencing experiments. Front Genet. 2013;4:178.

17. Rocke DM, Ruan L, Zhang Y, Gossett JJ, Durbin-Johnson B, Aviran S. Excess false positive rates in methods for differential gene expression analysis using RNA-Seq data. bioRxiv preprint. 2015. doi:http://dx.doi.org/10.1101/020784.

18. Rettiganti M, Nagaraja HN. Power analyses for negative binomial models with application to multiple sclerosis clinical trials. J Biopharm Stat. 2012;22(2):237–59.

19. Gordon A, Glazko G, Qiu X, Yakovlev A. Control of the mean number of false discoveries, Bonferroni and stability of multiple testing. Ann Appl Stat. 2007;1:179–190.

10

LW-FQZip 2: a parallelized reference-based compression of FASTQ files

Zhi-An Huang[1†], Zhenkun Wen[1†], Qingjin Deng[1], Ying Chu[1], Yiwen Sun[2] and Zexuan Zhu[1*]

Abstract

Background: The rapid progress of high-throughput DNA sequencing techniques has dramatically reduced the costs of whole genome sequencing, which leads to revolutionary advances in gene industry. The explosively increasing volume of raw data outpaces the decreasing disk cost and the storage of huge sequencing data has become a bottleneck of downstream analyses. Data compression is considered as a solution to reduce the dependency on storage. Efficient sequencing data compression methods are highly demanded.

Results: In this article, we present a lossless reference-based compression method namely LW-FQZip 2 targeted at FASTQ files. LW-FQZip 2 is improved from LW-FQZip 1 by introducing more efficient coding scheme and parallelism. Particularly, LW-FQZip 2 is equipped with a light-weight mapping model, bitwise prediction by partial matching model, arithmetic coding, and multi-threading parallelism. LW-FQZip 2 is evaluated on both short-read and long-read data generated from various sequencing platforms. The experimental results show that LW-FQZip 2 is able to obtain promising compression ratios at reasonable time and memory space costs.

Conclusions: The competence enables LW-FQZip 2 to serve as a candidate tool for archival or space-sensitive applications of high-throughput DNA sequencing data. LW-FQZip 2 is freely available at http://csse.szu.edu.cn/staff/zhuzx/LWFQZip2 and https://github.com/Zhuzxlab/LW-FQZip2.

Keywords: High-throughput sequencing, Sequencing data compression, Reference- based compression, Sequence alignment

Background

The rapid progress of high-throughput DNA sequencing techniques has dramatically reduced the costs of whole genome sequencing, which leads to revolutionary advances in gene industry [1, 2]. Genome studies have produced tremendous volume of data that poses great challenges to storage and transfer. Data compression becomes necessary as a silver-bullet solution to ease the dilemma [3–6]. Nevertheless, the popular generic compression tools, such as gzip (http://www.gzip.org/) and bzip2 (http://www.bzip.org), cannot obtain satisfactory performance on high-throughput DNA sequencing data, because they do not utilize the biological characteristics of the data like repeat fragments and palindromes. Efficient

compression methods oriented to high-throughput DNA sequencing data are highly demanded.

Many specialized compression methods have been proposed for sequencing data in raw FASTQ format [7–11], sequencing reads (DNA nucleotides only) [12–15] and/or aligned SAM/BAM format [16]. Depending on whether extra reference genomes are required, these compression methods normally can be classified into reference-based and reference-free methods.

Reference-based methods align the targeted sequences to some external reference sequence(s) for identifying the similarity between them. The variances of the alignment are encoded instead of the original targeted sequences. Reference-based methods generally obtain better compression ratios with more time consumption by involving a sequence alignment pre-processing. Representative state-of-the-art reference-based methods include CRAM [17], Quip [11], and DeeZ [18]. CRAM [17], working with BAM-based input, records the variances of reads and a reference genome with Huffman coding. Quip [11] with '-r'

* Correspondence: zhuzx@szu.edu.cn
†Equal contributors
[1]College of Computer Science and Software Engineering, Shenzhen University, Shenzhen 518060, China
Full list of author information is available at the end of the article

option compresses SAM/BAM data in a standard reference-based scheme and employs highly optimized statistical models for various SAM fields, thus leading to compromising compression rate. It is also applicable to reference-based FASTQ compression in '-a' mode where a *de novo* assembly procedure is introduced to construct references in the target data rather than using existing references. DeeZ [18] lowers the cost of representing common differences among the reads' mapping results by implicitly assembling the underlying donor genome in order to encode these variants only once.

Reference-free methods compress the raw sequencing data, mainly in FASTQ format, directly based on their natural statistics. For example, FQZcomp [8] uses an order-k context model to predict the nucleotide sequences in FASTQ format followed by arithmetic coding based compression. DSRC [10] partitions input FASTQ data into blocks enabling independent compression of them using LZ77 and Huffman encoding schemes. DSRC 2 [19] is an improvement of DSRC by introducing threaded parallelism and more efficient coding scheme. LFQC [20] preforms data transformation to the four fields of the sequencing records in an FASTQ file separately, followed by regular data compressor namely zpaq and lpaq8. LEON [21] first builds a reference as a probabilistic *de Bruijn graph* based on bloom filter, and then records the reads and quality scores as mapped paths in the graph using arithmetic encoding. SCALCE [22] reorganizes reads in an FASTQ file that share common 'core' substrings into groups, and then compacts the groups using gzip or LZ77-like tools. SeqDB [23] coordinates sequences bases and their corresponding quality scores into 2D byte arrays and compresses them with an existing multithreaded compressor Blosc. Quip [11], in addition to the reference-based compression mode, also provides reference-free compression using arithmetic coding based on high order Markov chains. Instead of exploiting the redundancy of homologous sequences, reference-free methods put more effort into predictive model and coding scheme, which tends to improve the time efficiency by sacrificing compression ratio to some degree [24].

This work focus on long-term archiving and space-sensitive scenarios, where superior compression ratio is pursued and reference-based methods are more favourable. In our previous work [25], we have proposed a self-contained reference-based method, namely LW-FQZip 1, to compress high-throughput DNA sequencing data in raw FASTQ format. LW-FQZip 1 introduces a light-weight mapping model to efficiently align short reads against the reference sequence based on a *k*-mer indexing strategy. The light-weight mapping model distinguishes LW-FQZip 1 from other reference-based methods for not relying on any external alignment software. Nevertheless, LW-FQZip 1 is far from satisfactory in terms of compression efficiency.

LW-FQZip 2 is an improved version of LW-FQZip 1 by introducing parallelism and more efficient coding schemes. Especially, LW-FQZip 2 is equipped with light-weight mapping model, bitwise prediction by partial matching (PPM), arithmetic coding, and multi-threading parallelism. It can support various FASTQ files generated from the most well-known high-throughput sequencing platforms and obtain superior compression ratios at reasonable time and memory space costs.

Implementation

The general framework of LW-FQZip 2 is shown in Fig. 1. Firstly, LW-FQZip 2 splits an input FASTQ file into three data streams (i.e., metadata, nucleotide sequences, and quality scores) and then the nucleotide sequences (also known as reads) are divided into equal-sized sub-blocks which are simultaneously fed to the light-weight mapping model implemented with multi-threading. After the sequence mapping, the matching results (i.e., the mapped position, palindrome flag, match length, and match type) and mismatch values are recorded in different intermediate files. Secondly, the metadata and quality scores are simultaneously proceeded by abridging the consecutive repeats with incremental encoding and run-length-limited encoding, respectively. Finally, the intermediate files generated from the metadata, nucleotide sequences, and quality scores streams are compacted with a combination of bitwise order-32 PPM model and arithmetic coder in parallel, except that the intermediate file storing mismatch values is compressed by a distinct bitwise order-28 arithmetic coder. In best compression ratio mode, i.e., LW-FQZip 2 with a '−g' option selected, the quality scores and mismatch values are compacted using the zpaq tool (http://mattmahoney.net/dc/zpaq.html) and the other intermediate files encoded with lpaq9m (http://mattmahoney.net/dc/text.html#1440). LW-FQZip 2 is available at http://csse.szu.edu.cn/staff/zhuzx/LWFQZip2 and https://github.com/Zhuzxlab/LW-FQZip2. The pseudo-code of LW-FQZip 2 is provided in Algorithm 1, Additional file 1. The key components of LW-FQZip 2 are described as follows.

Compression of metadata and quality scores

In LW-FQZip 2, the metadata are pre-processed with incremental encoding, with which the variances of one metadata to its previous neighbour is stored rather than the original data. The quality scores are pre-processed with run-length-limited encoding. More details of the incremental encoding and run-length-limited encoding are available in [25].

After pre-processing, the processed data can be compressed with lpaq9m (if '-g' option is selected) or a combination of PPM model and arithmetic coder. In the

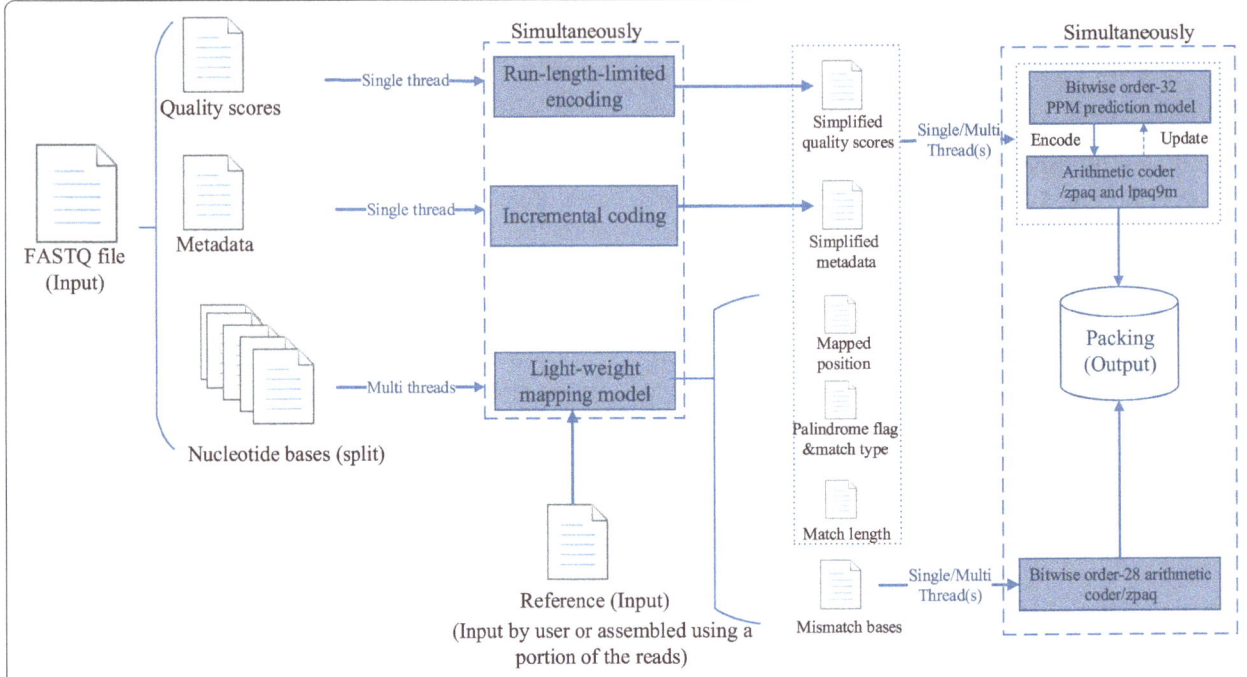

Fig. 1 The general framework of LW-FQZip 2. Firstly, the input FASTQ file is split into three data streams of metadata, bases, and quality scores. Secondly, the quality scores and metadata are compacted with run-length-limited encoding and incremental encoding, respectively. The nucleotide bases are partitioned and mapped to an external reference sequence based on the light-weight mapping model. Finally, the processed intermediate files from the three streams are compressed with arithmetic coder and/or other specific coding schemes

latter case, a binary tree is established to store the predictive context information. The pre-processed metadata and quality scores are transformed into binary streams to train an order-32 PPM model, i.e., prediction based on a 4-byte context (higher order can improve the compression ratios by promoting the predictive accuracy but it consumes more memory space and running time. A trade-off order 32 is adopted in this study). The binary quality scores are matched against the predicted results per bit by producing '0' or '1', and then the context prediction model is updated accordingly. Finally, the prediction results are recorded using arithmetic coder or zpaq (if '-g' option is selected). The pseudo-code of the compression with PPM prediction model and arithmetic coding/zpaq is provided in Algorithm 2, Additional file 1.

Reference-based compression of nucleotide sequences using light-weight mapping model

The target nucleotide sequences are mapped to an external reference sequence based on the light-weight mapping model [25] and the mapping results are recorded instead of the original sequences.

To make this article self-contained, the light-weight mapping model is briefly introduced in this subsection. The mapping model is designed to implement fast alignment by indexing the k-mer substrings within the reference. A hash table I_R is firstly established to save all

positions of k-mer substrings in the reference with some predefined prefix, e.g., 'CG'. On mapping a read of nucleotide bases X to the reference, the model identifies all k-mer substring included in I_R and selects the valuable k-mer substrings served as seeds, where some restrictions are predefined to eliminate the low-quality k-mer substrings (e.g., minimum seed length L, mismatch tolerance rate e). Based on the selected seeds, multiple local alignments are performed to identify the maximum matches. The mapping results of X including the mapped position, palindrome flag, match length, match type and mismatch values are recorded in some intermediate files. If no k-mer substring in X is identified in I_R, the palindrome of X undergoes the same mapping procedure. In the case neither X nor its palindrome is mapped to the reference, the plain X is output for encoding directly.

To improve the matching rate, the unmapped parts of reads are further partitioned into shorter segments and realigned against the reference where palindrome match is considered. The mismatched nucleotide bases are composed exclusively of four characters (i.e., {'A', 'C', 'G', 'T'}), which are much easier to encode than quality scores. Therefore, a simpler yet efficient model based on the bitwise order-28 arithmetic coder (http://cs.fit.edu/~mmahoney/compression/text.html#2212) or zpaq is adopted. If no proper reference is available, LW-FQZip 2 also provides an option '-a' to generate a reference by assembling a portion of reads that contain the predefined prefix.

Blocking and multithreading

Parallelism is introduced to LW-FQZip 2 to improve the computational efficiency using the Pthreads library. In the mapping procedure, the input FASTQ file is partitioned into b (empirically set to 10 in this study) equal-sized blocks. Accordingly, b threads are simultaneously executed with each running a light-weight mapping model for a corresponding block. Afterward, a new single thread is created to collect the mapping results of the previous b threads and dispatch the results into different intermediate files. After the three data streams, i.e., metadata, quality scores, and nucleotide sequences are properly processed, multiple threads are created to compress the six intermediate files with the corresponding encoding schemes as shown in Fig. 1.

In summary, LW-FQZip 2 improves LW-FQZip 1 by introducing multi-threading for the time-consuming read mapping and using more efficient encoding schemes based on PPM model, arithmetic coders, lpaq9m and/or zpaq. The implementation details of LW-FQZip 2 are provided in the Additional file 1.

Results and discussion

LW-FQZip 2 is verified using ten representative real-world FASTQ files (five short-read data and five long-read data) on a platform running 64-bit Red Hat 4.4.7-16 with four 8-core Intel(R) Xeon(R) E7-8837 CPUs (@2.67GHz with Hyper-Threading Technology). These data sets, generated from various well-known high-throughput DNA sequencing platforms, were downloaded from the Sequence Read Archive of the National Centre for Biotechnology Information (NCBI) [26]. Details of these data sets are provided in Table 1.

LW-FQZip 2 is compared with the other state-of-the-art lossless FASTQ compressors namely Quip [11], DSRC 2 [19], FQZcomp [8], CRAM [17], LFQC [20], LEON [21] and SCALCE [22]. The general-purpose compression tools, i.e., gzip and bzip2, as well as the original LW-FQZip 1 are also included in the comparison as baselines. All compared methods are configured to obtain the best compression ratio (the parameter settings of all methods and the software version information are provided in the Additional file 1). LW-FQZip 2 is evaluated on two modes, i.e., the normal mode ('LW-FQZip 2', using arithmetic coders to compress the intermediate files) and the best compression ratio mode ('LW-FQZip 2 (–g)', using lpaq9m and zpaq to compress the intermediate files). Quip is also executed in two modes, i.e., the reference-based compression ('quip -r') and the assembly-based compression ('quip -a', a reference is assembled with a portion of reads, and then a reference-based compression is conducted using the generated reference). Quip, CRAM, and LW-FQZip 1 fall within reference-based scheme. The other methods are reference-free methods.

The performance of the methods is evaluated in terms of compression ratio, speed, and memory consumption. The compression ratios of all compared methods on the ten FASTQ data sets are tabulated in Table 2. The average compression and decompression speeds of all methods are plotted in Fig. 2. The memory sizes consumed by the compared methods on each data set are reported in Table 3. More details of the data sets and experimental studies are provided in Additional file 1: Tables S1-S13. The average CPU utilization and version information of all compared methods are provided in Additional file 1: Tables S14 and S17, respectively.

From Table 2, it is shown that LW-FQZip 2 and LW-FQZip 2 (–g) successfully work on all test data sets.

Table 1 The ten real-world FASTQ data sets used for performance evaluation

	Datasets	Platforms	Species	Read length (bp)	Size (MB)	GC content
Long-read	SRR2916693	454GS	Pseudomonas moraviensis	67-1201	425	58.8%
	SRR2994368	Illumina Miseq	Escherichia coli	70-502	4688	49.7%
	SRR3211986	Pacbio RS	Homo sapiens	2-62746	1759	39.6%
	ERR739513	MinION	Phage	5-246140	871	47.9%
	SRR3190692	Illumina MiSeq	Escherichia coli	70-602	11379	52.3%
Short-read	ERR385912	Illumina Hiseq 2000	Escherichia coli	51	641	43.5%
	ERR386131	Ion Torrent PGM	Capsicum baccatum	151	1371	50.5%
	SRR034509	Illumina Analyzer II	Escherichia coli	101	5247	52.6%
	ERR174310	Illumina Hiseq 2000	Homo sapiens	202	105122	N.A.
	ERR194147	Illumina Hiseq 2000	Homo sapiens	101	202631	40.3%

Note: The long-read data sets have variable-length reads, while the short-read data sets have fixed-length reads

Table 2 The compression ratios of the compared methods on ten test data sets

		LW-FQZip 2	LW-FQZip 2 (−g)	LW-FQZip 1	Quip (−a)	Quip (−r)	DSRC 2	CRAM	FQZcomp	LFQC	LEON	SCALCE	bzip 2	gzip
Long-read	SRR2916693	16.7%	15.3%	18.1%	20.9%	20.5%	20.2%	21.9%	21.6%	**12.7%**	19.5%	17.2%[a]	24.2%	29.6%
	SRR2994368	17.3%	**16.0%**	17.9%	20.1%	N/A	23.2%	26.4%	N/A	N/A	23.1%	17.3%[a]	28.5%	34.2%
	SRR3211986	33.3%	**32.3%**	N/A	33.3%	N/A	N/A	33.9%	N/A	**32.3%**	N/A	33.4%[a]	36.4%	42.6%
	ERR739513	35.2%	**34.8%**	N/A	N/A	N/A	N/A	35.6%	N/A	34.9%	N/A	N/A	39.7%	45.4%
	SRR3190692	12.7%	**11.7%**	13.2%	16.5%	N/A	20.3%	22.3%	N/A	N/A	18.1%	12.7%[a]	24.4%	29.5%
Short-read	ERR385912	6.4%	**5.0%**	6.6%	7.2%	N/A	7.8%	N/A	N/A	5.8%	7.0%	6.6%[a]	13.9%	17.9%
	ERR386131	16.5%	16.0%	18.7%	17.7%	16.6%	16.8%	25.5%	24.6%	**15.5%**	N/A	16.6%[a]	21.5%	26.0%
	SRR034509	23.7%	**22.7%**	25.0%	25.1%	24.9%	26.1%	27.4%	26.1%	23.7%	27.9%	24.5%[a]	31.5%	36.9%
	ERR174310	21.0%	20.1%	N/A	**20.0%**	N/A	20.2%	N/A	N/A	N/A	25.3%	19.6%[a]	26.2%	31.7%
	ERR194147	20.1%	**14.3%**	N/A	20.0%	N/A	20.3%	N/A	N/A	N/A	20.3%	15.4%[a]	19.7%	23.6%

Compressed Ratio: the compressed file size divided by the original file size; 'N/A': the program cannot work on the data, some error occur in program, such as loses fidelity after decompression or decompression failed; [a]: the read order is changed after decompression; The best results are highlighted in bold

LW-FQZip 2 (–g) tends to obtain superior compression ratios to the other methods especially on long-read data. DSRC 2, Quip, FQZcomp, LEON, SCALCE and LW-FQZip 1 suffer from some issues like incompatibility and fidelity-loss on the long-read data generated from Pacbio RS and MinION platforms. LW-FQZip 1, CRAM, FQZcomp and LFQC also fail on some short-read data sets of large size. LFQC obtains comparable compression ratios to LW-FQZip 2 and LW-FQZip 2 (–g) in the data sets it works out. We made an extra comparison analysis

between LW-FQZip 2, LW-FQZip (–g) and LFQC in terms of compression ratio, memory usage, and time consumption in a radar chart in Fig. 3. The results show that LW-FQZip (–g) and LFQC attain slightly better compression ratios than LW-FQZip 2 at the cost of memory usage and time consumption, respectively. Nevertheless, LW-FQZip 2 achieves better compromise over all metrics than the other two methods.

In terms of compression and decompression speeds, LW-FQZip 2 outperforms other reference-based methods

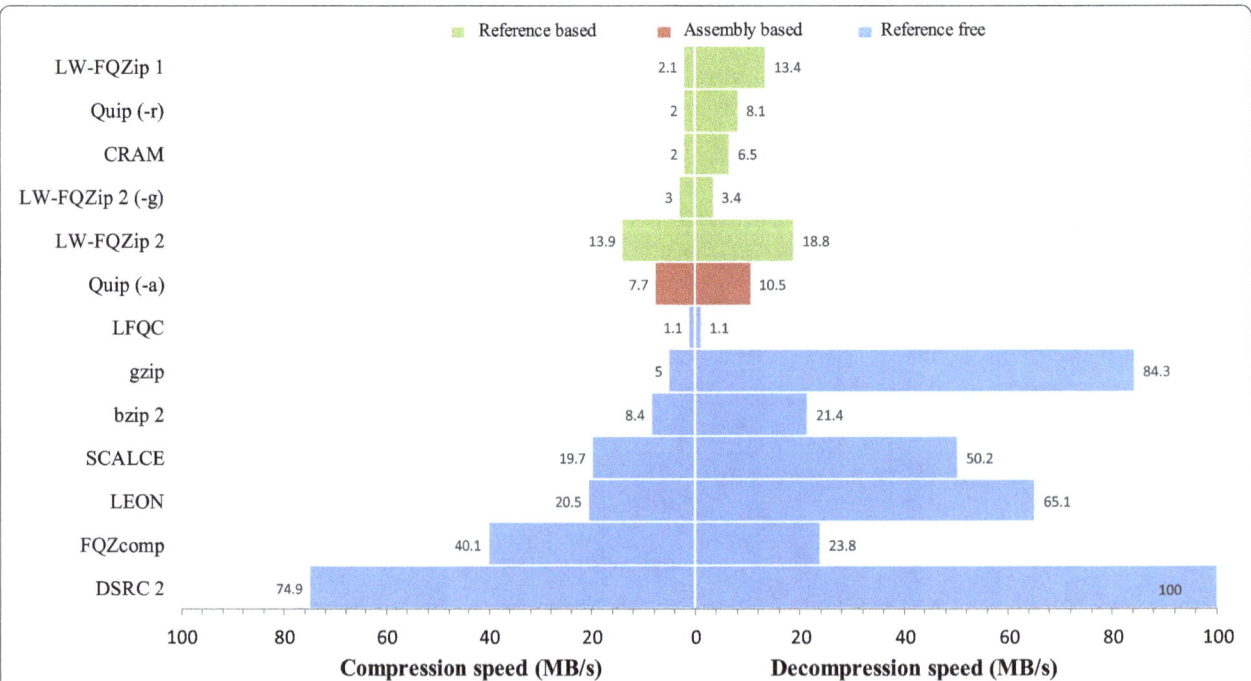

Fig. 2 The average compression and decompression speeds of the compared methods on ten test data sets. The compression speed is calculated as the original file size divided by the compression time. The decompression speed is calculated as the original file size divided by the decompression time

Table 3 The memory usage (MB) of the compared methods on ten test data sets

		LW-FQZip 2	LW-FQZip 2 (−g)	LW-FQZip 1	Quip (−a)	Quip (−r)	DSRC 2	CRAM	FQZcomp	LFQC	SCALCE	LEON	bzip 2	gzip
SRR2916693	compression	1605	4420	459	759	391	582	784	312	3965	1380	1722	7.6	**1.6**
	decompression	1598	4127	37	756	389	601	620	314	3932	1050	696	4.8	**1.5**
SRR2994368	compression	1582	14158	1048	2801	389	6175	1355	N/A	N/A	4073	5435	7.6	**1.6**
	decompression	1579	13154	69	2198	387	7234	652	N/A	N/A	1050	2623	4.8	**1.5**
SRR3211986	compression	1190	12935	N/A	1098	N/A	N/A	5777	N/A	4768	2158	N/A	7.6	**1.6**
	decompression	1528	5657	N/A	1109	N/A	N/A	2381	N/A	4320	1035	N/A	4.8	**1.5**
ERR739513	compression	1283	11511	N/A	N/A	N/A	N/A	3694	N/A	5108	N/A	N/A	7.6	**1.6**
	decompression	1403	11079	N/A	N/A	N/A	N/A	1455	N/A	4748	N/A	N/A	4.8	**1.5**
SRR3190692	compression	1726	14560	1058	3552	391	14157	1363	N/A	N/A	5219	6776	7.6	**1.6**
	decompression	1725	13329	69	2898	386	14794	661	N/A	N/A	1055	3217	4.8	**1.5**
ERR385912	compression	1603	2793	410	772	389	911	N/A	N/A	3140	1422	1717	7.6	**1.6**
	decompression	1603	2655	69	770	392	908	N/A	N/A	3060	1040	504	4.8	**1.5**
ERR386131	compression	1691	12443	1033	771	389	1844	1318	322	5175	1961	N/A	7.6	**1.6**
	decompression	1721	12165	39	768	384	1950	651	319	4835	1049	N/A	4.8	**1.5**
SRR034509	compression	1748	14670	1073	1531	383	6683	1351	324	5352	4151	4139	7.6	**1.6**
	decompression	1752	7042	71	1218	391	7736	653	309	4859	1050	1799	4.8	**1.5**
ERR174310	compression	1886	16270	N/A	5333	N/A	5558	N/A	N/A	N/A	5333	7797	7.6	**1.6**
	decompression	1865	11239	N/A	1156	N/A	18487	N/A	N/A	N/A	1045	5106	4.8	**1.5**
ERR194147	compression	1953	17771	N/A	782	N/A	20271	N/A	N/A	N/A	5380	7192	7.6	**1.6**
	decompression	1963	13908	N/A	780	N/A	24284	N/A	N/A	N/A	1057	5322	4.8	**1.5**

Note: The best results are highlighted in bold

as shown in Fig. 2. It is worth highlighting that LW-FQZip 2 outperforms LW-FQZip 1 in terms of compatibility, compression ratio, and speed, which suggests a substantial improvement of LW-FQZip 2 to LW-FQZip 1. As expected, reference-free methods tend to be faster than reference-based methods. Among the reference-free methods, DSRC 2 compresses the fastest with compromising compression ratio by taking full advantage of multi-threading. Quip and LEON manage to obtain some trade-offs between the three evaluation criteria.

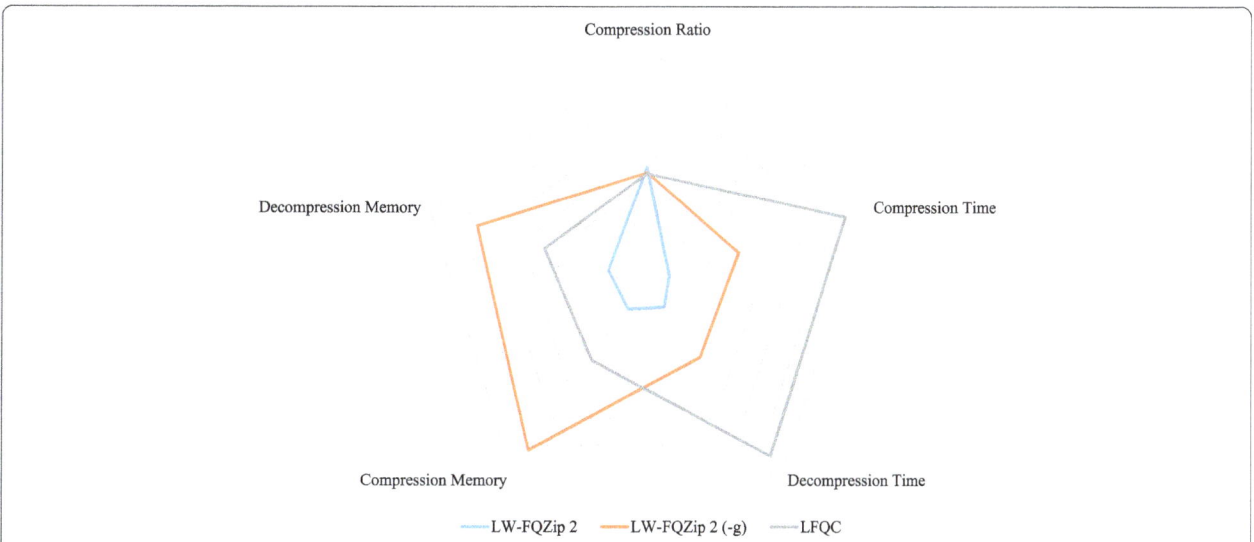

Fig. 3 Comparison between LW-FQZip 2, LW-FQZip 2 (−g) and LFQC in a radar chart in terms of average compression ratio, compression time, decompression time, compression memory usage, and decompression usage. In each criterion, the results of the three methods are normalized to the range of [0, 1] and a smaller value, i.e., closer to the centroid, indicates a better performance

SCALCE is a very efficient tool by introducing a boosting scheme based on locally consistent parsing (LCP) technique to sort the reads, which enables SCALCE to compress similar reads together and attain competitive compression ratios at high speed.

We also try to adopt the LCP technique in our method as the framework shown in Additional file 1: Figure S2. In this attempt, the successfully mapped reads are still compressed with the original LW-FQZip 2, whereas the unmapped reads undergo the LCP boosting and gzip compression. LCP is applied to only a small portion of the reads, yet the results shown in Additional file 1: Table S21, suggest that LCP can improve the compression ratio. Indeed, re-sorting the reads according to their similarity is really an efficient option to improve the compression ratio, especially for archiving-oriented applications. However, since the order of the reads is changed, it inevitably imposes extra cost if random access of the archive is the concern in the downstream analysis. LW-FQZip 2 is designed to preserve the original read order to facilitate the implementation of random access in the future extension of this tool.

The compared methods are also tested on the benchmark data sets suggested by the MPEG working group on genomic compression (https://github.com/sfu-comp bio/compression-benchmark/blob/master/samples.md). The results are presented in Additional file 1: Tables S18-S20, where the proposed method shows consistent efficiency.

The experimental results suggest that all specialized methods outperform the general-purpose tools in terms of compression ratio but use more memory space. The reference-based methods tend to be slower than the reference-free methods, due to the extra running time involved in the sequence alignment preprocessing. Different methods are designed with different strengths and can be used for different purposes. There is no single method dominates other methods in all criteria.

The effect of palindrome handling is investigated in the Additional file 1: Tables S15 and S16. With palindrome handling more reads can be mapped to the reference at a higher speed (the mapping is stopped once the first match is identified), thus the compression ratio and speed are improved slightly, while the extra memory consumption is eligible.

The effect of thread number is studied in Additional file 1: Figure S1. The compression speed is affected not only by the number of threads but also the disk I/O speed. As a result, the compression might not be speeded up proportionally as the number of threads increases (as shown in Additional file 1: Figure S1). Using faster disk system like solid state disk (SSD) can help to speed up both compression and decompression (see in the Additional file 1: Table S22).

Conclusions

This article presents a specialized compression tool LW-FQZip 2 for FASTQ files. LW-FQZip 2 shows superior compression ratios and compatibility with reasonable (de) compression speed and memory space consumption. It could serve as a candidate tool for archival or storage space-sensitive applications of sequencing data.

The emerging long-read technologies, e.g., Single Molecule Real Time (SMRT) sequencing [27] and Nanopore sequencing [28], produce much longer DNA sequences, reportedly providing a more complete picture of genome structure. They are deemed to be a complementary solution to overcome the shortages of short-read sequencing. The exponentially increasing long-read sequencing data poses new great challenges to the existing specialized compression methods. LW-FQZip 2 shows good compatibility to long-read sequencing data. The current work is hoped to provide insights into the storage problems of new sequencing data. More efficient alignment models for long-read data will be developed in the future work.

Additional file

Additional file 1: This document provides the implementation details of LW-FQZip 2 and the detailed experimental results of each compared algorithm in the related comparison study. **Algorithm 1.** The main procedure of LW-FQZip 2. **Algorithm 2.** Compression with PPM prediction model and arithmetic coding. **Table S1.** The performance of LW-FQZip 2. **Table S2.** The performance of LW-FQZip 2 (–g). **Table S3.** The performance of LW-FQZip 1. **Table S4.** The performance of Quip (–a). **Table S5.** The performance of Quip (–r). **Table S6.** The performance of DSRC 2. **Table S7.** The performance of CRAM. **Table S8.** The performance of FQZcomp. **Table S9.** The performance of LFQC. **Table S10.** The performance of LEON. **Table S11.** The performance of SCALCE. **Table S12.** The performance of bzip 2. **Table S13.** The performance of gzip. **Table S14.** The average number of CPU cores used by the compared methods. **Table S15.** The compression ratios and time consumptions of LW-FQZip 2 w/o complementary palindrome mapping. **Table S16.** The memory usage of the LW-FQZip 2 w/o complementary palindrome mapping. **Table S17.** The version information of all compared methods. **Table S18.** The compression ratios of the compared methods on benchmark data provided by MPEG working group on genomic compression. **Table S19.** The performance of LW-FQZip 2 on benchmark data provided by MPEG working group on genomic compression. **Table S20.** The performance of LW-FQZip 2 (–g) on benchmark data provided by MPEG working group on genomic compression. **Table S21.** The compression ratios of LW-FQZip2 + LCP, LW-FQZip2 -g + LCP, SCALCE, and LW-FQZip 2 on seven representative data sets. **Table S22.** The comparison of compression speed of LW-FQZip 2 using SSD and HDD disk systems. **Figure S1.** The compression speeds of LW-FQZip 2 using different number of threads on five representative data sets. **Figure S2.** The framework of LW-FQZip 2 with LCP technique.

Abbreviations
PPM: Prediction by partial matching; LCP: Locally consistent parsing; NCBI: National Centre for Biotechnology Information; SMRT: Single molecule real time; SSD: Solid state disk

Acknowledgements
We would like to thank the Editor and the reviewers for their precious comments on this work which helped improve the quality of this paper.

Funding

This work was supported in part by the National Natural Science Foundation of China (61471246, 61575125, and 61572328), Guangdong Foundation of Outstanding Young Teachers in Higher Education Institutions (Yq2013141 and Yq2015141), Guangdong Special Support Program of Top-notch Young Professionals (2014TQ01X273 and 2015TQ01R453), Guangdong Province Ordinary University Characteristic Innovation Project (2015KTSCX122), Shenzhen Scientific Research and Development Funding Program (ZYC201105170243A, JCYJ20150324141711587, CXZZ20140902102350474, JCYJ20150630105452814, CXZZ20140902160818443, CXZZ20150813151056544, and JCYJ20160331114551175), and China-UK Visual Information Processing Lab Foundation. The funding body played no role in the design or conclusions of the study.

Author's contributions

ZH & ZZ conceived the algorithm, developed the program, and wrote the manuscript. ZW and YS helped with manuscript editing, designed and performed experiments. QD&YC prepared the data sets, carried out analyses and helped with program design. All authors read and approved the final manuscript.

Competing interests

The authors declare that they have no competing interests.

Author details

[1]College of Computer Science and Software Engineering, Shenzhen University, Shenzhen 518060, China. [2]School of Medicine, Shenzhen University, Shenzhen 518060, China.

References

1. van Dijk EL, Auger H, Jaszczyszyn Y, Thermes C. Ten years of next-generation sequencing technology. Trends Genet. 2014;30(9):418–26.
2. Kozanitis C, Heiberg A, Varghese G, Bafna V. Using genome query language to uncover genetic variation. Bioinformatics. 2014;30(1):1–8.
3. Zhu Z, Zhang Y, Ji Z, He S, Yang X. High-throughput DNA sequence data compression. Brief Bioinform. 2015;16(1):1–15.
4. Numanagic I, Bonfield JK, Hach F. Comparison of high-throughput sequencing data compression tools. Nat Methods. 2016;13(12):1005–8.
5. Hosseini M, Pratas D, Pinho AJ. A survey on data compression methods for biological sequences. Information. 2016;7(4):56.
6. Zhu Z, Li L, Zhang Y, Yang Y, Yang X. CompMap: a reference-based compression program to speed up read mapping to related reference sequences. Bioinformatics. 2015;31(3):426–8.
7. Zhang, Y, Patel K, Endrawis T, Bowers A, Sun Y. A FASTQ compressor based on integer-mapped k-mer indexing for biologist. Gene. 2016;579(1):75-81.
8. Bonfield JK, Mahoney MV. Compression of FASTQ and SAM format sequencing data. PLoS ONE. 2013;8(3):e59190.
9. Tembe W, Lowey J, Suh E. G-SQZ: compact encoding of genomic sequence and quality data. Bioinformatics. 2010;26(17):2192–4.
10. Deorowicz S, Grabowski S. Compression of DNA sequence reads in FASTQ format. Bioinformatics. 2011;27(6):860–2.
11. Jones DC, Ruzzo WL, Peng X, Katze MG. Compression of next-generation sequencing reads aided by highly efficient de novo assembly. Nucleic Acids Res. 2012;40(22):e171.
12. Grabowski S, Deorowicz S, Roguski L. Disk-based compression of data from genome sequencing. Bioinformatics. 2015;31(9):1389–95.
13. Janin L, Schulz-Trieglaff O, Cox AJ. BEETL-fastq: a searchable compressed archive for DNA reads. Bioinformatics. 2014;30(19):2796–801.
14. Patro R, Kingsford C. Data-dependent bucketing improves reference-free compression of sequencing reads. Bioinformatics. 2015;31(17):2770–7.
15. Rozov R, Shamir R, Halperin E. Fast lossless compression via cascading Bloom filters. BMC Bioinformatics. 2014;15 Suppl 9:S7.
16. Li H, Handsaker B, Wysoker A, Fennell T, Ruan J, Homer N, Marth G, Abecasis G, Durbin R. The sequence alignment/map format and SAMtools. Bioinformatics. 2009;25(16):2078–9.
17. Fritz MH-Y, Leinonen R, Cochrane G, Birney E. Efficient storage of high throughput DNA sequencing data using reference-based compression. Genome Res. 2011;21(5):734–40.
18. Hach F, Numanagic I, Sahinalp SC. DeeZ: reference-based compression by local assembly. Nat Methods. 2014;11(11):1082–4.
19. Roguski L, Deorowicz S. DSRC 2–Industry-oriented compression of FASTQ files. Bioinformatics. 2014;30(15):2213–5.
20. Nicolae M, Pathak S, Rajasekaran S. LFQC: a lossless compression algorithm for FASTQ files. Bioinformatics. 2015;31(20):3276–81.
21. Benoit G, Lemaitre C, Lavenier D, Drezen E, Dayris T, Uricaru R, Rizk G. Reference-free compression of high throughput sequencing data with a probabilistic de Bruijn graph. BMC Bioinformatics. 2015;16:288.
22. Hach F, Numanagic I, Alkan C, Sahinalp SC. SCALCE: boosting sequence compression algorithms using locally consistent encoding. Bioinformatics. 2012;28(23):3051–7.
23. Howison M. High-throughput compression of FASTQ data with SeqDB. IEEE/ACM Trans Comput Biol Bioinform. 2013;10(1):213–8.
24. Kahn SD. On the future of genomic data. Science. 2011;331(6018):728–9.
25. Zhang Y, Li L, Yang Y, Yang X, He S, Zhu Z. Light-weight reference-based compression of FASTQ data. BMC Bioinformatics. 2015;16:188.
26. Leinonen R, Sugawara H, Shumway M. The sequence read archive. Nucleic Acids Res. 2010;39(suppl_1):D19-D21.
27. Flusberg BA, Webster DR, Lee JH, Travers KJ, Olivares EC, Clark TA, Korlach J, Turner SW. Direct detection of DNA methylation during single-molecule, real-time sequencing. Nat Methods. 2010;7(6):461–5.
28. Branton D, Deamer DW, Marziali A, Bayley H, Benner SA, Butler T, Di Ventra M, Garaj S, Hibbs A, Huang X. The potential and challenges of nanopore sequencing. Nat Biotechnol. 2008;26(10):1146–53.

11

Comprehensive evaluation of RNA-seq quantification methods for linearity

Haijing Jin[1], Ying-Wooi Wan[2] and Zhandong Liu[3]*

Abstract

Background: Deconvolution is a mathematical process of resolving an observed function into its constituent elements. In the field of biomedical research, deconvolution analysis is applied to obtain single cell-type or tissue specific signatures from a mixed signal and most of them follow the linearity assumption. Although recent development of next generation sequencing technology suggests RNA-seq as a fast and accurate method for obtaining transcriptomic profiles, few studies have been conducted to investigate best RNA-seq quantification methods that yield the optimum linear space for deconvolution analysis.

Results: Using a benchmark RNA-seq dataset, we investigated the linearity of abundance estimated from seven most popular RNA-seq quantification methods both at the gene and isoform levels. Linearity is evaluated through parameter estimation, concordance analysis and residual analysis based on a multiple linear regression model. Results show that count data gives poor parameter estimations, large intercepts and high inter-sample variability; while TPM value from Kallisto and Salmon shows high linearity in all analyses.

Conclusions: Salmon and Kallisto TPM data gives the best fit to the linear model studied. This suggests that TPM values estimated from Salmon and Kallisto are the ideal RNA-seq measurements for deconvolution studies.

Keywords: RNA-seq, Deconvolution, Linearity

Background

Next-generation sequencing based technology for RNA profiling (RNA-seq) has become the predominant method to quantify the transcript abundance in cells. Compared to microarray technology, RNA-seq offers broader quantification range and enables the detection of novel transcripts [1]. However, due to the fragmentation of sequencing material, there is greater complexity in quantification and analysis of RNA-seq data [2]. Current state-of-the-art quantification tools for RNA-seq data can be divided into two major categories [3]: alignment-based and alignment-free. Alignment-based quantification methods will first map each sequenced reads to a reference genome or transcriptome and then estimate the abundance of transcripts based on the alignment. Alignment-free quantification methods rely on light-weight pseudo-alignment in k-mer space to quantify the transcript abundance. An analytic challenge raised from these quantification methods is that different method generates abundance measurements in different units, including counts, FPKM (Fragments Per Kilobase of transcript per Million mapped reads), RPKM (Reads Per Kilobase of transcript per Million mapped reads), and TPM (Transcripts Per Million) [4]. Furthermore, various transformation strategies can be applied to quantification values in purpose of specific downstream analysis like differential gene expression analyses [5] or novel splicing site detection [6]. Although several studies have provided assessment of analysis tools for RNA-seq data, little consensus on the optimal analysis pipeline is obtained [4, 6–9].

*Correspondence: zhandonl@bcm.edu
[3]Department of Pediatrics-Neurology, Jan and Dan Duncan Neurological Research Institute, Baylor College of Medicine, 1250 Moursund St., Suite 1325, 77030 Houston, TX, USA
Full list of author information is available at the end of the article

Deconvolution is a mathematical process used to extract constituent elements from a mixture of multiple signals [10]. In the field of biomedical research, deconvolution is widely applied to retrieve cell-type or tissue specific gene expression profiles from heterogeneous tissue samples. Most deconvolution algorithms in the literature assume a linear model [10–17], in which the expression signal of the mixture is a weighted sum of the expression for its constitutive cell types. Previous analysis has shown the necessity of using anti-log expression microarray data to avoid unwanted bias introduced by non-linear transformation [18]. However, no study has assessed the linearity of transcript abundance in RNA-seq data. Therefore, in this study, we conducted a comprehensive comparison of seven RNA-seq quantification methods on the linearity of the estimated abundance using a deep sequencing dataset where RNA samples were mixed at known proportions. Our results will provide a good recommendation to researchers considering deconvolution on RNA-seq data.

Results
Data
We employed the benchmark dataset used to assess RNA-seq measurement performance in different application sites and platforms from the Sequencing Quality Control (SEQC) project [19]. In order to have minimal inter-sample variability in the linearity evaluation analyses, we included samples from the same platform (Illumina HiSeq 2000) and same sequencing center (NVS). Specifically, raw sequenced reads for four biological replicates of four types of samples (A, B, C, D) were obtained; where sample A is derived from universal human reference RNA, sample B is derived from human brain reference RNA, sample C is obtained by mixing A and B in ratio 3:1, and sample D is obtained by mixing A and B in ratio 1:3. Out of 12 samples from A, B and C, nine samples have about eighty million pairs of raw reads and three samples have double the depth. Overall, the mappability of all the samples is around 70–80%. A brief summary about the samples is given in Additional file 6: Table S1.

Quantification methods
We performed a literature survey and selected seven prevalent quantification methods for comparison. To increase the comparability of the estimated transcript abundance, all the alignment-based quantification methods were applied on mappings processed with Tophat2 [20]. HTSeq-count [21] provides the number of reads/fragments mapped unambiguously to a single feature, referred as count. Cufflinks [22] , which is also the most popular quantification method, uses comparative algorithm assembly to produce minimal set of transcript supported by the transcript alignment. The resulting transcript abundance is measured in FPKM. EdgeR [5]

models count data based on an overdisposed Poisson model and uses an empirical Bayes procedure to moderate the degree of overdispersion across genes, which is intended for downstream gene expression analysis. RSEM [3] uses a generative model of RNA-Seq reads and the expectation-maximization (EM) algorithm to estimate abundances of transcript feature. The two alignment-free based quantification methods, Kallisto [23] and Salmon [24] apply pseudo-alignment to find potential transcript origins of RNA-seq reads.

Data distribution
To our surprise, the distribution of the estimated abundance from each quantification method results in distinct distribution at both the gene level (Fig. 1a) and isoform level (Fig. 1b). Except Cufflinks FPKM, all distributions contain two sharp peaks for abundance at gene level (Fig. 1a). While count values are normalized to TPM or FPKM, the second peak is weakened and results in a smoother curve. Although the scaling/normalization factors based on library size and gene length used in TPM and FPKM will explain the reduced range of the quantifications, it cannot explain the reduced height and smoothened second peak in the distribution. For example, logarithm transformation reduces the range of HT-Count quantifications in scale, without weakening the second peak (Fig. 1a). The sharpness of the second peak observed at the gene level diminished at isoform level for all quantification methods (Fig. 1b). Nevertheless, the distribution pattern at both gene and isoform levels remains consistent for the same quantification method.

Assessment of linearity
To evaluate the linearity of the RNA-seq data produced by different quantification methods, we used a multiple linear regression model: $C \sim m \times A + n \times B + \epsilon$. Since sample C is derived by mixing sample A and B at ratio of 3:1, the expected values for the parameters m, n, and ϵ should be 0.75, 0.25, and zero respectively. We reason that if the RNA-seq data are linear, the fitted multiple linear regression model will provide precise estimation for the parameters m, n, and ϵ. Since there are four biological replicates for each sample types (A, B, C or A, B, D), there is a total of 64 possible models to be fitted. We fitted all 64 models and studied the performance based on the average of the parameters and predicted values. As shown in Fig. 2, the true value of C is linearly correlated to the fitted values from models with known mixture proportion 0.75 and 0.25 as parameters in all quantification methods. The linear relationships are especially pronounced at the gene level (Fig. 2a). Although there are more data points away from the diagonal at the isoform level (Fig. 2b), the high density along the diagonal represents strong linear relationships. Results on the estimated

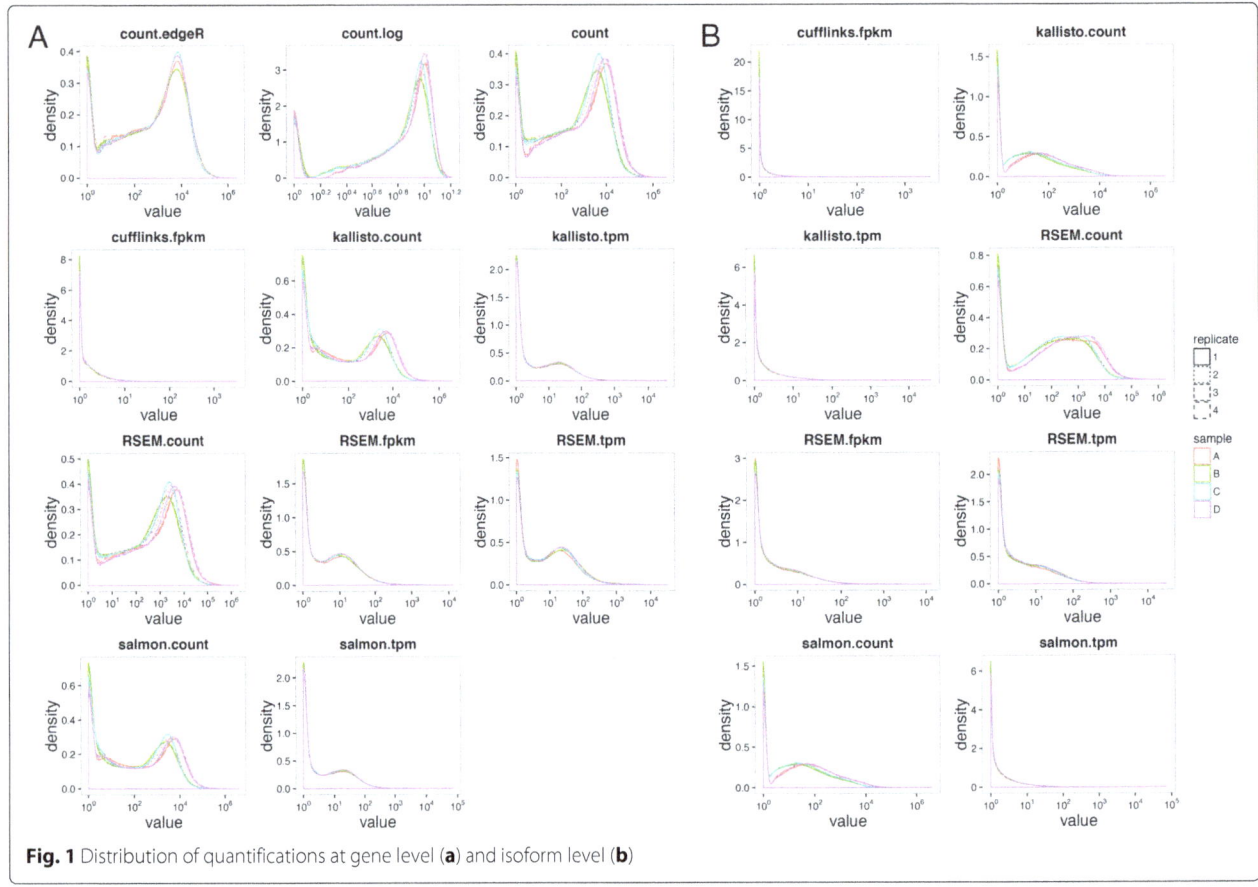

Fig. 1 Distribution of quantifications at gene level (**a**) and isoform level (**b**)

coefficients and intercepts from the 64 models show that count data produce large intercept values at both the gene level and isoform level (Fig. 3, top panel). The intercept values are also sensitive to the counts data from different samples fitted, resulted in highly variable results across models of different samples. FPKM values from Cufflinks give the largest variation in estimated m and n among all non-count quantifications. Overall, TPM and FPKM quantifications reported by Salmon, Kallisto and RSEM resulted in the best estimate of coefficients with small 95% confidence intervals. Similar analyses on isoform level abundance give the same results as observed in gene level data (Fig. 3b, middle and bottom panel).

We observed that the estimation of m resulted in two clusters from count data (Fig. 3, middle panel): one close to true value of 0.75 and another center around 0.4. We found that this phenomenon is due to the library size difference between samples. Specifically, results from cluster close to 0.75 are from 16 combinations with the second replicate of sample A. From Additional file 6: Table S1, we can find that the second replicate of A is the only sample among the four replicates to have similar library size with other samples of C and B. From this result, we could conclude that normalization of quantified abundance is

essential to eliminate the inter-sample variability to meet the linear assumption of deconvolution analysis. To further assess the linearity of the RNA-seq quantifications, we evaluated the fitted model's prediction through three analyses: 1) a concordance analysis between measured C and fitted value $\hat{C}$, 2) a receiver operating characteristic (ROC) curve-like analysis on the absolute residual values, and 3) residual analysis for rescaled model. Similar to parameter estimation analysis mentioned above, linear regression is performed on 64 models, which is constructed upon combinations of 4 replicates in each sample, then estimated value $\hat{C}$ and residuals are averaged for final analysis.

In the concordance analysis, we evaluated if the rank of genes from the fitted value C is consistent with the rank from the true observed C. Plots of the rank of C against the rank of $\hat{C}$ in Fig. 4a demonstrate that count data provide the best concordance while the Cufflink's FPKM result in the worst concordance. In addition, the log transformation of count data induces a slight underestimation. Furthermore, the concordance of abundance estimated from Kallisto and Salmon enhanced tremendously as the expression level increases. Although good concordance is observed in abundances quantified by some methods

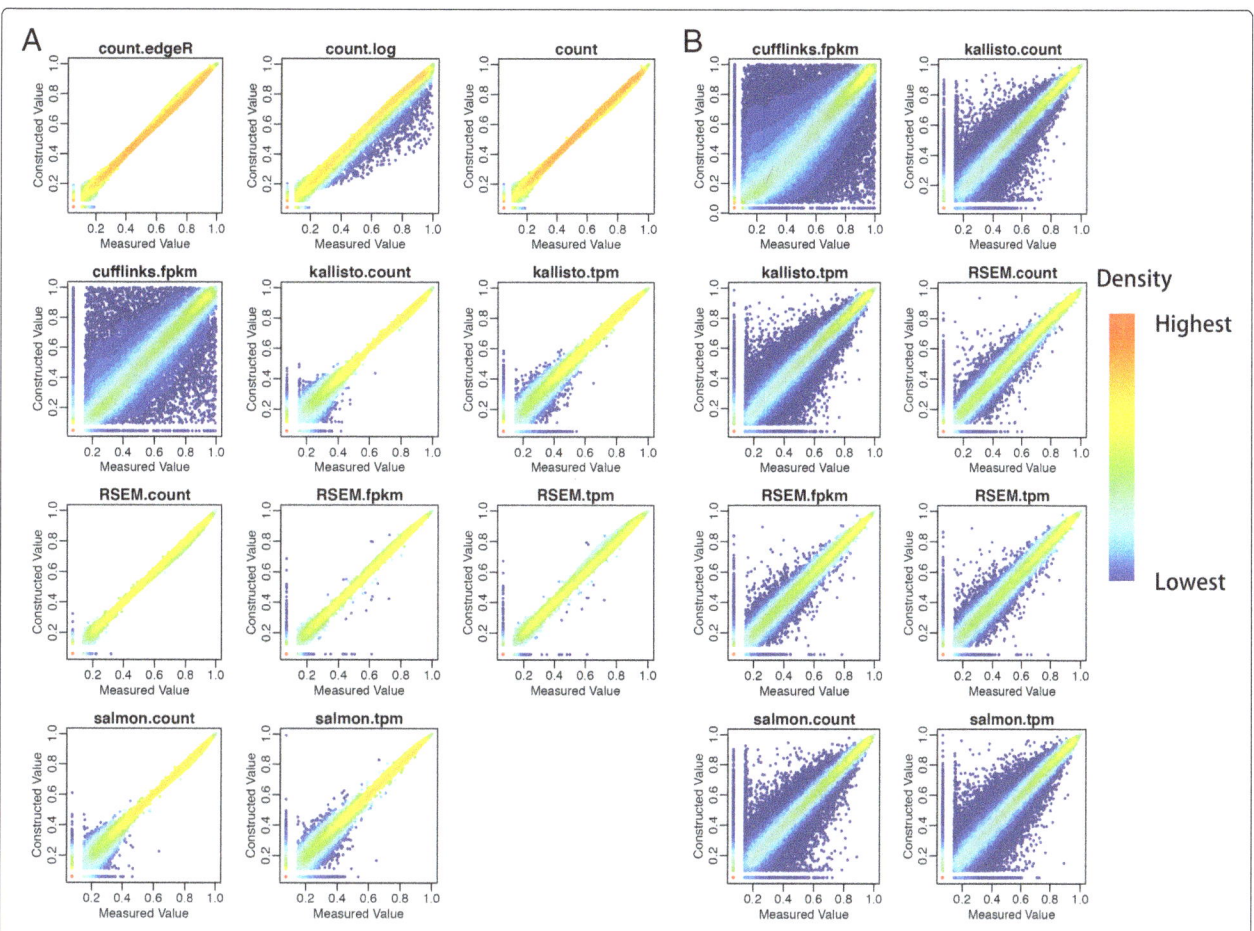

Fig. 2 Concordant analysis between rank of quantifications of $0.75 \times \bar{A} + 0.25 \times \bar{B}$(*Constructed Value*) and $\bar{C}$ (*Measured Value*) at gene level (**a**) and isoform level (**b**). Rankes were normalized by the number of quantifications in each plot

at gene level (Fig. 4a), concordance of abundance at iso-form levels is poor in all methods (Fig. 4b). The poor concordance of isoform level data might due to the bias introduced by isoform abundance quantification methods. Moreover, the concordance analysis (Fig. 2a) on measured C against constructed C based on ground truth $(0.75 \times A + 0.25 \times B)$represents consistent result to the concordance analysis between C against $\hat{C}$ (Fig. 4a). In the ROC-like analysis of absolute residual values, we evaluated the proportions of genes or isoforms with residual at a given threshold level t. Results on the abundance at the gene level show that all studied methods, except for log transformation of count, result in reasonable performance. Among which, TPM from Salmon and Kallisto perform the best (Fig. 5a). Similarly, TPM from Salmon and Kallisto gives the best performance at isoform level and count data perform poorly in general (Fig. 5b)

In the residual analysis, in order to make the residual comparable across quantification methods in different

ranges, we rescaled quantifications and use the model $\frac{C-\mu_C}{\sigma_C} \sim m \times \frac{A-\mu_A}{\sigma_A} + n \times \frac{B-\mu_B}{\sigma_B} + \epsilon$ to conduct multiple linear regression. From the residual plots, we could observe that all the methods except for Cufflinks and log transformation resulted in small residuals across fitted values (Fig. 6a).

Discussion

From the results in Fig. 3, we observed that the performance of linear model on count data is largely affected by the library size of samples fitted. Therefore, we suggest normalization as a compulsory preprocess step prior the deconvolution analysis. The poor estimation of isoform based data (Figs. 2, 3, 4, 5 and 6) might due to the challenge of isoform abundance quantification. In addition to the analyses presented in the "Results" section above, we also conducted the same set of analyses on sample D, using the model $D \sim m \times A + n \times B + \epsilon$. Since D is made up of A and B with proportion 1:3, the expected value for m

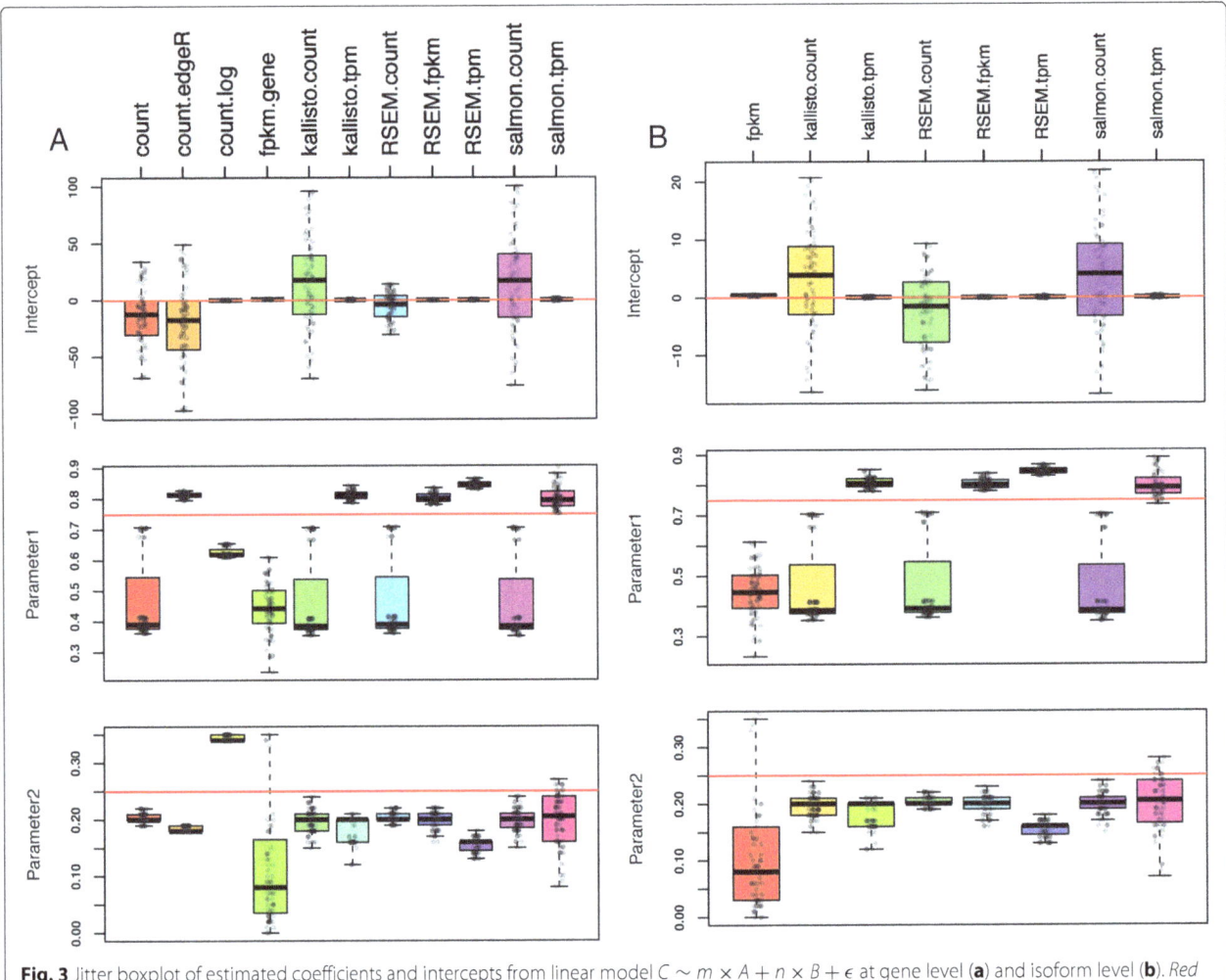

Fig. 3 Jitter boxplot of estimated coefficients and intercepts from linear model $C \sim m \times A + n \times B + \epsilon$ at gene level (**a**) and isoform level (**b**). *Red line* indicates expected estimates if C, A and B satisfy linear assumptiotn

and n is 0.25 and 0.75 respectively in this D-based model. All of the results from the D-based model are consistent with the findings observed from the C-based model discussed (Additional file 1: Figure S1, Additional file 2: Figure S2, Additional file 3: Figure S3, Additional file 4: Figure S4 and Additional file 5: Figure S5). Therefore, we can conclude that the evaluation of linearity of quantification data is not affected by the specific proportion of the original mixture. In summary, the main objective of this study is to provide reference to the researcher considering deconvolution as downstream analysis of RNA-Seq data. Although linear assumption is the priority of deconvolution analysis, it does not guarantee the performance. For future study, assessment of deconvolution performance is still required.

Conclusions

We conducted a comprehensive study to assess the linearity of gene and isoform abundance reported by different RNA-seq quantification methods based on the

performance how these quantifications fitted in a multiple regression linear model. From our analysis, we observed that abundance at both the gene and isoform level from different quantification methods exhibit distinct distribution patterns and thus give diverse results. Abundance reported in the units of counts gave poor linearity, demonstrated from the worst estimated parameters and intercept values of the model. This indicates the necessity of normalizing the abundance data prior the deconvolution analysis. In total, TPM reported by Salmon and Kallisto is the best abundance data for linear models as the estimated parameters are close to true mixture proportions and the fitted values are tightly linearly correlated to the sample's measured abundance. Moreover, when comparing within the same quantification method at both gene and isoform levels, the correlation between the measured and models' fitted values is lower at isoform level compared to gene level while the performance on estimating parameters of linear models are similar in both levels.

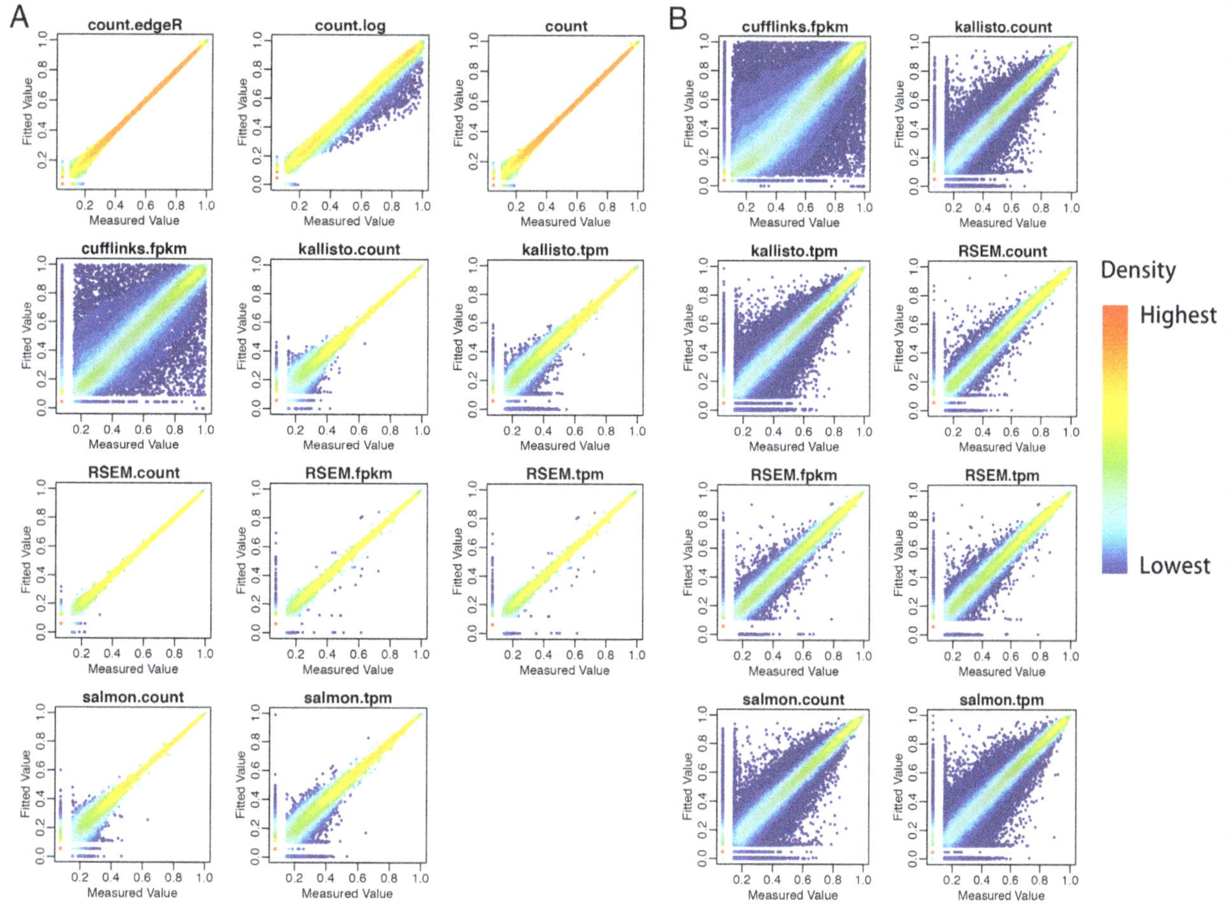

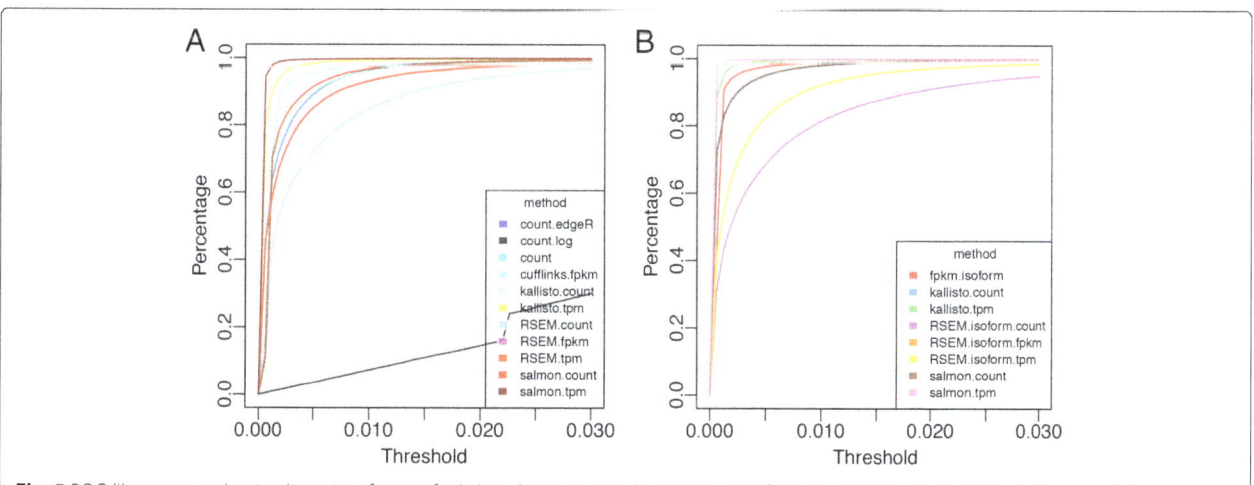

Fig. 4 Concordant analysis between rank of estimated quantifications and rank of measured abundance value at gene level (**a**) and isoform level (**b**). The *fitted value in the y-axis* is estimated from model $C \sim m \times A + n \times B + \epsilon$. Ranks were normalized by the number of quantifications in each plot

Fig. 5 ROC-like curve evaluating linearity of quantified abundance at gene level (**a**) and isoform level (**b**) based on residuals from model $C \sim m \times A + n \times B + \epsilon$. Proportion of variables with residuals *smaller* than a threshold is computed

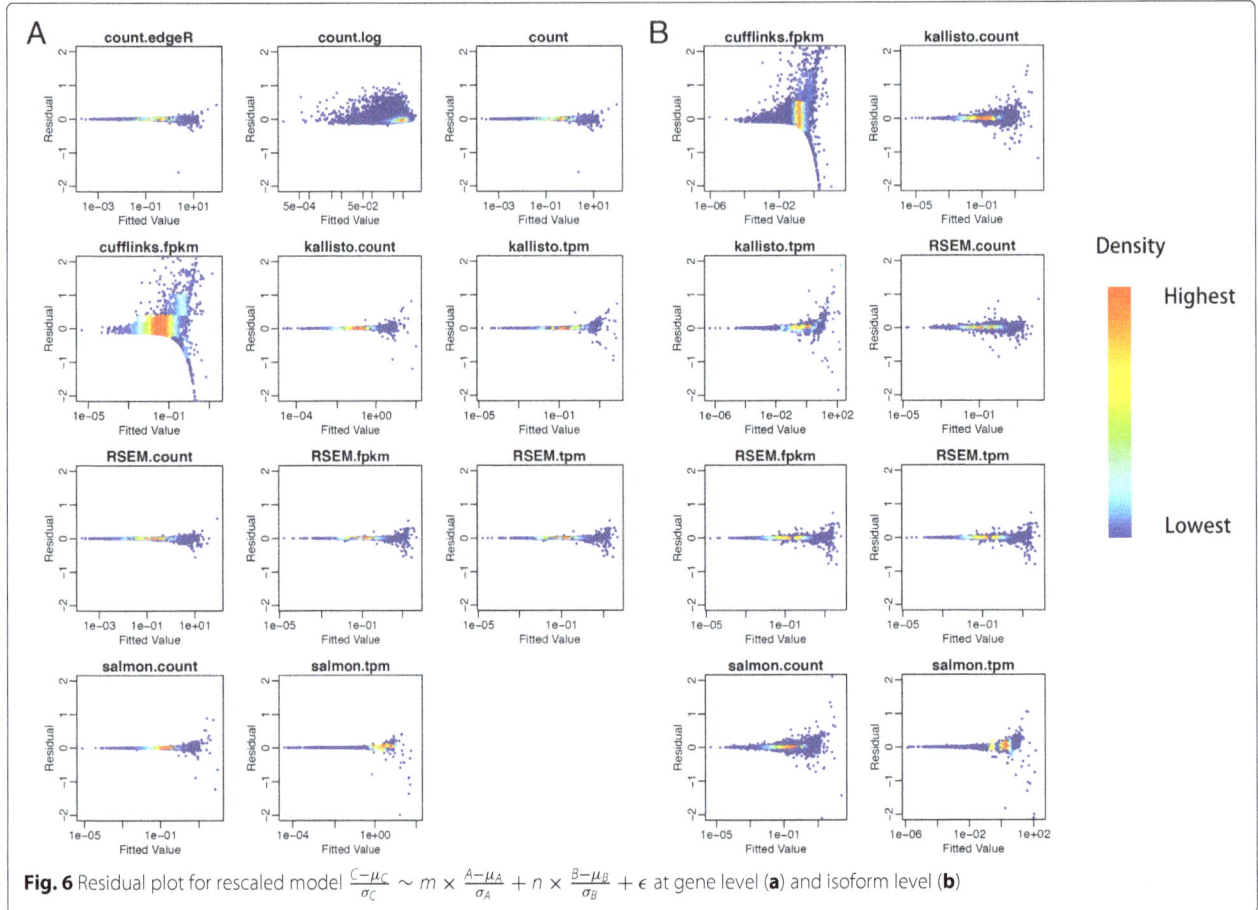

Fig. 6 Residual plot for rescaled model $\frac{C-\mu_C}{\sigma_C} \sim m \times \frac{A-\mu_A}{\sigma_A} + n \times \frac{B-\mu_B}{\sigma_B} + \epsilon$ at gene level (**a**) and isoform level (**b**)

Methods

Dataset

The dataset used in this study is the RNA-seq data generated by the Sequencing Quality Control (SEQC) project. Raw data in fastq format were obtained from GEO website with accession number GSE47774. List of downloaded samples and their details is specified in Additional file 6: Table S1.

Data quantification and analysis

Raw sequence reads from multiple lanes were first merged into one for each replicate of each sample. To prepare sequence map files for alignment-based quantification methods, the merged sequenced reads were mapped to human reference genome (hg19) using Tophat2 (version 2.1.0) with default parameters. The mapped files were then used by quantification methods HTSeq (version 0.6.1) to obtain count data, Cufflinks (versin 2.1.1) to obtain FPKM data. RSEM (version 1.2.31) will first mapped the raw reads using its default aligner Botwtie2 (version 2.1.0) onto human genome (hg19) and then estimate count, FPKM, and TPM. EdgeR (version 3.12.1) is used for count data transformation. Alignment-free quantification methods Kallisto (version 0.42.5) and Salmon

(version 0.6.0) estimate the count and TPM data based on index built from human transcriptome (GRCh37). All the quantification methods were run at both the gene and isoform levels. All the statistical analysis and plots were carried out in R environment (version 3.2.3).

Additional files

Additional file 1: Figure S1. Concordant analysis between rank of quantifications of $0.25 \times \bar{A} + 0.75 \times \bar{B}$(Constructed Value) and $\bar{D}$ (Measured Value) at gene level (a) and isoform level (b). Rankes were normalized by the number of quantifications in each plot.

Additional file 2: Figure S2. Jitter boxplot of estimated coefficients and intercepts from linear model $D \sim m \times A + n \times B + \epsilon$ at gene level (a) and isoform level (b). Red line indicates expected estimates if D, A and B satisfy linear assumption.

Additional file 3: Figure S3. Concordant analysis between rank of estimated quantifications and rank of measured abundance value at gene level (a) and isoform level (b). The fitted value in the y-axis is estimated from model $D \sim m \times A + n \times B + \epsilon$. Ranks were normalized by the number of quantifications in each plot.

Additional file 4: Figure S4. ROC-like curve evaluating linearity of quantified abundance at gene level (a) and isoform level (b) based on residuals from model $D \sim m \times A + n \times B + \epsilon$. Proportion of variables with residuals smaller than a threshold is computed.

Additional file 5: Figure S5. Residual plot for rescaled model
$\frac{D-\mu_D}{\sigma_D} \sim m \times \frac{A-\mu_A}{\sigma_A} + n \times \frac{B-\mu_B}{\sigma_B} + \epsilon$ at gene level (a) and isoform level (b).

Additional file 6: Table S1. Column 1 and 2 show the sample type and replicates index; Column 3 shows total read pairs, column 4-7 show the mapping rate from different quantification methods (Tophat, RSEM, Kallisto and Salmon).

Abbreviations

EM: Expectation maximization; FPKM: Fragments per kilobase of transcript per million mapped reads; NVS: Norvartis; ROC: Receiver operating characteristics; RPKM: Reads per kilobase of transcript per million mapped reads; SEQC:Sequencing qualify control; TPM: Transcripts per million

Acknowledgments

We thank Hari Krishna Yalamanchili and reviewers for comments that greatly improved this manuscript.

Funding

ZL, YW, and HJ are partially supported by the Houston Endowment, NSF DMS-1263932 and CPRIT RP170387. The publication cost of this article was funded by NSF DMS-1263932.

Authors' contributions

HJ performed the analysis of the data. The initial manuscript was drafted by HJ and ZL. YW and ZL conceived the study. All authors read and approved the final manuscript.

Competing interests

The authors declare that they have no competing interests.

About this supplement

This article has been published as part of *BMC Bioinformatics* Volume 18 Supplement 4, 2017: Selected original research articles from the Third International Workshop on Computational Network Biology: Modeling, Analysis, and Control (CNB-MAC 2016): bioinformatics. The full contents of the supplement are available online at https://bmcbioinformatics.biomedcentral .com/articles/supplements/volume-18-supplement-4.

Author details

[1]Graduate Program in Structural and Computational Biology and Molecular Biophysics, Baylor College of Medicine, One Baylor Plaza, 77030 Houston, TX, USA. [2]Department of Molecular and Human Genetics, Baylor College of Medicine, One Baylor Plaza, 77030 Houston, TX, USA. [3]Department of Pediatrics-Neurology, Jan and Dan Duncan Neurological Research Institute, Baylor College of Medicine, 1250 Moursund St., Suite 1325, 77030 Houston, TX, USA.

References

1. Wang Z, Gerstein M, Snyder M. RNA-Seq: a revolutionary tool for transcriptomics. Nature reviews Genetics. 2009;10(1):57–63. doi:10.1038/nrg2484.
2. Zhao S, Fung-Leung WP, Bittner A, Ngo K, Liu X. Comparison of RNA-Seq and microarray in transcriptome profiling of activated T cells. PLoS ONE. 2014;9(1):. doi:10.1371/journal.pone.0078644.
3. Teng M, Love MI, Davis CA, Djebali S, Dobin A, Graveley BR, Li S, Mason CE, Olson S, Pervouchine D, Sloan CA, Wei X, Zhan L, Irizarry RA. A benchmark for RNA-seq quantification pipelines. Genome Biol. 2016;17(1):74. doi:10.1186/s13059-016-0940-1.
4. Conesa A, Madrigal P, Tarazona S, Gomez-Cabrero D, Cervera A, McPherson A, Szcześniak MW, Gaffney DJ, Elo LL, Zhang X, Mortazavi A. A survey of best practices for RNA-seq data analysis. Genome Biol. 2016;17(1):13. doi:10.1186/s13059-016-0881-8.
5. Robinson MD, McCarthy DJ, Smyth GK. edgeR: a Bioconductor package for differential expression analysis of digital gene expression data. Bioinformatics. 2009;26(1):139–40. doi:10.1093/bioinformatics/btp616.
6. Germain PL, Vitriolo A, Adamo A, Laise P, Das V, Testa G. RNAontheBENCH: Computational and empirical resources for benchmarking RNAseq quantification and differential expression methods. Nucleic Acids Res. 2016;44(11):5054–67. doi:10.1093/nar/gkw448.
7. Kanitz A, Gypas F, Gruber AJ, Gruber AR, Martin G, Zavolan M. Comparative assessment of methods for the computational inference of transcript isoform abundance from RNA-seq data. Genome Biol. 2015;16(1):150. doi:10.1186/s13059-015-0702-5.
8. Chandramohan R, Wu PY, Phan JH, Wang MD. Benchmarking RNA-Seq quantification tools. In: Proceedings of the Annual International Conference of the IEEE Engineering in Medicine and Biology Society. EMBS; 2013. p. 647–50. doi:10.1109/EMBC.2013.6609583.
9. Fonseca NA, Marioni J, Brazma A. RNA-Seq gene profiling - a systematic empirical comparison. PLoS ONE. 2014;9(9). doi:10.1371/journal.pone.0107026.
10. Mohammadi S, Zuckerman N, Goldsmith A, Grama A. A critical survey of deconvolution methods for separating cell-types in complex tissues. arXiv:1510.04583 [cs.CE]. doi:10.1109/JPROC.2016.2607121.
11. Newman AM, Liu CL, Green MR, Gentles AJ, Feng W, Xu Y, Hoang CD, Diehn M, Alizadeh AA. Robust enumeration of cell subsets from tissue expression profiles. Nat Methods. 2015;12(2014):1–10. doi:10.1038/nmeth.3337.
12. Zhong Y, Wan YW, Pang K, Chow LML, Liu Z. Digital sorting of complex tissues for cell type-specific gene expression profiles. BMC Bioinforma. 2013;14:89. doi:10.1186/1471-2105-14-89.
13. Rahmani E, Zaitlen N, Baran Y, Eng C, Hu D, Galanter J, Oh S, Burchard EG, Eskin E, Zou J, Halperin E. Sparse PCA corrects for cell type heterogeneity in epigenome-wide association studies. Nat Methods. 2016;13(5):443–5. doi:10.1038/nmeth.3809.
14. Liebner DA, Huang K, Parvin JD. MMAD: microarray microdissection with analysis of differences is a computational tool for deconvoluting cell type-specific contributions from tissue samples. Bioinformatics. 2014;30(5):682–9. doi:10.1093/bioinformatics/btt566.
15. Abbas AR, Wolslegel K, Seshasayee D, Modrusan Z, Clark HF. Deconvolution of blood microarray data identifies cellular activation patterns in systemic lupus erythematosus. PLoS ONE. 2009;4(7). doi:10.1371/journal.pone.0006098. arXiv:1506.03733v1.
16. Gong T, Hartmann N, Kohane IS, Brinkmann V, Staedtler F, Letzkus M, Bongiovanni S, Szustakowski JD. Optimal deconvolution of transcriptional profiling data using quadratic programming with application to complex clinical blood samples. PLoS ONE. 2011;6(11). doi:10.1371/journal.pone.0027156.
17. Shen-Orr SS, Tibshirani R, Khatri P, Bodian DL, Staedtler F, Perry NM, Hastie T, Sarwal MM, Davis MM, Butte AJ. Cell type-specific gene expression differences in complex tissues. Nat Methods. 2010;7(4):287–9. doi:10.1038/nmeth.1439.
18. Zhong Y, Liu Z. Gene expression deconvolution in linear space. Nat Meth. 2012;9(1):8–9. doi:10.1038/nmeth.1830.
19. Su Z, Łabaj PP, Li SS, Thierry-Mieg J, Thierry-Mieg D, Shi W, Wang C, Schroth GP, Setterquist RA, Thompson JF, Jones WD, Xiao W, Xu W, Jensen RV, Kelly R, Xu J, Conesa A, Furlanello C, Gao HH, Hong H, Jafari N, Letovsky S, Liao Y, Lu F, Oakeley EJ, Peng Z, Praul CA, Santoyo-Lopez J, Scherer A, Shi T, Smyth GK, Staedtler F, Sykacek P, Tan XX, Thompson EA, Vandesompele J, Wang MD, Wang JJJ, Wolfinger RD, Zavadil J, Auerbach SS, Bao W, Binder H, Blomquist T, Brilliant MH, Bushel PR, Cai W, Catalano JG, Chang CW, Chen T, Chen G, Chen R, Chierici M, Chu TM, Clevert DA, Deng Y, Derti A, Devanarayan V, Dong Z, Dopazo J, Du T, Fang H, Fang Y, Fasold M, Fernandez A, Fischer M, Furió-Tari P, Fuscoe JC, Caimet F, Gaj S, Gandara J, Gao HH, Ge W, Gondo Y, Gong B, Gong M, Gong Z, Green B, Guo C, Guo L-WL, Guo L-WL, Hadfield J, Hellemans J, Hochreiter S, Jia M, Jian M, Johnson CD, Kay S, Kleinjans J, Lababidi S, Levy S, Li QZ, Li L, Li P, Li Y, Li H, Li J, Li SS, Lin SM, López FJ, Lu X, Luo Y, Ma X, Meehan J, Megherbi DB, Mei N, Mu B, Ning B, Pandey A, Pérez-Florido J, Perkins RG, Peters R, Phan JH, Piroozznia M, Qian F, Qing T, Rainbow L, Rocca-Serra P, Sambourg L, Sansone SA, Schwartz S, Shah R, Shen J, Smith TM, Stegle O,

Stralis-Pavese N, Stupka E, Suzuki Y, Szkotnicki LT, Tinning M, Tu B, van Delft J, Vela-Boza A, Venturini E, Walker SJ, Wan L, Wang W, Wang JJJ, Wang JJJ, Wieben ED, Willey JC, Wu PY, Xuan J, Yang Y, Ye Z, Yin Y, Yu Y, Yuan YC, Zhang J, Zhang KK, Zhang WW, Zhang WW, Zhang Y, Zhao C, Zheng Y, Zhou Y, Zumbo P, Tong W, Kreil DP, Mason CE, Shi L. A comprehensive assessment of RNA-seq accuracy, reproducibility and information content by the Sequencing Quality Control Consortium. Nat Biotechnol. 2014;32(9):903–14. doi:10.1038/nbt.2957. NIHMS150003.

20. Kim D, Pertea G, Trapnell C, Pimentel H, Kelley R, Salzberg SL. TopHat2: accurate alignment of transcriptomes in the presence of insertions, deletions and gene fusions. Genome Biol. 2013;14(4):36. doi:10.1186/gb-2013-14-4-r36.

21. Anders S, Pyl PT, Huber W. HTSeq-A Python framework to work with high-throughput sequencing data. Bioinformatics. 2015;31(2):166–9. doi:10.1093/bioinformatics/btu638.

22. Trapnell C, Williams BA, Pertea G, Mortazavi A, Kwan G, van Baren MJ, Salzberg SL, Wold BJ, Pachter L. Transcript assembly and quantification by RNA-Seq reveals unannotated transcripts and isoform switching during cell differentiation. Nat Biotechnol. 2010;28(5):511–5. doi:10.1038/nbt.1621. 171.

23. Bray NL, Pimentel H, Melsted P, Pachter L. Near-optimal probabilistic RNA-seq quantification. Nat Biotechnol. 2016;34(5):525–7. doi:10.1038/nbt.3519. 1505.02710.

24. Patro R, Duggal G, Kingsford C. Salmon: accurate, versatile and ultrafast quantification from RNA-seq data using lightweight-alignment. bioRxiv. 2015021592. doi:10.1101/021592. 1505.02710.

12

A time series driven decomposed evolutionary optimization approach for reconstructing large-scale gene regulatory networks based on fuzzy cognitive maps

Jing Liu[1*], Yaxiong Chi[1], Chen Zhu[1] and Yaochu Jin[2]

Abstract

Background: Reconstructing gene regulatory networks (GRNs) from expression data plays an important role in understanding the fundamental cellular processes and revealing the underlying relations among genes. Although many algorithms have been proposed to reconstruct GRNs, more rapid and efficient methods which can handle large-scale problems still need to be developed. The process of reconstructing GRNs can be formulated as an optimization problem, which is actually reconstructing GRNs from time series data, and the reconstructed GRNs have good ability to simulate the observed time series. This is a typical big optimization problem, since the number of variables needs to be optimized increases quadratically with the scale of GRNs, resulting an exponential increase in the number of candidate solutions. Thus, there is a legitimate need to devise methods capable of automatically reconstructing large-scale GRNs.

Results: In this paper, we use fuzzy cognitive maps (FCMs) to model GRNs, in which each node of FCMs represent a single gene. However, most of the current training algorithms for FCMs are only able to train FCMs with dozens of nodes. Here, a new evolutionary algorithm is proposed to train FCMs, which combines a dynamical multi-agent genetic algorithm (dMAGA) with the decomposition-based model, and termed as dMAGA-FCM$_D$, which is able to deal with large-scale FCMs with up to 500 nodes. Both large-scale synthetic FCMs and the benchmark DREAM4 for reconstructing biological GRNs are used in the experiments to validate the performance of dMAGA-FCM$_D$.

Conclusions: The dMAGA-FCM$_D$ is compared with the other four algorithms which are all state-of-the-art FCM training algorithms, and the results show that the dMAGA-FCM$_D$ performs the best. In addition, the experimental results on FCMs with 500 nodes and DREAM4 project demonstrate that dMAGA-FCM$_D$ is capable of effectively and computationally efficiently training large-scale FCMs and GRNs.

Keywords: Gene regulatory networks, Fuzzy cognitive maps, Big data, Big optimization, Multi-agent genetic algorithm, Decomposition

Background

In the age of big data, there is an urgent need to develop effective and computationally efficient methods to convert data into knowledge. When we extract knowledge from big data, we often need to solve big optimization problems, which may involve thousands of, even millions of, decision variables. For example, in the *"Optimization of Big Data Competition"* organized at the *IEEE 2015 Congress on Evolutionary Computation*, a set of big optimization benchmark problems were formulated, which have up to 4864 decision variables extracted from EEG data analyses [1, 2].

In biology, techniques of high-throughput experiment such as DNA microarrays bring the "big data" in molecular biology and system biology, which is made up with a large number of molecular entities and their products.

---

* Correspondence: neouma@mail.xidian.edu.cn
[1]Key Laboratory of Intelligent Perception and Image Understanding of
Ministry of Education, Xidian University, Xi'an 710071, China
Full list of author information is available at the end of the article

Finding the interactions between these entities is a key step to understand the governing mechanism of biological systems. The so-called gene regulatory networks, or GRNs in short, has been proved to be the most widely used model to analyze the dynamic behavior of biological systems. GRNs model the molecular entities as networks which consist of a group of nodes (representing genes, proteins and small molecules) which influence each other. And the objective of GRNs is to capture the interconnections among these genes. By reconstructing the complex interconnections within these GRNs, we can highlight inhibitory or excitatory interactions, as well as how intracellular or extracellular factors (environmental and drug-induced effects) affect gene products or deregulate cellular process. The reconstruction of a GRN based on expression data is also called reverse engineering or network inference.

In recent years, various algorithms, especially evolutionary algorithms (EAs) have been proposed to infer GRNs by analyzing of gene expression data, such as genetic algorithm (GA) [3], genetic programming (GP) [4], evolution strategy (ES) [5], and ant colony optimization (ACO) [6]. However, the GRNs modeled by the above algorithms only consist of a limited number of genes. How to reconstruct the large scale GRNs is still an underdetermined problem in biology.

Various models have been used to model GRNs. The simplest models are based on Boolean networks. In the reverse engineering, Boolean networks are used to infer both the underlying topology and the Boolean functions at the nodes from observed gene expression data [7].

In addition, continuous networks, an extension of the Boolean networks, are also widely used to model GRNs. Each gene is considered to be a node in the network, while the edge represents the relationship between genes. In biological systems, the activity level of genes is considered in a continuous range, continuous representation captures this feature, while the Boolean model is not work. Many methods based on continuous networks have been proposed to the inference of GRNs. Such as linear regression based, and the mutual information (MI) based methods. ARACNE algorithm [8], the classical MI based method, calculate the value of MIs of all gene pairs. If the calculated value exceeds the given threshold, then one regulatory interaction is inferred [9].

Many probabilistic graphical models have been proposed by researchers to measure the high-order dependency among distinct gene expression patterns. The Bayesian network is one of the most popular methods used in the inference of GRNs. In the Bayesian network, directed acyclic graphs are used to indicate the conditional dependency among random variables [10].

Despite that plenty of proposed models using gene expression data to infer GRNs [11], these models only adopt scaled real values or Boolean values to indicate the levels of gene expression [12]. FCMs use fuzzy values, which integrate the benefits of both real values and Boolean values, to figure out the real relationships in GRNs [12]. Many researches [13] have validated that FCMs are efficient and powerful tool when it comes to modeling complex regulatory network systems.

Thus, within this context, we focus on developing a new evolutionary approach framework to train GRNs, which is based on fuzzy cognitive maps, or FCMs in short. As a type of effective computation tool, fuzzy cognitive maps (FCMs) were proposed by Kosko in [14] for the first time. Many work have been demonstrated the effective ability of FCMs in modeling complex systems. An FCM is a network model to describe the relationship between different concepts in the real world. Nodes in the network stand for the concepts and weighted edges are used to quantify the relationship between concepts. Compared to traditional modeling techniques, FCMs are more reasonable and intuitive description of human reasoning and have been successfully used in numerous practical application areas, such as medical diagnosis [15], metabolism network modeling [16], process control [17], military science [18], and modeling of software development [19, 20].

In recent years, lots of work has been carried out in the research of train FCMs from data. In fact, given an initial state of an FCM, which is represented by a set of state values of constituent nodes, a trained FCM can simulate the evolution of state values over time to predict the future state values. Thus, the objective of these learning algorithms is to determine weights in FCMs so that the response time sequences of each node in the trained FCMs can fit the observed time sequences as far as possible.

Therefore, viewed from methodology, the learning objective of reconstructing FCM models from time series data can be described as an optimum formula. And the learning algorithm is to minimize the optimum formula as far as possible, which is simulating the observed time sequence. This will become a typical big optimization problem when the scale of FCM expands to a high level, since the number of decision variables, namely, the number of weights needs to be determined, increases quadratically with the number of nodes in FCMs. For example, if we need to train an FCM with 500 nodes, 250 000 decision variables need to be optimized is a typical large-scale optimization problem.

Evolutionary algorithms (EAs) are a subset of generic population-based optimization algorithms inspired by the biological evolution. In real applications, EAs often performs well to a wide range of problems. EAs and other population based metaheuristics have been shown to be powerful in training FCMs, including genetic algorithms (GAs) [21, 22], particle swarm optimizations (PSOs) [23], simulated annealing (SA) [24], ant colony optimization (ACO) algorithms [6, 25], and memetic algorithms [26].

Stach et al. [21] proposed a method to optimize FCMs using real-coded genetic algorithm (RCGA). The aim of RCGA is to model the system structure from the time sequence which was observed from the complex system. Since time series data only at two adjacent time point can be used as a training sample, the whole time sequence can be broken down into several pairs according to time point t and $t + 1$. The estimated values of time point $t + 1$ can be calculated by the values of time point t in each pair. The objective of learning algorithm is to find a FCM that can reproduce the time sequence which is the same as the observed time sequence. The RCGA first established the error between the output data of estimated FCM and the observed time sequence as an optimum formula, then, the operators of RCGA were used to solve the optimum formula which is actually minimize the error. The experiments results in [21] show that RCGA have a big advantage over ES.

Subsequently, in order to train large scale FCMs, Stach et al. in [22] applied a scalable divide and conquer strategy to speed up RCGA. In divide and conquer RCGA, the whole training data are divided into several subsets. Then, a FCM model can be learned from each subset by RCGA. And these learned FCM models are averaged as the final solution.

Particle swarm optimization (PSO) was also extended to solve the optimization problem of training FCM by Parsopoulos et al. in [23]. The learning algorithm aims to reduce the searching space continually and find a small region of candidate FCM models which is close to the real FCM model.

In addition of the above learning algorithms, other population-based algorithms were also used to learning FCMs. For example, a simulated annealing (SA) based FCM learning algorithm was proposed by Alizadeh et al. in [24]. In the learning algorithm, the optimal object is the same as the one in [21], except that the optimization algorithm is changed to SA. Chen et al. [25] proposed an ant colony optimization (ACO) algorithm for training FCMs with no more than 40 nodes, where the weights were discretized. Additionally, FCMs were also applied to solve many practical problems. In a recent work by Chen et al., biological GRNs were modeled as FCMs, and the author proposed a hybrid method which combines inherently continuous ACO algorithm with a decomposition-based optimization strategy to reconstruct biological GRNs with 100 genes from gene expression data [6].

Acampora et al. [26] introduced a memetic algorithm to generate FCM models from historical data without a priori knowledge. Extensive comparative studies performed on both synthetic and real-world data verified the competitive performance of the memetic algorithm.

However, FCMs learned by most existing algorithms are always small-scale with dozens of nodes, and only the ACO_{RD} proposed in [6] used FCMs to reconstruct gene regulatory networks (GRNs) with 100 nodes. Thus, there is a demand to develop methods capable of training large-scale FCMs based on time series data.

In our previous work, multi-agent systems and GAs are combined to form a new algorithm named as a multi-agent genetic algorithm (MAGA) for large-scale global numerical optimization [27]. The results shown MAGA performed well even for the optimization problems with 10 000 decision variables. In [28, 29], MAGA was also used as the learning algorithm to solve constraint satisfaction problems and combinatorial optimization problems, and the experiment results show a good performance. MAGA was extended to successfully solve the big optimization problem extracted from EEG data analysis mentioned above [30, 31]. Moreover, in [32], a new version of MAGA, termed as dynamic MAGA (dMAGA), was proposed, which can effectively train FCMs with 200 nodes (40 000 variables). However, dMAGA formulated the FCM training problem as one single optimization problem, where all weights are determined simultaneously. Such formulation prevents dMAGA from being able to efficiently handle even larger FCMs, such as FCMs with 500 nodes.

In fact, to learn an FCM can be considered as to learn how each node in the FCM is affected by other nodes. Thus, in order to make a training algorithm for FCMs scalable to a large number of nodes, we can decompose an FCM learning problem into multiple optimization problems, where each optimization problem corresponds to the training of a single node. That is to say, for training an FCM with N nodes, the original optimization problem having $N \times N$ weights (decision variables), will be decomposed into N optimization problems with each having only N decision variables. A similar idea has also been reported in [6, 33] in reconstructing GRNs based on FCMs, which successfully reduces the size of the optimization algorithm and favors a distributed implementation. Nevertheless, efficiently solving N optimization problems with N decision variables remains to be challenging, in particular when N becomes large.

To take full advantages of both dMAGA and the above-mentioned decomposition-based optimization approach, this work proposes a new algorithm for training larger FCMs by combining the decomposition based approach with dMAGA, which is termed as dMAGA-FCM_D. FCMs with various sizes are used to verify the performance of the proposed dMAGA-FCM_D. The results show that dMAGA-FCM_D is able to effectively train FCMs with 500 nodes and significantly enhance the performance of the original dMAGA. Moreover, dMAGA-FCM$_D$ is employed to reconstruct a biological GRN, which is a challenging real-world problem. The proposed dMAGA-FCM$_D$ is shown to outperform a few state-of-the-art algorithms on the benchmark problem DREAM4 [34], which the *Data Error* is as low as 0.2, where the results of other algorithms are around 0.4. DREAM4 project is a GRNs inference challenge which aims at reconstructing network structure from simulated steady-state and time series data.

Methods
Decomposition-based FCM for GRNs reconstruction
The relations between different concepts in a complex system can be described as a directed graph which consists of nodes and weighted edges. To adequately describe the concepts, it is necessary to use real number to quantity the expression level of each concept. For an FCM consists of N concept nodes, the state values of these nodes are described as a vector C:

$$C = [C_1, C_2, ..., C_N] \tag{1}$$

where, $C_i \in [0, 1]$, $i = (1, 2, ..., N)$. Once the state value of each node is described, we need to describe the relations between nodes. Here, we use a an $N \times N$ weight matrix w to define the relations, which is also the candidate solution to the FCM learning problem,

$$w = \begin{bmatrix} w_{11} & w_{12} & ... & w_{1N} \\ w_{21} & w_{22} & ... & w_{2N} \\ \vdots & \vdots & \ddots & \vdots \\ w_{N1} & w_{N2} & ... & w_{NN} \end{bmatrix} \tag{2}$$

A weight is assigned to an edge connecting any two nodes to quantify the strength and the type of the relations between the two nodes. The absolute value of a weight represents how strong the source node affects the target nodes, while the positive or negative sign of weights denotes an excitatory or inhibitory relation [25]. In (2), all weights w_{ij} are in the range of [-1, 1], representing the causal relations between nodes i and j, $i, j = 1, 2, ..., N$. Fig. 1a shows a simple FCM with 5 nodes and 7 edges, and the corresponding weight matrix is presented in Fig. 1b, where, e.g., $w_{12} = +0.34$ indicates that there is an excitatory edge pointing from node 1 to node 2 with a strength of 0.34. $w_{54} = 0$ means there is no causal relation between nodes 5 and 4. Similarly, $w_{44} = -0.9$ suggests that node 4 has a negative effect of feedback regulatory on itself. The higher the absolute values of the weights is, the stronger the relations.

In fact, an FCM with N nodes can be decomposed into N single sub-maps in the following way. For simplicity, we focus on the input information of each node. Each node and its input nodes can be regarded as a sub-map. Every single map corresponds to a column vector in the weight matrix. In other words, the weight matrix in

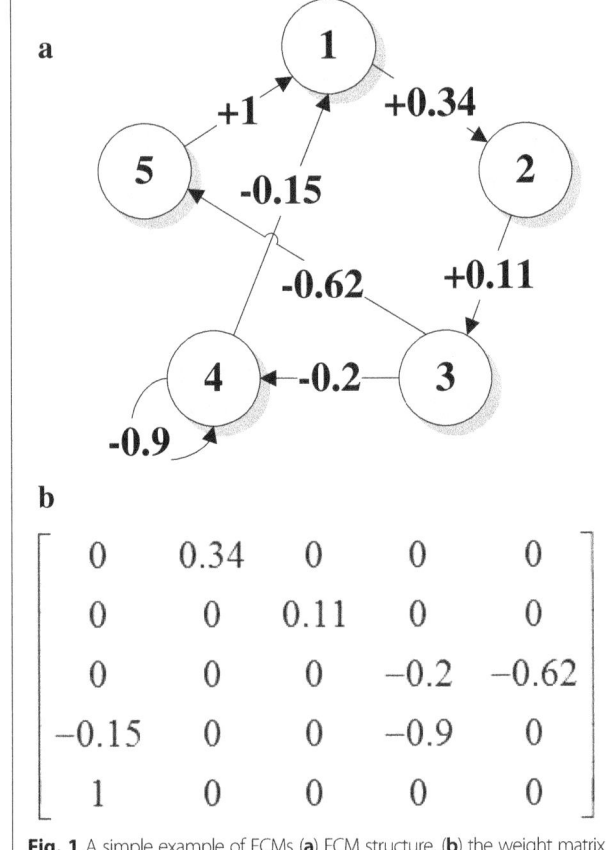

Fig. 1 A simple example of FCMs (**a**) FCM structure, (**b**) the weight matrix

(2) can be represented as $w = [w_1, w_2, ..., w_N]$, where $w_i = (w_{i1}, w_{i2}, ..., w_{iN})$, $i = 1, 2, ..., N$. For example, the FCM in Fig. 1 can be decomposed into 5 sub-maps shown in Fig. 2a and b shows the corresponding weight relations of each single sub-map.

When the activation degree of nodes are produced, we use the following equation to predict the activation degree C_i^{t+1} based on the known values C_i^t,

$$C_i^{t+1} = g\left(\sum_{j=1}^{N} w_{ji} C_j^t\right), \forall i \in \{1, 2, ..., N\} \tag{3}$$

where C_i^t is the state value node i at the t-th iteration in one response, and $g(\cdot)$ is a sigmoid activation function that maps the activation level to the unit interval [0, 1],

$$g(x) = \frac{1}{1 + e^{-\lambda x}} \tag{4}$$

where parameter λ decides the slop of the function, and a small value lead to highly nonlinear model. Logistic transformation function is the most commonly used function in FCMs and offers significantly greater advantages than other functions [35, 36]. According to the recommendation in [37], we set 5 as the value of λ in this paper.

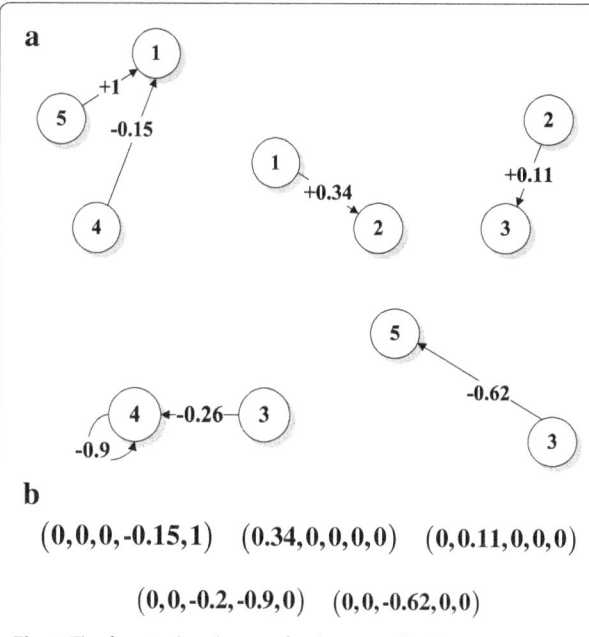

a

b

$(0,0,0,-0.15,1)$ $(0.34,0,0,0,0)$ $(0,0.11,0,0,0)$

$(0,0,-0.2,-0.9,0)$ $(0,0,-0.62,0,0)$

Fig. 2 The five single sub-maps for the example FCM in Fig. 1 (**a**) 5 single sub-maps, (**b**) the corresponding weight relations

FCM learning algorithms aims to find the relations between different concepts from response sequences. In computation terms, the objective is to learn the best inter-connection matrix which performs the best on simulating the response sequences. Specifically, the error between the responses generated by candidate FCM and observed response sequences are formulated as an optimization expression, and learning algorithms are used to minimize it. And the error mentioned above is labeled as *Data_Error*,

$$Data\_Error(w) = \frac{1}{NN_s(N_t-1)} \sum_{n=1}^{N} \sum_{s=1}^{N_s} \sum_{t=1}^{N_t-1} \left(C_n^t(s)-\hat{C}_n^t(s)\right)2 \quad (5)$$

where N_s is the number of response sequences, N_t is the number of time instants in the response sequences, $C_n^t(s)$ is the t-th desired state value of node n in the s-th response sequence, and $\hat{C}_n^t(s)$ is the t-th state value of node n in the s-th generated response sequence.

From the decomposition point of view, *Data_Error* in (5) is actually averaged over the data error of each node and can be re-formulated as

$$Data\_Error(w) = \frac{1}{N} \sum_{n=1}^{N} \left(\frac{1}{N_s(N_t-1)} \sum_{s=1}^{N_s} \sum_{t=1}^{N_t-1} \left(C_n^t(s)-\hat{C}_n^t(s)\right)^2 \right)$$

$$= \frac{1}{N} \sum_{n=1}^{N} Data\_Error_n(w_n) \quad (6)$$

Thus, the data error of node i can be represented as follows,

$$Data\_Error_i(w_i) = \frac{1}{N_s(N_t-1)} \sum_{s=1}^{N_s} \sum_{t=1}^{N_t-1} \left(C_i^t(s)-\hat{C}_i^t(s)\right)^2 \quad (7)$$

When we calculate $\hat{C}_i^t(s)$, the state values of input nodes to node i in the desired response sequences are used. In this way, the data error of each node can be calculated independently. In the following text, Eq. (7) is used as the objective function for optimizing the weights of the i-th sub-map.

Decomposition-based dMAGA for training FCMs

Different from the dMAGA proposed in [32], which is used to optimize the whole weight matrix simultaneously, dMAGA-FCM$_D$ proposed in this paper optimizes the column vectors of the weight matrix independently. In the following, we first define the agents used in dMAGA, and then introduce the genetic operators to be performed on the agents. Finally, the detailed implementation of dMAGA-FCM$_D$ is provided.

Definition of the agents

In dMAGA-FCM$_D$, each candidate weight column vector w_i is regarded as an agent, where i means the i-th single sub-map.

Definition 1

For a FCM with N nodes, each agent is represented as an N-dimensional vector. Once the representation of agent is determined, the most important thing is to define the expression of energy of each agent. In this work, the energy is the negative value of *Data_Error* defined in (7). The dMAGA-FCM$_D$ aims to increase the energy of each agent as much as possible in order to survive in the environment, which is equivalent to decrease the *Data_Error*. To realize the local perceptivity of the agents, the environment is organized in a lattice-like structure, in which each agent competes or cooperates with their neighbors. An agent interacts with its neighbors, exchanging information with each other, thereby achieving global information share. The lattice in which all agents live is defined as follows.

Definition 2

All agents live in a lattice-like environment, L, which is called an agent lattice. The size of L is $L_{size} \times L_{size}$, where L_{size} is an integer. Each agent is fixed on a lattice-point and can only interact with its neighbors. The agent located at (a, b) is represented as $L_{a,b}$, a, $b = 1, 2, ..., L_{size}$, then the neighbors of $L_{a,b}$, $Neighbors_{a,b}$ are defined as follows.

$$Neighbors_{a,b} \subseteq \{L_{a',b}, L_{a,b'}, L_{a,b''}, L_{a'',b}\} \quad (8)$$

where

$$a' = \begin{cases} a-1, & a\neq1 \\ \boldsymbol{L}_{size}, & a=1 \end{cases}, \quad b' = \begin{cases} b-1, & b\neq1 \\ \boldsymbol{L}_{size}, & b=1 \end{cases},$$

$$a'' = \begin{cases} a+1, & a\neq\boldsymbol{L}_{size} \\ 1, & a=\boldsymbol{L}_{size} \end{cases}, \text{ and } b'' = \begin{cases} b+1, & b\neq\boldsymbol{L}_{size} \\ 1, & b=\boldsymbol{L}_{size} \end{cases}.$$

As shown in Fig. 3, there are four agents at most in the neighborhood for each agent. In the standard MAGA, each agent has to compete with the other four agents around it, in other words, the four agents around it are all its neighbors, so the agents with low energy are eliminated early, which make MAGA lost the diversity of the population under the huge selection pressure. Thus, in dMAGA-FCM$_D$, in order to tune the selection pressure, each agent can select its neighbors dynamically according to the energy. Namely, there is no need to compete with all the agents around it. Thus, $\boldsymbol{Neighbors}_{a,b}$ in is a subset of (8).

To dynamically determine the neighbor of each agent, at the beginning of each generation, all agents will be sorted according to a decreasing order of their energy and evenly divided into four levels. And the agents in the first/second/third/fourth level can select four/three/two/one neighbor(s) randomly from (8). In this way, an agent with a lower level of energy will have a smaller number of neighbors, thereby improving its chance to survive than those in MAGA.

Genetic operators for agents

Four genetic operators are used to guide the evolutionary search, which are the neighborhood competition, the crossover, the mutation, and the self-learning operator, of which the neighborhood competition and crossover operators account for competition and cooperator among agents, while the mutation and self-learning operators are dedicated to improving the agents' ability of local search and making use of knowledge to survive. Suppose the four operators are performed on agent $\boldsymbol{L}_{a,b} = (l_1, l_2, ..., l_N)$, and $\boldsymbol{Max}_{a,b} = (m_1, m_2, ..., m_N)$ is the agent with the maximum energy in neighbors of $\boldsymbol{L}_{a,b}$.

Neighborhood competition operator

If $\boldsymbol{L}_{a,b}$ satisfies $\textbf{Energy}(\boldsymbol{L}_{a,b}) > \textbf{Energy}(\boldsymbol{Max}_{a,b})$, which means $\boldsymbol{L}_{a,b}$ defeats all the neighbors and keep the current solution in lattice; otherwise, $\boldsymbol{L}_{a,b}$ lost the change to survive and have to be replaced by $\boldsymbol{Max}_{a,b}$ which is the energy with maximum energy in the neighborhood. There are also two strategies determine how the agent $\boldsymbol{L}_{a,b}$ would be replaced. That is, if $U(0, 1) < P_o$, strategy 1 is adopted; otherwise, strategy 2 is used. Here, P_o is a predefined probability, and $U(0, 1)$ is a random number uniformly distributed in the interval of [0, 1].

Strategy 1

A new agent $\boldsymbol{New}_{a,b} = (e_1, e_2, ..., e_N)$ is generated as follows:

$$e_i = \begin{cases} \text{-1}, & \omega < \text{-1} \\ 1, & \omega > 1 \\ \omega, & \text{otherwise} \end{cases}, \text{ where} \qquad (9)$$

$$\omega = m_i + U(-1, 1) \times (m_i - l_i), \quad i = 1, 2, ..., N$$

Since there might be still some useful information in $\boldsymbol{L}_{a,b}$ even if it is a loser, Strategy 1 uses both $\boldsymbol{L}_{a,b}$ and $\boldsymbol{Max}_{a,b}$ to generate a new agent, which is used to replace $\boldsymbol{L}_{a,b}$.

Strategy 2

Randomly select two integers k and s satisfying $1 < k < s < N$, then a new agent is generated as follows to replace $\boldsymbol{L}_{a,b}$,

$$\boldsymbol{New}_{a,b} = (m_1, m_2, ..., m_{k-1}, m_s, m_{s-1}, ..., m_{k+1}, m_k, m_{s+1}, ..., m_N)$$

$$(10)$$

Crossover and mutation operators

The orthogonal crossover and Gaussian mutation operators are applied on $\boldsymbol{L}_{a,b}$ and $\boldsymbol{Max}_{a,b}$. The reader are referred to [32] for the details of these operators.

Self-learning operator

In MAGA, a small scale MAGA is introduced as the self-learning operator for agents to be able to use knowledge, which however, is still a sort of random-based local search strategy. Therefore, to make use of the properties of FCMs, in dMAGA, a one dimensional

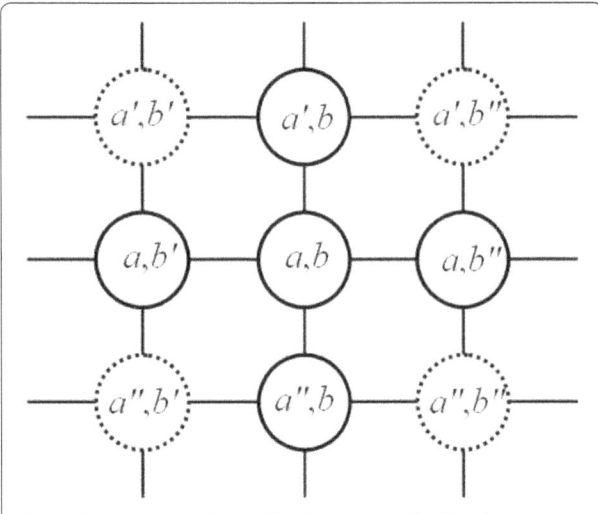

Fig. 3 The topology of the defined agent neighborhood

search strategy is adopted to implement the self-learning operator. In this work, we also perform this self-learning operator on $L_{a,b}$, and more details can be found in [32].

Implementation of dMAGA-FCM$_D$

dMAGA-FCM$_D$ optimizes each sub-map sequentially. For each sub-map, the neighbors of each agent is first determined, then each agent compete with its neighbors according to their energy, and the agent with a higher level of energy survives in the population. Once the competition is performed, the crossover and mutation operators are performed on each agent with a probability P_c and P_m, respectively. Then, the best agent in the current generation improves its energy through self-learning operator.

Algorithm 1 shows more details of dMAGA-FCM$_D$, where L_i^t represents the agent lattice at the t-th generation for the i-th single sub-map, $CBest_i^t$ represents the best agent in the t-th generation for the i-th single sub-map, and $Best_i$ represents the best agent found for the i-th single sub-map.

Results

Experiments on synthetic FCMs

In this section, the dMAGA-FCM$_D$ is tested on synthetic FCMs. For fully test the performance of the dMAGA-FCM$_D$, the scale of FCMs is varying from 5 to 500 nodes. In order to show the improvement of dMAGA-FCM$_D$ over dMAGA, extensive comparisons are made between these two algorithms. In addition, dMAGA-FCM$_D$ is compared with three representative existing methods based on evolutionary algorithms, namely, RCGA [21], ACO$_{RD}$ (an improved variant of ACO) [6] and differential evolution (DE) [38], where the results of RCGA and DE are taken from the literature. ACO$_{RD}$ and dMAGA are run under the same experimental settings as those of dMAGA-FCM$_D$. All experiments on dMAGA-FCM$_D$ are conducted for the same parameter settings, which are given in [32].

Data_Error, which is presented in (5), is an important measure to evaluate the performance of different learning algorithms. Unlike the *Data_Error* is used to evaluate the fitting ability of time sequences, *Model_Error* is a direct comparison between the weights of the learned model and the target model,

$$Model\_Error = \frac{1}{N^2} \sum_{i=1}^{N} \sum_{j=1}^{N} \left| w_{ij} - \hat{w}_{ij} \right| \qquad (11)$$

where $\hat{w}_{ij}$ is the learned weight of between nodes i to j.

In order to evaluate the performance of the algorithms under comparison in predicting the existence of edges, the problem of training FCMs is extended and transformed into a binary classification problem. That is, the target FCM model and the learned FCM model are transformed into binary networks according to a predefined threshold. The absolute weights that are larger than the threshold are transformed to 1; otherwise 0. Once the transformational rule is determined, we need to choose a value of the predefined threshold. In this paper, we choose 0.05 as the value of the threshold which is the same value used in [37]. In FCMs, we usually think that the causal relation with strength less than 0.05 usually has no significance in practical problems.

The author of [37] also gives the definition of positive and negative edges. According to the definition, the edges with absolute weights larger than 0.05 are identified as negative edges; otherwise, they are identified as positive edges. The *SS mean* is employed to evaluate the performance,

$$SS\ mean = \frac{2 \times Sensitivity \times Specificity}{Sensitivity + Specificity} \qquad (12)$$

where

$$Sensitivity = \frac{N_{TP}}{N_{TP} + N_{FN}} \qquad (13)$$

$$Specificity = \frac{N_{TN}}{N_{TN} + N_{FP}} \qquad (14)$$

where *TP*, *FN*, *TN*, and *FP* are defined in Table 1. *T* (true) means the correctly identified edges. *F* (false) means the edge is identified as the opposite character. *P* (positive) means positive edge and *N* (negative) means negative edge which has defined above (12). For example, N_{FP} is the number of false positive edges, which means the number of negative edges that are incorrectly identified as positive edges.

In this paper, the artificial response time sequences used to train FCM models are generated by a two-step method proposed in [25]. First, random real numbers which should be within the interval [-1, 1] are assigned to the weights of the interconnection matrix, an additional file shows this in more detail (see Additional file 1). However, according to the viewpoint in [25], which is if the weight between two nodes is smaller than 0.05, the relation between these two nodes can be ignored in practical application, we check each weight whether its absolute value is smaller than 0.05. If so, the weight will be set to 0.

Table 1 Definition of *TP*, *FN*, *TN*, and *FP*

		Target networks	
		0	1
Learned networks	0	TP	FP
	1	FN	TN

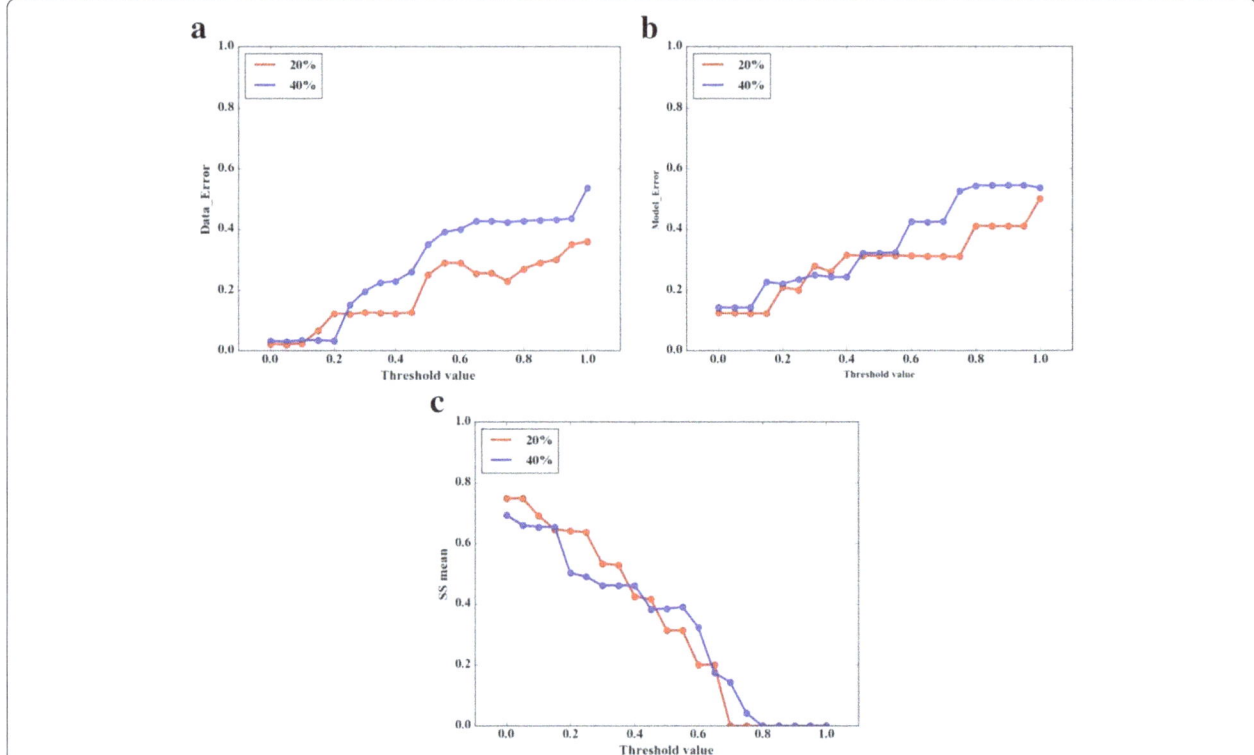

Fig. 4 Comparison in terms of (**a**) *Data_Error*, (**b**) *Model_Error* and (**c**) *SS mean* on FCMs (10 nodes) with various of threshold value which ranges from 0 to 1. The red line and blue line show the comparison experiments for FCMs with 20% density and 40% density, respectively

Second, the state value of response time sequences at the initial time point are assigned by random value ranging from 0 to 1. Then, the subsequent time sequences can be generated according to (3) based on the FCM model and initial state values.

Here, the threshold value (0.05) used in this work is always a default parameter in many work [6, 12, 25] about FCMs. However, different threshold value will affect the algorithm performance. In order to explore the impact of the threshold value on algorithm performance, we

conduct the following experiment. When the network (10 Node, 20% density and 40% density) was learned, we set the value of threshold ranges from 0 to 1 in the step of 0.05 and get a series of networks. Then we calculate the *Data_Error*, *Model_Error* and *SS mean* for networks with different threshold values. From Fig. 4, we can see that the value of threshold affect the algorithm performance greatly, no matter what the density of the network, the performance of the algorithm decreases with the increase of threshold value. When the threshold value is

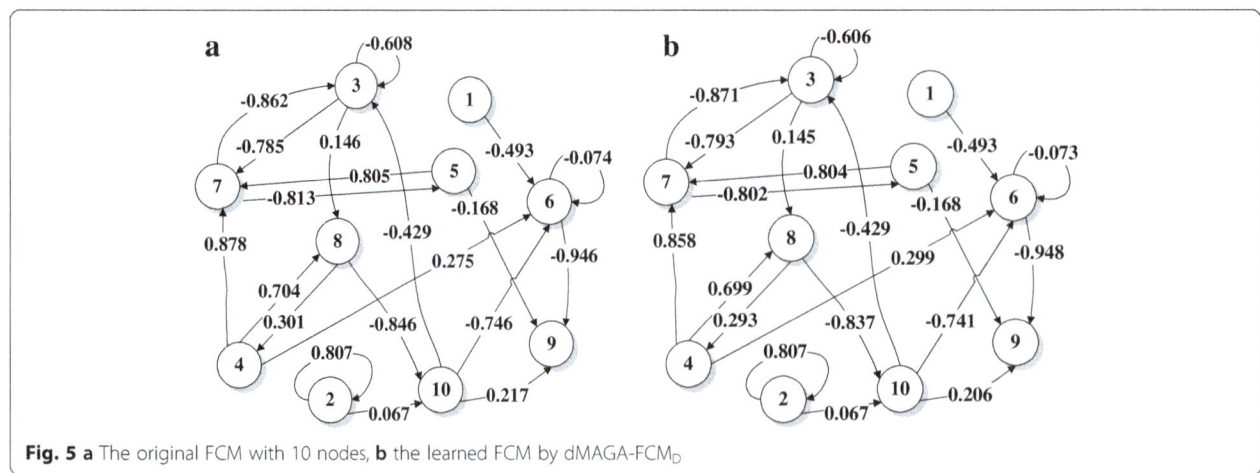

Fig. 5 a The original FCM with 10 nodes, **b** the learned FCM by dMAGA-FCM$_D$

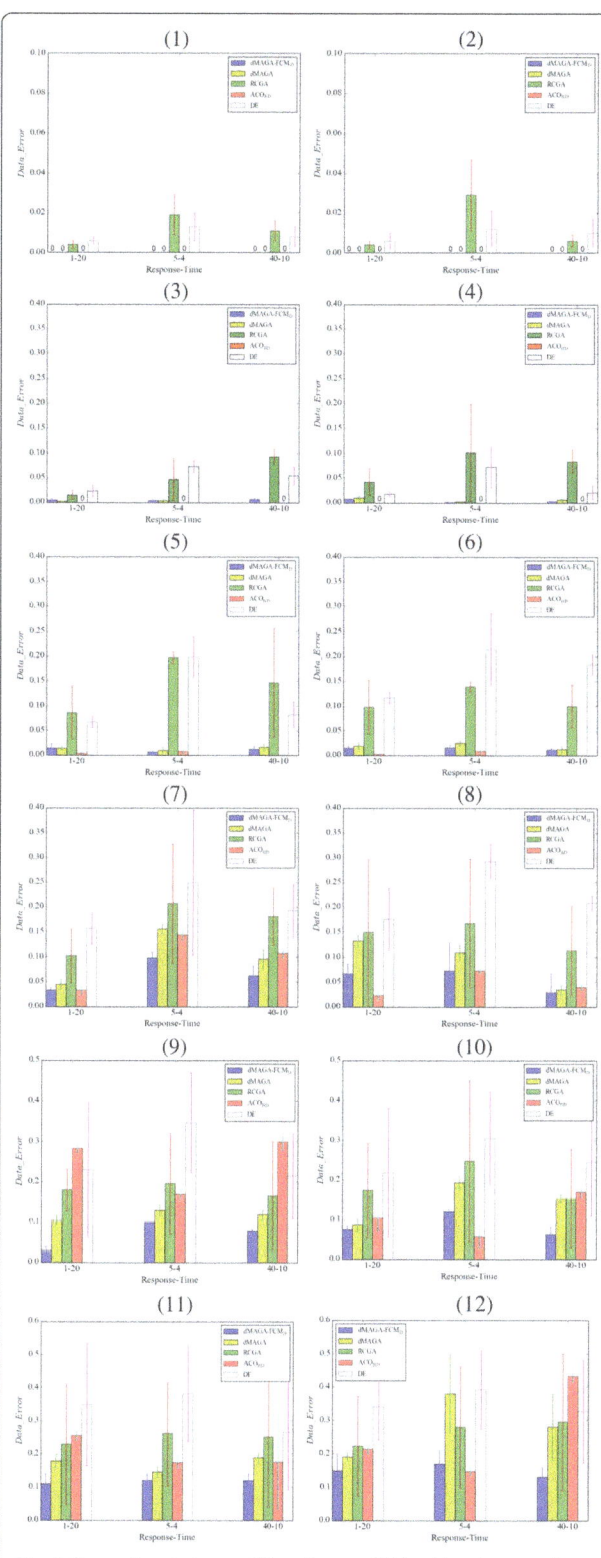

Fig. 6 Comparison in terms of *Data_Error* on FCMs with various number of nodes. (1) 5 nodes, density = 20%. (2) 5 nodes, density = 40%. (3) 10 nodes, density = 20%. (4) 10 nodes, density = 40%. (5) 20 nodes, density = 20%. (6) 20 nodes, density = 40%. (7) 40 nodes, density = 20%. (8) 40 nodes, density = 40%. (9) 100 nodes, density = 20%. (10) 100 nodes, density = 40%. (11) 200 nodes, density = 20%. (12) 200 nodes, Density = 40%

bigger than 0.6, the value of *SS mean* is even 0 which means there is no existence of edge in this network is predicted correctly. Thus, it is important to select an appropriate threshold for the performance of the algorithm.

In real-world, the network structures of FCMs are always sparse. There are only a fraction of links between nodes in FCMs. For this reason, we need to control the density of the random FCM model. When we generate the FCM models by the method mentioned above, a large part of weights are assigned to 0 and the others are set to nonzero random real numbers. For example, if the edge density of 20% is expected for an FCM with 10 nodes, 20 random edges (there are $10 \times 10 = 100$ edges with the full connected graph, then 20% of 100 edges is 20) will be selected and random values will be assigned to these edges.

The size of generated FCM varies from 5 to 500 ($N = 5$, 10, 20, 40, 100, 200, 300, and 500), and for each size, the edge density of 20% and 40% are used. In order to compare with other algorithms, experiments on three scenarios are conducted. The first scenario has one response sequence with 20 time points each ($N_s = 1$, $N_t = 20$), the second scenario has five response sequences with four time points each ($N_s = 5$, $N_t = 4$), and the third scenario has 40 response sequences with 10 time points each ($N_s = 40$, $N_t = 10$). Figure 5a shows an original FCM with 10 nodes with 20% edge density and Fig. 5b shows the corresponding FCM learned by dMAGA-FCM$_D$. By comparing the original and learned FCMs, we see that the network structure is fully correctly learned and the errors between the original and learned weights are smaller than 0.02. Comprehensive comparative results in terms of *Data_Error* on FCMs with 5 ~ 200 nodes are presented in Fig. 6, and a detailed comparison of the FCMs with 300 and 500 nodes is reported in Table 2. The comparison in terms of *Model_Error* is given in Fig. 7 and Table 3. The results are averaged over 30

Table 2 Comparison in terms of *Data_Error* on larger synthetic FCMs (Average ± Standard deviation)

#Nodes	Edge density	N_s-N_t	dMAGA-FCM$_D$	dMAGA
300	20%	1–20	0.105 ± 0.021	0.297 ± 0.102
		5–4	0.125 ± 0.053	0.247 ± 0.031
		40–10	0.118 ± 0.037	0.336 ± 0.025
	40%	1–20	0.121 ± 0.082	0.214 ± 0.112
		5–4	0.063 ± 0.019	0.351 ± 0.051
		40–10	0.139 ± 0.070	0.298 ± 0.042
500	20%	1–20	0.145 ± 0.107	0.368 ± 0.114
		5–4	0.186 ± 0.034	0.420 ± 0.027
		40–10	0.156 ± 0.072	0.395 ± 0.064
	40%	1–20	0.130 ± 0.056	0.348 ± 0.108
		5–4	0.164 ± 0.081	0.416 ± 0.086
		40–10	0.173 ± 0.022	0.404 ± 0.053

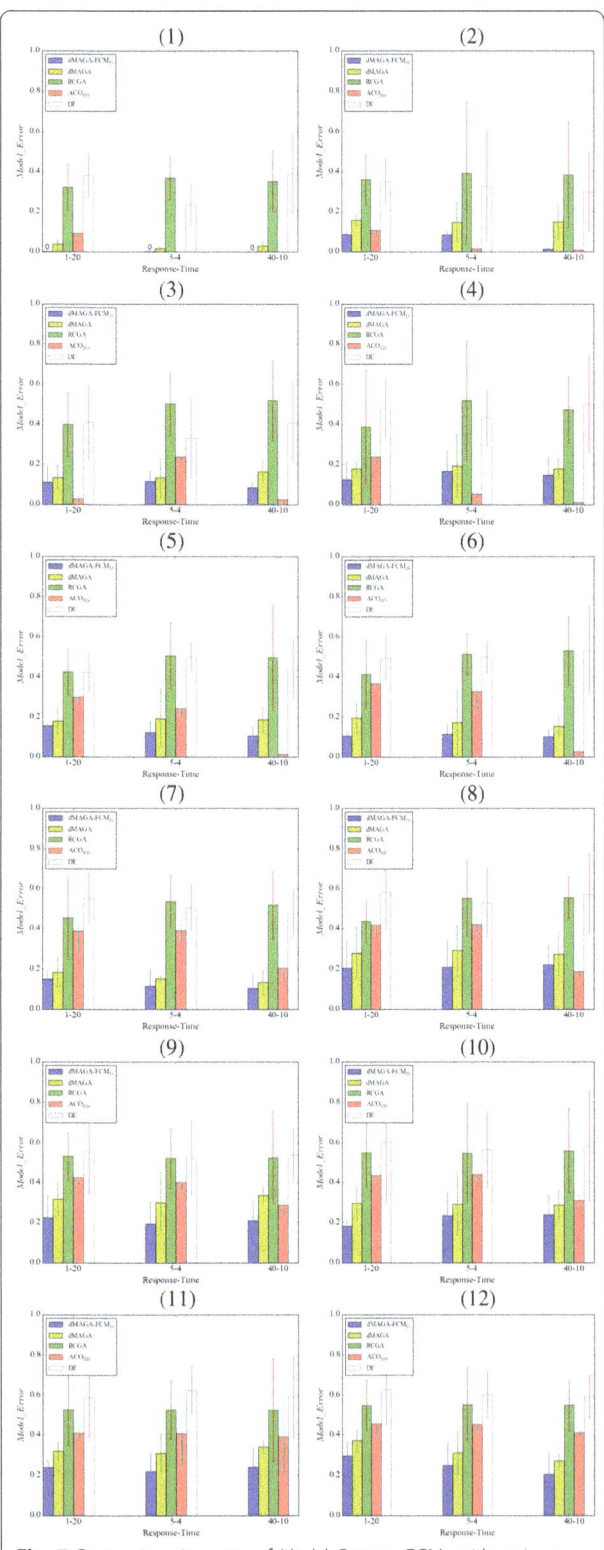

Fig. 7 Comparison in terms of *Model_Error* on FCMs with various number of nodes. (*1*) 5 nodes, density = 20%. (*2*) 5 nodes, density = 40%. (*3*) 10 nodes, density = 20%. (*4*) 10 nodes, density = 40%. (*5*) 20 nodes, density = 20%. (*6*) 20 nodes, density = 40%. (*7*) 40 nodes, density = 20%. (*8*) 40 nodes, density = 40%. (*9*) 100 nodes, density = 20%. (*10*) 100 nodes, density = 40%. (*11*) 200 nodes, density = 20%. (*12*) 200 nodes, density = 40%

Table 3 Comparison in terms of *Model_Error* on larger synthetic FCMs (Average ± Standard deviation)

#Nodes	Edge density	N_s-N_t	dMAGA-FCM$_D$	dMAGA
300	20%	1–20	0.305 ± 0.024	0.371 ± 0.062
		5–4	0.321 ± 0.072	0.382 ± 0.028
		40–10	0.241 ± 0.010	0.298 ± 0.091
	40%	1–20	0.352 ± 0.016	0.386 ± 0.083
		5–4	0.257 ± 0.051	0.296 ± 0.024
		40–10	0.310 ± 0.027	0.390 ± 0.109
500	20%	1–20	0.323 ± 0.032	0.401 ± 0.102
		5–4	0.317 ± 0.017	0.414 ± 0.023
		40–10	0.321 ± 0.045	0.374 ± 0.127
	40%	1–20	0.285 ± 0.081	0.339 ± 0.134
		5–4	0.379 ± 0.015	0.430 ± 0.085
		40–10	0.392 ± 0.029	0.407 ± 0.141

independent runs for FCMs with 5 ~ 200 nodes and 10 independent runs for FCMs with 300 and 500 nodes.

As can be seen, in terms of *Data_Error*, dMAGA-FCM$_D$, dMAGA and ACO$_{RD}$ all reach 0.000 for FCMs with 5 nodes, no matter whether the edge density is 20% or 40%. For FCMs with 10 and 20 nodes, although *Data_Error* of dMAGA-FCM$_D$ is larger than that of ACO$_{RD}$, it is much smaller than those of RCGA and DE. For FCMs with 40 ~ 200 nodes, dMAGA-FCM$_D$ outperforms ACO$_{RD}$ on 13 out of 16 FCMs and outperforms all other methods with different edge densities and N_s-N_t combinations. Moreover, Table 2 shows that dMAGA-FCM$_D$ clearly outperforms dMAGA on FCMs with 300 and 500 nodes.

The comparison in terms of *Model_Error* shows that for FCMs with 5 ~ 200 nodes, dMAGA-FCM$_D$ always outperforms dMAGA, RCGA, and DE. For FCMs with 10 nodes, ACO$_{RD}$ performs better than dMAGA-FCM$_D$, but for FCMs with 20 and 40 nodes, dMAGA-FCM$_D$ outperforms ACO$_{RD}$ on 9 out of all the 12 cases regardless the edge density and N_s-N_t combination. Moreover, all the averaged *Model_Error* of dMAGA-FCM$_D$ for FCMs with 300 and 500 nodes is smaller than 0.4,which is consistently smaller than that of dMAGA.

In addition, the above results show that the performance of dMAGA-FCM$_D$ in terms of *Data_Error* and *Model_Error* is not very sensitive to the number of nodes, the number of response sequences, and the number of time points in a certain range, where at least four time-points are required and the number of nodes are limited less than 500. It indicates that dMAGA-FCM$_D$ is robust and is scalable to the size of FCMs to a certain extent.

Note also that, the experimental results of dMAGA-FCM$_D$ are obtained with less than 1.5×10^6 fitness function evaluations (see Fig. 8), whereas the results of compared algorithms use 3×10^6 fitness function evaluations, which are much larger than that of dMAGA-

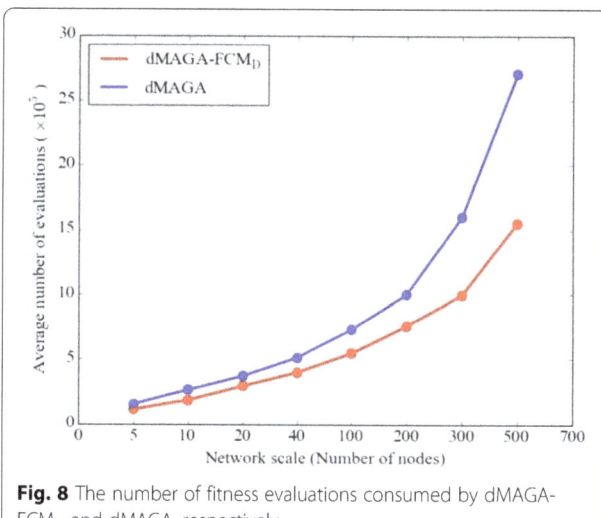

Fig. 8 The number of fitness evaluations consumed by dMAGA-FCM$_D$ and dMAGA, respectively

FCM$_D$. As can be seen from Fig. 8, dMAGA-FCM$_D$ efficiently reduces the computational cost compared with dMAGA, especially for large-scale FCMs. Therefore, dMAGA-FCM$_D$ has exhibited better performance with lower computational cost compared to the state-of-the-art, demonstrating that the algorithm is very competitive for solving large-scale problems.

Figure 9 report the performance of dMAGA-FCM$_D$ in terms of *SS Mean*. As we can see, for FCMs with $5 \sim 20$ nodes, dMAGA-FCM$_D$ consistently outperforms dMAGA, RCGA and DE, and also outperforms ACO$_{RD}$ on 12 out of all the 18 cases. For FCMs with $40 \sim 200$ nodes, dMAGA-FCM$_D$ performs better than all the other learning algorithms on all cases. Table 4 shows the comparative results in terms of *SS mean* of dMAGA-FCM$_D$ and dMAGA for FCMs with 300 and 500 nodes. As can be seen, compared with dMAGA, dMAGA-FCM$_D$ significantly enhances the ability of dMAGA in training larger FCMs.

Experiments on DREAM4 in silico network challenge

Gene regulatory networks (GRNs) have been widely used to model, analyze and predict the behavior of biological organisms. A GRN aims to build the relationships between a set of molecular entities and is often modeled as a network composed of nodes (representing genes, proteins or metabolites) and edges (representing molecular interactions such as protein–DNA and protein–protein interactions or rather indirect relationships between genes) [39]. In this section, we employ the proposed dMAGA-FCM$_D$ to reconstruct a biological GRN based on gene expression time series data, known as DREAM4 [34], a widely used benchmark for evaluating reverse engineering methods [40]. The gene expression time series data were generated based on the network structures of *Escherichia coli* and *Saccha-romyces cerevisiae*. Time series database contains a variety of network sizes with

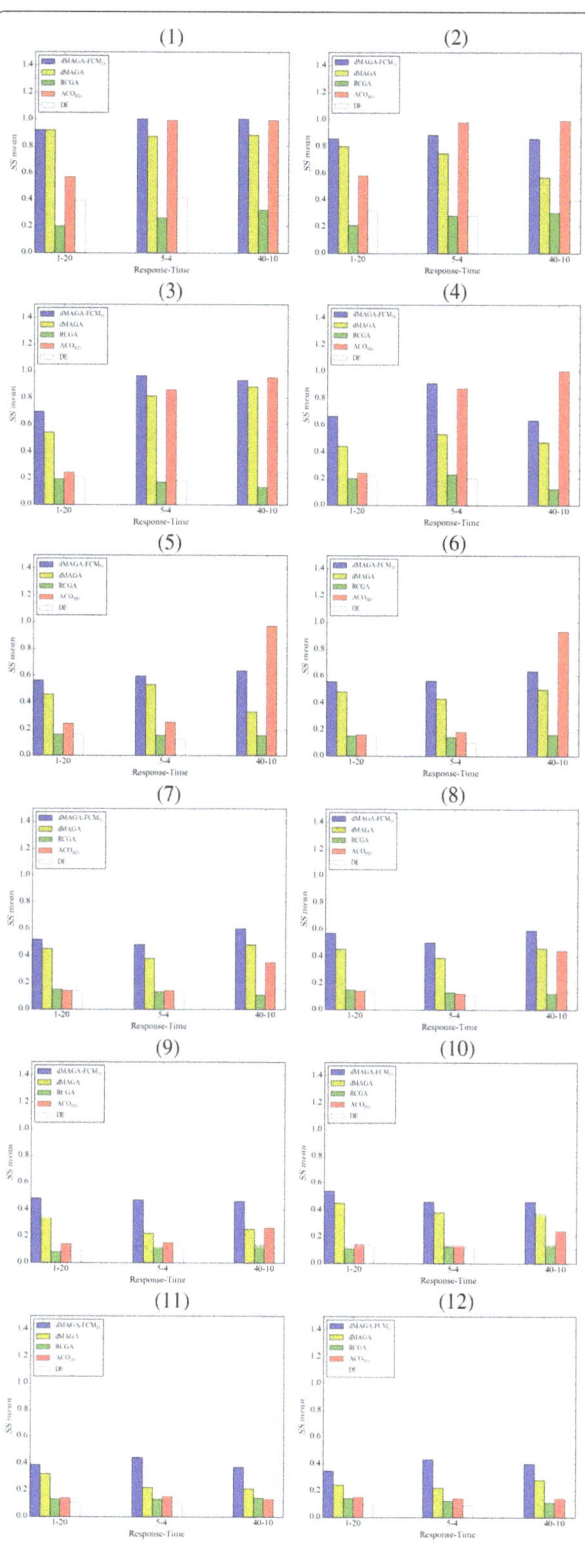

Fig. 9 Comparison in terms of *SS Mean* on FCMs with various number of nodes. (*1*) 5 nodes Density = 20%, (*2*) 5 nodes Density = 40%, (*3*) 10 nodes, density = 20%, (*4*) 10 nodes, density = 40%, (*5*) 20 nodes, density = 20%, (*6*) 20 nodes, density = 40%, (*7*) 40 nodes, density = 20%, (*8*) 40 nodes, density = 40%, (*9*) 100 nodes, density = 20%, (*10*) 100 nodes, density = 40%, (*11*) 200 nodes, density = 20%, (*12*) 200 nodes, density = 40%

Table 4 Comparison in terms of *SS mean* on larger synthetic FCMs

#Nodes	Edge density	N_s-N_t	dMAGA-FCM$_D$	dMAGA
300	20%	1–20	0.329	0.309
		5–4	0.303	0.215
		40–10	0.425	0.302
	40%	1–20	0.374	0.290
		5–4	0.384	0.296
		40–10	0.455	0.388
500	20%	1–20	0.237	0.163
		5–4	0.243	0.177
		40–10	0.341	0.139
	40%	1–20	0.290	0.193
		5–4	0.246	0.186
		40–10	0.342	0.151

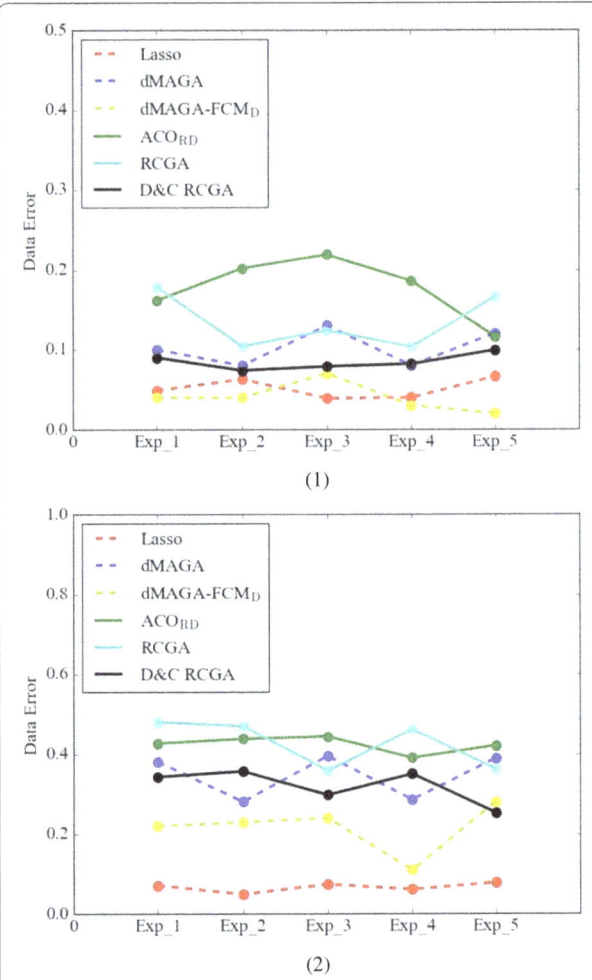

Fig. 10 Experimental results in terms of *Data_Error* for GRNs with (*1*) 10 genes and (*2*) 100 genes. (*1*) Five experiments for GRNs with 10 genes. (*2*) Five experiments for GRNs with 100 genes

10 and 100 genes. Perturbation and noise on expression profiles are generated by differential equations. There are five separate networks for each type of networks.

In the DREAM4 data, under the perturbations of internal noise and measurement noise, there are five time series for each gene. Each time series contains 21 time points. The first 10 time points evaluate the response of gene networks in the presence of perturbations. The perturbations are revoked at the 11-th time instant. So the last 11 time points reflect the response of gene networks after the wild type network is restored. In this experiment, we use the last 11 time points and test the performance of dMAGA-FCM$_D$ in terms of *Data_Error* and *SS mean*.

In the experiments, we perform several efficient evolutionary-based algorithms, dMAGA, RCGA, D&C RCGA and DE, one DREAM data. We also perform Lasso [41] for comparison. The experimental results are shown in Fig. 10. As can be seen, *Data_Error* of dMAGA-FCM$_D$ perform the best for GRNs with 10 genes except one case (Exp_3). For GRNs with 100 genes, Lasso is better than dMAGA-FCM$_D$ in terms of *Data_Error*, and dMAGA-FCM$_D$ performs better than other evolutionary-based algorithms. We can also see that the decomposition strategy is always useful, because the dMAGA-FCM$_D$ is always better than dMAGA, D&C RCGA, which is RCGA with decomposition strategy, is always better than RCGA no matter the number of genes is 10 or 100.

Discussion

The proposed algorithm dMAGA-FCM$_D$ enhances prediction, which is valuable for machine learning. As the decomposition strategy, dMAGA-FCM$_D$ has reduced the computational cost. Reconstructed FCMs can be modeled as a big data optimization problem. Previous studies indicate that the number of variables needs to be

optimized increases quadratically with the scale of FCMs, resulting an exponential increase in the number of candidate solutions. Currently, most learning algorithms can only handle small scale FCMs which number of nodes is smaller than 100. In this work, we show that the dMAGA-FCM$_D$ can handle large FCMs with 300 and 500 nodes (see Table 2, 3 and 4). The experiment results also show that dMAGA-FCM$_D$ efficiently reduces the computational cost compared with our previous work dMAGA, especially for large-scale FCMs, which demonstrating that the algorithm is very competitive for solving large-scale problems.

In the DREAM4 data, the dMAGA-FCM$_D$ performs better than the other evolutionary-based learning algorithms. However, in the five experiments for GRNs with 100 genes, the Lasso is better than dMAGA-FCM$_D$. The performed Lasso takes into account the sparseness of the network. Therefore, the properties of Lasso algorithm provide a good direction for us to improve the dMAGA-FCM$_D$.

Conclusions

In this paper, we propose a new algorithm, termed as dMAGA-FCM$_D$, to train large-scale GRNs based on FCMs using time series data by introducing the decomposition based optimization approach into the dynamical multi-agent genetic algorithm. The dMAGA-FCM$_D$ has shown to be well suited for learning causal relations of complex systems. Extensive experiments are conducted on both synthetic FCMs and DREAM4, a challenge to reverse GRNs from simulated time series data. Experimental results show that dMAGA-FCM$_D$ is able to effectively train FCMs with up to 500 nodes and improves ability of reconstructing GRNs with high accuracy, outperforming the compared state-of-the-art learning algorithms. Our results also indicate that dMAGA-FCM$_D$ is promising for solving larger-scale GRNs, which will be considered in our future work.

Abbreviations
dMAGA-FCM$_D$: Dynamical multi-agent genetic algorithm with the decomposition-based model; FCM: Fuzzy cognitive maps; GRN: Gene regulatory networks

Acknowledgements
Not applicable.

Funding
This work is partially supported by the Outstanding Young Scholar Program of National Natural Science Foundation of China (NSFC) under Grant 61522311 and the Overseas, Hong Kong & Macao Scholars Collaborated Research Program of NSFC under Grant 61528205.

Authors' contributions
JL conceived the new algorithm. YC and CZ conducted all the experiments. YJ advised on the design of experiments. All authors contributed to writing the manuscript All authors read and approved the final manuscript.

Competing interests
The authors declare that they have no competing interests.

Author details
[1]Key Laboratory of Intelligent Perception and Image Understanding of Ministry of Education, Xidian University, Xi'an 710071, China. [2]Department of Computer Science, University of Surrey, Guildford GU2 7XH, UK.

References
1. Abbass HA. Calibrating independent component analysis with Laplacian reference for real-time EEG artifact removal. 21st International Conference on Neural Information Processing. 2014. p. 68–75.
2. Goh SK, Abbass HA, Tan KC. Artifact removal from EEG using a multi-objective independent component analysis model. 21st International Conference on Neural Information Processing. 2014. p. 570–577.
3. Repsilber D, Liljenström H, Andersson SGE. Reverse engineering of regulatory networks: simulation studies on a genetic algorithm approach for ranking hypotheses. Biosystems. 2002;66(1):31–41.
4. Eriksson R, Olsson B. Adapting genetic regulatory models by genetic programming. Biosystems. 2004;76(1):217–27.
5. Fomekong-Nanfack Y, Kaandorp JA, Blom J. Efficient parameter estimation for spatio-temporal models of pattern formation: case study of Drosophila melanogaster. Bioinformatics. 2007;23(24):3356–63.
6. Chen Y, Mazlack LJ, Lu LJ. Inferring fuzzy cognitive map models for gene regulatory networks from gene expression data. IEEE International Conference on Bioinformatics and Biomedicine. 2012. p. 589–601
7. Kauffman SA. The origins of order: self organization and selection in evolution. J Evol Biol. 1992;13(1):133–44.
8. Margolin AA, Wang K, Lim WK, Kustagi M, Nemenman I, Califano A. Reverse engineering cellular networks. Nat Protoc. 2006;1(2):662–71.
9. Butte AJ, Kohane IS. Mutual information relevance networks: functional genomic clustering using pairwise entropy measurements. In Pac Symp Biocomput. 2000;5:418–29.
10. Friedman N, Linial M, Nachman I, Pe'er D. Using bayesian networks to analyze expression data. J Comput Biol. 2000;7(3–4):601–20.
11. Marbach D, Costello JC, Kuffner R, Vega NM, Prill RJ, Camacho DM, et al. Wisdom of crowds for robust gene network inference. Nat Methods. 2012;9(8):796–804.
12. Chen Y, Mazlack LJ, Ali AM, Long LJ. Inferring causal networks using fuzzy cognitive maps and evolutionary algorithms with application to gene regulatory network reconstruction. Appl Soft Comput. 2015;37:667–79.
13. Papageorgiou EI. Learning algorithms for fuzzy cognitive maps - a review study. IEEE Trans Syst Man Cybern Part C. 2012;42(1):150–63.
14. Kosko B. Fuzzy cognitive maps. Int J Man Mach Stud. 1986;24:65–75.
15. Georgopoulos VC, Malandraki GA, Stylios CD. A fuzzy cognitive map approach to differential diagnosis of specific language impairment. J Artif Intel Med. 2003;29(3):261–78.
16. Dickerson JA, Cox Z, Wurtele ES, Fulmer AW. Creating metabolic and regulatory network models using fuzzy cognitive maps. North Am Fuzzy Inform Proc Conf (NAFIPS). 2001;4:2171–6.
17. Papageorgiou E, Groumpos P. A weight adaptation method for fuzzy cognitive maps to a process control problem. Berlin: Lecture Notes in Computer Science, Springer; 2004. p. 3037.
18. Bakken BT, Gilljam M. Training to improve decision-making-system dynamics applied to higher level military operations. In: 20th International System Dynamics Conference. Palermo; 2002.
19. Stach W, Kurgan L. Modeling software development project using fuzzy cognitive maps. Proc. 4th ASERC Workshop on Quantitative and Soft Software Engineering (QSSE'04). 2004. p. 55–60.
20. Stach W, Kurgan L, Pedrycz W, Reformat M. Parallel fuzzy cognitive maps as a tool for modeling software development project. Proc. 2004 North American Fuzzy Information Processing Society Conf. (NAFIPS'04), Banff; 2004. p. 28–33
21. Stac W, Kurgan L, Pedrycz W, Reformat M. Genetic learning of fuzzy cognitive maps. Fuzzy Sets Syst. 2005;153:371–401.
22. Stach W, Kurgan L, Pedrycz W. A divide and conquer method for learning large fuzzy cognitive maps. Fuzzy Sets Syst. 2010;161:2515–32.
23. Papageorgiou EI, Parsopoulos KE, Stylios CS, Groumpos PP, Vrahatis MN. Fuzzy cognitive maps learning using particle swarm optimization. J Intell Inf Syst. 2005;25:95–121.
24. Ghazanfari M, Alizadeh S, Fathian M, Koulouriotis DE. Comparing simulated annealing and genetic algorithm in learning FCM. Appl Math Comput. 2007;192:56–68.
25. Chen Y, Mazlack LJ, Lu LJ. Learning fuzzy cognitive maps from data by ant colony optimization. In: Proceedings of Genetic and Evolutionary Computation Conference (GECCO). 2012. p. 9–16.
26. Acampora G, Pedrycz W, Vitiello A. A competent memetic algorithm for learning fuzzy cognitive maps. IEEE Trans Fuzzy Systems. 2015;23(6):2397–411.
27. Zhong W, Liu J, Xue M, Jiao L. A multiagent genetic algorithm for global numerical optimization. IEEE Trans Syst Man Cybern Part B. 2004;34(2):1128–41.
28. Liu J, Zhong W, Jiao L. A multiagent evolutionary algorithm for constraint satisfaction problems. IEEE Trans Syst Man Cybern Part B. 2006;36(1):54–73.
29. Liu J, Zhong W, Jiao L. A multiagent evolutionary algorithm for combinatorial optimization problems. IEEE Trans Syst Man Cybern Part B. 2010;40(1):229–40.
30. Zhang Y, Zhou M, Jiang Z, Liu J. A multi-agent genetic algorithm for big optimization problems. IEEE Congr Evol Comput (CEC). 2015;703–7.
31. Zhang Y, Liu J, Zhou M, Jiang Z. A multi-objective memetic algorithm based on decomposition for big optimization problems. Memetic Comput. 2016;8(1):45–61.
32. Liu J, Chi Y, Zhu C. A dynamic multi-agent genetic algorithm for gene regulatory network reconstruction based on fuzzy cognitive maps. IEEE Trans Fuzzy Systems. 2016; 24(2):419–31.
33. Chi Y, Liu J. Reconstructing gene regulatory networks with a memetic-neural hybrid based on fuzzy cognitive maps. Natural Computing. 2016. In press.
34. Greenfield A, Madar A, Ostrer H, Bonneau R. DREAM4: combining genetic and dynamic information to identify biological networks and dynamical models. PLoS One. 2010;5:e13397.

35. Aguilar J. A survey about fuzzy cognitive maps papers. Int J Comput Cognition. 2005;3(2):27–33.

36. Bueno S, Salmeron JL. Benchmarking main activation functions in fuzzy cognitive maps. Expert Syst Appl. 2009;36(3):5221–9.

37. Stach W. Learning and aggregation of fuzzy cognitive maps – an evolutionary approach. PhD Dissertation: University of Alberta; 2010.

38. Papageorgiou EI, Groumpos PP. Optimization of fuzzy cognitive map model in clinical radiotherapy through the differential evolution algorithm. Biomed Soft Comput Human Sci. 2004;9(2):25–31.

39. Hecker M, Lambeck S, Toepfer S, Someren E, Guthke R. Gene regulatory network inference: data integration in dynamic models-a review. Biosystems. 2009;96(1):86–103.

40. Thomas SA, Jin Y. Reconstructing gene regulatory networks: where optimization meets big data. Evol Intel. 2014;7(1):29–47.

41. Tibshirani R. Regression shrinkage and selection via the lasso. J Royal Stat Soc Series B (Methodological). 1996;58:267–88.

13

EnCOUNTer: a parsing tool to uncover the mature N-terminus of organelle-targeted proteins in complex samples

Willy Vincent Bienvenut[*], Jean-Pierre Scarpelli, Johan Dumestier, Thierry Meinnel and Carmela Giglione

Abstract

Background: Characterization of mature protein N-termini by large scale proteomics is challenging. This is especially true for proteins undergoing cleavage of transit peptides when they are targeted to specific organelles, such as mitochondria or chloroplast. Protein neo-N-termini can be located up to 100–150 amino acids downstream from the initiator methionine and are not easily predictable. Although some bioinformatics tools are available, they usually require extensive manual validation to identify the exact N-terminal position. The situation becomes even more complex when post-translational modifications take place at the neo-N-terminus. Although N-terminal acetylation occurs mostly in the cytosol, it is also observed in some organelles such as chloroplast. To date, no bioinformatics tool is available to define mature protein starting positions, the associated N-terminus acetylation status and/or yield for each proteoform. In this context, we have developed the EnCOUNTer tool (i) to score all characterized peptides using discriminating parameters to identify bona fide mature protein N-termini and (ii) to determine the N-terminus acetylation yield of the most reliable ones.

Results: Based on large scale proteomics analyses using the SILProNAQ methodology, tandem mass spectrometry favoured the characterization of thousands of peptides. Data processing using the EnCOUNTer tool provided an efficient and rapid way to extract the most reliable mature protein N-termini. Selected peptides were subjected to N-terminus acetylation yield determination. In an *A. thaliana* cell lysate, 1232 distinct proteotypic N-termini were characterized of which 648 were located at the predicted protein N-terminus (position 1/2) and 584 were located further downstream (starting at position > 2). A large number of these N-termini were associated with various well-defined maturation processes occurring on organelle-targeted proteins (mitochondria, chloroplast and peroxisome), secreted proteins or membrane-targeted proteins. It was also possible to highlight some protein alternative starts, splicing variants or erroneous protein sequence predictions.

Conclusions: The EnCOUNTer tool provides a unique way to extract accurately the most relevant mature proteins N-terminal peptides from large scale experimental datasets. Such data processing allows the identification of the exact N-terminus position and the associated acetylation yield.

Keywords: N-terminal modifications, Protein maturation, Acetylation, Quantitation, Processing tool, Organelle proteins, Transit peptide, Cleavage site

* Correspondence: Willy.Bienvenut@i2bc.Paris-Saclay.fr
Institute for Integrative Biology of the Cell (I2BC), CEA, CNRS, Univ. Paris-Sud, Université Paris Saclay, 91198 Gif-sur-Yvette Cedex, France

Background

N-terminal acetylation (NTA) is one of the major protein modifications of the eukaryotic cytosol and occurs mainly co-translationally [1, 2]. In plants, most chloroplast proteins are encoded in the nucleus, translated in the cytosol and targeted to the chloroplast by a transit peptide that is cleaved upon arrival inside the organelle. Large scale analyses show that 20–40% of these proteins are N-acetylated in their mature chloroplastic form [3, 4]. The determination of the associated cleavage site of the transit peptide (TP) are still challenging. The cleavage positions of the mitochondrial or plastid TP (mTP or cTP) can be predicted using TargetP or ChloroP softwares [5, 6], but the predictions are not always reliable [7]. Although experimental data provide useful information, it still remains difficult to identify the true N-terminal peptides amid the multitude of internal peptides identified in a large scale experiment. In addition, the determination of NTA yield is a difficult task with the tools currently available. As an exemple, Mascot Distiller (MD) allows NTA quantitation using peptides N-terminally labeled with d_3- (heavy, H) or d_0- (light, L) acetyl [8]. Although this tool was used to define Lys ε-acetylation yield [9], it is originally dedicated to provide protein differential quantitation. The determination of the NTA yield for each proteoform (especially for the mitochondrial and the plastidic mature proteins) is not easily available and requires some additional processing [10].

Therefore, the development of a new tool designed to perform the extraction of the data computed by Mascot and Mascot-Distiller is required. The combination of the outputs must provide a list of the mature N-termini and the associated accurate NTA yields. Although some alternative tools could be able to perform H/L ratio quantitation such as MaxQuant, the EnCOUNTer script is not able, presently, to handle other input file format than the Mascot and Mascot-Distiller ones.

The EnCOUNTer tool (Extraction and Calculation Of Unbiased N-Termini) uses a stepwise approach. First, the characterized peptides are scored to discriminate between protein N-termini at position 1–2 and downstream N-termini (DNT) of the protein sequence. This determination is based on a curated experimental dataset. Second, EnCOUNTer recalculates the average NTA yield taking into account the first residue of the characterised mature proteins. Finally, it provides an exhaustive list of the processed N-termini with the recalculated unbiased NTA yield. The EnCOUNTer tools was trained using a manually validated dataset (Additional file 1: Table S1). As a proof of concept, the optimized parameters were used against a complex *Arabidopsis thaliana* experimental dataset obtained after an enrichment of the mature protein N-termini using the SILProNAQ approach [2]. Such experimental data set provides 584 DNT peptides (related to 383 distinct proteins) of which 338 were quantified for NTA yield. Some of these N-termini (112 hits), were experimentally validated and their positions well correlated with known cleavage sites of signal peptides, mTPs or cTPs (based on UniProtKB/Swiss-Prot annotations). Some others (224 hits) were in accordance with transit peptide cleavage site predictions within a range of ± 2 residues. In addition, 648 protein N-termini were characterized at positions 1 or 2 (on the initiator methionine or after its excision) of which 303 were also quantified for protein NTA yield (Additional file 2: Table S2).

Methods

Sample preparation and raw data aquisition

Proteins extracted from *A. thaliana* Col. 0 seedling were used to perform N-terminus enrichment using SCX chromatography. Rapidly, 1 mg of protein was denatured and reduced followed by cysteine alkylation with iodoacetamide. After cold acetone precipitation, proteins were resuspended in 50 mM NH_4HCO_3 and digested by 1/100 (w/w) of TPCK treated porcine trypsin (Sigma-Aldrich) for 1.5 h at 37 °C, twice. Peptides were desalted with Sep-Pak columns and the retained material was eluted with 80% acetonitrile (ACN), 0.1% TFA and then evaporated to dryness. The collected material was resuspended in Strong Cation eXchange (SCX) LC buffer (5 mM KH_2PO_4, 30% ACN and 0.05% formic acid) and injected into an Alliance HPLC system using a fluorimeter detector (Waters) equipped with polysulfoethyl A column (200 × 2.1 mm, 5 μm 200 Å; PolyLC, Colombia, MD). Peptides were eluted with a KCl gradient (SCX-LC buffer B: 350 mM KCl in SCX-LC buffer A; 0–5 min, 0% B; 15–40 min, 5–26% B; 40–45 min, 26–35% B). Fractions were collected every 2 min for 40 min and the solvent was evaporated to dryness before storage at −20 °C. Fractions eluted from SCX columns with retention times of 3 to 22 min were analyzed as previously described [1] with an Easy Nano-LC II (Thermo Scientific) coupled to a LTQ-Orbitrap™ Velos (Thermo Scientific). Finally, data processing usually combines a few acquisition files, i.e. 10 files related to each individual SCX fraction (1 h analysis) and 6 files related to combined fractions (for more details, see [2]). Furthermore, acquisition files obtained from SCX fraction 5 and 6 were used as training dataset and testing dataset, respectively.

Mascot Distiller/Mascot data processing and *.xml exports

Regardless of the number of Orbitrap-MS acquired files, Mascot Distiller (Ver. 2.5.1, Matrix Science, London, UK) combines all acquired files together for a unique processing event (Fig. 1). The EnCOUNTer script was first used to extract data from the raw files followed by protein identification using the Mascot protein

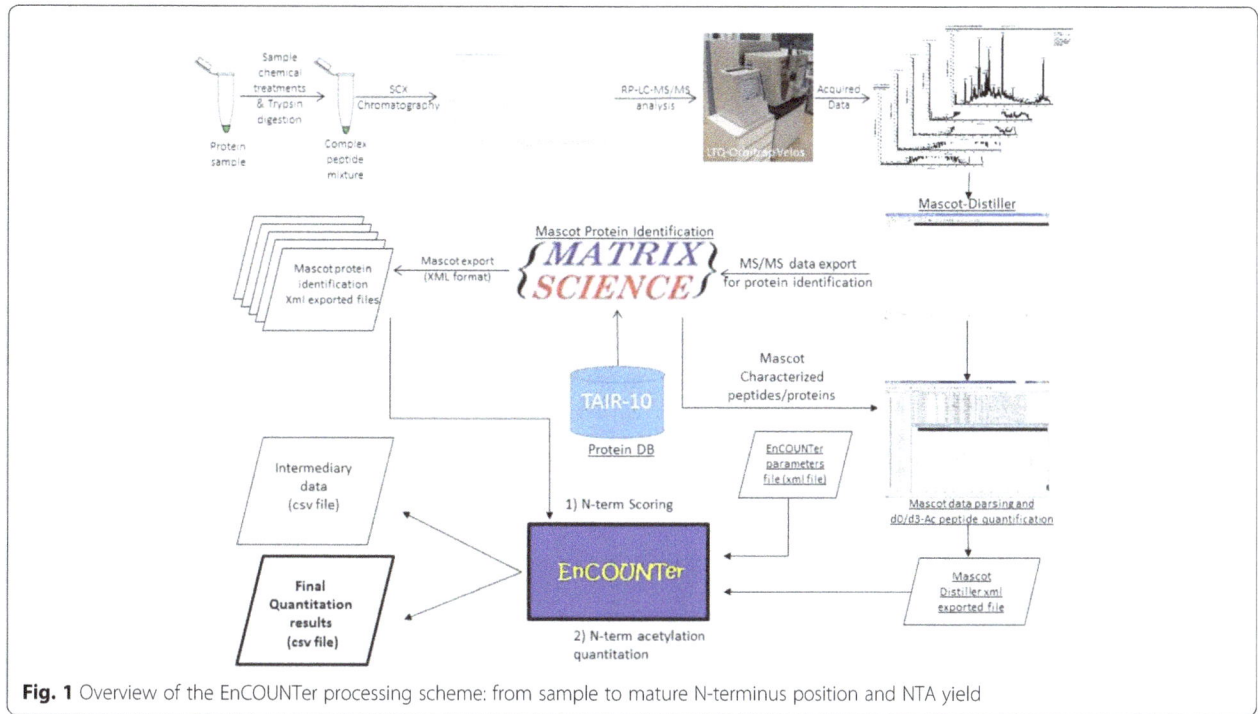

Fig. 1 Overview of the EnCOUNTer processing scheme: from sample to mature N-terminus position and NTA yield

identification tool (Ver 2.4, Matrix Science, London, UK). Precursor detection was obtained after peak re-gridding of 100 points per Da for a peak half width of 0.02 Da. Precursor ion charge was determined directly from the Orbitrap survey scan and limited at 1 to 5. The "time domain aggregated MS/MS spectra" was re-gridded at 20 points per Da for a peak half width of 0.2 Da and MS/MS spectra containing more than 10 peaks were retained. MS/MS spectra of similar precursor masses were not combined at any time and considered separately for peptide identification. Additional filtering parameters such as correlation threshold and minimal signal over noise ratio were defined at 0.7 and 1, respectively for the MS and MS/MS in the relevant mass range (400–20000 Da and 50–10000 Da, respectively) with a maximum peak iteration of 500. Alternative high resolution mass spectrometers could be used but the MD parameters applied for raw data extraction should be optimized accordingly.

MD extracted data were submitted to Mascot 2.4 software for protein identification and post-translational modification characterization. The database used was "The Arabidopsis Information Resource" (TAIR ver. 10; www. arabidopsis.org [11]). The parent and fragment mass tolerance were 5 ppm and 0.4 Da, respectively. Additionally, carbamidomethylcysteine and d_3-acetyl on Lys were defined as fix modifications and methionine oxidation as variable modification. Semi-trypsin was defined for the enzyme cleavage rule with up to 6 missed cleavages. Peptide N-terminus acetylation status, i.e. d3-

Acetyl (chemically induced modification) or d0-Acetyl (endogenous modification) were investigated using the Mascot quantification option (associated to the MD parameters). These parameters ("Acetylation [MD]" quantification method) are available in Additional file 3. Then, MD uploaded the Mascot processing results and parsed them using relaxed parameters (minimum peptide identification score was set at 25, 0.2 for the P-value, 0.1 for the peak correlation coefficient, the area fraction coefficient and the precursor standard error). Irrelevant and false positive peptide hits generated at this step were filtered out at the final stage of the EnCOUNTer process.

EnCOUNTer also required protein identification data generated by Mascot. These data were automatically exported in xml files with the same MD parameters for the P-value and the Mascot score threshold. Additionally, the "MudPIT Scoring" and "Bold Red peptides" option were selected. These exported files contain all "Protein Hit Information" except pI and Taxonomy ID and all "Peptide Match Information" except the frame number and the unassigned queries.

EnCOUNTer processing

Basically, the EnCOUNTer tool requires the MD exported file, the associated Mascot results and a parameter file. Although the tool could be used with the default parameters provided (Additional file 4), an optimization of the scoring parameters using a relevant training dataset has been performed. During the scoring parameter optimization, the EnCOUNTer tool required an additional files containing the list of "curated N-termini" (True / False N-termini; Additional file 1: Table S1). At the end of the optimization, a file containing all optimized values was generated (*.json). This file could be applied on other experimental datasets (from a similar origin) without the optimization of the scoring scheme.

Parsing function

The EnCOUNTer tool parses the pre-processed data exported from MD and Mascot identification tool. Mascot matched queries (only Mascot first-ranked peptide sequences) and associated protein AC were extracted from the MD xml file. Each of these entries were enriched with information, e.g. peptide sequence, starting position, MD processing results such as H/L ratio and signal quality coefficients. Then, the collected results were complemented with data extracted from the Mascot exported files such as the peptide identification score, identification E value... Of note, some peptides were not proteotypic [12] and shared with few distinct proteins or, alternatively, to different translational isoforms of the same protein (especially for TAIR database). The redundancy is noted and these data could be easily removed at will. Also, the shared peptides were distinctively labelled in the final result list.

N-terminus scoring function

The EnCOUNTer tool should discriminate internal peptides from the mature protein N-termini. Biological details associated to nuclear encoded mitochondrial/plastidic proteins TP such as sequence composition and average length [13–16] (also observed from experimental dataset [3, 17, 18]), highlighted some features useful to define relevant scoring coefficients (Additional file 5: Figure S1 and Additional file 6: Figure S2). To this end, we defined a scoring function based on six distinct coefficients related to i) peptide "starting position", ii) residues around the "starting position", iii) characterized N-terminal modifications, iv) alternative start positions at the vicinity of the "starting position", v) matched peptide redundancies and finally iv) the "Localization" score. Some of these features could be optimized from the training dataset (such as "starting position" or the "residues around the starting position") whereas some other should be defined by the users to valorize/penalize experimental observations (such as "data redundancy" or "multiple transit peptide cleavage sites").

Peptide "starting position" score (Bound Score)

Based on the experimental training dataset, EnCOUNTer determines the optimal range (OptiMin and OptiMax) where "true" N-termini are the most frequently distributed.

The Matthews Correlation Coefficient (MCC) was determined for all possible combinations of positions between the two endpoints of the N-terminal distribution range of the "True" hits for the DNT candidates (defined as ExpMin and ExpMax). The optimum range defined with the higher MCC provides the optimum endpoints (OptiMin and OptiMax). This positional range is associated with a scoring weight of 2 to favor the characterization of these N-termini. This calculation was associated to a "K fold cross validation" (using 10 randomized fractions) to determine the robustness of the prediction and the results of the investigation were exported in the *.bound file (specifically for the "bound" K fold test) and *.json (all optimized values).

Nevertheless, some relevant candidates (Experimental "True" N-termini) were still present outside of these optimal values, i.e. in between ExpMin/OptiMin and OptiMax/ExpMax. Since the experimental dataset may be slightly different compared to the training dataset (considering the ExpMin and ExpMax values extracted from the training dataset), the ExpMin and ExpMax values were pondered by the standard deviation observed during the "K fold cross correlation" as an estimation of the dataset variability (defined as Min and Max respectively). Both ranges, i.e. Min/OptiMin and OptiMax/Max, were associated with a scoring weight of 1 (neutral effect on the result) that prevented their elimination at this stage. All others positions are associated with a scoring weight below 1 (e.g. 0.1) to penalize such less biologically relevant positions. Starting positions 1–2 were subjected to a special scoring detailed below.

Residues around the starting position ("Spec" Score)

Based on the training dataset, EnCOUNTer is able to provide a scoring matrix associated to the amino acid presents around the experimentally characterized starting position. For each position located between P_n to P_{-n}, a binary classification was performed for each of the 21 possible amino acids. Such investigation provides a distribution of True Positive (True N-termini candidate has the defined residue at P_i), True Negative (False N-termini candidate has not the defined residue at P_i), False Positive (True N-termini candidate has not the defined residue at P_i) and False Negative (False N-termini candidate has the defined residue at P_i). The MCC was calculated for each of the 21 residues for each of the defined position between P_n to P_{-n}. The result of the MCC provided an overview of the "abundance" for each residue at a specific position (P_i) based on the training dataset. In accordance with our scoring scheme (1 for the neutral value), the MCC results were translated by 1 unit (tMCC). This tMCC matrix (defined for each of the 20 amino acids on a "±n" residues around the starting position of the peptides present in the training dataset) was used to determine the "Spec" score for each experimental candidate (with a starting position higher than 2) as a product of the tMCCs values for each position between P_{-n} to P_n (Eq. 1). The size of the screening window (±n) was also optimized automatically to determine the optimum MCC value.

$$Spe\,Score = \prod_{i=-n}^{n} tMCC(Xxx; Pi) \qquad (1)$$

Determination of the "Spec" score base on the tMCC determined for each possible residue (Xxx) at the define position P_i (in the P_{-n} - P_n range).

A "K fold cross validation" (subdivided in 10 subsets) was applied after the optimization step to determine the robustness of the prediction. The "K fold cross validation" result was exported in the *.spec (specifically for the "Spec" K fold test) and *.json (all optimized values).

N-terminal modifications (Acetyl Score)
Due to the MD processing applied during the peptide identification step, peptide's N-terminal modifications are restricted to d0/d3-NTA. Three different situations could occur (d0-NTA, d3-NTA and d0/d3-NTA). It could be interesting to segregate differentially such peptides especially for GAP test [19] where the main goal is to identify the N-terminal acetylated (NTAed) proteins and to rate them differently with values higher than 1 to valorize the modification or below 1 to penalize it. Characterization of the pair d0/d3-Ac reinforces the legacy of such N-termini the MS/MS spectra related to d0-Ac and d3-NTA could be considered as two independent events) and a score higher than 1 could be applied.

Alternative start positions (Prox score)
Despite proteins processing sites are usually considered to be unique, experimentally-based results tend to display a different reality involving multiple vicinal cleavage positions [7, 8, 17]. This provides clear and highly valuable distinctive criteria compared to internal protein fragments. The number of potential cleavage sites was determined in a defined window around the investigated position. Both the window size and the coefficient-weight could be defined in the parameters file (Additional file 4), and the "prox score" for this coefficient was obtained using Eq. 2. Initial parameters should be defined in the configuration file and optimized using the reference dataset but usually these multiple and vicinal cleavages [3, 17] are observed in a windows of ± 5–10 residues (defined at will in the parameter file).

$$Prox.\,Score. = R^m \qquad (2)$$

With R = user defined weight and m = number of alternative cleavages sites experimentally characterized in the defined window (±5 residues range defined in the "Default parameters");

Peptide redundancy (Rep Score)
Multiple characterization of the same peptide strengthens the probability to match a real event. Since each analysis provides thousands of acquired independent MS/MS spectra, the identification of the same peptide from different MS/MS acquisition could be considered as independent event and strongly increase the probability to match a "real event" or "true peptide". To take advantage of this redundancy, the number of occurrences of the same peptide (not considering variable charge states or possible associated modifications such as Met oxidation) was used in the "Rep Score". Nevertheless, the number of duplicates matches could reach few hundreds to few thousands of hits for the same peptide especially if multiple LC-MS acquisitions are processed together. To maintain the weight of this coefficient within the range compatible with the others scoring coefficients, the number of occurrences for identical peptides was logarithm pondered in Eq. 3.

$$Rep\,Score = K^{\log(q)} \qquad (3)$$

Where K is the score associated to such event (K = 2 is defined in the "Default parameters") and q = number of experimental occurrences of the investigated starting position;

Localization score (Loc Score)
It is experimentally infrequent [2, 20, 21] to characterize mature protein N-termini both at the N-terminal side of the predicted protein (Pos 1–2) and further downstream in the same sample. Thus, it could be interesting to take advantage of such information to penalize/favor DNT peptides. The weight applied to DNT hits should be defined at will in the configuration file.

Protein N-terminal scoring at position 1–2
Since a negative dataset could not be defined for the N-termini at position 1 and 2, automated optimization of the score is not possible. The "Spec" score for these peptides is set at the optimized "Spec-score-threshold (automatically defined during parameters optimization) to favor the final NTA quantification of these peptides. To note, the other scoring coefficients (i.e. N-terminal modification characterized and peptide redundancy) were applied for these positions. Then, the final EnCOUNTer score for these peptides (Position 1–2) could not be compared with the DNT associated scores.

Scoring parameters optimization and calculation

Since few parameters such as sample preparation or species influencing the type and number of downstream N-termini (True or False hits), a test sample dataset should be used to optimize the parameters. Alternatively, default parameters are provided for the *A. thaliana* samples.

The EnCOUNTer tool is able to optimize few scoring coefficients (i.e. the optimum downstream N-termini range and the "vicinal cleavage site" scoring matrix) using a reference files. Some other parameters are not optimized automatically and could be subject to modification in the parameter file. Each of them (Prox, Rep and Loc scores) should be defined before the EnCOUNTer optimization to determine automatically the EnCOUNTer threshold. This threshold is optimized using an MCC approach at the end of the optimization step. Since the EnCOUNTer score is the product of the six previously defined coefficients (Eq. 4), each of them could be neutralized using the unit value ("1") in the parameter files except for the "Spec" scoring coefficient which is the backbone of this approach. Scoring calculation was applied for each Mascot characterized peptide.

$$
\begin{aligned}
\text{EnCOUNTer Score } = \ & \text{Bound} \times \text{Spec} \\
& \times \text{Acetyl} \times \text{Prox} \\
& \times \text{Rep} \times \text{Loc}
\end{aligned} \qquad (4)
$$

This optimization finishes with a "K fold cross validation" to provide some insights about the prediction robustness and the results are saved in the *.score (specifically for the final EnCOUNTer score K fold test), *.json (all optimized values) and *.param (all parameters resumed) file.

NTA quantification function

Mascot Distiller is an interesting tool to determine d0/d3-Acetylation yield for each characterized peptide. Although, final quantitative values are provided per protein and not per protein starting position, the quantitative data remain available in the MD xml exported file. The EnCOUNTer script re-organizes them to provide N-terminal acetylation yield for each distinct proteoform. Since MD processing was performed using relaxed parameters (see Mascot Distiller data processing section), EnCOUNTer filters those data to retain only the most relevant ones for the NTA quantitation based on MS signal quality coefficients defined by MD. EnCOUNTer tool uses the "Correlation Coefficient" (related to the fitting between the theoretical and experimental isotopic distribution; higher than 0.8), the "Fraction Coefficient" (defining the fraction of the signal of the peak of interest over all signal; usually higher than 0.5), the SigQual coefficient is associated to the H/L standard deviation

(defined by the least squares fit to the heavy vs. light component intensities from the scans in the XIC peak; lower than 0.05 Da), the E-value (lower than 0.05), the Mascot score associated to the matched query (higher than 30; highly dependent of the database used) and finally the EnCOUNTer score threshold (automatically defined as previously described). These coefficients could not be defined by default and are strongly related to the raw MS signal quality. They should be adapted accordingly to the instrument (MS and LC) used for sample separation and analysis. The characterised peptides passing those criteria were used to determine the final H/L ratio for each distinct protein positions based on a logarithmic means. Jointly, the logarithmic deviation (σ) of the NTA yield was determined to provide the minimum/maximum NTA range when more than one ratios were determined. Finally, the average NTA yield was obtained from the average H/L ratio using the Eq. 5 and the confidence interval (Min and Max NTA percentages) was defined by the Eqs. 6 and 7 respectively using the logarithmic divergence coefficient.

$$\% \, \mathbf{NTA} = 1 \, / \, (1 \, + \, < \mathrm{H/L \ ratio} >) \qquad (5)$$

$$\% \, \mathbf{NTA_{Min}} = 1 \, / \, (1 \, + \, < \mathrm{H/L \ ratio} > \, \times \sigma) \qquad (6)$$

$$\% \, \mathbf{NTA_{Max}} = 1 \, / \, (1 \, + \, < \mathrm{H/L \ ratio} > /\sigma) \qquad (7)$$

EnCOUNTer data export

The final results were exported in a *.csv file providing protein AC's, the proteotypicity, the starting position, the N-terminal modifications characterised, the mature N-terminal sequence (first 10 residues after the starting position), the EnCOUNTer score, the < H/L > ratio (and deviation), the N-terminus acetylation yield (Average, Min and Max values). An additional file was also exported containing all collected and processed data (EnCOUNTer Intermediary file).

Training and testing dataset

An experimental dataset collected during a large scale *A. thaliana* N-terminome characterization was used for the optimizing and the testing steps. Out of the 16 acquired files, data associated to fraction 5 and 6 were used as training and testing datasets, respectively. First, the acquired data were processed for protein/peptide identification as described in "Distiller/Mascot data processing" section. For the peptides associated to a unique gene-ID but few different translation versions, the lowest TAIR extension number was retained in the final list. Non-proteotypic peptides (*i.e.*, peptide matching several distinct gene-IDs) were removed from the final list. Each characterized peptide was manually checked to identify mature N-termini. Information from specialized databases such as AT_Chloro [22, 23], PPDB [24], SUBA [25], TopFind [26], MASCP-Gator [27] and various prediction tools such as TargetP / ChloroP / SignalP [5, 6]) or Mitofate [28] were used to assess N-termini relevancy and protein sub-cellular localization for each candidate (Additional file 1: Table S1). A total of 784 and 1006 checked peptides were dispatched in few different subcategories (Table 1 and Additional file 1: Table S1) in Fraction 5 and 6, respectively.

EnCOUNTer Launch

The EnCOUNTer script should be launched in a prompt windows associated with the required files (fully described in the help support and the user manual). First, EnCOUNTer determined the optimized scoring parameters using the training dataset (MD and Mascot exported files) and the reference N-terminal list. A few files are exported at the end of the optimization including the optimized "scoring parameter" (*.json file) and the detailed results of the optimization and "K fold cross validation" (*.bound, *.param, *.score and *.spec). Second, the experimental datasets (MD and Mascot exported files) were scored using the previously optimized parameters to discriminate and quantify the mature N-termini and associated NTA yield. At the end of the process, the EnCOUNTer script provided two distinct files, i.e. the intermediary and final EnCOUNTer results. The intermediary file provided the detailed values used to determine the EnCOUNTer score and the individual NTA quantitation, whereas the final Encounter file provided the aggregated results per distinct proteoforms (EnCOUNTer score and the final NTA yied).

Results and discussion
Training and testing datasets

Two experimental samples were defined as training and testing dataset i.e. fraction 5 and fraction 6 respectively. The peptides characterized after the Mascot identification step are filtered using few different Mascot–associated values using the peptide E-value and the minimum Mascot score defined in the configuration files. These thresholds should be adapted to reach 1% of False Discovery Rate (FDR) at the peptide level. Applying these thresholds, false positive identifications for the expected

N-terminal peptide (position 1 and 2) were infrequent (Table 1). As an example, no false candidate was characterized in Fraction 5 and only one probable false hit was listed in Fraction 6 (Additional file 1: Table S1). For these starting positions, the associated localizations were mainly the cytosol (49 hits), the membrane/vacuole (17 hits), the peroxisome (6 hits) or the mitochondria (without mTP, 5 hits). Only one plastidic protein (AT2G44640.1) was characterized with a mature N-termini at position 2. This infrequent but not unusual chloroplastic N-terminus [29] was confirmed experimentally and reported in PPDB [21]. The characterized N-termini at position 1–2 corresponded well to the expected cytoplasmic localizations.

Additionally, 595 peptides were characterized with downstream starting position (Start position > 2). These hits were sorted between True N-termini (mature protein N-termini; 203 hits), False N-termini (erroneous mature N-termini; 329 hits) and ambiguous N-termini (mainly poor MS/MS spectra quality or inconsistencies with previous biological and experimental facts; 63 hits). Only the True/False candidates were used during the EnCOUNTer training step. The main subcellular localization is the chloroplast with 73% of the candidates (149 hits) for the "True" dataset. Other locations such as cytosol, membrane or mitochondria were also found (21, 7 and 5%, respectively). At the contrary, the "False" dataset exhibits random location and similar distributions were also observed in Fraction 6 dataset (Table 1 and Additional file 1: Table S1). These two manually curated datasets (Fraction 5 and 6) were used during the EnCOUNTer training and testing steps, respectively.

N-terminus scoring optimization
Residues around the starting position ("Spec" Score)

Residues close to the N-terminal position are, sometimes, associated to artifact modifications and/or (un)-expected endoproteolytic cleavages. As an example, hydroxylated residues (Ser, Thr or Tyr could be modified with a d3-NTA during sample preparation. Such modification located at P_{1-3} (see [30] for detailed positional nomenclature) could be wrongly associated to d3-NTA. The specificity of the endoproteinase used during sample preparation could also create a bias in the characterized

Table 1 Distribution of the manually checked peptides for the training and testing datasets (Fraction 5 and Fraction 6 respectively; based on Additional file 1: Table S1)

Starting position	Classification	Hits for Fraction 5	Hits for Fraction 6
Position 1 and 2	True Protein N-termini	189	261
	False Protein N-termini	0	1
Position > 2	True downstream N-termini	202	232
	Ambiguous downstream N-termini	63	61
	False downstream N-termini	329	451

peptides. Then, Arg residue at P'_1 could be due to trypsin endoproteolytic cleavage (Additional file 7: Figure S3B) and may not be relevant as True maturation site. Along a streamline, the endoproteinases or the associated contaminants could generate numerous unexpected peptides. As an example, the presence of pseudotrypsin [31] or chymotyrypsin could generate some alternative N-termini with a Phe or Leu at P_{-1} [32]. Typically, it is well know that some positions ahead of the TP cleaving position could be specific such as Ala at P_{-1} and Val/Ile at P_{-3} for the plastidic proteins (Additional file 6: Figure S2A-B and [3, 21]) or the Arg at P_{-2} and P_{-3} for mitochondrial proteins (Additional file 6: Figure S2C-D and [16]). Interestingly, different residues appear to be predominant for other protein subclasses, for example Leu at position P_{-9} or Asp at P_{1-3} for proteins carrying a signal peptide (Fig. 2 and Additional file 6: Figure S2). The tMCC profiles (distribution of tMCC values at position P_{-n} to P_n for each distinct residue) clearly reflect the importance of these residues (Fig. 2 and Additional file 7: Figure S3) used in the "Spec" score.

The "Spec" coefficient is determined using a weight matrix of based on the distribution of specific residues around the cleavage site using the MCCs. MCC reflects the presence/lack of specific residues around the cleavage position. This determination can be performed for both the "True" and the "False" reference dataset (Additional file 7: Figure S3A and B) compared to a random distribution of transit peptide cleavage position (Additional file 7: Figure S3C). "Spec" score is the main basis

of EnCOUNTer scoring scheme and could be used alone to determine the final EnCOUNTer score. The Spec-associated matrix was determined for both the "True" and the "False" subsets from Fraction 5 dataset. Based on the "True" hits, EnCOUNTer allows a discrimination at 94.0% accuracy and 97.6% specificity with 4.3% FDR whereas the optimization based on the "False" dataset reached only 88.5% accuracy and 91.6% specificity with 5.6% FDR (Table 2 and Additional file 8: Table S3). Only the optimization using the true hits is retained for the final scoring scheme.th=tlb=

Finally, a K fold cross validation (k = 10) was performed to determine the robustness of this approach. The accuracy reach 88.5 ± 4.1% and 94.9 ± 4.2% sensitivity with 9.3 ± 7.7% FDR (Table 2 and Additional file 8: Table S3). Although additional features should be used to prevent the loss of "True" hits, the results obtained using only the "Spec" score are extremely promising.

Peptide "starting position" score (Bound Score)

For most proteins, the mature N-term position is located on the first two residues of the protein sequence (position 1–2). Nevertheless, some proteins N-termini could be located further downstream (Position > 2). For example, the mTP cleavage position is expected between positions 20–70 whereas for the position for the cTP of *A. thaliana* nuclear encoded proteins is expected between positions 40–70 [16, 33]. In our training datasets (Additional file 1: Table S1), the validated downstream starting positions were distributed from position 3 to

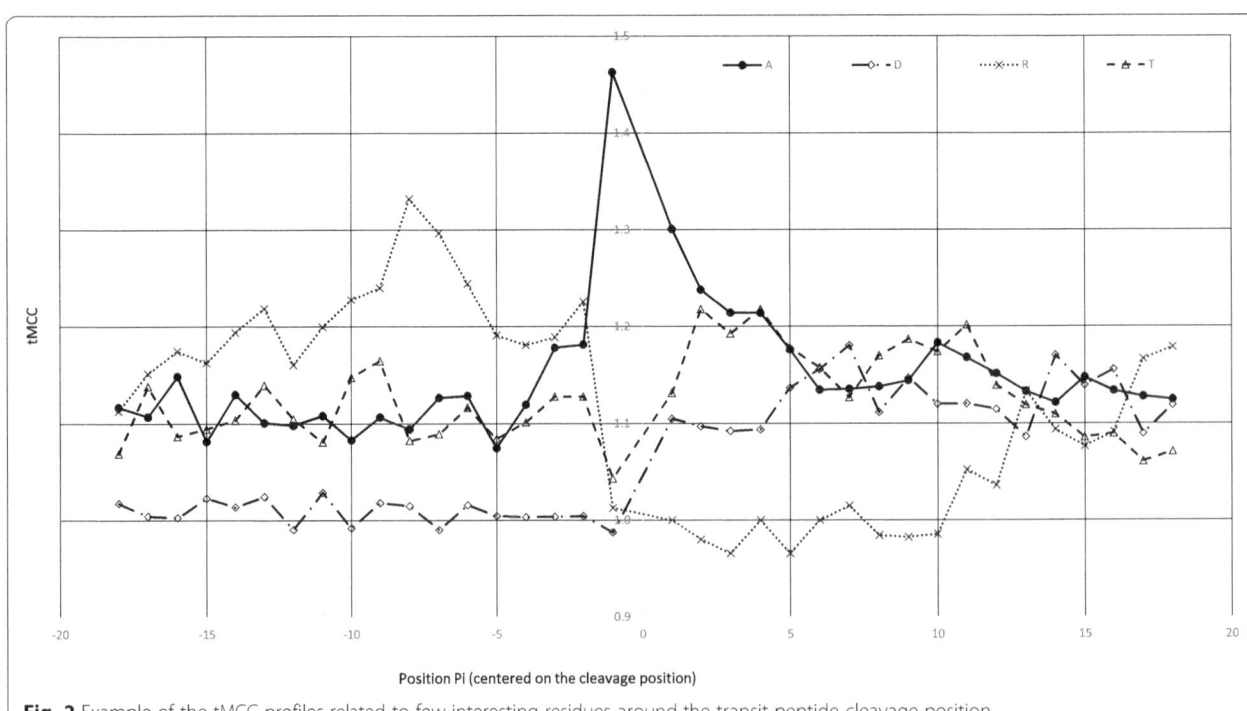

Fig. 2 Example of the tMCC profiles related to few interesting residues around the transit peptide cleavage position

Table 2 Results of the automated optimisation of the Bound and Spec parameters using the K fold cross validation result ($n = 10$) and the final scoring scheme using the same validation approach

Investigated parameters	Dataset or Scoring Scheme	ExpMin Position	ExpMax Position	EnCOUNTer or Spec threshold	True Positive	True Negative	False Positive	False Negative	Accuracy	Sensitivity	Specificity	False Discovery Rate	MCC
Spec (True dataset)	Training	-	-	> 62.9 ± 2.0	162 ± 3	288 ± 3	7 ± 2	21 ± 2	94.0 ± 0.3%	88.5 ± 1.0%	97.4 ± 0.5%	4.6 ± 0.9%	0.87 ± 0.01
	Validation	-	-	N.R.	16 ± 2	31 ± 3	2 ± 2	4 ± 2	88.5 ± 4.1%	78.7 ± 7.3%	94.9 ± 4.2%	9.3 ± 7.7%	0.67 ± 0.11
Bound (True dataset)	Training	17 ± 4	80 ± 6	-	164 ± 4	241 ± 6	56 ± 6	19 ± 3	84.3 ± 1.5%	89.8 ± 0.7%	81.2 ± 1.8%	25.3 ± 1.6%	0.71 ± 0.01
	Validation	N.R.	N.F.	-	18 ± 3	26 ± 4	7 ± 3	3 ± 1	82.9 ± 5.5%	87.7 ± 4.3%	80.0 ± 7.8%	26.6 ± 8.7%	0.67 ± 0.09
	All data together	14	78	63.2	183	266	63	20	84.4%	90.1%	80.9%	25.6%	0.69
Spec / Bound / Prox (True dataset)	Training	17 ± 4	80 ± 6	> 129.9 ± 0.6	167 ± 4	293 ± 5	3 ± 1	16 ± 2	96.1 ± 0.6%	91.2 ± 0.6%	99.1 ± 0.2%	1.6 ± 0.3%	0.92 ± 0.01
	Validation	N.R.	N.F.	N.R.	19 ± 4	33 ± 5	0 ± 1	2 ± 2	95.9 ± 2.9%	91.1 ± 5.3%	98.7 ± 2.3%	1.9 ± 3.2%	0.91 ± 0.06
Spec / Bound / Prox (False dataset)	Training	86 ± 9	300 ± 1	< 69.3 ± 6.1 (*)	272 ± 5	108 ± 4	74 ± 3	24 ± 4	79.5 ± 0.5%	92.0 ± 1.2%	59.1 ± 1.7%	21.4 ± 0.6%	0.59 ± 0.01
	Validation	N.R.	N.F.	N.R.	30 ± 3	12 ± 3	9 ± 3	3 ± 3	78.0 ± 4.1%	90.5 ± 6.7%	58.1 ± 10.1%	22.0 ± 5.4%	0.55 ± 0.08
Fraction 5 dataset	True dataset (Spec only)	14	78	65.1	179	321	8	24	94.0%	88.2%	97.6%	4.3%	0.872
	True dataset (Spec, Bound)	14	78	112.8	180	326	3	23	95.1%	88.7%	99.1%	1.6%	0.897
	True dataset (Spec, Bound, Prox)	14	78	130.1	185	326	3	18	96.1%	91.1%	99.1%	1.6%	0.917
	False dataset (Spec Only)	79	300	72.2	285	185	17	44	88.5%	86.6%	91.6%	5.6%	0.767
	False dataset (Spec, Bound, Prox)	79	300	66.7	304	118	84	25	79.5%	92.4%	58.4%	21.6%	0.556
	False dataset (Stringent params)	14	78	133.8	186	326	3	17	96.2%	91.6%	99.1%	1.6%	0.921
Fraction 6 dataset	True dataset (Spec, Bound, Prox)	14	78	130.1	179	442	8	51	91.3%	77.8%	98.2%	4.3%	0.806

(*) for the prediction based on the False dataset, the EnCOUNTer score must be below the determined Threshold for the True hits

106 (defined as ExpMin and ExpMax) for the validated candidates ("True" dataset) vs. 4 to 1104 for the irrelevant candidates (False dataset).

Interestingly, few distinct TP regions (Additional file 5: Figure S1) could be highlighted and are associated with proteins carrying a signal peptide (positions between 20 to 35 [34, 35]), mitochondrial TP (between positions 25 to 65 [16, 28]) and plastidic TP (between positions 30 to 95 [18]). Comparatively, the starting positions of the "False" candidates were evenly distributed. Then, it is interesting to favor/penalize selected regions depending of the protein training set. This allows the EnCOUNTer tool to define the optimum range where mature N-terminal positions are characterized from the training dataset. The optimum range for Fraction 5 dataset is in between positions 14–78 with 84.4% accuracy and 80.9% specificity with 25.6%. FDR. The associated "K fold cross correlation" (K = 10) highlights the robustness of this determination (Additional file 8: Table S3). This parameter cannot be used alone but always in combination with the Spec score (at least). When combining "Spec" and "Bound" score, 95.1% sensitivity and 99.1% accuracy with 1.6% FDR are reached on Fraction 5 training dataset. Such combination clearly improved the EnCOUNTer discrimination power compared to the "Spec" score alone.

Influence of the other scoring coefficients
By default, reliable predictions are reached using the Spec score and the Bound score together (95.1% accuracy at 1.6% FDR on the training dataset). Nevertheless, it could be possible to improve the prediction specificity or sensitivity using the additional coefficients Acetyl, Prox, Rep and/or Loc. Depending on the coefficient applied, it was possible to improve the sensitivity or the specificity of the EnCOUNTer tool (data not shown). As an example, the combination of Spec, Bound and Prox coefficients provides a final 96.1% accuracy and 99.1% sensitivity with 1.6% FDR. The associated K fold cross validation (k = 10) was performed and provided 95.9 ± 2.9% accuracy and 98.7 ± 2.3% sensitivity at 1.9 ± 3.2% FDR (Table 2 and Additional file 8: Table S3).

Although, the overall accuracy could be improved using different scoring combinations this was usually detrimental to the sensitivity. Depending on the goal (sensitivity, accuracy, specificity), scoring coefficient combinations could be adapted to reach better result than those provided in Table 2 i.e. better accuracy or better sensitivity... In hour hands, the combination of the scoring coefficients Spec, Bound and Prox provides a good starting compromise (Table 2) that could be optimized at will. These optimized parameters were applied to the Fraction 6 training dataset and provided the

discrimination of the N-termini at 91.3% of sensitivity and 98.2% of specificity (4.3% of FDR).

Protein N-terminal Acetylation quantitation
As previously mentioned, MD could provide protein NTA quantitation regardless of the multiple protein proteoforms. This is the example for the protein At1g16080.1 of which four distinct N-terminal positions could be characterized (positions 42, 44, 45 and 48; Additional file 1: Table S1). MD gave a single NTA yield of 35.5% (Min = 0.4%; Max = 98.98%) whereas EnCOUNTer provided four distinct NTA yield, i.e. 100.0% (Min: 99.8%; Max + 100.0%), 29.5%, 42.4% (Min: 42.0%; Max + 42.8%) and 2.1% respectively for each proteoforms. Another frequent MD processing error is the aggregation of H/L value associated to internal peptides. As an example, the MD quantification of At2g16600.1 protein combines the NTA yield associated to position 2 and 21 for a final NTA yield of 99.2% (Min = 26.5%; Max = 100.0%) whereas EnCOUNTer quantify only the N-terminus at position 2 with 99.8% NTA (Min = 98.5%; Max = 100.0%). Furthermore, the EnCOUNTer score of the peptide starting at position 21 is below the EnCOUNTer threshold and is not considered as a significant N-terminus. It is clear that EnCOUNTer discriminates the different N-termini and provides the most accurate NTA yield for each of them with an error range below 1% in average (Additional file 1: Table S1).

Example of application
As an example of application, our whole experimental dataset (N-terminus enriched fraction from A. *thaliana* leave lysate [2]) was processed using the optimized EnCOUNTer parameters. The parameters used where based on the results obtained during the optimization phase (Table 2), i.e. the combination of the Spec, Bound and Prox coefficients. 3964 potential N-termini were listed of which 1554 have an EnCOUNTer score higher than the threshold (EnCOUNTer Threshold = 130.1). After the removal of the non-proteotypic peptides, 1257 probable mature N-termini were listed of which 649 were located at position 1–2 and 608 at positions downstream of the protein N-terminus (Position >2). The NTA yield was determined for 594 N-termini (excluded none proteotypic N-termini) of which 275 were located at position 1–2 and 319 were associated to DNT (Additional file 2: Table S2).

As previously observed [2], 73% of the characterised N-termini at position 1–2 were found fully acetylated (NTA > 95%), 18% not acetylated (NTA < 5%) and 9% were partially acetylated (Fig. 3a). These N-termini peptides were mainly located (Fig. 3c) in the cytosol (39%), the nucleus (26%) and also in the mitochondria (6%) and the plastid (5%). Protein located in these last two

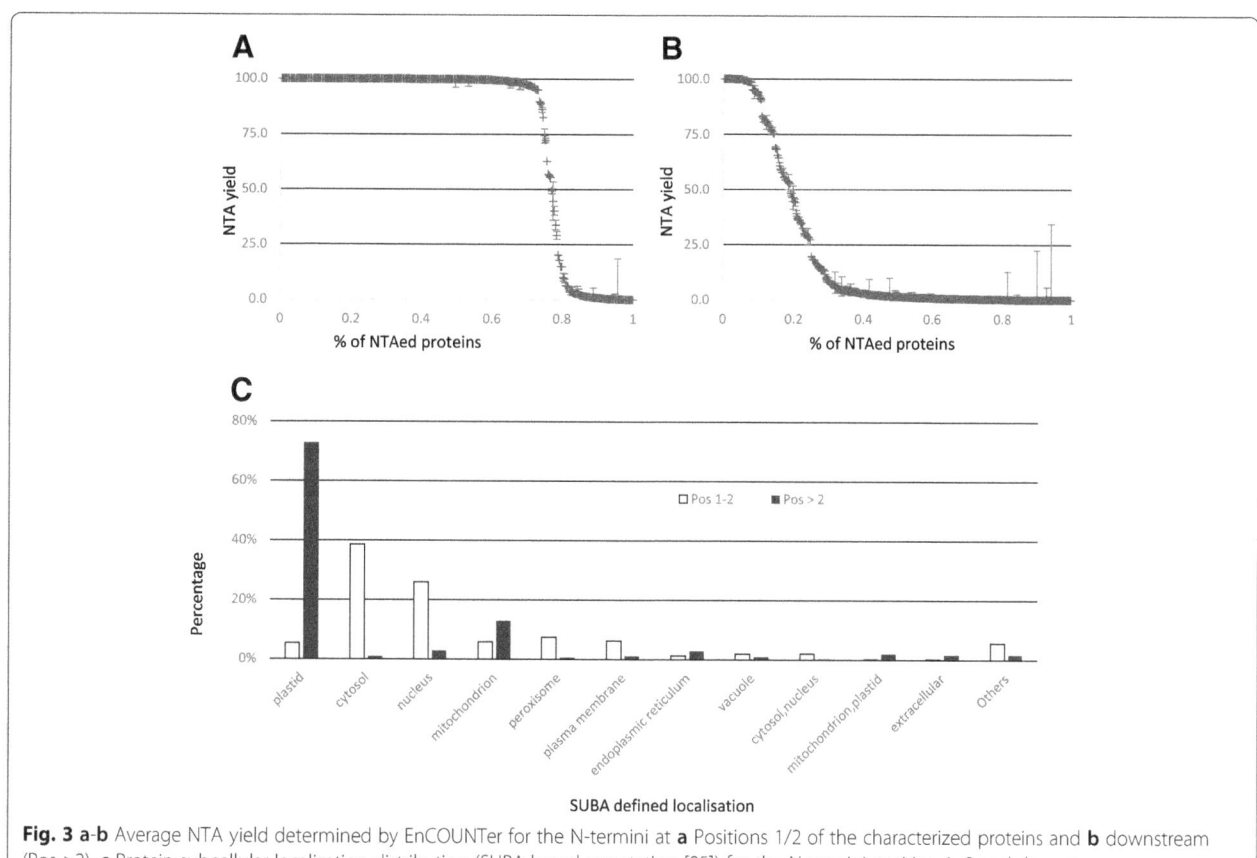

Fig. 3 a-b Average NTA yield determined by EnCOUNTer for the N-termini at **a** Positions 1/2 of the characterized proteins and **b** downstream (Pos >2). **c** Protein subcellular localization distribution (SUBA based annotation [25]) for the N-termini position 1–2 and downstream

compartments are frequently associated to TP excision. Nevertheless, most of the characterised mitochondrial proteins are outer membrane proteins and the characterisation of a starting position 1 and 2 is biologically relevant. Similarly, 17 out of the 35 plastid proteins are coded by the plastid genome. These proteins, expressed directly in the plastid, do not undergo transit peptide excision and were expected in this subset. Considering the other 18 nucleus-encoded proteins which were annotated in SUBA (The SUBcellular localisation database for *A. thaliana* [25]) as being translocated to the chloroplast, some of them are known to be chloroplastic such as At2g44640.1 or At4g28440.1 [21] and the absence of cTP could be explained by some previously described alternative import mechanisms [29]. Other candidates are erroneously associated to plastidic localisation such as At5g24650.1 (TIM17) which is clearly a mitochondrial protein.

In addition to these expected N-termini, 608 downstream N-termini could be characterized with an EnCOUNTer score higher than the threshold of which 319 were quantified for NTA. The pattern of the DNT-NTA yield with 8% of the downstream N-termini fully acetylated (>95%), 25% partially NTAed and 67% not acetylated (<5%) was clearly different (Fig. 3b) from

protein NTA profile (Fig. 3a). The subcellular distribution (Fig. 3c) was also strongly modified and the main localisation for the downstream N-termini was for 73% associated to plastidic proteins. Additionally, DNT also revealed mitochondrial N-termini (13%) resulting from mTP excision and alternative maturation of peroxisomal proteins (e.g. At2g33150.1 [36]), membrane proteins (e.g. At3g06035.1 or At5g19250.1 [37]) or vacuolar proteins (At5g60360.1 [38] or At2g23000). As previously observed for Pos 1–2, some of the SUBA subcellular localisation were erroneous, e.g. cytosolic localisation for At1G12900 or At4g26300 while they are localized in the chloroplastic stroma [39]. Some of the DNTs could also be a consequence of an alternative splicing or alternative start position (e.g. At1g66240 [40]), or errors on the gene starting position (At1g23820). Most of the 608 downstream N-termini highlighted by the EnCOUNTer tool were clearly due to protein maturation processes. This result confirms the added-value of EnCOUNTer to highlight mature proteins N-termini in complex peptide mixtures.

Conclusions

Throughout few thousands experimentally characterised N-termini, the EnCOUNTer tool is able to parse the most relevant mature protein N-termini with 96.1% accuracy and 99.1% specificity on the training dataset (91.3% sensitivity and 98.2% specificity using Fraction 6 dataset). Furthermore, the EnCOUNTer tool is able to provide reliable NTA yield for each distinct proteoform at the expected protein N-terminus (Pos 1–2) but also downstream.

Applied to a large experimental dataset, the EnCOUNTer tool was able to characterize more than 1200 N-termini of which almost 600 were quantified for NTA yield. Those characterised DNT could be associated to different maturation processes including nuclear encoded proteins targeting to various organelles (e.g. mitochondria, chloroplast or peroxisome), cytosolic maturations involving transient targeting peptides (e.g. membrane or secreted proteins) or erroneously assigned protein starts. This tool provides a unique way to determine the experimental position of the protein mature N-terminus and NTA acetylation yield for few hundreds up to thousands of candidates. This tool is especially interesting to determine accurately and rapidly the influence of various stresses on protein N-terminal status and N-terminal modification yield.

Additional files

Additional file 1: Table S1. List of the manually validated dataset (Fraction 5 and fraction 6) and curated subcellular localization.

Additional file 2: Table S2. Results obtained after EnCOUNTer processing applied to a large *A. thaliana* LC-MS/MS dataset

Additional file 3: Mascot Distiller processing method for protein N-terminal acetylation quantitation.

Additional file 4: EnCOUNTer parameters file.

Additional file 5: Figure S1. Distribution of the transit peptide cleavage position for fraction 5 validated dataset (True and False) and few specific subset (Mitochondria, plastid…).

Additional file 6: Figure S2. Web logos and heatmaps of 231 of nuclear coded plastidic N-termini (A-B), 30 mitochondrial N-termini (C-D) and 35 N-termini associated to protein carrying an excised signal peptide (E-F). The sequences around the transit peptide cleavage are compared to the average distribution of *A. thaliana* using the stand alone ICELogo tool [41].

Additional file 7: Figure S3. tMCC profiles of the Spec matrix for each possible residue for the A) True subset, B) False subset and C) from random A. thaliana proteins.

Additional file 8: Table S3. Result of the K fold cross validation ($n = 10$) for the optimization of bound, Spec and the global scoring using fraction 5 training dataset (True vs. False dataset).

Abbreviations
ACN: Acetonitrile; cTP: Plastid transit peptide; d0-Ac: endogeneous acetyl group; d3-Ac: deuterated acetyl group; DNT: Downstream N-termini; FDR: False Discovery Rate; H: Heavy (d3-Ac); L: Light (d0-Ac); LC: Liquid chromatography; MCC: Matthews Correlation Coefficient; MD: Mascot Distiller; MS: Mass spectrometry; mTP: Mitochondrial transit peptide; NTA: N-terminal acetylation; NTAed: N-terminal acetylated; SCX: Strong Cation eXchange; TFA: Trifluoroacetic acid; tMCC: translated (by 1 unit upper) Matthews Correlation Coefficient; TP: Transit peptide

Acknowledgement
This study has benefited from the facilities and expertise of the SICaPS platform of I2BC (Institute for Integrative Biology of the Cell). The authors also thank A. Estreicher and Y Vandenbrouck for their advices, counsels, and expertise.

Funding
This work was supported by the "French National Research Agency (grant No. ANR-13-BSV6-0004) and is directly associated to the development of the EnCOUNTer tool. JD has benefited from the support of the "LabEX Saclay Plant Sciences-SPS" (ANR-10-LABX-0040-SPS) for its gratification. The Lidex BIG (https://www.universite-paris-saclay.fr/fr/recherche/projet/lidex-big) finance experimental analyses to provide raw material used forthe development and the optimisation of the EnCOUNTer tool.

Authors' contributions
WB was involved in the acquisition of data, the conception and the design of the EnCOUNTer tool, the analysis and the interpretation of data. JPS and JD developed the python script, WIB, CG and TM were involved in drafting the manuscript. All authors read and approved the final manuscript.

Competing interests
The authors declare that they have no competing interests.

References
1. Giglione C, Fieulaine S, Meinnel T. N-terminal protein modifications: Bringing back into play the ribosome. Biochimie. 2015;114:134–46.
2. Linster E, Stephan I, Bienvenut WV, Maple-Grodem J, Myklebust LM, Huber M, Reichelt M, Sticht C, Geir Moller S, Meinnel T, et al. Downregulation of N-terminal acetylation triggers ABA-mediated drought responses in Arabidopsis. Nat Commun. 2015;6:7640.
3. Bienvenut WV, Espagne C, Martinez A, Majeran W, Valot B, Zivy M, Vallon O, Adam Z, Meinnel T, Giglione C. Dynamics of post-translational modifications and protein stability in the stroma of Chlamydomonas reinhardtii chloroplasts. Proteomics. 2011;11(9):1734–50.
4. Rowland E, Kim J, Bhuiyan NH, van Wijk KJ. The Arabidopsis Chloroplast Stromal N-Terminome: Complexities of Amino-Terminal Protein Maturation and Stability. Plant Physiol. 2015;169(3):1881–96.
5. Emanuelsson O, Brunak S, von Heijne G, Nielsen H. Locating proteins in the cell using TargetP, SignalP and related tools. Nat Protoc. 2007;2(4):953–71.
6. Emanuelsson O, Nielsen H, von Heijne G. ChloroP, a neural network-based method for predicting chloroplast transit peptides and their cleavage sites. Protein Sci. 1999;8(5):978–84.
7. Bienvenut WV, Sumpton D, Martinez A, Lilla S, Espagne C, Meinnel T, Giglione C. Comparative large scale characterization of plant versus mammal proteins reveals similar and idiosyncratic N-alpha-acetylation features. Mol Cell Proteomics. 2012;11(6):M111. 015131.
8. Bienvenut WV, Giglione C, Meinnel T. Proteome-wide analysis of the amino terminal status of Escherichia coli proteins at the steady-state and upon deformylation inhibition. Proteomics. 2015;15(14):2503–18.
9. ElBashir R, Vanselow JT, Kraus A, Janzen CJ, Siegel TN, Schlosser A. Fragment ion patchwork quantification for measuring site-specific acetylation degrees. Anal Chem. 2015;87(19):9939–45.
10. Knudsen AD, Bennike T, Kjeldal H, Birkelund S, Otzen DE, Stensballe A. Condenser: a statistical aggregation tool for multi-sample quantitative proteomic data from Matrix Science Mascot Distiller. J Proteomics. 2014;103: 261–6.
11. Huala E, Dickerman AW, Garcia-Hernandez M, Weems D, Reiser L, LaFond F, Hanley D, Kiphart D, Zhuang M, Huang W, et al. The Arabidopsis Information Resource (TAIR): a comprehensive database and web-based information retrieval, analysis, and visualization system for a model plant. Nucleic Acids Res. 2001;29(1):102–5.
12. Mallick P, Schirle M, Chen SS, Flory MR, Lee H, Martin D, Ranish J, Raught B, Schmitt R, Werner T, et al. Computational prediction of proteotypic peptides for quantitative proteomics. Nat Biotechnol. 2007;25(1):125–31.
13. Kunze M, Berger J. The similarity between N-terminal targeting signals for protein import into different organelles and its evolutionary relevance. Front Physiol. 2015;6:259.
14. Patron NJ, Waller RF. Transit peptide diversity and divergence: A global analysis of plastid targeting signals. Bioessays. 2007;29(10):1048–58.

15. Shi LX, Theg SM. The chloroplast protein import system: from algae to trees. Biochim Biophys Acta. 2013;1833(2):314–31.

16. Huang S, Taylor NL, Whelan J, Millar AH. Refining the definition of plant mitochondrial presequences through analysis of sorting signals, N-terminal modifications, and cleavage motifs. Plant Physiol. 2009;150(3):1272–85.

17. Vaca Jacome AS, Rabilloud T, Schaeffer-Reiss C, Rompais M, Ayoub D, Lane L, Bairoch A, Van Dorsselaer A, Carapito C. N-terminome analysis of the human mitochondrial proteome. Proteomics. 2015;15(14):2519–24.

18. Bionda T, Tillmann B, Simm S, Beilstein K, Ruprecht M, Schleiff E. Chloroplast import signals: the length requirement for translocation in vitro and in vivo. J Mol Biol. 2010;402(3):510–23.

19. Dinh TV, Bienvenut WV, Linster E, Feldman-Salit A, Jung VA, Meinnel T, Hell R, Giglione C, Wirtz M. Molecular identification and functional characterization of the first Nalpha-acetyltransferase in plastids by global acetylome profiling. Proteomics. 2015.

20. Bienvenut WV, Sumpton D, Lilla S, Martinez A, Meinnel T, Giglione C. Influence of various endogenous and artefact modifications on large scale proteomics analysis. Rapid Commun Mass Spectrom. 2013;27:443–50.

21. Zybailov B, Rutschow H, Friso G, Rudella A, Emanuelsson O, Sun Q, van Wijk KJ. Sorting signals, N-terminal modifications and abundance of the chloroplast proteome. PLoS One. 2008;3(4):e1994.

22. Bruley C, Dupierris V, Salvi D, Rolland N, Ferro M. AT_CHLORO: A Chloroplast Protein Database Dedicated to Sub-Plastidial Localization. Front Plant Sci. 2012;3:205.

23. Ferro M, Brugiere S, Salvi D, Seigneurin-Berny D, Court M, Moyet L, Ramus C, Miras S, Mellal M, Le Gall S, et al. AT_CHLORO, a comprehensive chloroplast proteome database with subplastidial localization and curated information on envelope proteins. Mol Cell Proteomics. 2010;9(6):1063–84.

24. Sun Q, Zybailov B, Majeran W, Friso G, Olinares PD, van Wijk KJ. PPDB, the Plant Proteomics Database at Cornell. Nucleic Acids Res. 2009;37(Database issue):D969–74.

25. Heazlewood JL, Verboom RE, Tonti-Filippini J, Small I, Millar AH. SUBA: the Arabidopsis Subcellular Database. Nucleic Acids Res. 2007;35(Database issue):D213–8.

26. Fortelny N, Yang S, Pavlidis P, Lange PF, Overall CM. Proteome TopFIND 3.0 with TopFINDer and PathFINDer: database and analysis tools for the association of protein termini to pre- and post-translational events. Nucleic Acids Res. 2015;43(Database issue):D290–7.

27. Joshi HJ, Hirsch-Hoffmann M, Baerenfaller K, Gruissem W, Baginsky S, Schmidt R, Schulze WX, Sun Q, van Wijk KJ, Egelhofer V, et al. MASCP Gator: an aggregation portal for the visualization of Arabidopsis proteomics data. Plant Physiol. 2011;155(1):259–70.

28. Fukasawa Y, Tsuji J, Fu SC, Tomii K, Horton P, Imai K. MitoFates: Improved Prediction of Mitochondrial Targeting Sequences and Their Cleavage Sites. Mol Cell Proteomics. 2015;14(4):1113–26.

29. Miras S, Salvi D, Piette L, Seigneurin-Berny D, Grunwald D, Reinbothe C, Joyard J, Reinbothe S, Rolland N. Toc159- and Toc75-independent import of a transit sequence-less precursor into the inner envelope of chloroplasts. J Biol Chem. 2007;282(40):29482–92.

30. Schechter I, Berger A. On the Size of the active site in proteases. I) Papain. Biochem Biophys Res Comm. 1967;27:157–62.

31. Keil-Dlouha VV, Zylber N, Imhoff J, Tong N, Keil B. Proteolytic activity of pseudotrypsin. FEBS Lett. 1971;16(4):291–5.

32. Hedstrom L, Szilagyi L, Rutter WJ. Converting trypsin to chymotrypsin: the role of surface loops. Science. 1992;255(5049):1249–53.

33. Jarvis P. Targeting of nucleus-encoded proteins to chloroplasts in plants. New Phytol. 2008;179(2):257–85.

34. Kall L, Krogh A, Sonnhammer EL. A combined transmembrane topology and signal peptide prediction method. J Mol Biol. 2004;338(5):1027–36.

35. Dalbey RE, Wang P, van Dijl JM. Membrane proteases in the bacterial protein secretion and quality control pathway. Microbiol Mol Biol Rev. 2012; 76(2):311–30.

36. Carrie C, Murcha MW, Millar AH, Smith SM, Whelan J. Nine 3-ketoacyl-CoA thiolases (KATs) and acetoacetyl-CoA thiolases (ACATs) encoded by five genes in Arabidopsis thaliana are targeted either to peroxisomes or cytosol but not to mitochondria. Plant Mol Biol. 2007;63(1):97–108.

37. Borner GH, Lilley KS, Stevens TJ, Dupree P. Identification of glycosylphosphatidylinositol-anchored proteins in Arabidopsis. A proteomic and genomic analysis. Plant Physiol. 2003;132(2):568–77.

38. Sohn EJ, Kim ES, Zhao M, Kim SJ, Kim H, Kim YW, Lee YJ, Hillmer S, Sohn U, Jiang L, et al. Rha1, an Arabidopsis Rab5 homolog, plays a critical role in the vacuolar trafficking of soluble cargo proteins. Plant Cell. 2003;15(5):1057–70.

39. Huang M, Friso G, Nishimura K, Qu X, Olinares PD, Majeran W, Sun Q, van Wijk KJ. Construction of plastid reference proteomes for maize and Arabidopsis and evaluation of their orthologous relationships; the concept of orthoproteomics. J Proteome Res. 2013;12(1):491–504.

40. Puig S, Mira H, Dorcey E, Sancenon V, Andres-Colas N, Garcia-Molina A, Burkhead JL, Gogolin KA, Abdel-Ghany SE, Thiele DJ, et al. Higher plants possess two different types of ATX1-like copper chaperones. Biochem Biophys Res Commun. 2007;354(2):385–90.

41. Colaert N, Helsens K, Martens L, Vandekerckhove J, Gevaert K. Improved visualization of protein consensus sequences by iceLogo. Nat Methods. 2009;6(11):786–7.

14

NEArender: an R package for functional interpretation of 'omics' data via network enrichment analysis

Ashwini Jeggari[3] and Andrey Alexeyenko[1,2]*

Abstract

Background: The statistical evaluation of pathway enrichment, i.e. of gene profiles' confluence to the pathway level, allows exploring molecular landscapes using functionally annotated gene sets. However, pathway scores can also be used as predictive features in machine learning. That requires, firstly, increasing statistical power and biological relevance via a network enrichment analysis (NEA) and, secondly, a fast and convenient procedure for rendering the original data into a space of pathway scores. However, previous implementations of NEA involved multiple runs of network randomization and were therefore slow.

Results: Here, we present a new R package NEArender which can transform raw 'omics' features of experimental or clinical samples into matrices describing the same samples with many fewer NEA-based pathway scores. This is done via a parametric estimation of the null binomial distribution and is thus much faster and less biased than randomization procedures. Further, we compare estimates from these two alternative procedures and demonstrate that the summarization of individual genes to pathways increases the statistical power compared to both the default differential expression analysis on individual genes and the state-of-the-art gene set enrichment analysis. The package also contains functions for preparing input, modeling null distributions, and evaluating alternative versions of the global network.

Conclusions: Beyond the state-of-the-art exploration of molecular data through pathway enrichment, score matrices produced by NEArender can be used in larger bioinformatics pipelines as input for phenotype modeling, predicting disease outcomes etc. This approach is often more sensitive and robust than using the original data. The package NEArender is complementary to the online NEA tool EviNet (https://www.evinet.org) and, unlike of the latter, enables high performance of computations off-line.

The R package NEArender version 1.4 is available at CRAN repository
https://cran.r-project.org/web/packages/NEArender/

Keywords: Enrichment, Network analysis, Network benchmark, R package

* Correspondence: andrej.alekseenko@scilifelab.se
[1]National Bioinformatics Infrastructure Sweden, Science for Life Laboratory, Stockholm, Sweden
[2]Department of Microbiology, Tumor and Cell biology, Karolinska Institutet, Stockholm, Sweden
Full list of author information is available at the end of the article

Background

NEA employs network topology to evaluate functional impact of experimentally determined genes and gene sets by detecting enrichment of previously characterized gene sets, such as pathways. NEA became a natural extension of the gene set enrichment analysis, GSEA [1] into the network domain [2, 3]. Performance and applicability of GSEA have been limited by the following: 1) only a minority of genes possesses specific pathway annotations, 2) it can be applied only to genes altered in a specific way, detectable by the given platform (typically transcriptomics) and not to those regulated via other mechanisms, and 3) statistical power of the analysis is limited by gene set size. NEA largely overcomes these limitations due to one key difference: while GSEA counts the number of genes shared between an experimental list and a pathway, NEA considers network edges between any genes of the two sets in the global network. On the other hand, the earlier NEA versions had an own drawback: the error rate was estimated in multiple, time-consuming instances of randomized networks. Being a non-parametric approach, the randomization allows solving a wide range of higher-order topological problems. However, for many applications this would be prohibitively slow. In the present R package, we implement a much faster parametric estimation of connectivity rate expected by chance and show that this also eliminates bias on small gene sets. Finally, we provide an illustration of how NEA increases robustness of non-replicated analyses of gene differential expression.

Results

Software functionality

NEArender possesses both core and ancillary functions for network enrichment analysis. It is implemented as an R package and described in details, beyond the official manual, in an R vignette https://cran.r-project.org/web/packages/NEArender/vignettes/NEArender_vignette.pdf

The input shall contain three components: 1) one or multiple user-defined (experimental or theoretical) gene sets which have to be functionally characterized, called altered gene sets (AGS); 2) a collection of functional gene sets (FGS) which would enable functional characterization through their known functions, and 3) a global network of functional coupling (NET). Input can be provided as either text files or pre-processed R lists and matrices. Figure 1 presents the relationship between the major components and steps of NEArender.

In fact, although we denote input in terms of "genes", a range of functional nodes of a biological network can be analyzed with this algorithm, such as protein molecules, genomic regions that encode proteins, microRNAs, promoters, and enhancers etc. Nodes listed in AGS, FGS, and NET should employ the same ID format. The package performs network enrichment analysis through the fast binomial test as described in Methods and quantifies enrichment in each AGS-FGS pair with a number of statistics: chi-squared score, z-score, p-value, q-value (the p-value adjusted for multiple testing, a.k.a. false discovery rate [4], number of network edges that exist between any nodes of AGS and FGS (but neither within AGS nor within FGS), and respective number of AGS-FGS edges expected by chance, calculated with the binomial formula.

In order to run enrichment analysis, the user should prepare the input components. Since the AGS is the most dynamic and user-specific part of the input, functionality for AGS compilation and processing is most developed. First, AGSs can be prepared in advance externally, e.g. as a list of genomic variations reported in a given genome. Alternatively, the package functions can create AGSs from sample columns of an R matrix. This can be done with a number of different algorithms (two of these we use below in the section 'Robustness of results obtained from different replicates'. Then the number of genes included in each AGS would be either data-driven (all significant genes) or pre-defined by user (top ranking ones regardless of significance). In addition, the algorithm 'toprandom' generates random AGSs of a user-set size. Finally, a special function allows direct creation of AGSs as full sample-specific sets of mutated genes. FGSs are usually imported from a file by listing all members of each functional set. Due to the high network density and, hence, statistical power of NEA, there is a special option, which is not available in GSEA: single genes can be treated as FGS as well. A full list of such single-gene FGSs can be created from all network nodes of NET with parameter `as_genes_fgs`, so that each FGS item in the output list contains just one gene. It is practical to use a large pre-compiled FGS collection, such as all ontology terms, pathways, or a union of resources, e.g. MSigDB database [5]. Alternatively, users can create custom single- or multi-gene FGS collections of their own. We note however that using relatively few FGSs and/or AGSs in one analysis (so that the total number of AGS-FGS tests is below a few hundreds) would not allow estimating the q-values properly.

While higher order topological biases are, as discussed above, of arguable importance for NEA, another network feature is vital for the enrichment evaluation used in this package. Namely, the parametric algorithm would produce unbiased estimates only in scale-free networks, i.e. where node connectivity values follow the power law distribution [6]. Being an almost ubiquitous feature in the full scale biological networks, this is still not the case in networks that are artificially constructed from e.g. ChIP-seq based collections of transcription factor binding events [7] or from computationally predicted microRNA-transcript targeting data [8]. If such network

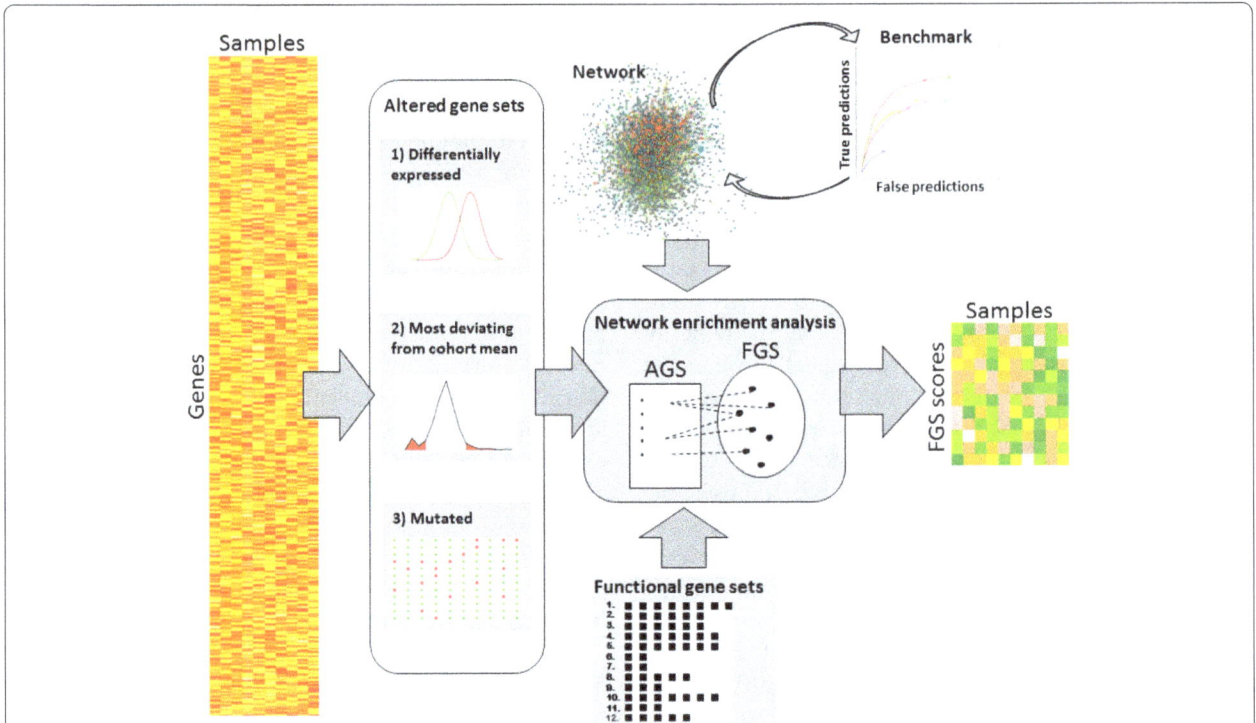

Fig. 1 Analysis flow in NEArender. The original matrix of 'omics' (mutation, methylation, expression etc.) data described a limited number of samples (patients etc.) with a much larger number of gene feature rows. At the first, preparatory step each sample was described via a characteristic sample-specific altered gene set (AGS). In parallel, a collection of functional gene sets (FGS) that share certain functionally annotations (within each set) was downloaded or prepared otherwise. A global gene/protein network (NET) was also provided (possibly selected from a number of alternatives based on benchmark results). In the course of network enrichment analysis (NEA) each AGS received as many NEA scores as there were FGSs, i.e. obtained coordinates in the multidimensional FGS space. This created an output matrix of the same number of sample columns but many fewer rows

components are desirable, we recommend employing software that involves network permutation tests, i.e. the randomization, such as in [2, 3]. Therefore, the package enables evaluating network topology for scale-freeness and second-order dependencies (described below) as well as benchmarking alternative NETs using either standard or custom FGSs, as described in [9]. Briefly, the benchmark consists of as many test cases as there are FGS members in total (multiple occurrences of the same genes in different FGS are treated separately). For each such gene, the procedure tests the null hypothesis of the gene not being an FGS member. The true positive or false negative result is assigned if the gene receives an NEA score above or below a certain threshold, respectively. In parallel, randomly picked genes with close node degree values are tested against the same FGS to estimate specificity via the false positive versus true negative ratio. The counts of alternative test outcomes TP, TN, FP, and FN at variable NEA thresholds are used for plotting ROC curves.

Comparison to randomization based algorithm

We compared the p-value distributions between the two methods (Fig. 2). The both were capable of adjusting p-values for multiple testing, but we omit the adjusted

value analysis here since our GS test set of 330 FGS (mostly KEGG pathways with addition of GO terms and other sets related to cancer, cytokine signaling, intercellular communications etc. [10] abounded with highly functionally similar GS pairs and thus the fraction of significant q-values was ~30%. We instead focus on the p-values in order not to miss important distribution details. It is apparent that the estimates from network randomization become more consistent with the growing number of randomization runs, from $N = 3$ to $N = 300$. The third column of Fig. 2 displays strong and asymptotically increasing correlation between p-values from the network randomization z-test (NRZ) and chi-squared binomial formula (CSB) (Spearman $r = \{0.83; 0.94; 0.97; 0.98; 0.99\}$ for $N = \{3; 10; 30; 100; 300\}$, respectively). Since CSB employed a deterministic procedure, we assume that all dispersion of p-value points in the NRZ-CSB space should be attributed to sampling errors by the stochastic NRZ algorithm. CSB and NRZ p-values converged sufficiently well only at $N > 30$. According to the quantile analysis with Q-Q plots (2nd column), CSB p-values were more conservative than those of NRZ. Next, the latter was somewhat less sensitive than CSB in regard of small GS, especially of those

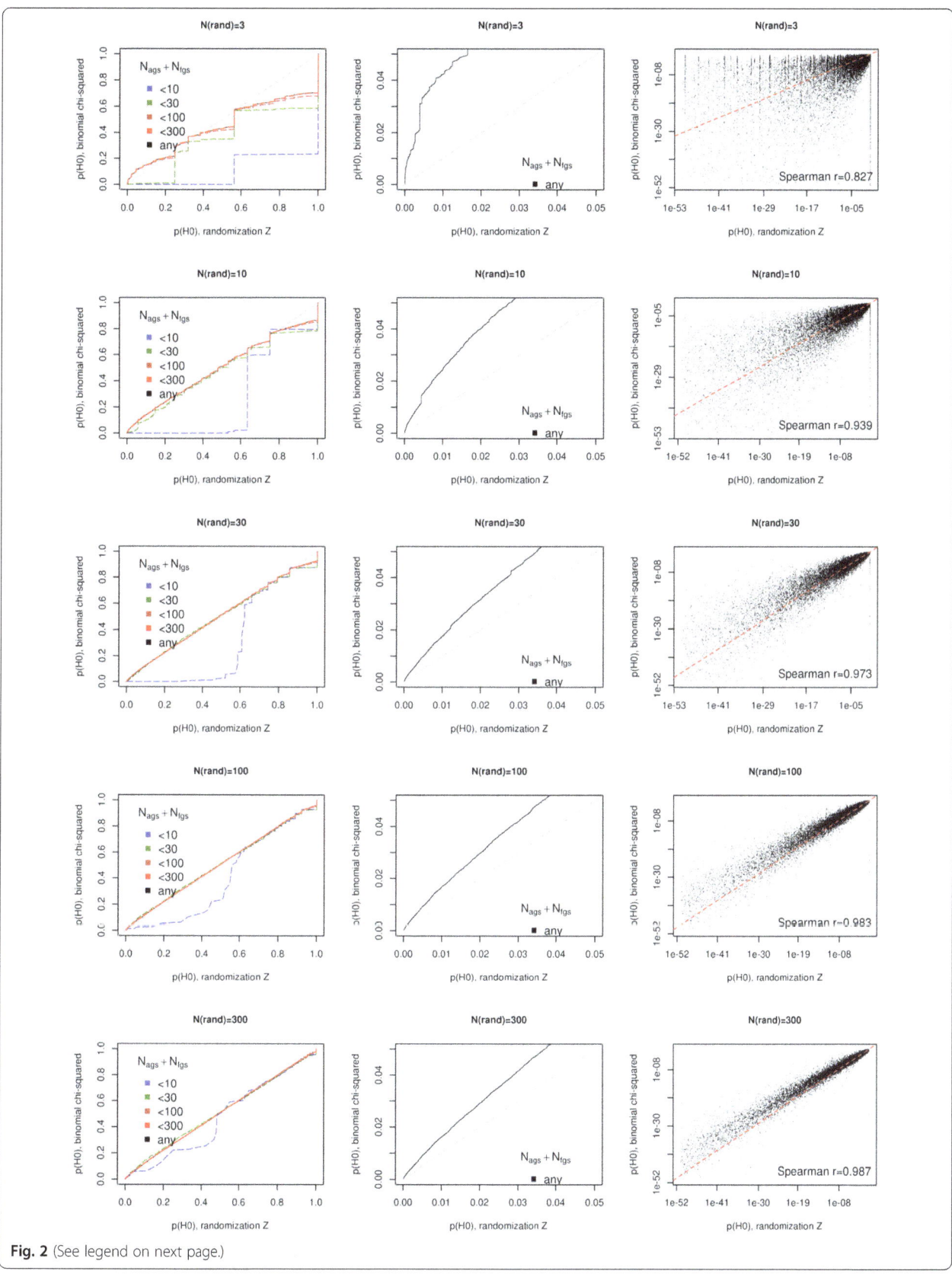

Fig. 2 (See legend on next page.)

(See figure on previous page.)
Fig. 2 Comparative sensitivity and sources of bias in randomization-based versus binomial calculation of network enrichment. Network enrichment between all vs. all 330 gene sets was analyzed with both NRZ and CSB methods. P-value distributions were compared using Q-Q plots (columns 1 and 2) and scatter plots of log (p) values (column 3). Q-Q plots in column 1 display both the total distributions (black lines), i.e. regardless of GS size, and distribution fractions that correspond to smaller GS ($N_{AGS} + N_{FGS}$, color lines). The QQ-plots in column 2 are insets of the black, un-stratified Q-Q plots of column 1. Identity lines (x = y) are plotted in dotted grey and dotted red in the QQ-plots and scatter plots, respectively. Analogous plots in regard of other factors biasing NRZ p-values ($C_{AGS} + C_{FGS}$ and N_{edges}) are provided in Additional file 1: Figure S1 (note that plot columns 2 and 3 are the same as in the present figure)

where both GSs were small, below 30 or 10 nodes altogether (green and blue lines in the first column). Again, NRZ p-values from the computationally feasible but insufficient $N = \{3; 10; 30\}$ exhibited more disagreement with respective CSB values. In addition to the sums $N_{AGS} + N_{FGS}$, we stratified the Q-Q plots by the sum of node degrees $C_{AGS} + C_{FGS}$ and by the actual number of edges between two GSs, N_{edges} (Additional file 1: Figure S1A and B). The sum $C_{AGS} + C_{FGS}$ appeared a strongly biasing factor (although N and C expectedly correlated in the test set). The influence of N_{edges} was almost non-existent.

Since there was no gold standard for network enrichment, we could not unambiguously conclude that CSB was more precise than NRZ. However it seems to be preferable because of its, in general, more conservative estimates and, in particular, higher sensitivity on small GS as well as the convergence of NRZ at higher values of N. The NRZ-specific bias on small GSs likely occurred because between-GS connectivity values could not take on negative values, hence the left tail of the distribution of N_{edges} was compressed and standard deviation must be underestimated.

Thus while using NRZ, the deviation from CSB values and bias due to small GS size were considerably large at $N < 100$. However, performing many randomizations might make the procedure prohibitively slow. While using NRZ, the computation time grows linearly with the number of randomizations. For example on a server running 2.6.32-642.el6.x86_64 (OS Scientific Linux),one randomization of the given network took around 3 min plus 20 min for counting connectivity in the given FGS collection, so that the total time for 100 randomizations was 100*(20 + 3) = 2300 min. For the most direct comparison, the same perl software [9] in the CSB mode, i.e. without network randomization, run 15 min. In the both cases, the perl process occupied between 300 and 400 MB RAM. Running R package NEArender required less than 5 min and below 200 MB RAM. We note that in many practical tasks the number of either AGS or FGS could be significantly reduced. However as stated above, the current package is mostly meant for generating NEA scores that could be used as predictive features in machine learning - and those AGSxFGS matrices are supposed to be large.

Robustness of results obtained from different replicates

Statistically underpowered experimental designs are not welcomed in the scientific community but, due to the lack of resources, are still practiced. In particular, it is common to represent patients in large cohort screens with single samples [11, 12]. Both the GSEA and NEA analyses measure enrichment in order to summarize signals from individual genes' to the level of pathways and biological processes. This feature suggests a potential increase in robustness of conclusions in experiments that lack replicates. We decided to investigate this robustness of using single-gene expression values versus GSEA and NEA under different scenarios using cell transcriptome data from the FANTOM5 CAGE RNA sequencing set [13]. In order to make results comparable between NEA and GSEA, we use the binomial version of the latter [1], which allowed applying the analyses to the same input gene sets. Commonly, statistical power is analyzed via extrapolating variance estimates [14, 15]. Since GSEA and NEA do not provide such estimates, we instead watched consistency of conclusions drawn from individual replicates.

The samples, in 3 to 5 biological (distinct healthy individuals) replicates, originated from different fibroblast, epithelial, and smooth muscle cells. In total, 43 replicates from 11 cell types quantified RNA transcript tags mapped to 16620 genes. This provided a broad range of degrees of dissimilarity between type-specific transcriptomes. The dissimilarity was evaluated as a fraction of significantly DE genes in fully replicated designs and ranged from 2.4% ("fibroblast.periodontal" vs. "fibroblast.gingivial") to 54.8% ("smooth.muscle.umbilical.artery" vs. "gingival.epithelial") DE genes. This set-up allowed us to model situations of analyzing DE of each gene g using single samples, e.g. g_{Ai} vs. g_{Bj} on samples i and j from cell types A and B, and then using all available replicates for A and B (N_A, N_B): $g_{A\{1...N_A\}}$ vs. $g_{B\{1...N_B\}}$. Note that calculating DE p-values was thus possible only in the latter case. Otherwise, DE could only be ranked by values of fold change between A_i and B_j. We quantified DE in all the 55 possible $A_i - B_j$ pairs between the 11 cell types. As an example, if there were $N_A = 3$ and $N_B = 5$ replicates, then we could model DE measurements on 5*3 = 15 pairs of single samples plus one analysis $g_{A\{1...N_A\}}$ vs. $g_{B\{1...N_B\}}$. In the following, we review preservation of DE estimates across different cell samples contrasts analyzed with three different methods: the default DE analysis on individual genes, GSEA, and NEA. The preservation was evaluated in the form of Spearman rank correlation coefficients between: 1) fold change values of individual genes for gene-wise analysis "as is",2) P-values of FGS enrichment scores obtained with binomial GSEA, and 3) P-values of FGS enrichment scores obtained with CSB NEA.

The example scatterplots at Fig. 3 (a, c, and e) display agreement between the same replicate pairs of the same contrast between fibroblast gingivial vs. tenocyte cells. The fold change values and p-values, although being different in their nature and scale, allowed us to rank the results in the same way as a biological researcher would prioritize them. We note here that in GSEA and NEA neither p-values nor enrichment scores (being mutually rank-invariant) explicitly provide sampling errors similar to between-sample variance in the DE analysis. All such comparisons between sample pairs were quantified by Spearman rank correlations and used as individual points in the plots on the right (b, d, and f). One observation from the plots C, and E was that regardless of the correlation strength, NEA possessed a much higher statistical power to detect enrichment – which has been discussed in details by Alexeyenko et al. in 2012 [2]. Indeed, no GSEA p-values appeared significant after the Bonferroni adjustment for multiple testing in either replicate pair (brown dotted vertical and horizontal lines). For comparison, almost 1/3rd of the NEA scores were significant in the both replicate pairs.

GSEA and NEA were each tested using six alternative variants of AGS, generated for each contrast: a) lists of top 100, 300, and 1000 DE genes ranked by fold change and b) lists of genes with absolute $\log_2$ (fold change) values exceeding 1, 2, and 4. AGS sizes of the form (b) were not fixed and depended on the magnitude of the difference between the two transcriptomes. When using all available replicates in the analyses (a), the genes were ranked by p-value with functions lmFit and eBayes in R package limma [16]. In (b) we required that, in addition to the fold change cutoff, the adjusted p-values did not exceed 0.05.

Overall, the scores from NEA agreed with each other much better than those from GSEA (Fig. 3, g and h). By investigating the AGS- and method-specific scatterplots for AGSs of type "abs ($\log_2$ (fold change)) > 2" (Fig. 3, b, d, and f; respective plots for all the six AGS types are available as Additional file 1: Figure S3), we observed that the level of correlation between the analyses (Y-axis) depended on strength of the pairwise difference between cell type transcriptomes, expressed as the fraction of DE genes (X-axis). Ideally, we would observe independence of the difference strength (so that ranks

fully correlate regardless of fold change, p-value, FDR etc.) and perfect correlation between DE results from both individual sample pairs and replicated analyses. However, it was unrealistic to expect large and efficient AGSs from few (e.g. 5...7%) significantly DE genes. Still, we can conclude that NEA output was considerably closer to the desirable pattern than those of the gene-wise and GSEA analyses. Indeed, in addition to the stronger overall correlation, NEA performed particularly well on weakly different cell types pairs (see the regression lines' intercepts with Y-axes and the rank r values in the bottom right corners in Fig. 3, b, d, and f). The only exception could be found for AGSs top100 (Additional file 1: Figure S3), where GSEA appeared superior over NEA. However, this option would be practically unusable in GSEA because of its low statistical power on small gene sets and, hence, few significant enrichment values.

We also could compare consistency of results obtained via i) individual sample pairs against each other ("g_{Ai} vs. g_{Bj}" against "g_{Am} vs. g_{Bn}") and ii) results from individual sample pairs against the replicated analysis ("g_{Ai} vs. g_{Bj}" against "$g_{A\{1...N_A\}}$ vs. $g_{B\{1...N_B\}}$"). One can also see that the option (ii), i.e. the comparison of non-replicated to replicated analyses (colored points), often exhibited poorer correlation than comparisons (i) (black points), especially for GSEA (the same pattern can be seen in the boxplots of Additional file 1: Figure S2).

Discussion

We demonstrated that estimates of enrichment from the network randomization procedure were biased. They could sufficiently converge to respective CSB values only at very large, impractical numbers of randomizations runs.

While considering the "lack-of-replicates" scenario, we assumed that the former shall always be preferred. However, our analysis demonstrated the higher (although still imperfect) robustness of NEA scores in the full absence or shortage of replicates. This conclusion about superiority of NEA over gene-wise analyses is, of course, only relevant when the research problem can be approached with an exploratory analysis at a pathway level and not when it requires identifying individual consistently deregulated genes. In addition, this advantage of NEA explains the earlier observed higher robustness of its scores, compared to individual gene profiles, as descriptors of experimental models of cancer-inhibitory fibroblasts [10]. Importantly, biological replicates of patients' samples are rarely available in clinical cohorts, so that we anticipate efficient usage of NEA in such setups.

NEA can also be performed using the online tool available at https://www.evinet.org. The latter is more interactive, user-friendly, and focused on visualization, while the package NEArender enables high performance, possibly with parallel computation. In

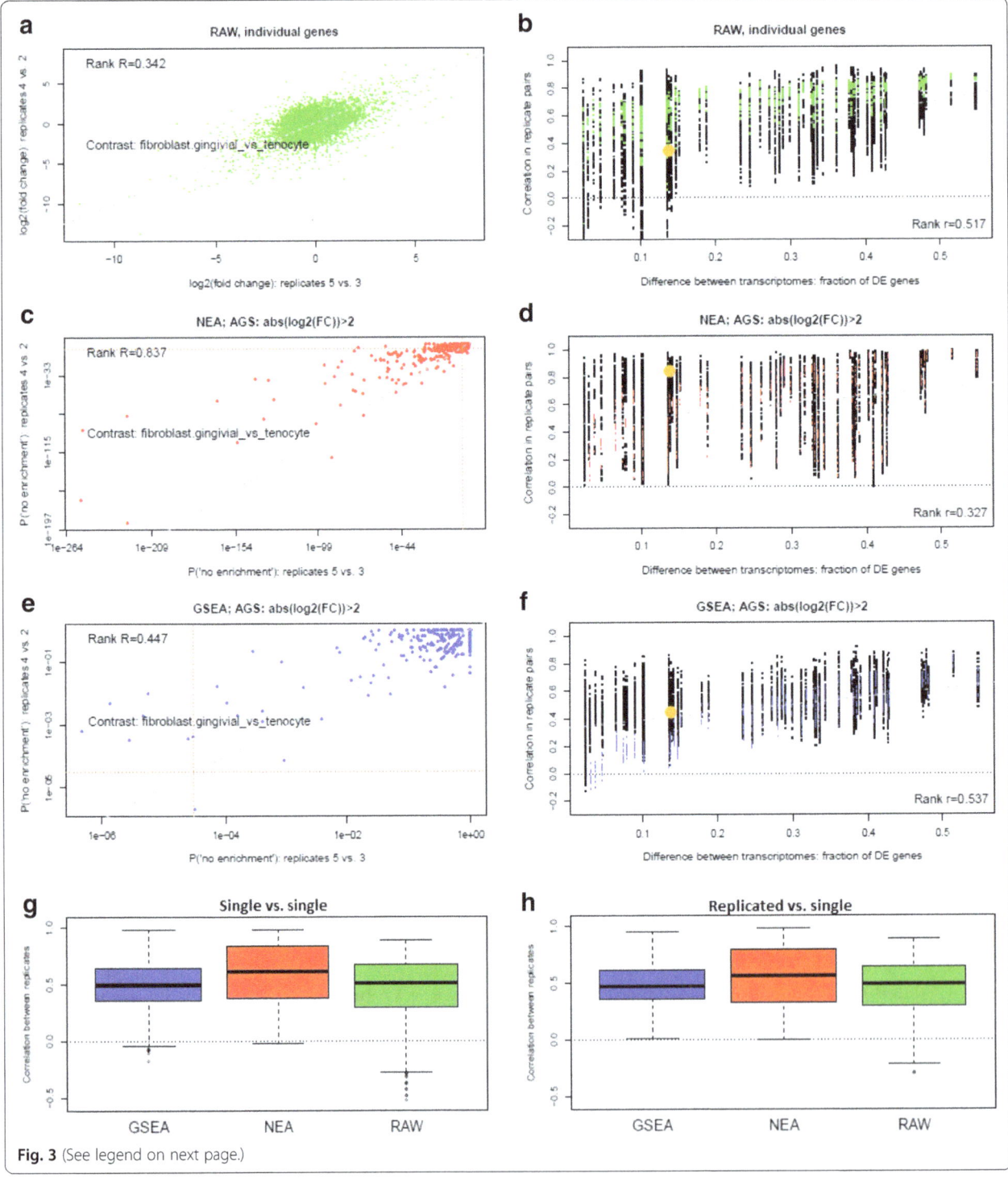

Fig. 3 (See legend on next page.)

Fig. 3 Agreement between biological replicates in alternative approaches to differential expression analysis. **a**, **c**, **e**: Examples of Spearman rank correlations between DE analyses without replicates on gingivial epithelial versus tenocyte cells using samples from two different donor pairs (gingivial epithelial: #4 and #5; tenocyte: #2 and #3). Since p-values in the non-replicated DE analyses were not available, the plot A (RAW) represents raw fold change values for 16620 genes, while plots C and E represent p-values for 330 FGSs. In C and E, the vertical and horizontal brown dotted lines delineate p-values significant after the Bonferroni correction. Dashed grey line: the linear regression fit of Y on X (the R values shown in the corners were calculated using the rank formula and are thus independent of the fits). NEA and GSEA values were obtained for AGSs representing DE genes in each analysis that satisfied the criterion `abs(log₂(fold change))>2`, i.e. the 4-fold change in either direction. These example rank R values from A, C, and E are plotted at respectively B, D, and F as big orange dots. Thus in B, D, and F, the Spearman coefficients from A, C, and E (top left corner) as well as from all other pairwise comparisons are plotted as a function of relative difference strength between respective transcriptomes (X axis). As an example, according to the results of fully powered DE analysis for plots A, C, and E, 13.8% genes were DE with adjusted p-value < 0.05. Hence the value 0.138 is used as X-coordinate for the orange dots. The grey dotted linear regression line and the Spearman rank R value quantify the relations between X and Y. Black and colored points correspond to 1) correlation values of non-replicated DE analyses with each other and 2) correlations where one of the two analyses was replicated, respectively. The boxplots G and H summarize results across all cell types and AGS versions for non-replicated vs. half-replicated options (1) and (2). All pairwise contrasts of GSEA and RAW against NEA were significant with p(H0) < 0.001. In Additional file 1: Figure S3, six plots for NEA and GSEA represent results of using all the six alternative AGS versions

combination with the ancillary functions for input preparation and benchmarking NEArender is meant to become a practically useful part of bioinformatics and biostatistics R pipelines.

Conclusions

Output of NEArender allows using sample- or patient-specific pathway scores as predictive features of phenotype or clinical variables. Compared to GSEA, this raises statistical power and robustness to practically acceptable levels. Compared to previous implementations of NEA, the analysis runs much faster and removes bias due to smaller gene set size. Package NEArender also contains previously not available ancillary functions for preparing and benchmarking input.

Methods

Significance estimation

NEA considers topological properties of the network via node degrees of GS members as well as the total number of edges in the whole network. For a long time, confidence of network enrichment scores has been mostly computed using network randomizations in order to estimate the connectivity value (i.e. number of edges between two GSs expected by chance) as well as its standard deviation [2, 3, 17]. Then the network enrichment statistic was computed as a z-score:

$$z = \frac{n_{AGS\text{-}FGS} - \hat{n}_{AGS\text{-}FGS}}{\sigma_{AGS\text{-}FGS}}$$

where $n_{AGS\text{-}FGS}$ was the actual number of edges that connect any nodes of AGS and any nodes of FGS, while $\hat{n}_{AGS\text{-}FGS}$ and $\sigma_{AGS\text{-}FGS}$ were the average number of AGS-FGS edges and respective standard deviation found in a number of randomized instances of the actual network.

The randomization algorithm by Maslov and Sneppen [17] was based on rewiring each original edge between nodes i *and* j, so that i and j instead get connected to randomly sampled (without replacement) nodes k and l. This allowed preserving individual node degrees (i.e. the total number of edges for a given node) as well as the global topological properties of the network (first of all the scale-freeness). However, McCormack et al. [3] demonstrated that in this randomization procedure higher-order topological properties, such as the propensity of high-degree nodes to avoid connections with other high-degree nodes, could still be biased. They suggested another algorithm for cases of relevance. In our view, the removal of higher order topological biases is not always justified and the decision to apply this over-randomization should depend on a particular research question.

The focus of our present work is on creating a software implementation which is fast and independent of GS sizes and numbers of edges connecting them. In order to do that, we evaluate enrichment of AGS versus FGS using the binomial formula:

$$\chi^2 = \frac{(n_{AGS\text{-}FGS} - \hat{n}_{AGS\text{-}FGS})^2}{n_{AGS\text{-}FGS}} + \frac{(!n_{AGS\text{-}FGS} - !\hat{n}_{AGS\text{-}FGS})^2}{!n_{AGS\text{-}FGS}},$$

The respective number of links expected by chance is calculated simply as:

$$\hat{n}_{AGS\text{-}FGS} = \frac{N_{AGS} * N_{FGS}}{2 * N_{total}},$$

where $!n$ denotes "other than n", N_{AGS} and N_{FGS} report the sums of connectivities of individual nodes (genes) in AGS and FGS, respectively, and N_{total} is the number of edges in the whole network. We note here that this simplified calculation is legitimate if only direct AGS-FGS edges are of interest (which is typically the case of NEA using sufficiently dense NETs, i.e. when in practically all AGS-FGS pairs both expected and actual connectivity is expressed with positive values) and if higher-order topology issues may be neglected.

The Gaussian z-scores are consecutively "reverse-engineered" by `nea.render` from X^2 scores via p-values of the latter. Since X^2, unlike of z, is only defined on the non-negative domain, z values are coerced negative in cases of depletion (as opposed to enrichment).

Topology analysis

The package includes auxiliary functions for visual inspection of both second order biases (`topology2nd`) and scale-freeness (`connectivity`). The vignette to the package provides examples of various deviations found in nine example networks of different provenance.

Parallel computation

At the most computationally intense step, which is the counting of actual network edges in each AGS-FGS pair, the package can employ parallel jobs enabled with R package `parallel` (https://stat.ethz.ch/R-manual/R-devel/library/parallel/doc/parallel.pdf).

Gene set enrichment analysis

Either together with or instead of NEA, users can also perform the conventional binomial enrichment analysis GSEA (note that here the binomial analysis is applied to the gene quantities rather than to network edge counts as in NEA). It accepts the same input as `nea.render` (excluding NET), and produces similarly arranged output from Fisher's exact test: odds ratio estimate, p-value, q-value, and the number of genes shared by AGS and FGS.

Included example datasets

In comparison to GSEA, NEA requires an extra component: a global network of functional coupling between genes and/or proteins. The package contains the following data sets: a small and a large version of NET `net.-kegg` [18] and `net.merged` [9], a collection of FGSs `can.sig.go` (2406 distinct genes in 34 KEGG pathways [18] and GO terms [19] as well as three inputs for creating AGSs: somatic point mutations `tcga.gbm` [20] and two subsets of FANTOM5 transcriptomics data `fantom5.43samples` and `fant.carc` [13].

Additional files

Additional file 1: Figure S1. Sensitivity and sources of bias in randomization-based versus binomial calculation of network enrichment. **Figure S2.** Rank correlation coefficients between results of differential expression analysis on different sample pairs and groups (Additional file 2). **Figure S3.** Agreement between biological replicates in alternative approaches to differential expression analysis.

Additional file 2: Supplementary File Boxplots.Rfree.P_based.

Abbreviations

AGS: Altered genes set; CSB: Chi-squared binomial formula; DE: Differential gene expression; FDR: False discovery rate; FGS: Functional gene set; FN: False negative; FP: False positive; GS: Gene set; GSEA: Gene set enrichment analysis; NET: Global network of functional coupling; NRZ: Network randomization z-test; TN: True negative; TP: True positive

Acknowledgements

The support by NBIS (National Bioinformatics Infrastructure Sweden) is gratefully acknowledged, as well as assistance of the Science for Life Laboratory IT team.

Funding

The authors would like to thank Vera and Emil Kornells Stiftelse for the financial support. The publication of this article was funded from this source.

Authors' contributions

AA wrote software, the manuscript, and helped in preparing the documentation. AJ tested software, prepared the package and documentation, and wrote and edited the manuscript. All the authors have read and approved the final manuscript.

Competing interests

The authors declare that they have neither financial nor non-financial competing interests.

About this supplement

This article has been published as part of BMC Bioinformatics Volume 18 Supplement 5, 2017: Selected works from the Joint 15th Network Tools and Applications in Biology International Workshop and 11th Integrative Bioinformatics International Symposium (NETTAB / IB 2015). The full contents of the supplement are available online at https://bmcbioinformatics.biomedcentral.com/articles/supplements/volume-18-supplement-5.

Author details

[1]National Bioinformatics Infrastructure Sweden, Science for Life Laboratory, Stockholm, Sweden. [2]Department of Microbiology, Tumor and Cell biology, Karolinska Institutet, Stockholm, Sweden. [3]Department of Cell and Molecular Biology, Karolinska Institutet, Stockholm, Sweden.

References

1. Bayerlová M, Jung K, Kramer F, Klemm F, Bleckmann A, Beißbarth T. Comparative study on gene set and pathway topology-based enrichment methods. BMC Bioinformatics [Internet]. 2015 Dec [cited 2015 Oct 31];16(1). Available from: http://www.biomedcentral.com/1471-2105/16/334
2. Alexeyenko A, Lee W, Pernemalm M, Guegan J, Dessen P, Lazar V, et al. Network enrichment analysis: extension of gene-set enrichment analysis to gene networks. BMC Bioinformatics. 2012;13:226.
3. McCormack T, Frings O, Alexeyenko A, Sonnhammer ELL. Statistical assessment of crosstalk enrichment between gene groups in biological networks. PLoS One. 2013;8(1):e54945.
4. Storey JD, Tibshirani R. Statistical significance for genomewide studies. Proc Natl Acad Sci U S A. 2003;100(16):9440–5.
5. Liberzon A, Birger C, Thorvaldsdóttir H, Ghandi M, Mesirov JP, Tamayo P. The Molecular Signatures Database (MSigDB) hallmark gene set collection. Cell Syst. 2015;1(6):417–25.
6. Barabási A-L, Oltvai ZN. Network biology: understanding the cell's functional organization. Nat Rev Genet. 2004;5(2):101–13.
7. Bovolenta LA, Acencio ML, Lemke N. HTRIdb: an open-access database for experimentally verified human transcriptional regulation interactions. BMC Genomics. 2012;13:405.
8. Griffiths-Jones S. miRBase: microRNA sequences, targets and gene nomenclature. Nucleic Acids Res. 2006;34(90001):D140–4.
9. Merid SK, Goranskaya D, Alexeyenko A. Distinguishing between driver and passenger mutations in individual cancer genomes by network enrichment analysis. BMC Bioinformatics. 2014;15:308.
10. Alexeyenko A, Alkasalias T, Pavlova T, Szekely L, Kashuba V, Rundqvist H, et al. Confrontation of fibroblasts with cancer cells in vitro: gene network

analysis of transcriptome changes and differential capacity to inhibit tumor growth. J Exp Clin Cancer Res CR. 2015;34(1):62.

11. Cancer Genome Atlas Research Network. Comprehensive genomic characterization defines human glioblastoma genes and core pathways. Nature. 2008;455(7216):1061–8.

12. Cancer Genome Atlas Research Network. Integrated genomic analyses of ovarian carcinoma. Nature. 2011;474(7353):609–15.

13. Forrest ARR, Kawaji H, Rehli M, Kenneth Baillie J, de Hoon MJL, Haberle V, et al. A promoter-level mammalian expression atlas. Nature. 2014;507(7493):462–70.

14. Gregory R. Warnes, Peng Liu, Fasheng Li. ssize [Internet]. Bioconductor. [cited 2016 Jul 19]. Available from: http://bioconductor.org/packages/ssize/

15. Guerrero-Bosagna C. High type II error and interpretation inconsistencies when attempting to refute transgenerational epigenetic inheritance. Genome Biol [Internet]. 2016 Dec [cited 2016 Jul 17];17(1). Available from: http://genomebiology.biomedcentral.com/articles/10.1186/s13059-016-0982-4

16. Ritchie ME, Phipson B, Wu D, Hu Y, Law CW, Shi W, et al. limma powers differential expression analyses for RNA-sequencing and microarray studies. Nucleic Acids Res. 2015;43(7):e47.

17. Maslov S, Sneppen K. Specificity and stability in topology of protein networks. Science. 2002;296(5569):910–3.

18. Kanehisa M, Goto S, Kawashima S, Nakaya A. The KEGG databases at GenomeNet. Nucleic Acids Res. 2002;30(1):42–6.

19. Ashburner M, Ball CA, Blake JA, Botstein D, Butler H, Cherry JM, et al. Gene ontology: tool for the unification of biology. The Gene Ontology Consortium. Nat Genet. 2000;25(1):25–9.

20. International Cancer Genome Consortium, Hudson TJ, Anderson W, Artez A, Barker AD, Bell C, et al. International network of cancer genome projects. Nature. 2010;464(7291):993–8.

15

TIGER*i*: modeling and visualizing the responses to perturbation of a transcription factor network

Namshik Han[1,3*], Harry A. Noyes[2] and Andy Brass[3*]

Abstract

Background: Transcription factor (TF) networks play a key role in controlling the transfer of genetic information from gene to mRNA. Much progress has been made on understanding and reverse-engineering TF network topologies using a range of experimental and theoretical methodologies. Less work has focused on using these models to examine how TF networks respond to changes in the cellular environment.

Methods: In this paper, we have developed a simple, pragmatic methodology, TIGER*i* (Transcription-factor-activity Illustrator for Global Explanation of Regulatory interaction), to model the response of an inferred TF network to changes in cellular environment. The methodology was tested using publicly available data comparing gene expression profiles of a mouse p38α (Mapk14) knock-out line to the original wild-type.

Results: Using the model, we have examined changes in the TF network resulting from the presence or absence of p38α. A part of this network was confirmed by experimental work in the original paper. Additional relationships were identified by our analysis, for example between p38α and HNF3, and between p38α and SOX9, and these are strongly supported by published evidence. FXR and MYC were also discovered in our analysis as two novel links of p38α. To provide a computational methodology to the biomedical communities that has more user-friendly interface, we also developed a standalone GUI (graphical user interface) software for TIGERi and it is freely available at https://github.com/namshik/tigeri/.

Conclusions: We therefore believe that our computational approach can identify new members of networks and new interactions between members that are supported by published data but have not been integrated into the existing network models. Moreover, ones who want to analyze their own data with TIGER*i* could use the software without any command line experience. This work could therefore accelerate researches in transcriptional gene regulation in higher eukaryotes.

Keywords: Machine Learning, Transcriptional regulatory network, Transcription factor binding site, Gene expression

Background

Integrated functional genomics attempts to utilize the vast wealth of data produced by modern large scale genomic and post-genomic projects to understand the functions of cells and organisms [1]. The rapidly increasing amount of high throughput sequencing data makes it essential to develop new analytical tools that can systematically process and integrate those datasets. This presents both challenges and opportunities to the computer science community.

Transcription factor (TF) proteins bind to promoter elements on genomic DNA at TF binding sites (TFBS), to help control the transfer of genetic information from gene to mRNA [2]. Understanding the mechanisms underlying mRNA transcription is one of the "grand challenges" in modern biology. Experimental techniques allow direct measurement of individual gene transcription, but the contribution of multiple TFs is hard to determine [3–5]. Measuring the concentration of TF proteins and their affinity for the promoter region of genes is difficult because

* Correspondence: namshik.han@gurdon.cam.ac.uk;
andy.brass@manchester.ac.uk
[1]Gurdon Institute, University of Cambridge, Cambridge, UK
[3]School of Computer Science and School of Health Sciences, University of Manchester, Manchester, UK
Full list of author information is available at the end of the article

concentrations are low and protein-DNA interactions are subject to multiple controls, resulting in measurement artifacts [6–8]. Post transcriptional regulation compounds these difficulties because other molecules modify mRNA stability and hence the signals from the TFs [3, 9–11]. In such a complex environment, *in-silico* techniques can provide insights and hypotheses into the underlying TF regulatory activity, although they clearly have limitations.

Reverse-engineering of TF network and TFBS information

A number of techniques are available to uncover the topology of the TF network—the networks of complex reactions and interactions in the cell that control transcript levels [12]. One strategy is to use principals of reverse-engineering and use gene expression data to infer regulatory interactions [13, 14]. Various reverse-engineering methods can reduce the dimensionality of the classic combinatorial search problem and utilize genome sequence data to enhance the sensitivity and specificity of predictions. However, they have difficulties in describing regulatory control by mechanisms other than TFs. Reverse-engineering of TF networks in the lower eukaryotes has been well developed [15–17]. However, the problems in mapping the regulatory mechanisms in cells of higher eukaryotes have made such global studies either impossible or impractical. Some recent studies have begun to address this issue [18–20], but have tended to focus only understanding which TFs bind to which genes—not looking in detail at the nature of the TF/ TFBS interaction. A recent study [21] identified key biological features in transcriptional changes, however this method has difficulties in inferring the dynamics of the interactions. Furthermore, TF concentrations were not considered during the identification of the features.

To date, various reverse-engineering methods can reduce the dimensionality of the reverse-engineering problem and utilize genome sequence data to enhance the sensitivity and specificity of predicted interactions. However, they have difficulties in describing regulatory control by mechanisms other than TFs. To address this issue TFBSs information is required to complement the gene expression data. We used a list of 132,654 TFBSs between 20,920 genes and 174 TFs that had been identified by searching an alignment of five mammal species for conserved 5' and 3' regions [22]. Connectivity data is notorious for high false positive rates; however, our connectivity data is robust against the problem because it extracts binding information from well conserved upstream regions. A more detailed explanation is addressed in the Methods and Results section and a schematic diagram of the connectivity data is presented in Fig. 1.

Identifying regulation type by combining TFA and TFC analysis

Transcription factor activities (TFAs) are the intensity of the interactions between a certain transcription factor

(TF) and its targets at a certain experimental point [23]. Thus, the estimated strength of TFAs between each TF and its target gene are useful to know which TF is acting on which gene at a given time point or experiment condition. However, simply knowing the regulatory activities under a single experimental condition provides limited information about the transcriptional network. To understand the mechanism of regulatory interactions, we developed a method that identifies statistically significant differences in TFAs under two different conditions. The significant differences indicate the changing level of TFAs between two conditions, so varying trends of TFAs in whole experimental process are easily detected and can be used to identify TF-specific regulatory patterns (up and down-regulation).

A highly concentrated TF induces more gene expressions rather than a lower concentrated TF. High-affinity binding sites induce the gene expressions at any level of TF concentration (TFCs), but low-affinity binding sites require high level of TF concentration for induction [24]. Thus, we might assume that TF concentration level is an important factor for investigating TFAs, and TFA investigation with considering TFC provides more reliable and accurate results closer to the complex reality of biology. To address this problem we proposed a probabilistic variational inference method to infer the concentration of each TF protein (TFC) and the regulatory intensities (TFA) of each TF and gene pair [4].

Aside from the method, there have been some notable attempts to infer TFAs based on integrating gene expression data and TFBSs information. The approaches use various well-known statistical inference techniques such as network component analysis [25], support vector machine [26], multivariate regression plus backward variable selection [27] and partial least squares [28]. However, the TFAs, which are inferred by these methods, do not contain any information on the strength and the sign of the physical interaction between a TF and its target genes. Moreover, the regulatory interactions can change easily in response to changing experimental conditions and over time. Since the methods are not fully probabilistic, they are not ideal for investigating the stochastic interactions. A linear regression based probabilistic method to model the full probability distribution of each TFA on each gene was developed [23]. The limitation of this method, however, is that it does not infer the TFAs and TFCs separately. This is a serious problem in subsequent analysis and prediction.

Methods

Transcription regulatory circuits and mathematical model

Transcription regulatory circuits can be thought of as having *trans-* and *cis-*inputs that are transformed into genetic information at mRNA level [29]. These circuits

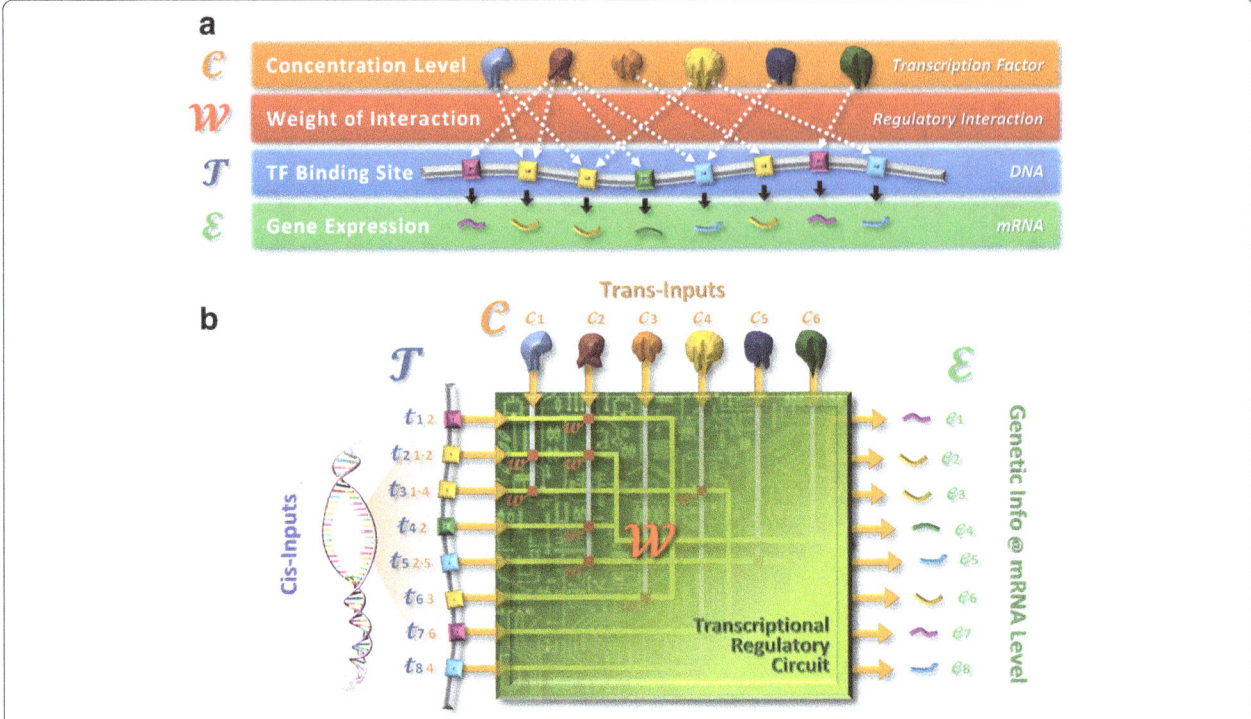

Fig. 1 Schematic diagrams for understanding basic concepts of this study **a** The four components of transcription. The transcriptional regulators interact with their targets genes to regulate gene expression at the mRNA level. The cellular environment controls the concentration of TFs, **C**. The TFs bind to specific sites close to the target genes, described in model by the connection matrix, **T**. The TFs bind to their different target genes with varying strengths to regulate transcription. The strength of each of these pair-wise interactions is described by a weight matrix **W**. This all finally results in the transcription of mRNA at particular concentration, **ε**. **b** A schematic of a transcriptional regulatory circuit. The circuit takes *trans*- and *cis*-inputs to transform the genetic information at mRNA level. The four components for transcription (as described above) are the key elements for the circuit

are a key component in the regulation of mRNA levels in the cell, and have a number of components (shown in Fig. 1): TFs, whose concentration can change, bind to TFBSs upstream of genes with a strength that is a function of the particular TF-gene interaction, to control the concentration of mRNA produced. A number of mathematical models have been developed which attempt to describe these interactions [15–17, 19, 21, 30]. For example Sanguinetti et al. [4] model the log gene expression in the form:

$$\underline{e} = \underline{\underline{T}}\,\underline{\underline{W}}\,\underline{c} + \underline{v} \qquad (1)$$

Where:

i) $\underline{e}$ is a set of logged gene expression measurements.
ii) $\underline{\underline{T}}$ is a binary matrix capturing the connection topology—the specific set of TFBS upstream of genes and the TFs that bind to them. If TF f binds upstream of gene g then $\underline{\underline{T}}_{gf} = 1$.
iii) $\underline{\underline{W}}$ is a weight matrix that captures the nature of the interaction strengths between TF-gene pairs in regulating expression of a specific gene.

iv) $\underline{c}$ is the vector of concentrations of each of the TFs.
v) $\underline{v}$ is a vector of independent and identically distributed variables modeling the noise in the system. The model assumes that a spherical Gaussian term could explain all noise on gene expression profiling data.

Typically, we have knowledge of $\underline{e}$ (from gene expression profiling experiments, such as microarray or RNA-seq) and would like to infer the set of TFC $\underline{c}$, and TFA $\underline{\underline{W}}$ giving rise to this signal. Given $\underline{\underline{T}}$ and $\underline{e}$ Sanguinetti et al. [4] then show how it is possible to solve for $\underline{c}$ and $\underline{\underline{W}}$ using a discrete time state space model (Eq. 1) with expectation-maximization (EM) algorithm. In the model, elements of the $\underline{c}$ matrix indicate the concentration level of a given TF protein (TFC) at a specific time. Elements of the $\underline{\underline{W}}$ matrix represent the regulatory intensity (TFA) between a given TF protein and its binding affinity to its target genes. The baseline expression level is the mean vector. The measurement noise $\underline{v}$ follows zero-mean i.i.d. Gaussian noise. To estimate the $\underline{c}$ and $\underline{\underline{W}}$ matrices, the model used posterior estimation of

Bayer's theorem. During this estimation, EM algorithm allowed the model to efficiently approximate the log likelihood. However, it is rare to have a complete knowledge of $\underline{\underline{T}}$ —we simply do not know the binding sites for all TFs in a typical higher eukaryotic cell. Recent experimental techniques, such as ChIP-chip and ChIP-seq can provide useful data to help construct the connection topology $\underline{\underline{T}}$ [31], however they have clear limitations if we are looking for a complete topology [32, 33]. A number of theoretical techniques are also available to uncover the connection topology [34–36]. The techniques generally use principals of reverse-engineering and use gene expression and genome sequence data to infer regulatory interactions.

Gene expression data
Gene expression datasets were downloaded from Gene Expression Omnibus (accession number GSE7342 for p38α and GSE36890 for STAT5) [37, 38]. The expression profiling data of GSE7342 dataset was normalized by the robust microarray average (RMA) method. The read counts of GSE36890 dataset was normalized to the reads per kilobase of exon per mega-base of library size (RPKM).

Generating $\underline{\underline{T}}$, the connection topology
In this paper we have taken a conservative strategy for generating $\underline{\underline{T}}$ which looks at upstream region of genes that are well-conserved in multiple mammalian genomes. We used a published catalogue of common regulatory motifs that were overrepresented in gene upstream regions [22]. These motifs were identified by constructing genome-wide alignments for four mammalian species in promoter regions and 3' UTRs relating to well-annotated genes from the RefSeq database. The same TFs were assumed to bind the same TFBSs in mice since the TFBS had been discovered in an alignment of human, mouse, rat and dog promoter regions. TFBS upstream of human 13,330 RefSeq genes were predicted. Mouse genes corresponding to the published list of human genes [22] were identified using Ensembl mouse gene annotation.

Estimation of statistically significant changes
We were specifically interested in any TF activity that exhibits statistically significant changes between the two conditions. In particular, we are interested in changes that may be due to a change in activity of the TF, and not just in its concentration. We therefore scaled the TFA by the predicted TFC as a measure for changes in activity [24]. A joint analysis of TFA and TFC should provide more robust predictions of those TFs whose activity has changed for reasons beyond those of a simple change in concentration. To compare two different

conditions, the normalized TFA by TFC in wild-type condition were subtracted by the normalized TFA in knock-out condition. We therefore determined those interactions for which:

$$\left[\left\|\frac{W_{gf}}{c_f}\right\|\right]_{WT} - \left[\left\|\frac{W_{gf}}{c_f}\right\|\right]_{KO} > \textbf{\textit{Cutoff}} \qquad (2)$$

The value of the **Cutoff** was chosen such that all differences at the 95% confidence interval were considered significant (±2 standard deviations). ±2SD limit is widely chosen as a normal limit because it fits well into two important categories: (1) confident interval and (2) testing hypothesis.

Gene Ontology (GO) analysis
GO analysis was performed by using DAVID [39]. The sets of genes showing significant changes identified in Eq. 2 were submitted to DAVID using the default parameters in order to obtain the GO term classifications of each gene. Our computational pipeline utilized the results to investigate the functionality of genes and their regulatory TFs. The detailed methods and the result figures of GO analysis are supplied in Additional file 1: Supplementary Text and Figure S1–S3.

Results
Estimating the responses to perturbation of transcription networks
We have developed a strategy which used forward-engineering to construct the connection topology (see Fig. 1, Methods, and Additional file 1: Supplementary Text and Figure S1–S3), based on a previous study of regions upstream of genes conserved in multiple mammalian genomes [22]. The structure of this network of transcriptional regulatory interactions between TFs and the genes whose transcription they control is described by a binary matrix $T \in \Re^{n \times m}$, where n is the number of TFs and m is the number of genes; An element (i, j) of the matrix is '1' if TF i binds to the upstream control region of gene j, '0' otherwise. We have then employed a mathematical model to integrate the connection topology data and a gene expression dataset from a higher eukaryote in which we are interested in modeling the changes that occur in the TF network in response to a change in the cellular environment (Fig. 2). Our approach could be seen as complementary to 'Integrative methods', as defined in [40], as it provides a strategy for creating an approximate connection topology if more detailed information is not available. The connection topology that is being used for this analysis contains many approximations and is certainly incomplete. However, it should be noted that we are looking at the differences between the models, for example

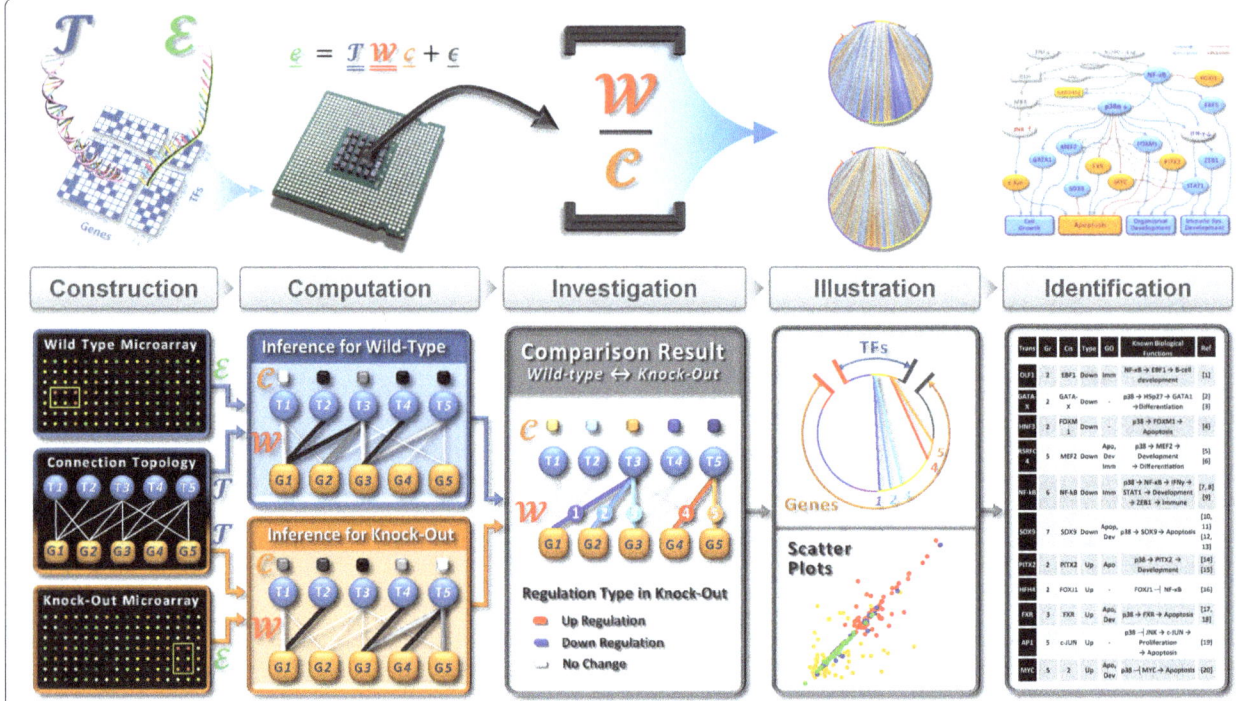

Fig. 2 Overview of our strategy and work-flow of our computational pipeline with a plain example. Our strategy uses a computational pipeline based on a reverse-engineering technique. The pipeline takes as inputs the results of transcription (gene expression data ε and connectivity information T and outputs the sources of transcription (strengths W and concentrations C). The pipeline is composed of five parts: Construction: RMA normalization of gene expression profiling data ε and a binary matrix containing connection topology T is constructed using by forward-engineering strategy. Computation: The gene expression profiling data and connectivity data are utilized to infer TF-gene interaction strengths W and TF concentration levels C. Investigation: Once the strengths and concentrations are inferred, the actual TF activities are estimated by normalizing the strengths on the concentrations. The statistically significant changes in the TF-gene interactions strength, TF concentration levels, and TF activities are calculated. Illustration: The changes are illustrated in round limpet-like plot or in the scattered plots that shows the changes between individual TF and genes. Identification: The candidate TFs are identified, and Gene Ontology (GO) analysis are performed on the genes that are regulated by the candidate TFs. The literature is reviewed to find the supporting evidence, and the individual links between the candidate TFs and their potential biological functions are identified and summarized in a table. Based on the table, we finally construct the comprehensive TF network for p38α

between a wild-type and knock-out state, and those differences will be in parts of the model for which we do have data.

The results of our approach provide a set of TFs and their target genes which are related by significant up- or down-regulation in transcription. It provides a clear indication of the changes in TFA and TFC of TFs that are controlling transcriptional regulatory mechanisms in response to a specific stimulus. We therefore showed an "integrated" approach for network inference, based on a forward-engineered connection topology, can produce plausible and testable hypotheses about the responses to perturbation of transcription networks in higher eukaryotes.

Illustrating interpretable images of complex data
The visualization tools then make patterns apparent that would be difficult to detect in numerical data (Fig. 2). To distinguish regulation patterns between different experimental conditions, recognizing at a glance is

important. However, computing results are formed in large numerical matrix, thus it is not only difficult to navigate through the whole matrix but also impossible to present the results in one page.

Figure 3 shows a graphical representation of the significant changes in TFA matrices $\underline{\underline{W}}$ (n by m) and TFC vectors $\underline{c}$ (n) obtained from this analysis. The patterns of the responses to perturbation of TF networks are readily observed in this single-shot image that presents approximately 2000 significant changes of varying TF activities on the 132,654 TFBSs after deleting p38α. In upper part of the plots, the TFs place in the functional group order. The genes, which have at least one significant interaction with TFs, locate in bottom part of the plots. A line in the plots presents a regulatory interaction (normalized TFA by TFC) between a TF and its target gene, and line color indicates a significant difference between the strengths of the regulatory interaction of two conditions. For example, we can easily find in visualized format (Fig. 3) that TF group three has

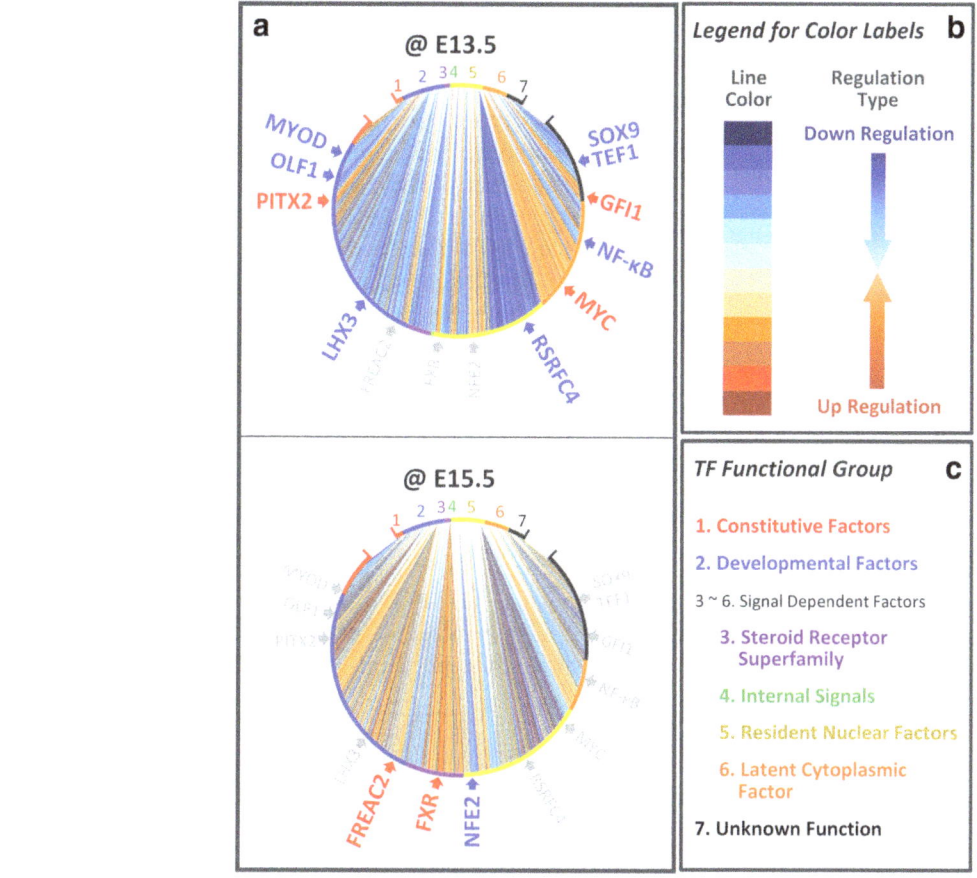

Fig. 3 Global view of the significant changes in TF activities. Our visualization tools make it possible to distinguish specific features and trends in each condition. **a** The changes in TF activities underlying absence of p38α are presented in the limpet-like plots. In the upper part of the limpet plots, the TFs are placed in order of functional group (Fig. 3c). Genes that have at least one significant change are located in the bottom of the plots. A line presents how much the TF activity of a certain gene is changed between the wild-type mice and the knock-out mice. If a value of the change is greater than zero, it is displayed in blue indicating that the TF-gene pair has significantly higher TF activation in the wild-type mice (Down-regulation after deleting p38α); while, if the change is less than zero, it is displayed in red indicating that the pair has significantly higher TF activation in the knock-out mice (Up-regulation in deleting p38α). **b** The legend for the line color is present. **c** The perimeters of the plots are broken into different colored regions corresponding to different functional groups listed in the key

distinct patterns (down-regulation at E13.5, up-regulation at E15.5) between two time points.

Modelling the changes of transcription factor network in p38α deficient mice

The computational pipeline as highlighted in Fig. 2 was applied to a published study of the effect p38α knock-out in mouse embryos [37]. This study developed four gene expression profiling datasets (Gene Expression Omnibus, accession number GSE7342) comprising of two time points at days 13.5 and 15.5 of embryonic development (E13.5 and E15.5) for p38α knock-outs and their wild-type controls. This data set was chosen for this study as it includes experimental measurements of gene expression in the wild-type and knock-out mice and showed that p38α deficient mice have significantly different phenotype. Thus, the experimental datasets were used as positive controls for our theoretical study.

The TF-gene interaction strengths (TFAs) $\underline{\underline{W}}$ and TF concentration levels (TFCs) $\underline{c}$ in each of these four data sets were then inferred to produce four weight matrices of TFAs:

$$\left[\underline{\underline{W}}\right]_{WT@E13.5}, \left[\underline{\underline{W}}\right]_{WT@E15.5}, \left[\underline{\underline{W}}\right]_{KO@E13.5}, \left[\underline{\underline{W}}\right]_{KO@E15.5}$$

and four concentration vectors of TFCs:

$$[\underline{c}]_{WT@E13.5}, [\underline{c}]_{WT@E15.5}, [\underline{c}]_{KO@E13.5}, [\underline{c}]_{KO@E15.5}$$

From the TFA weight $\underline{\underline{W}}$ and connection topology $\underline{\underline{T}}$ matrices, the average strength of TFA, S_{fi} for each TF in the datasets was calculated:

$$\mathbf{s}_f = \frac{\Sigma_g |\mathbf{w}_{gf}|}{\Sigma_g |\mathbf{\tau}_{gf}|} \qquad (3)$$

By comparing the average strengths between wild-type and knock-out mice, it is possible to see which of the TFs have significantly changed as a consequence of the removal of p38α.

Figure 4a, b show the changes in TFA strengths $\mathbf{s}_f$ between wild-type and knock-out mice at E13.5 and E15.5. It can be seen that a number of TFs show a significant signal (>2 s.d.) in this data. These are shown with more detail in Table 1. Figure 4c, d show the inferred TFCs $\underline{c}$ obtained for the E13.5 and E15.5 time points. Again, from this graph it is possible to see that a number of TFs appear to be responding to the p38α status. These are shown in more detail in Table 2.

Transcriptional regulatory network for p38α

Gene Ontology (GO) analysis on the target genes of the TFs with strongly changed activity showed enrichment for three GO terms and provided insight into the

functional role of the TFs (see Methods, Tables 1 and 2, and Additional file 1: Supplementary Text and Figure S1–S3). The three GO terms are the regulations of the apoptosis (programmed cell death), the downward spiral of the developmental process, and the immune system development. The JNK-c-Jun pathway stimulates the apoptosis, and the I-kB kinase/NF-kB cascade acts as a suppressor of the JNK-c-Jun pathway [41]. Inhibition of p38α MAPK retards another JNK-c-Jun pathway inhibitor NF-kB cascade, but promotes JNK-c-Jun pathway which induces the apoptosis by expressing the Bcl2 protein family [20, 42]. On the other hand, developmental process related genes are down-regulated in the p38α knock-out mice. The study of p38α MAPK [37] reported that the p38α knock-out mice die within days after birth. We do not have enough gene expression profiling data (either other time points in the embryonic period or postnatal period) to investigate TFAs in whole developmental process of the p38α knock-out mice; we cannot confirm but suppose that it might be the reason of the death of the knock-out mice. Further, the genes interact with the TFs which are reported as crucial TFs in the developmental

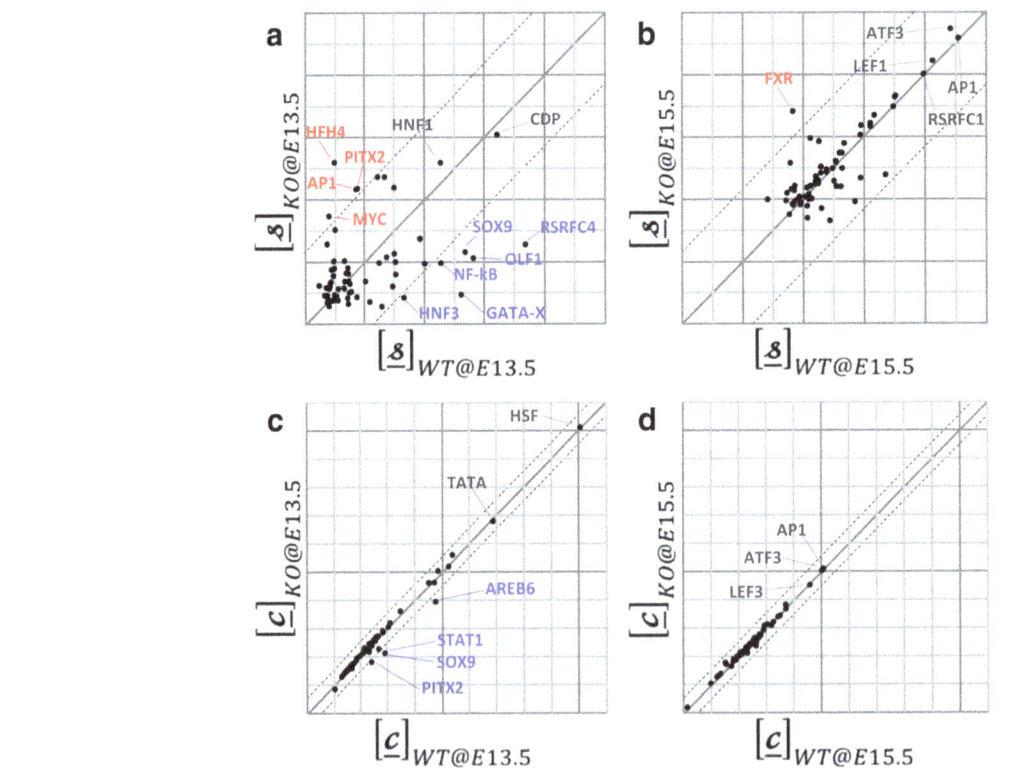

Fig. 4 a, b The average strength of changes in TF-gene interaction **S** and TF concentration levels **c** between wild-type and knock-out mice. The figures clearly show not only which TFs have strong interaction strengths or high concentrations (*gray*-colored TFs) but also which TFs have significant changes in their interaction pattern or concentration (*blue*- or *red*-colored TFs). The dotted lines indicate the standard deviation (=2) centered on the median value of the straight lines. (figure a for time point E13.5 and figure b for time point E15.5). Five TFs (shown in *red*) interact particularly strongly with their target genes in the p38α knout-mice. In contrast, six TFs (shown in *blue*) interact less-strongly in the p38α knock-out than wild-type. **c, d** The TF concentration levels 켙 of wild-type and knock-out mice. TF concentration levels are plotted. The strongest signal was observed in E13.5, only. Deleting p38α induces a down regulation of AREB6, PITX2, STAT1 and SOX9

Table 1 TFs showing significant changes in interaction strength between wild-type and knock-out mice

Trans name	TF group	Cis name	Regul-ation type	GO analy-sis	Known biological functions	Ref.
OLF1	1.Dev	EBF1	Down	Imm	NF-κB → **EBF1** → B-cell development	[52]
GATA-X	1. Dev	GATA1 ~6	Down	-	p38 → HSp27 → **GATA1** → Differentiation → **GATA1** → IL9 → Asthma	[53, 54]
HNF3	1. Dev	FOXM1	Down	-	p38 → **FOXM1** → Apoptosis	[45]
RSRFC4	5. Res	MEF2	Down	Apo, Dev Imm	p38 → **MEF2** → Development → Differentiation	[46, 55]
NF-κB	6. Lat	NF-κB	Down	Imm	p38 → **NF-κB** → IFNγ → STAT1 → Development → ZEB1 → Immune	[56–58]
SOX9	7. Unk	SOX9	Down	Apop, Dev	p38 → **SOX9** → Apoptosis	[59–62]
PITX2	1. Dev	PITX2	Up	Apo	p38 → **PITX2** → Development → Apoptosis	[63, 64]
HFH4	1. Dev	FOXJ1	Up	-	**FOXJ1** ⊣ NF-κB	[65]
FXR	3. Ste	FXR	Up	Apo, Dev	p38 → **FXR** → Apoptosis	[66, 67]
AP1	5. Res	c-JUN	Up	-	p38 ⊣ JNK → **c-JUN** → Proliferation → Apoptosis	[37]
MYC	5. Res	MYC	Up	Apo, Dev	p38 ⊣ **MYC** → Apoptosis	[68]

The TFs predicted to have significantly different behaviors between the wild-type and knock-out mice. "Trans Name"—the official gene symbol of the TF. "Cis Name"—the name given to the binding site of the TF. TFs were characterized into different functional groupings (see Fig. 3c for details: Dev—cell-type specific developmental TFs, Res—signal dependent resident nuclear factors, Lat—signal dependent latent cytoplasmic factors, Ste—signal dependent steroid receptor group, Unk—unknown). "Regulation Type", the way in which the TF regulates its target genes in the absence of p38α. "GO Analysis", provides more functional classification for the identified TFs (see Methods and Additional file 1: Figure S1 to S3 for details). The abbreviations of the GO terms are: Apo, Apoptosis; Dev, Developmental Process; Imm, Immune System Development. "Known Biological Functions" summarizes the findings from the recent biological literature as shown in the "References"
The boldface ones are the main node in Fig. 5

process and the immune system development. Our results are therefore in broad accordance with the experimentally validated results, so it confirmed that our pipeline produces reliable results.

Combining the data in Tables 1 and 2 with those obtained from the literature it is possible to build a putative model for the effects of p38α knock-out (Fig. 5). This figure shows TFs with a strong response in our analysis as nodes, with links that demonstrate regulatory interactions between them. The TF network therefore comprehensively shows the biological consequences of p38α knock-out at transcriptional level.

Discussion

We have developed a novel strategy for discovering changes in transcriptional regulatory networks of higher eukaryotes. It integrates methods for inferring TF-gene interaction strengths (TFAs) and TF concentration levels (TFCs); identifying statistically significant changes in TFAs and TFCs; analyzing the changes; classifying TFs into functional

groups; and visualizing the changes. To our knowledge, this is the first ensemble approach for characterizing the transcriptional function of TF proteins and their target genes in higher eukaryotes. Reverse-engineering of TF networks has been well developed in the lower eukaryotes [15, 17]. However, the problems in mapping the regulatory mechanisms in cells of higher eukaryotes have made such global studies either impossible or impractical. Some recent studies have begun to address this issue [16, 19, 30], but have tended to focus only on understanding which TFs bind to which genes—not looking in detail at the nature of the TF-gene interaction. Other studies [5, 21] identified key biological features in transcriptional changes, however the methods have difficulties in inferring the dynamics of the interactions. A recent review [40] has categorized techniques for network inference and listed their limitations.

We validated our computational pipeline using the p38α gene expression profiling data and our connectivity data. The study of p38α MAPK [37] used various experimental methods including a gene expression profiling analysis to

Table 2 TFs showing significant changes in concentration between wild-type and knock-out mice

Trans name	TF group	Cis name	Level	GO analy-sis	Known biological functions	Ref.
AREB6	1.Dev	ZEB1	Down	-	p38 → IFNγ → **ZEB1** → Immune	[58]
PITX2	1.Dev	PITX2	Down	Apo	p38 → **PITX2** → Development → Apoptosis	[63, 64]
STAT1	6. Lat	STAT1	Down	-	**STAT1** → Development → Immune	[57, 69]
SOX9	7. Unk	SOX9	Down	Apo, Dev	p38 → **SOX9** → Apoptosis	[59–62]

TFs changing their concentration levels significantly between wild-type and knock-out mice. "Level", the changes of TF concentration level in the absence of p38α. Other column headings and abbreviations are the same as those in Table 1
The boldface ones are the main node in Fig. 5

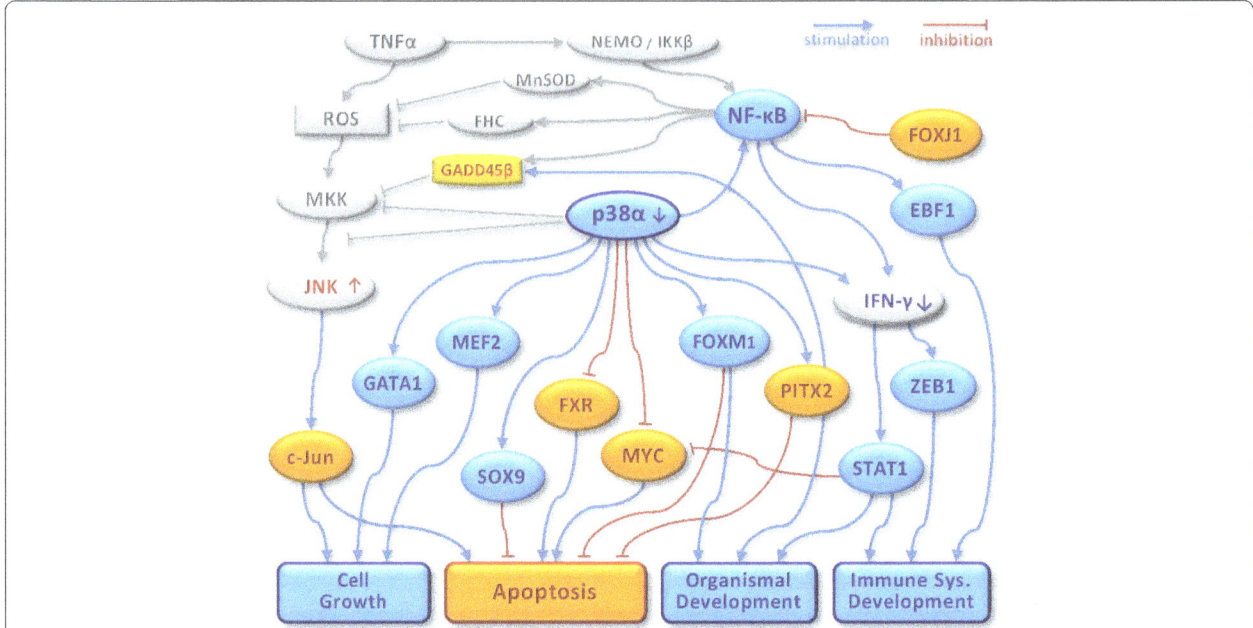

Fig. 5 A comprehensive transcriptional regulatory network for the p38α. The TFs depicted in gray is already known to p38α [41, 51]. The TFs identified in our analysis were then added to the figure and colored blue if their activity was down regulated in the absence of p38α, and colored orange if their activity was up regulated in the absence of p38α. Lines in the figure represent interactions that are known in the literature (listed in more detail in Tables 1 and 2)

show that p38α negatively regulates cell proliferation by antagonizing the JNK-c-Jun pathway. We utilized the published gene expression profiling dataset from their study, to demonstrate that our computational pipeline is able to infer from the gene expression profiling data the same *in silico* conclusions that the authors obtained from their invitro experiments. Therefore, our analysis focused on the JNK-c-Jun pathway to validate the accuracy, robustness and reliability of our strategy. The results are consistent with the experimentally validated inhibitory effect of p38α on transcriptional networks [37]. Their published data confirmed that the most important TF involved in the response to the knock-out was c-Jun, with a clear change observed in both its activation and concentration. In our theoretical work, we also showed a significant change in TFA of c-Jun, but we did not see any corresponding change in the predicted TFC, which is disappointing.

The p38α MAPK is one of many signal transduction pathways and works in both cell-type specific and cell-context specific manner. It plays a pivotal role in converting extra-cellular signal into a wide range of cellular response [43]. We classified a set of TFs that responded to the deletion of p38α into functional groups (Tables 1 and 2, Fig. 3c), that are either developmental factors (group 2) or extra-cellular signal dependent factors (group 3). Developmental factors are also dependent on extra-cellular signals because cells may require such signals to generate developmental

factors [44]. In Fig. 3a, it can be seen that the main factors that responded to the knock-out are the extra-cellular signal dependent factors. None of the TFs that significantly respond in the knock-out are constitutive factors. Our results are consistent with recent publications on the JNK-c-Jun pathway (see citations in Tables 1 and 2).

Our analyses generated a comprehensive transcriptional regulatory network for p38α. The network and a detailed description are shown in Fig. 5. The nodes in the graph were generated from our analysis of responding TFs. The edges in this network were derived from the literature or GO analysis (citations in Tables 1 and 2). The edges or links in the network of p38α regulated TFs have mostly been previously reported, but none of the reports had integrated all these p38α related TFs into a single comprehensive network diagram. Together these results predict a set of TFs that are in some way regulated by p38α, a set somewhat larger than that identified in the original paper. For example, we predict that Foxm1 (HNF3) responds to the p38α status. Recent papers, published since the original study, provide some support for this hypothesis [45, 46]. Most parts of the network are reported in numerous biological studies. However, our network reveals novel links such as p38α—FXR and p38α—MYC. The inferred links are supported by direct experimental evidence so validating the approach, but that in addition novel links have been proposed that are now testable.

The data shown in Tables 1 and 2 and visualized in Fig. 5, provide evidence that the methodology described in this paper is capable of generating plausible hypotheses about linkage between p38α and a range of different TFs. The hypotheses presented in this table have been generated solely from our input data (the connection topology data and gene expression profiling data), but are well-supported by the literature. The methodology has therefore demonstrated that it can produce plausible and testable hypotheses, even if the specific details of those interactions may not be completely accurate. This is not surprising given the fact that we only have an incomplete model of the transcription process. Any *in-silico* techniques which uses predicted TF/TFBSs interactions can provide only a limited view of the complete complexity of transcription control due to the nature of the binding between the TF and the TFBS and the complex effect of gene expression on the TFBS—for example dependent on the epigenetic factors, such as the pattern of histones or DNA methylation at the binding site—as well as the state and concentration of the TF itself. Analysis is complicated by the fact that there are other processes in the cell that act to control mRNA concentration. Such as the rate of RNAi regulated mRNA degradation [9, 10] or susceptibility to attack by RNAses [3, 11]. TFBS can hidden by histones [7, 8], or made more accessible by genomic uncoiling [6]. Furthermore, most TF binding may be cell or species specific not all sites are functional even if occupied, and many functional sites have low levels of conservation [47]. This rather undermines the commonly accepted assumption that TFBSs can be discovered by conservation [22]. However although the exact binding sites may not be conserved the set of TFs that bind a gene somewhere probably is.

p38α deficient mice showed significantly different phenotype which indicates its role is critical. The p38α study also provided the gene expression profiling dataset of wild-type mice as well as p38α deficient mice, so that we could apply our pipeline on the dataset to investigate TFAs and TFCs. It allowed us to directly compare our *in-silico* results to the experimental in-vitro results, and it validated our findings. However, the experiment was done on two time-points that could limit our validation. Thus, we tested our pipeline on a larger dataset from a recent STAT5 transcription factor study [38] which is consisted of 18 samples in five time-points. This study showed the critical role of STAT5-tetramer in immune system. To do this, the authors made STAT5-tetramer deficient mice by generating STAT5A-STAT5B double-knockin mice. Interleukin 2 (IL2) and IL15 are two of well-known upstream regulators of STAT5A-STATB, so they measured IL2- and IL15-induced gene expression profiling in both wild-type mice and STAT5-tetramer deficient mice. We downloaded the RNA-seq gene

expression dataset from this study and analyzed with our pipeline. TF activities were decreased in STAT5-tetramer deficient mice (both IL2- and IL15-induced), particularly at 4, 24, 48 h (Fig. 6). This general trend is well-corresponded to the experimental findings as the author reported IL2- and IL15-induced gene expression were both down-regulated in STAT5-tetramer deficient

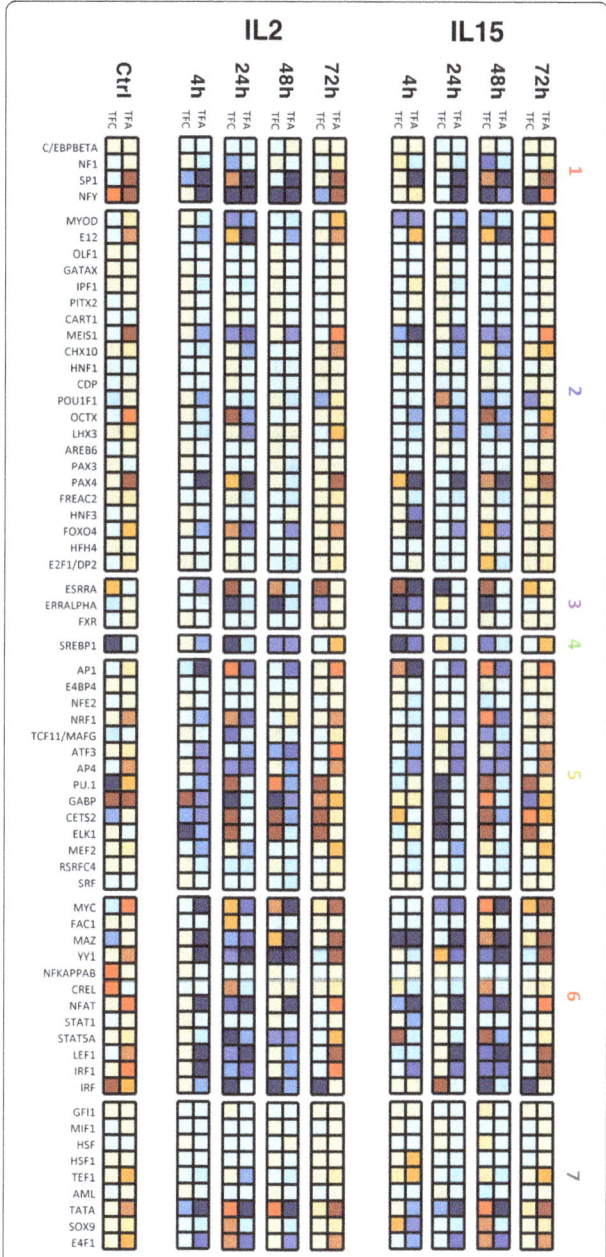

Fig. 6 TFA and TFC changes in STAT5-tetramer deficient mice. TFA and TFC of 65 TFs were estimated from IL2- and IL5-induced RNA-seq datasets and compared between wild-type and STAT5-tetramer deficient mice. Thus, TFA or TFC of a given TF is shown in *red* color if it is higher in STAT5-tetramer deficient mice than wild-type mice. If the level of TFA or TFC is higher, the color is darker. The numbers in right-side of heat-map indicates TF functional group (please see legend in Fig. 3c)

mice. However, most TFs were re-activated at the last time-points which is also exactly same observation in the experimental result. In particular, STAT5A is in our TF list so we closely investigated its activation patterns. STAT5A showed weak activities and concentration level at control sample, but it dramatically de-activated at 4, 24, 48 h and then re-activated at 72 h. More interestingly, all TFs in cytoplasmic factor group (marked as number 6 in Fig. 6) including STAT5A have same activation patterns with STAT5. Moreover, there are a few interesting TFs which are not classified as cytoplasmic factor but also followed same up- and down-regulated patterns with STAT5 (e.g. SP1, NFY, E12, MEIS1, PAX4, AP1, NRF1, TCF11, AP4, GABP, TATA, E4F1). We considered these are new findings and could lead new insight and testable hypotheses.

Conclusions

Our objective was to develop an effective computational pipeline which produces reliable and explicit models of transcriptional regulatory networks. Even though the TFBS information is incomplete due to the difficulties in identifying them, our pipeline predicts new biological hypotheses on a genome-wide scale by combining TFBS and gene expression information. TIGERi is publicly available as a stand-alone GUI software, so ones who have their own gene expression profiling data could easily use the TIGERi software to analyze their data on their fingertips. It would facilitate transcriptional gene regulation researches in the biomedical community.

Our approach can be applied to other gene expression datasets to provide a display of the transcriptional regulatory networks and identify novel candidate genes and TFs underlying specific phenotypes. For example, our methodology has been successfully applied to three recent studies [48–50]. The pipeline would be particularly valuable if it were run on large-scale multi-time point genomic data. It is also the case that we might expect the method to become increasingly predictive with improved connection topologies created from large scale experimentally validated TF/TFBS datasets as opposed to those generated from simple conservation data.

Acknowledgements
Our thanks to Magus Rattray and Guido Sanguinetti for many constructive discussions.

Funding
This work was supported by European Research Council CRIPTON Grant (RG59701), and Institutional funding to the Gurdon Institute by Wellcome Trust Core Grant (092096) and Cancer Research UK Grant (C6946/A14492). The publication charges for this article was funded by European Research Council CRIPTON Grant.

Authors' contributions
Conceptualization, NH, HAN and AB; Methodology, NH; Software, NH; Analysis, NH and AB; Writing, NH, HAN and AB; Supervision, AB; Funding Acquisition, NH and AB All authors have read and approved the final manuscript.

Competing interests
The authors declare that they have no competing interests.

About this supplement
This article has been published as part of BMC Bioinformatics Volume 18 Supplement 7, 2017: Proceedings of the Tenth International Workshop on Data and Text Mining in Biomedical Informatics. The full contents of the supplement are available online at https://bmcbioinformatics.biomedcentral.com/articles/supplements/volume-18-supplement-7.

Author details
[1]Gurdon Institute, University of Cambridge, Cambridge, UK. [2]School of Biological Sciences, University of Liverpool, Liverpool, UK. [3]School of Computer Science and School of Health Sciences, University of Manchester, Manchester, UK.

References

1. Olivera DJ, Nikolaub B, Wurtelea ES. Functional genomics: high-throughput mRNA, protein, and metabolite analyses. Metab Eng. 2002;4(1):98–106.
2. Scott MP. Development: the natural history of genes. Cell. 2000;100(1):27–40.
3. Amrani N, Sachs MS, Jacobson A. Early nonsense: mRNA decay solves a translational problem. Nat Rev Mol Cell Biol. 2006;7:415–25.
4. Sanguinetti G, Lawrence ND, Rattray M. Probabilistic inference of transcription factor concentrations and gene-specific regulatory activities. Bioinformatics. 2006;22(22):2755–81.
5. Cheng C, Yan X, Sun F, Li LM. Inferring activity changes of transcription factors by binding association with sorted expression profiles. BMC Bioinf. 2007;8:452.
6. Strick TR, Croquette V, Bensimon D. Single-molecule analysis of DNA uncoiling by a type II topoisomerase. Nature. 2000;404:901–4.
7. Jagannathan I, Cole HA, Hayes JJ. Base excision repair in nucleosome substrates. Chromosome Res. 2006;14:27–37.
8. Sonehara H, Nagata M, Aoki F. Roles of the first and second round of DNA replication in the regulation of zygotic gene activation in mice. J Reprod Dev. 2008;54(5):381–4.
9. Keene JD. RNA regulons: coordination of post-transcriptional events. Nat Rev Genet. 2007;8:533–43.
10. Anderson P, Kedersha N. RNA granules: post-transcriptional and epigenetic modulators of gene expression. Nat Rev Mol Cell Biol. 2009;10:430–6.
11. Brogna S, Wen J. Nonsense-mediated mRNA decay (NMD) mechanisms. Nat Struct Mol Biol. 2009;16:107–13.
12. Marko-Varga G. Pathway proteomics: global and focused approaches. Am J Pharmacogenomics. 2005;5(2):113–22.
13. Gardner TS, Faith JJ. Reverse-engineering transcription control networks. Phys Life Rev. 2005;2(1):65–88.
14. Bansal M, Belcastro V, Ambesi-Impiombato A, Bernardo D. How to infer gene networks from expression profiles. Mol Syst Biol. 2007;3(78):78–87.
15. Herrgård MJ, Lee B-S, Portnoy V, Palsson BØ. Integrated analysis of regulatory and metabolic networks reveals novel regulatory mechanisms in Saccharomyces cerevisiae. Genome Res. 2006;16(5):627–35.
16. Ramsey SA, Klemm SL, Zak DE, Kennedy KA, Thorsson V, Li B, Gilchrist M, Gold ES, Johnson CD, Litvak V, et al. Uncovering a macrophage transcriptional program by integrating evidence from motif scanning and expression dynamics. PLoS Comput Biol. 2008;4(3):e1000021.
17. Ye C, Galbraith SJ, Liao JC, Eskin E. Using network component analysis to dissect regulatory networks mediated by transcription factors in yeast. PLoS Comput Biol. 2009;5(3):e1000311.
18. Pham H, Ferrari R, Cokus SJ, Kurdistani SK, Pellegrini M. Modeling the regulatory network of histone acetylation in saccharomyces cerevisiae. Mol Syst Biol. 2007;3:153.
19. Gatta GD, Bansal M, Ambesi-Impiombato A, Antonini D, Missero C, Bernardo D. Direct targets of the TRP63 transcription factor revealed by a combination of gene expression profiling and reverse engineering. Genome Res. 2008;18(6):939–48.
20. Cai B, Chang SH, Becker EBE, Bonni A, Xia Z. p38 MAP kinase mediates apoptosis through phosphorylation of BimEL at ser-65. J Biol Chem. 2006;281(35):25215–22.
21. Oliveira AP, Patil KR, Nielsen J. Architecture of transcriptional regulatory circuits is knitted over the topology of bio-molecular interaction networks. BMC Syst Biol. 2008;2:17.

22. Xie X, Lu J, Kulbokas EJ, Golub TR, Mootha V, KLindblad-Toh V, Lander ES, Kellis M. Systematic discovery of regulatory motifs in human promoters and 3' UTRs by comparison of several mammals. Nature. 2005;434:338–45.

23. Sanguinetti G, Lawrence ND, Rattray M. A probabilistic dynamical model for quantitative inference of the regulatory mechanism of transcription. Bioinformatics. 2006;22(14):1753–9.

24. Driever W, Thoma G, Nüsslein-Volhard C. Determination of spatial domains of zygotic gene expression in the drosophila embryo by the affinity of binding sites for the bicoid morphogen. Nature. 1989;340:363–7.

25. Liao JC, Boscolo R, Yang Y-L, Tran LM, Sabatti C, Roychowdhury VP. Network component analysis: reconstruction of regulatory signals in biological systems. Proc Natl Acad Sci. 2003;100(26):15522–7.

26. Alter O, Golub GH. Integrative analysis of genome-scale data by using pseudoinverse projection predicts novel correlation between DNA replication and RNA transcription. Proc Natl Acad Sci. 2004;101:16577–82.

27. Gao F, Foat BC, Bussemaker HJ. Defining transcriptional networks through integrative modeling of mRNA expression and transcription factor binding data. BMC Bioinforma. 2004;5(31):31.

28. Boulesteix A-L, Strimmer K. Predicting transcription factor activities from combined analysis of microarray and ChIP data: a partial least squares approach. Theor Biol Med Model. 2005;2(23):23.

29. Kim HD, Shay T, O'Shea EK, Regev A. Transcriptional regulatory circuits - predicting numbers from alphabets. Science. 2009;325:429–32.

30. Honkela A, Girardot C, Gustafson EH, Liu Y-H, Furlong EEM, Lawrence ND, Rattray M. Model-based method for transcription factor target identification with limited data. Proc Natl Acad Sci U S A. 2010;107(17):7793–8.

31. Gerstein MB, Kundaje A, Hariharan M, Landt SG, Yan K-K, Cheng C, Mu XJ, Khurana E, Rozowsky J, Alexander R, et al. Architecture of the human regulatory network derived from ENCODE data. Nature. 2012;489(7414):91–100.

32. Buck MJ, Lie JD. ChIP-chip: considerations for the design, analysis, and application of genome-wide chromatin immunoprecipitation experiments. Genomics. 2004;83(3):349–60.

33. Park PJ. ChIP-seq: advantages and challenges of a maturing technology. Nat Rev Genet. 2009;10(10):669–80.

34. Zhu Z, Shendure J, Church GM. Discovering functional transcription-factor combinations in the human cell cycle. Genome Res. 2005;15:848–55.

35. Chang L-W, Nagarajan R, Magee JA, Milbrandt J, Stormo GD. A systematic model to predict transcriptional regulatory. Genome Res. 2006;16:405–13.

36. Davies SR, Chang L-W, Patra D, Xing X, Posey K, Hecht J, Stormo GD, Sandell LJ. Computational identification and functional validation of regulatory motifs in cartilage-expressed genes. Genome Res. 2007;17:1438–47.

37. Hui L, Bakiri L, Mairhorfer A, Schweifer N, Haslinger C, Kenner L, Komnenovic V, Scheuch H, Beug H, Wagne EF. p38alpha suppresses normal and cancer cell proliferation by antagonizing the JNK–c-Jun pathway. Nat Genet. 2007;39:741–9.

38. Lin JX, Li P, Liu D, Jin HT, He J, Ata Ur Rasheed M, Rochman Y, Wang L, Cui K, Liu C, et al. Critical role of STAT5 transcription factor tetramerization for cytokine responses and normal immune function. Immunity. 2012;36:586–99.

39. Huang DW, Sherman BT, Lempicki RA. Systematic and integrative analysis of large gene lists using DAVID bioinformatics resources. Nat Protoc. 2009;4(1):44–57.

40. Smet RD, Marchal K. Advantages and limitations of current network inference methods. Nat Rev Microbiol. 2010;8(10):717–29.

41. Perkins ND. Integrating cell-signalling pathways with NF-kappaB and IKK function. Nat Rev Mol Cell Bio. 2007;8:49–62.

42. Brichese L, Cazettes G, Valette A. JNK is associated with Bcl-2 and PP1 in mitochondria: paclitaxel induces its activation and its association with the phosphorylated form of Bcl-2. Cell Cycle. 2004;3(10):1312–9.

43. Wagner EF, Nebreda ÁR. Signal integration by JNK and p38 MAPK pathways in cancer development. Nat Rev Cancer. 2009;9(8):537–49.

44. Brivanlou AH, Jr JED. Signal transduction and the control of gene expression. Science. 2002;295(5556):813–8.

45. Behren A, Muhlen S, Sanhueza GAA, Schwager C, Plinkert PK, Huber PE, Abdollahi A, Simon C. Phenotype-assisted transcriptome analysis identifies FOXM1 downstream from Ras-MKK3-p38 to regulate in vitro cellular invasion. Oncogene. 2010;29(10):1519–30.

46. Sanchez-Calderon H, Rosa LR-d, Milo M, Pichel JG, Holley M, Varela-Nieto I. RNA microarray analysis in prenatal mouse cochlea reveals novel IGF-I target genes: implication of MEF2 and FOXM1 transcription factors. PLoS One. 2010;5(1):e8699.

47. Schmidt D, Wilson MD, Ballester B, Schwalie PC, Brown GD, Marshall A, Kutter C, Watt S, Martinez-Jimenez CP, Mackay S, et al. Five-vertebrate ChIP-seq reveals the evolutionary dynamics of transcription factor binding. Science. 2010;328(5981):1036–40.

48. McMaster A, Jangani M, Sommer P, Han N, Brass A, Beesley S, Lu W, Berry A, Loudon A, Donn R, et al. Ultradian cortisol pulsatility encodes a distinct, biologically important signal. PLoS One. 2011;6(1):e15766.

49. Han N, Dol Z, Vasieva O, Hyde R, Liloglou T, Raji O, Brambilla E, Brambilla C, Martinet Y, Sozzi G, et al. Progressive lung cancer determined by expression profiling and transcriptional regulation. Int J Oncol. 2012;41(1):242–52.

50. Darieva Z, Han N, Warwood S, Doris KS, Morgan BA, Sharrocks AD. Protein kinase C regulates late cell cycle-dependent gene expression. Mol Cell Biol. 2012;32(22):4651–61.

51. Papa S, Bubici C, Zazzeroni F, Pham CG, Kuntzen C, Knabb JR, Dean K, Franzoso G. The NF-kappaB-mediated control of the JNK cascade in the antagonism of programmed cell death in health and disease. Cell Death Differ. 2006;13:712–29.

52. Guo F, Tanzer S, Busslinger M, Weih F. Lack of nuclear factor-kappa B2/p100 causes a RelB-dependent block in early B lymphopoiesis. Blood. 2008;112(3):551–9.

53. Thonel A, Vandekerckhove J, Lanneau D, Selvakumar S, Courtois G, Hazoume A, Brunet M, Maurel S, Hammann A, Ribeil JA, et al. HSP27 controls GATA-1 protein level during erythroid cell differentiation. Blood. 2010;116(1):85–96.

54. Stassen M, Klein M, Becker M, Bopp T, Neudorfl C, Richter C, Heib V, Klein-Hessling S, Serfling E, Schild H, et al. p38 MAP kinase drives the expression of mast cell-derived IL-9 via activation of the transcription factor GATA-1. Mol Immunol. 2007;44(5):926–33.

55. Angelis L, Zhao J, Andreucci JJ, Olson EN, McDermott GCJC. Regulation of vertebrate myotome development by the p38 MAP kinase–MEF2 signaling pathway. Dev Biol. 2005;283(1):171–9.

56. Ramsauer K, Sadzak I, Porras A, Pilz A, Nebreda AR, Decker T, Kovarik P. p38 MAPK enhances STAT1-dependent transcription independently of Ser-727 phosphorylation. Proc Natl Acad Sci. 2002;99(20):12859–64.

57. Bradham C, McClay DR. p38 MAPK in development and cancer. Cell Cycle. 2006;5(8):824–8.

58. Li S, Gallup M, Chen Y-T, McNamara NA. Molecular mechanism of proinflammatory cytokine-mediated squamous metaplasia in human corneal epithelial cells. Invest Ophthalmol Vis Sci. 2010;51(5):2466–75.

59. Tew SR, Hardingham TE. Regulation of SOX9 mRNA in human articular chondrocytes involving p38 MAPK activation and mRNA stabilization. J Biol Chem. 2006;281(51):39471–9.

60. Zhang R, Murakami S, Coustry F, Wang Y, Crombrugghe B. Constitutive activation of MKK6 in chondrocytes of transgenic mice inhibits proliferation and delays endochondral bone formation. Proc Natl Acad Sci. 2006;103:352–70.

61. Akiyama H, Chaboissier M-C, Martin JF, Schedl A, Crombrugghe B. The transcription factor Sox9 has essential roles in successive steps of the chondrocyte differentiation pathway and is required for expression of Sox5 and Sox6. Genes Dev. 2002;16:2813–28.

62. Seymour PA, Freude KK, Tran MN, Mayes EE, Jensen J, Kist R, Scherer G, Sander M. SOX9 is required for maintenance of the pancreatic progenitor cell pool. Proc Natl Acad Sci. 2007;104(6):1865–70.

63. Acharya A, Lingenfelter DJ, Huang L, Gage PJ, Walter MA. Human PRKC apoptosis WT1 regulator is a novel PITX2-interacting protein that regulates PITX2 transcriptional activity in ocular cells. J Biol Chem. 2009;284(50):34829–38.

64. Toro R, Saadi I, Kuburas A, Nemer M, Russo AF. Cell-specific activation of the atrial natriuretic factor promoter by PITX2 and MEF2A. J Biol Chem. 2004;279(50):52087–94.

65. Li C-S, Chae S-C, Lee J-H, Zhang Q, Chung H-T. Identification of single nucleotide polymorphisms in FOXJ1 and their association with allergic rhinitis. J Hum Genet. 2006;51(4):292–7.

66. Gottardi AD, Dumonceau J-M, Bruttin F, Vonlaufen A, Morard I, Spahr L, Rubbia-Brandt L, Frossard J-L, Dinjens WN, Rabinovitch PS, et al. Expression of the bile acid receptor FXR in Barrett's esophagus and enhancement of apoptosis by guggulsterone in vitro. Mol Cancer. 2006;5(48):48.

67. Fujino T, Murakami K, Ozawa I, Minegishi Y, Kashimura R, Akita T, Saitou S, Atsumi T, Sato T, Ando K, et al. Hypoxia downregulates farnesoid X receptor via a hypoxia-inducible factor-independent but p38 mitogen-activated protein kinase-dependent pathway. FEBS J. 2009;276(5):1319–32.

68. Desbiens KM, Deschesnes RG, Labrie MM, Desfosses Y, Lambert H, Landry J, Bellmann AK. c-Myc potentiates the mitochondrial pathway of apoptosis by acting upstream of apoptosis signal-regulating kinase 1 (Ask1) in the p38 signalling cascade. Biochem J. 2003;372:631–41.

69. Shuai K, Liu B. Regulation of JAK-STAT signalling in the immune system. Nat Rev Immunol. 2003;3(11):900–11.

metaX: a flexible and comprehensive software for processing metabolomics data

Bo Wen[1,2], Zhanlong Mei[1,2], Chunwei Zeng[1,2] and Siqi Liu[1,2]*

Abstract

Background: Non-targeted metabolomics based on mass spectrometry enables high-throughput profiling of the metabolites in a biological sample. The large amount of data generated from mass spectrometry requires intensive computational processing for annotation of mass spectra and identification of metabolites. Computational analysis tools that are fully integrated with multiple functions and are easily operated by users who lack extensive knowledge in programing are needed in this research field.

Results: We herein developed an R package, metaX, that is capable of end-to-end metabolomics data analysis through a set of interchangeable modules. Specifically, metaX provides several functions, such as peak picking and annotation, data quality assessment, missing value imputation, data normalization, univariate and multivariate statistics, power analysis and sample size estimation, receiver operating characteristic analysis, biomarker selection, pathway annotation, correlation network analysis, and metabolite identification. In addition, metaX offers a web-based interface (http://metax.genomics.cn) for data quality assessment and normalization method evaluation, and it generates an HTML-based report with a visualized interface. The metaX utilities were demonstrated with a published metabolomics dataset on a large scale. The software is available for operation as either a web-based graphical user interface (GUI) or in the form of command line functions. The package and the example reports are available at http://metax.genomics.cn/.

Conclusions: The pipeline of metaX is platform-independent and is easy to use for analysis of metabolomics data generated from mass spectrometry.

Keywords: Metabolomics, Pipeline, Workflow, Quality control, Normalization

Background

Biochemicals (metabolites) with low molecular masses are the ultimate products of biological metabolism, while a metabolome represents the total composite in a given biological system and reflects the interactions among an organism's genome, gene expression status and the relevant micro-environment [1]. The most prevalent technology used in analysis of metabolomics is non-targeted mass spectrometry (MS) coupled with either liquid chromatography (LC-MS) or gas chromatography (GC-MS) [2, 3]. Generally, these techniques generate a set data of mass spectra with chromatography that includes retention time, peak intensity and chemical masses. Data analysis involves stepwise procedures including peak picking, quality control, data cleaning, preprocessing, univariate and multivariate statistical analysis and data visualization. A number of software packages are available for MS-based metabolomics data analysis as listed in Table 1, including propriety commercial, open-source, and online workflows. The MS manufacturers generally provide propriety software, like SIEVE (Thermo Scientific), MassHunter (Agilent Technologies) and Progenesis QI (Waters), which are often limited in scope and function. Open-source software, such as XCMS [4], CAMERA [5], MAIT [6], MetaboAnalyst [7] and Workflow4Metabolomics [8], usually cover limited processing steps. There is no such comprehensive pipeline that is used across the metabolomics community [9, 10]. Referring to the capabilities of the tools mainly used (as shown in Table 1), an automatic and comprehensive open source pipeline is urgent in

* Correspondence: siqiliu@genomics.cn
[1]BGI-Shenzhen, Shenzhen 518083, China
[2]China National GeneBank-Shenzhen, BGI-Shenzhen, Shenzhen, Guangdong 518083, China

Table 1 Qualitative assessment of metaX compared to other existing metabolomics tools

No.	1	2	3	4	5	6	7	8	9
Feature	metaX	MAT	Workflow4Metabolomics	MetMSLine	metaMS	MetaboNexus	MetaboAnalyst	XCMSOnline	MeltDB
Year	2015	2014	2014	2013	2013	2014	2009	2012	2008
Language	R, Java	R	R, Perl, Python, Java	R	R	R	R, Java	R	perl, JavaScript and R
Platform independent	√	√	√	√	√	Windows only	√	√	√
Open source	√	√	√	√	√	√	√	√	project- and user-specific access
Usable offline	√	√	√	√	√	√	√	-	-
Power analysis	√	-	-	-	-	-	√	-	-
Automatic outlier samples finding	√	-	√	√	-	-	-	-	-
PCA	√	√	√	√	-	√	√	√	√
Cluster analysis	√	√	√	√	-	√	√	-	√
PLS-DA	√	√	√	-	-	√	√	-	√
ROC analysis	√	-	-	-	-	√	√	-	-
Normalization	Sum, PQN, VSN, QC-RSC, ComBat, SVR, quantiles	-	Linear or local polynomial regression fitting	QC-LSC	-	Internal standard or quantile normalization	Normalized by sum/median, Normalized by reference sample/feature, sample specific normalization and quantile normalization	-	Normalized by specific compound or feature
Biomarker analysis	√	-	-	-	-	√	√	-	-
Correlation network analysis	√	-	-	-	-	-	-	-	-
Metabolite identification	√	√	√	√	√	√	-	√	√
Functional analysis	√	-	-	-	-	√	√	-	√
Quality assessment	√	√	-	-	-	-	-	√	-
Peak picking	√	√	√	√	√	√	√	√	√
HTML-Based report	√	-	-	-	-	-	(PDF)	-	(PDF)

Table 1 Qualitative assessment of metaX compared to other existing metabolomics tools (*Continued*)

No.	14	15	10	11	12	13	14
Feature	Mzmine	Mzmatch	apLCMS	EigenMS	MetaC	Metabomxtr	Metabolomics
Year	2006	2011	2009	2014	2011	2014	2014
Language	JAVA	JAVA, R	R	R	R	R	R
Platform independent	√	√	√	√	√(windows & MacOS)	√	√
Open source	√	√	√	√	√	√	√
Usable offline	√	√	√	√	√	√	√
Power analysis	-	-	-	-	-	-	-
Automatic outlier samples finding	-	-	-	-	-	-	-
PCA	√	-	-	-	-	-	√
Cluster analysis	√	-	-	-	-	-	√
PLS-DA	-	-	-	-	-	-	-
ROC analysis	-	-	-	-	-	-	-
Normalization	Linear normalizaiton, normalized by internal standards	Normalized by Reference sample	-	combination of ANOVA and singular value decomposition	interral standard, medium, biomass(divides the irtensity of each metabolite in a specific sampe by the value of the biomass meas red for this specific sample)	normalized using a mixture model with batch-specific thresholds and run order correction	normalized by sum,mean or media of each sample;normalized by specific reference;normalized by internal standards or optimal selection of multiple internal standards;
Biomarker analysis	-	-	-	-	-	-	-
Correlation network analysis	-	-	-	-	-	-	-
Metabolite identification	√	√	√	-	-	-	-
Functional analysis	-	-	-	-	-	-	-
Quality assessment	-	-	-	-	-	-	-
Peak picking	√	√	√	-	-	-	-
HTML-Based report	-	-	-	-	-	-	-

bioinformatics analysis of metabolomics. Basically, the pipeline aims for users to easily perform end-to-end metabolomics data analysis with a flexible combination of different methods to efficiently integrate new modules and to build customized pipelines in multiple ways.

We herein developed a comprehensive workflow for analysis of metabolomics data, termed metaX. At the present time, R [11] is a popular statistical programming environment and provides a convenient environment for statistical analysis of metabolomic and other -omics data [12, 13]. We thus designed metaX as an R package that automates analysis of untargeted metabolomics data acquired from LC/MS or GC/MS and offers a user-friendly web-based interface for data quality assessment and normalization evaluation. This workflow, which is open source and rich in functions, encourages experienced programmers to improve the relevant functions or to build their own pipeline within the R framework. Overall, metaX aims to be a tool array that utilizes an end-to-end statistical analysis of metabolomics data.

Implementation

A stepwise overview of data processing using metaX is illustrated in Fig. 1.

Peak picking and inputs

In general, metaX can take mzXML files as input or a peak table file as input. If taking mzXML files as input, metaX will use the R package XCMS [4] to detect peaks, then use the CAMERA [5] package to perform peak annotation. If a peaks table file is an input, metaX transforms the table data from a peak detection software, such as Progenesis QI (exported comma separated value (csv) format file), into an R object compatible with the subsequent workflow.

Pre-processing of raw peak data metabolite

The raw peak intensity data was pre-processed in metaX. Firstly, if a metabolite feature is detected in < 50% of quality control (QC) samples or detected in < 20% of experimental samples, it is removed from data analysis [14]. Secondly, a missing value after the first filtering is retained and imputed. In metaX, four methods are implemented to perform missing value imputation: k-nearest neighbor (KNN), Bayesian principal component analysis replacement (BPCA), svdImpute and random forest imputation (missForest) [15].

Data scaling and transformation

Five different scaling approaches are offered in metaX: Pareto scaling, vast scaling, range scaling, autoscaling

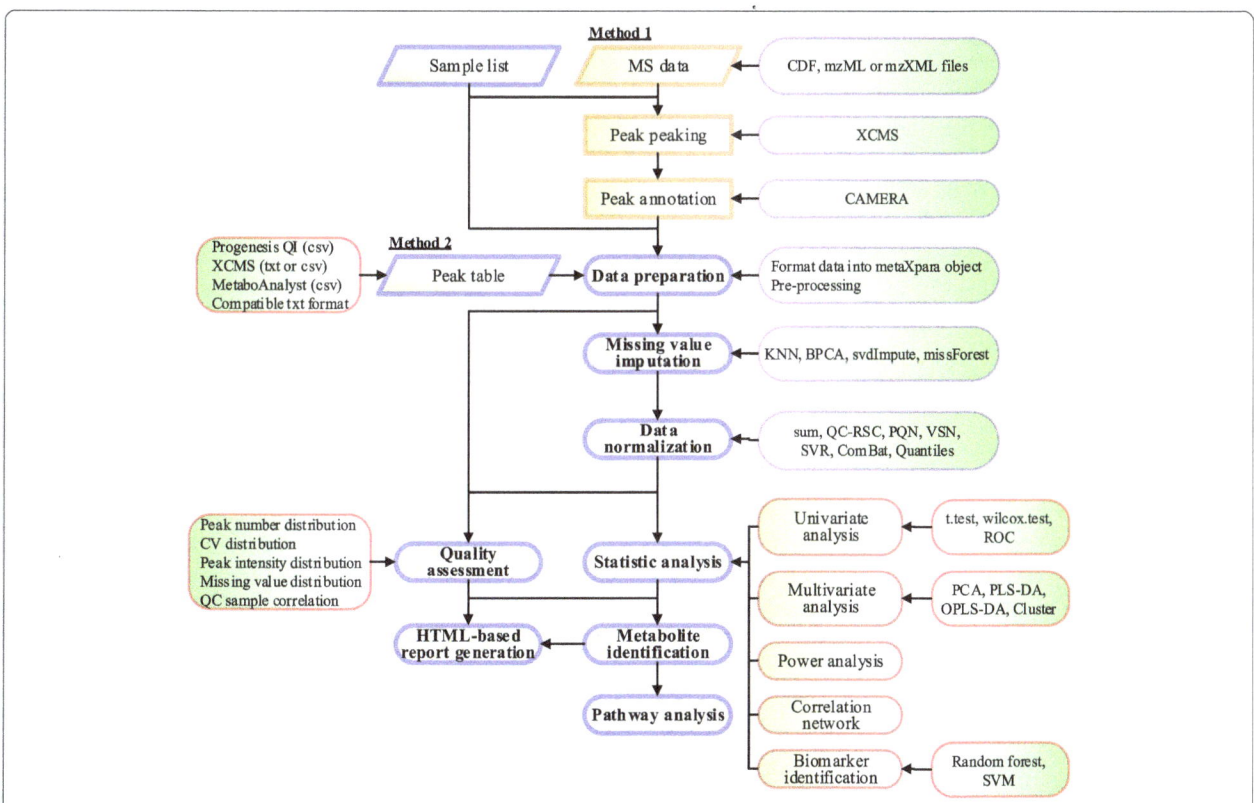

Fig. 1 Overview of metaX. This figure summarizes the main modules, functions and features of metaX. The input data and the functions are included in the figure

and level scaling [16]. The formulas of these scaling approaches are described in detail elsewhere [16]. In addition, three transformation approaches are offered in metaX: log, generalized logarithm (glog) and cube root transformation.

Removal of outliers

metaX provides the ability to automatically remove the outlier samples in the pre-processed data based on expansion of the Hotelling's T2 distribution ellipse [17]. A sample within the first and second component principal component analysis (PCA) score plot beyond the expanded ellipse is removed, and then the PCA model is recalculated. In default mode, three rounds of outlier removal are performed.

Normalization

A metabolomics dataset usually contains unwanted variations introduced by signal drift/attenuation and multiplicative noise across the dynamic range. These effects can detrimentally impact the significant signal discovery and MS features that are required for rigorous quality assurance [14, 18]. In metaX, two types of normalization methods are provided: 1) Sample-based normalization is used to correct different concentrations of samples, such as normalization to total sum, probabilistic quotient normalization (PQN), variance stabilizing normalization (VSN) and quantile-based methods. 2) Peak-based normalization is implemented to correct data within batch experiment analytical variation and batch-to-batch variation in large-scale studies [19]. In this normalization, if a study contains QC samples, the QC-robust spline batch correction (QC-RSC) can be used to alleviate the effects of peak area attenuation [19]. During normalization, the degree of smoothening is controlled by a parameter that sets the proportion of points for smoothening at each point, while in metaX, this parameter is automatically assigned by using leave-one-out cross validation. On the basis of QC samples, a metabolite feature with a coefficient of variation (CV) over the predetermined value is excluded after normalization. The CV threshold could be set by users; generally, CV values ≤ 30% are recommended. Support vector regression (SVR) [20] and ComBat [21] normalization methods are also implemented in metaX. A user-friendly web-based interface (http://metax.genomics.cn) was offered for rapid evaluation of the data normalization methods for a specified dataset.

Assessment of data quality

Pre- and post-normalization, the data quality is visually assessed in several aspects, 1) the peak number distribution, 2) the number of missing value distribution, 3) the boxplot of peak intensity, 4) the total peak intensity distribution, 5) the correlation heatmap of QC samples if available, 6) the metabolite m/z (or mass) distribution, 7)

the plot of m/z versus retention time, and 8) the PCA score or loading plot of all samples. There are two ways to perform data quality assessment in metaX, the command line mode and the user-friendly web-based interface at http://metax.genomics.cn/.

Univariate and multivariate statistical analysis

metaX offers both univariate and multivariate statistical analysis. For univariate statistical analysis, the parametric statistical test (Students t-test), non-parametric statistical test (Mann-Whitney U test), and classical univariate receiver operating characteristic (ROC) curve analysis are implemented. For multivariate statistical analysis, metaX offers functionalities for cluster analysis, multivariate modelling, including PCA, partial least squares-discriminant analysis (PLS-DA) and orthogonal partial least squares-discriminant analysis (OPLS-DA), with numerical and graphical results and diagnostics (optimal number of components estimated by cross-validation, R^2, Q^2, variable importance in projection (VIP), statistical significance of the model by permutation testing) [22]. In terms of the univariate test analysis, metaX also offers the false discovery rate (FDR)-corrected p-value by using the Benjamini-Hochberg FDR algorithm [23]. The PLS-DA was implemented based on the functions from the pls package [24], and the OPLS-DA was performed using the functions from the ropls package [25].

Power and sample size analysis

metaX offers an easy-to-use function to perform the power and sample size analysis. This function is based on the Bioconductor package SSPA [26] and outputs a figure to show the distribution curve of sample size versus the estimated power.

Metabolite correlation network analysis

metaX offers two types of network analysis. One is the correlation network analysis without regard for experimental groups information, and the other is differential correlation network analysis, which aims to identify metabolite correlation differences in a physiological state. The former was implemented using the cor function from the stats package to calculate the correlation coefficient, and the latter was implemented using the function comp.2.cc.fdr from the DiffCorr package [27] to calculate the significantly differential correlations. The igraph package [28] was used for network analysis and visualization. In addition, the network can be exported as a file in formats such as gml and pajek, which can be imported into Cytoscape [29] and Gephi [30] for network analysis and visualization. Both of the correlation network analyses aim to describe the correlation patterns among metabolites across samples, in which nodes represent metabolites and edges represent the correlation between different metabolites. The network analysis offers

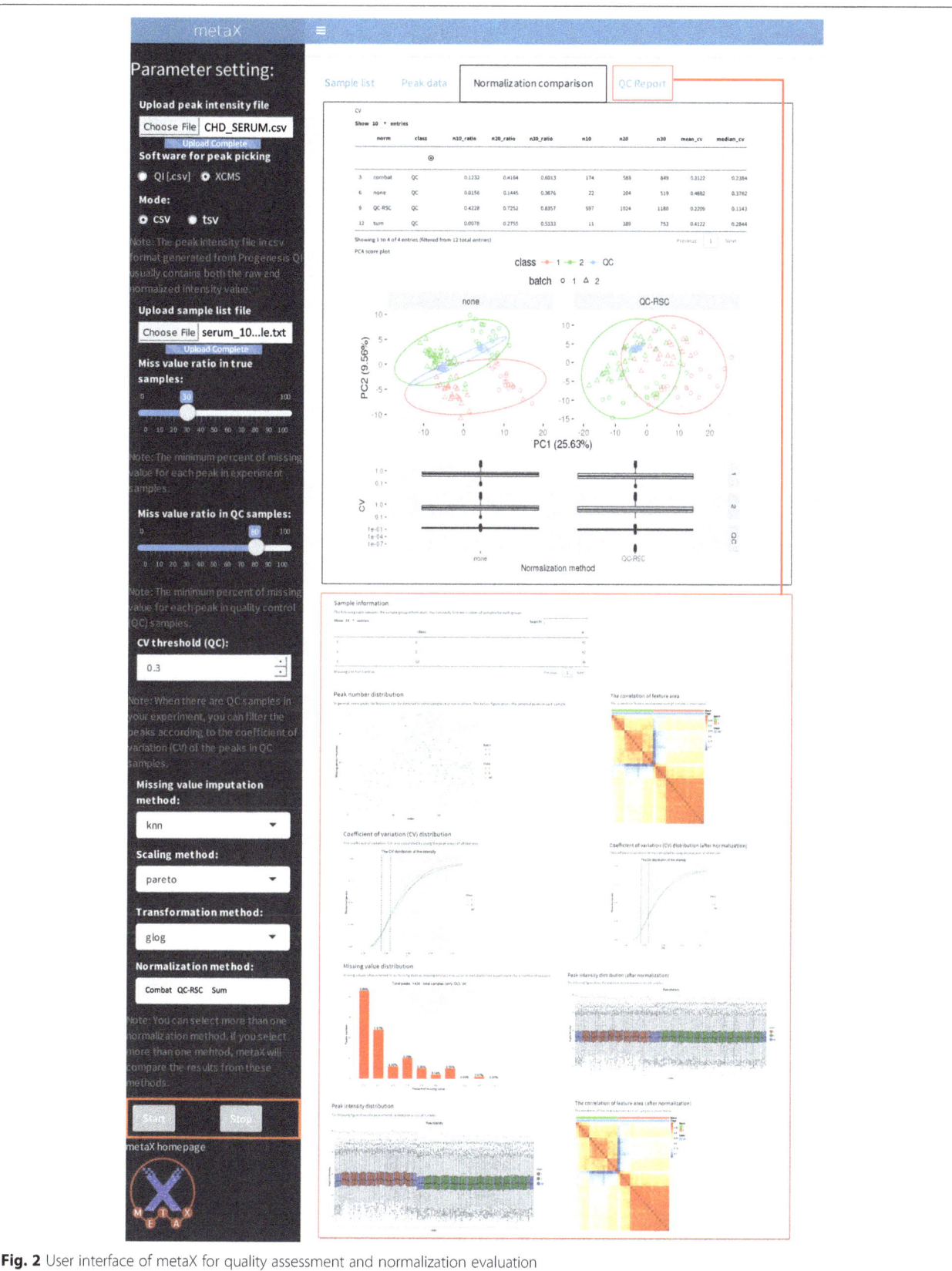

Fig. 2 User interface of metaX for quality assessment and normalization evaluation

a complementary method to univariate and multivariate statistical analysis methods.

Metabolite identification

Currently, metaX provides a function for metabolite identification based on the Human Metabolome Database (HMDB) [31], KEGG [32, 33], MassBank [34], PubChem [35], LIPID MAPS [36], MetaCyc [37] and PlantCyc (www.plantcyc.org). Moreover, metaX can easily be extended to support the other databases. The metabolites having molecular weights within a specified tolerance to the query m/z or molecular weight value are retrieved from the databases as putative identifications. The information of adducts and isotopes is utilized

to assist in metabolite identification if it is present. The default tolerance is 10 ppm.

Functional analysis

At present, metaX provides a function for metabolite pathway analysis based on IMPaLA [38].

Biomarker analysis

metaX uses functions from the R package "caret" to perform the biomarker selection, model creation and performance evaluation [39]. Currently, two methods, random forest [40] and support vector machine (SVM), are implemented to automatically select the metabolites which show the best performance. After the best set features are

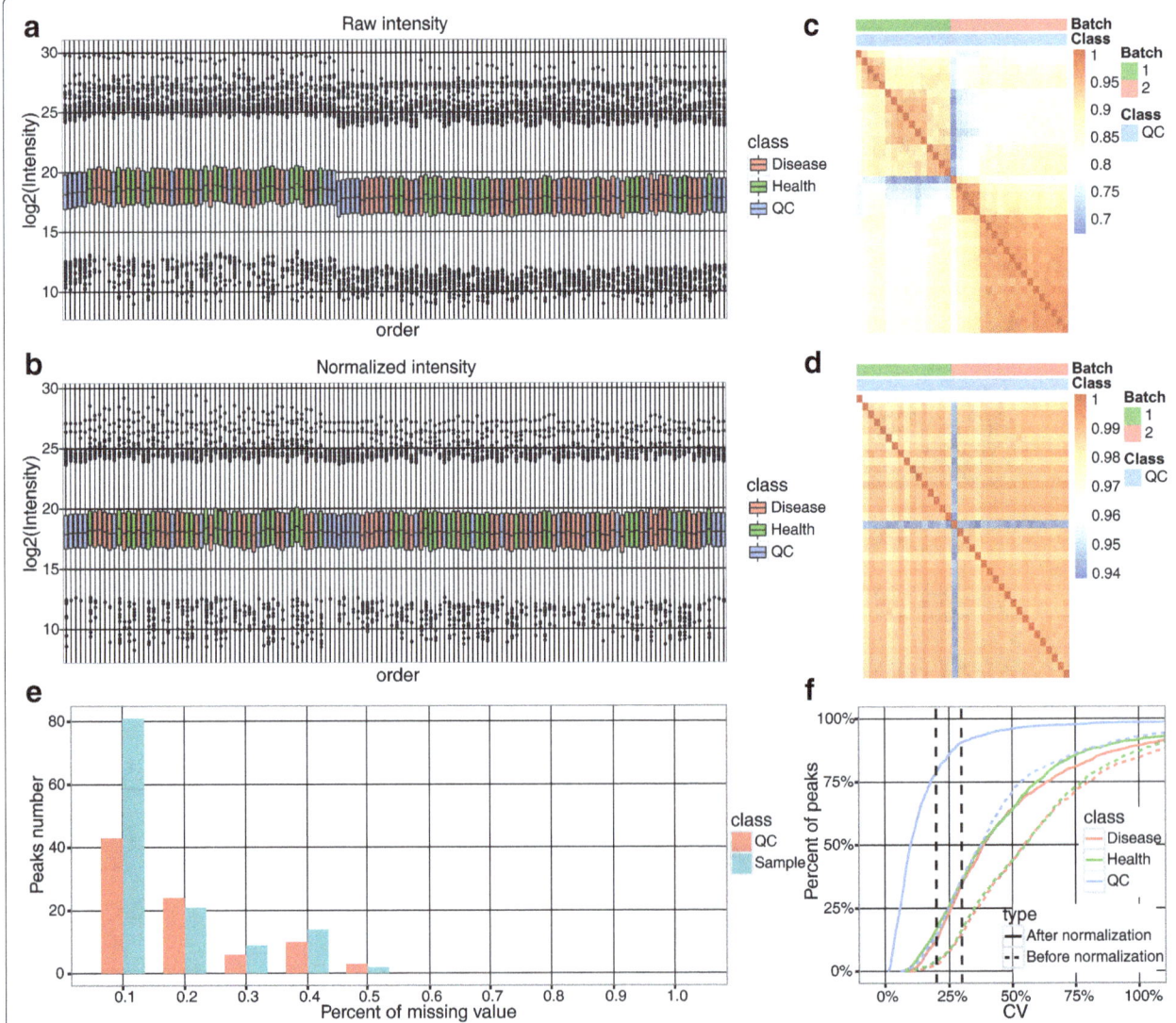

Fig. 3 QC charts generated by metaX. **a** The intensity of feature distribution before normalization. **b** The intensity of feature distribution after normalization. **c** The correlation plot of QC samples before normalization. **d** The correlation plot of QC samples after normalization. **e** The missing value distribution in experimental and QC samples. **f** The CV distribution of all features before and after normalization for each group

selected, a randomForest model can be created and the ROC curve can be plotted.

HTML-based report generation

metaX outputs an HTML-based report by using the Nozzle package [41], which contains quality assessment plots and other analysis results.

Results and discussion

To illustrate the applications of metaX, a published non-targeted LC-MS metabolomics dataset from a coronary heart disease (CHD) study was used [42, 43]. The dataset consisted of two batches of 138 plasma samples (59 CHD patients, 43 healthy controls and 36 QC samples) acquired in positive ion mode on an LTQ Orbitrap Velos instrument (Thermo Fisher Scientific, MA, USA). LC-MS raw data files were converted to mzXML format using ProteoWizard (version 3.0.5941) [44] and then were processed by XCMS [4] and CAMERA [5] for peak picking and peak annotation, respectively. In total, 1438 features were retained for downstream analysis. The mzXML files can be downloaded from the Dryad Digital Repository [43]. It merits to note that the study focus is mainly on the software application

and its capabilities, not on the biological interpretation of the generated results.

Quality assessment of metabolomics data using metaX

In metabolomics studies, data quality checks are crucial prerequisites to achieve reliable results. metaX offers a quick and easy data quality check of metabolomics data. This can be done using the R function in metaX or a user-friendly web interface at the website http://metax.geno-mics.cn/ as shown in Fig. 2. The mainly QC charts generated by metaX for the CHD dataset are illustrated in Figs. 3 and 4. The number of features detected per sample over the analysis time (injection order) is illustrated in Fig. 4c, revealing that the peaks acquired from any group, disease, healthy and QC, are randomly distributed. The intensities of all features per samples before and after normalization over the analysis time (injection order) are illustrated in Fig. 3a and b, respectively. The missing value distribution is shown in Fig. 3e, which gives an overview of the percent of missing values of all features in both the QC and experiment samples. According to Chawade's view, the total missing value plot and the total intensity plot derived from raw data and treated with/without normalization

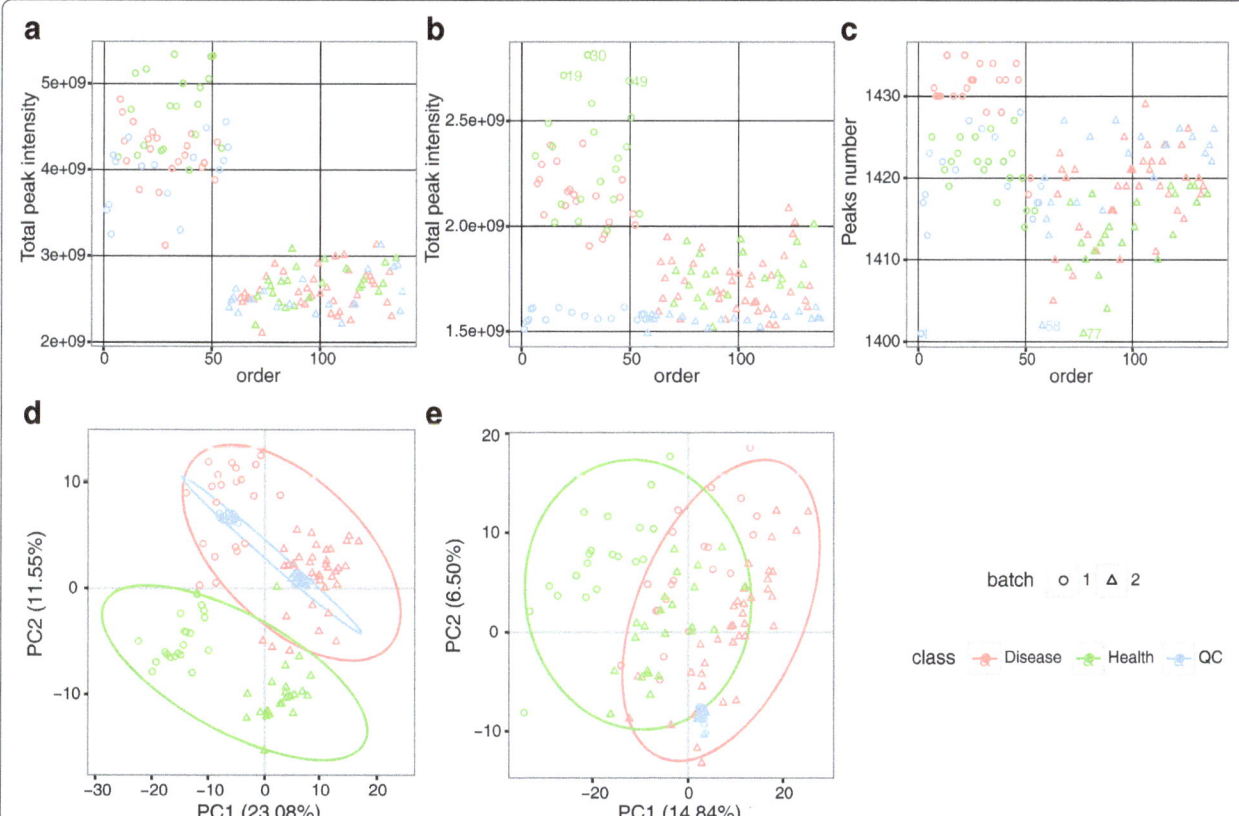

Fig. 4 QC charts generated by metaX. **a** The sum intensity of all features per sample before normalization over the analysis time (injection order). **b** The sum intensity of all features per sample after normalization over the analysis time (injection order). **c** The number of features per sample over the analysis time (injection order). **d** The score plot of PCA for the raw feature intensity data. **e** The score plot of PCA for the normalized data

could be used to identify sample outliers [45]. Our analysis supported this. The correlation plots of QC samples before and after normalization by SVR are illustrated in Fig. 3c and d and indicate that the lowest correlation efficiency is enhanced from approximately 0.7 to 0.9. The CV distribution of all features before and after normalization for each group is displayed in Fig. 3f, implying that after normalization, the signal quality is obviously improved. The sum intensity of all features per sample before and after normalization over the analysis time (injection order) is illustrated in Fig. 4a and b, suggesting that normalization could narrow the signal variation. The score plots of PCA for the raw feature intensity data and the normalized data are shown in Fig. 4d and e, respectively, which give an overview of the dataset and showing trends, groupings and outliers before data normalization and after data normalization. The score plot of PCA (Fig. 4d) for the non-normalized data provided a simple and easily interpretable visual check of the presence of batch effects. In Fig. 4d, the two data batches appear as two separated groups upon PCA analysis without normalization, whereas in Fig. 4e, after normalization the batch effect was reduced and all of the QC samples were clustered tightly, which provides an initial evaluation of the data quality. Overall, these QC charts demonstrate the necessity of normalization for metabolomics data, while metaX enables overview of the data quality with different charts.

Evaluation of normalization methods using metaX

A systematic bias in high-throughput metabolomics data is often introduced by various steps of sample processing and data generation. Data normalization can reduce systematic biases. A question related to this issue is how to select a proper normalization method. metaX provides a user-friendly web-based Shiny application (http://metax.genomics.cn) for this purpose. To select the optimal normalization approach for the CHD dataset, seven methods are evaluated using metaX. Figure 5 shows the score plots of PCA using different normalization methods. They indicate that after normalization using QC-RSC, ComBat or SVR, all of the QC samples are clustered more tightly, and the batch effect is effectively reduced compared with other methods. Table 2 presents the quantitative comparison metrics acquired by the different methods. From the results it is clear that all normalization methods performed better than non-normalization used in most of the metrics. Specifically, SVR detects the largest number of features (1293) with CV ≤ 30% in QC samples, followed by QC-RSC (1191). For the average CV of features in QC samples, SVR achieved the best performance, followed by QC-RSC. This is similar to the findings in a previous study [20]. However, QC-RSC could detect the largest number of differentially expressed features (178), followed by SVR (170). Taken together, for this data set, SVR could be an

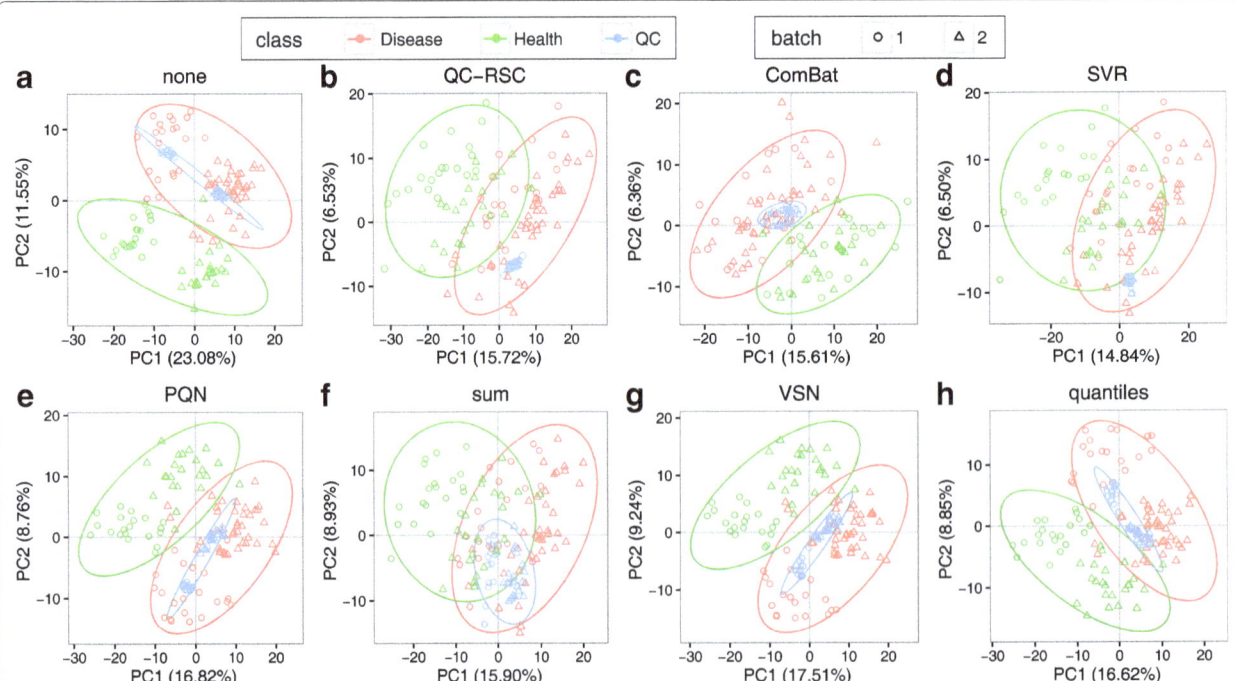

Fig. 5 Comparison of different normalization methods from PCA. **a** none, **b** QC-RSC, **c** ComBat, **d** SRV, **e**) PQN, **f** sum, **g** VSN and **h** quantiles. The different points in the figures refer to different samples, and the samples were color-coded according to their group information and shape-coded according to their batch information

Table 2 The comparison of different normalization methods

Methods	NO. of peaks	NO. of peaks (CV ≤ 30%)[a]	DEF[b]	Mean (CV) CHD [d]	Mean (CV) Health [d]	Mean (CV) QC [e]
ComBat	1438	930	127	0.4261	0.3816	0.1636
none	1438	527	65	0.4865	0.4739	0.2114
QC_RSC	1438	1191	178	0.5108	0.4664	0.1098
SVR	1438	1293	170	0.4853	0.4583	0.1081
PQN	1438	793	125	0.4945	0.4681	0.1777
Quantiles	1438	740	118	0.4911	0.4646	0.1895
sum	1438	761	119	0.5044	0.4733	0.1979
VSN	1438	772	120	0.5014	0.4761	0.1912

Note:
[a] After normalization, the number of peaks with CV ≤ 30% in QC samples
[b] DEF: differentially expressed features with q-value $< = 0.05$, fold change $> = 1.5$ or fold change $< = 0.667$ and VIP $> = 1$
[c] Mean (CV) CHD: The average CV of peaks in CHD disease group
[d] Mean (CV) Health: The average CV of peaks in health group
[e] Mean (CV) QC: The average CV of peaks in QC group

optimal normalization method, thus it was chosen as the default normalization method for the downstream analysis.

Univariate and multivariate statistical analysis
Data for the QC samples are removed from the dataset prior to univariate and multivariate analysis in metaX. For univariate analysis, Mann-Whitney U test and Students t-test are performed to compare disease and health groups, followed by false discovery correction using the Benjamini-Hochberg method using metaX. The results, along with the fold change of the disease group versus health group, are presented in Additional file 1: Table S1. In total, 171 features (13.22% of total features) are detected under the criteria of the corrected p-value (Mann-Whitney U test) ≤ 0.05, fold change ≥ 1.5 or ≤ 0.667 and VIP $> =1$, and 170 features (13.15% of total features) are detected under the criterion of the

corrected p-value (Students t-test) ≤ 0.05, fold change ≥ 1.5 or ≤ 0.667 and VIP $> = 1$. The result is comparable with that of the previous study [42].

For multivariate analysis, PCA, PLS-DA and OPLS-DA are performed by metaX. In PCA analysis, the normalized peak intensity matrix is glog transformed, followed by Pareto scaling and centering, and then two components are selected. The PCA score and loading plots are shown in Fig. 6a and b, respectively. The score plot indicates that there is an apparent difference between the disease and health groups. For PLS-DA and OPLS-DA, the normalized peak intensity matrix is also glog transformed, followed by Pareto scaling and centering. Two components are selected for PLS-DA and two components (one orthogonal and one predictive) for OPLS-DA. The score and loading plots for PLS-DA and OPLS-DA are shown in Fig. 7a and c, respectively. The R^2Y and Q^2Y values of the PLS-DA model, which are

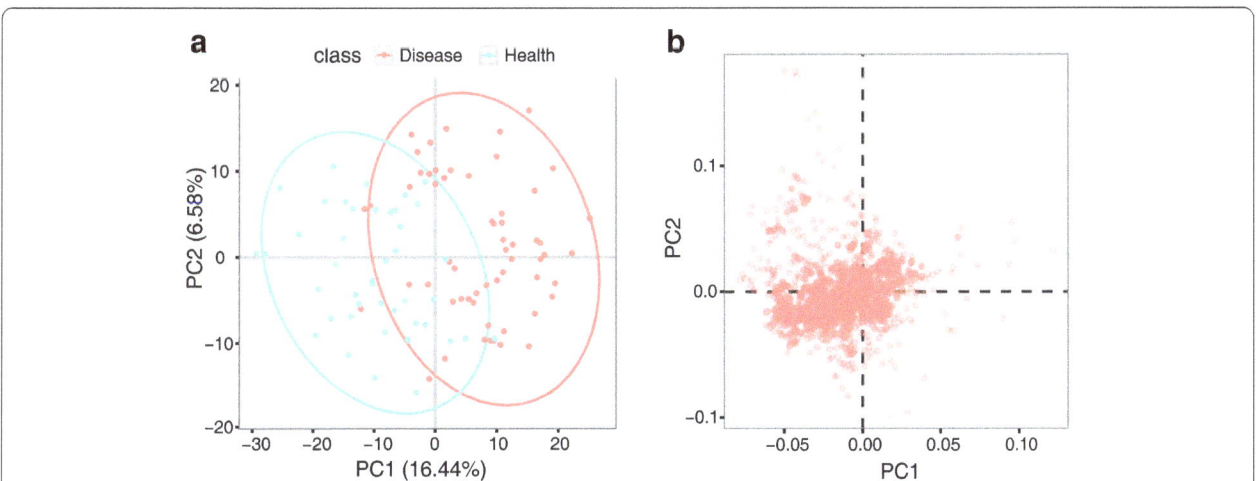

Fig. 6 The score and loading plots of PCA. **a** Score plot of PCA and (**b**) Loading plot of PCA. The different points in the figures refer to different samples, and the samples are color-coded according to their group information. The QC samples were removed before performing the PCA analysis

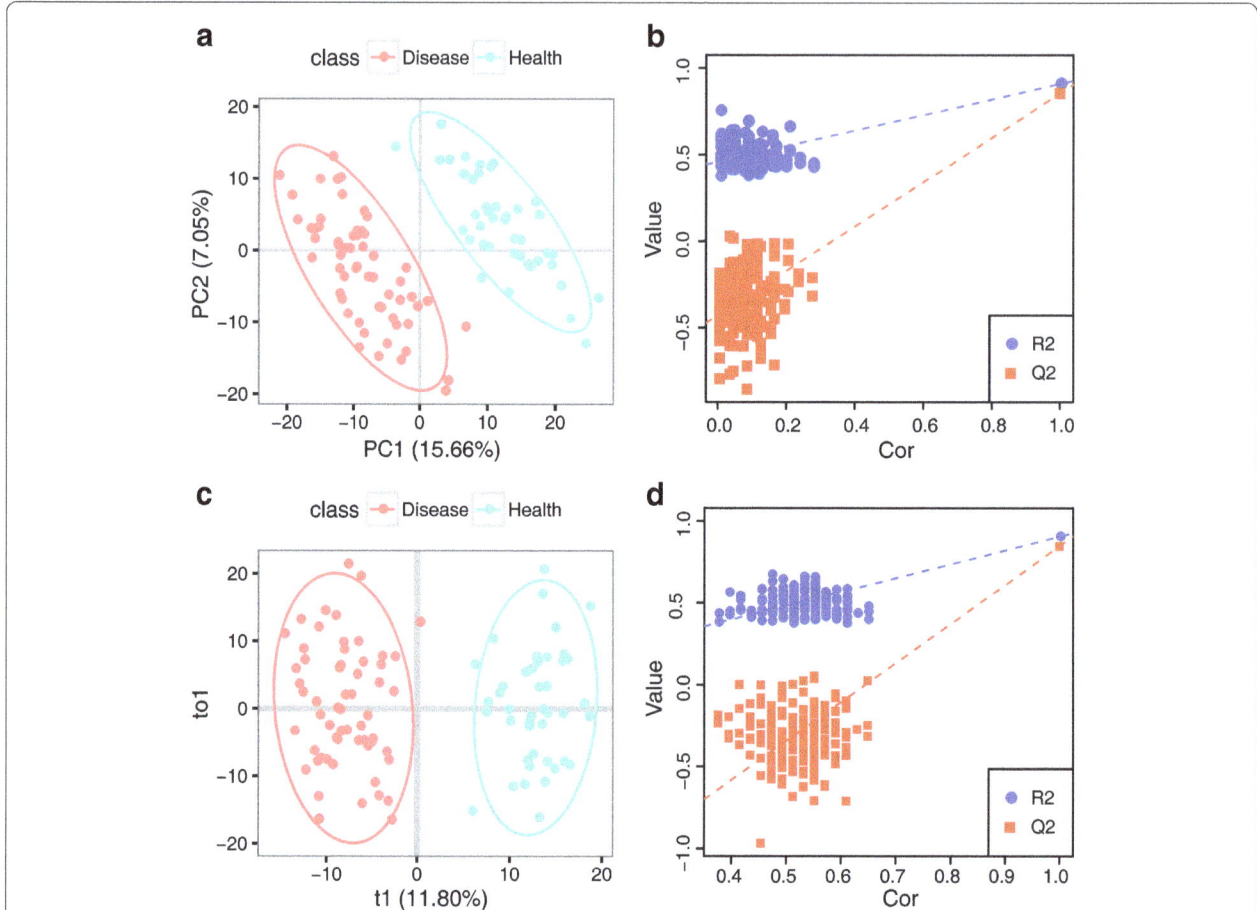

Fig. 7 The score and permutation test plots of PLS-DA and OPLS-DA. **a** Score plot of PLS-DA. R^2Y: 0.908, Q^2Y: 0.854. **b** Permutation test plot of PLS-DA, p-value < = 0.05. **c** Score plot of OPLS-DA. R^2Y: 0.905, Q^2Y: 0.847. **d** Permutation test plot of OPLS-DA, p-value < = 0.05. The different points in the score plots (A and C) refer to different samples, and the samples are color-coded according to their group information. The number of permutations for the permutation test is 200

0.908 and 0.854, respectively, indicate that the model has good goodness of fit and predictive ability. The R^2Y and Q^2Y values of the OPLS-DA model, which are 0.905 and 0.847, respectively, indicate that the model also has good goodness of fit and predictive ability. Overall, the two multivariate data analysis methods, PLS-DA and OPLS-DA, give similar results. To test the validity of the models of PLS-DA and OPLS-DA, a permutation test (n = 200) is performed. As shown in Fig. 7b and d, the test indicated that the two models are valid, and the good predictive ability of the model is not because of over-fitting with a p-value less than 0.05. Taken together, the results of PCA and PLS-DA (or OPLS-DA) show a distinct separation between the disease and health groups.

Table 3 The biomarkers selected by metaX

MZ	RT (min)	Mass	HMDB	Name	Delta (ppm)	Chemical formula
308.0498	10.46	285.0629	HMDB14387	Cladribine	−8.18	C10H12ClN5O3
424.3412	11.94	423.3349	HMDB06469	Linoleyl carnitine	−2.31	C25H45NO4
155.0281	2.81	116.066	HMDB32411	2-Methyl-1-methylthio-2-butene	−8.77	C6H12S
130.0499	3.43	129.0426	HMDB00267	Pyroglutamic acid	0.15	C5H7NO3
174.9913	2.30	NULL	NULL	NULL	NULL	NULL
309.0533	10.47	270.0892	HMDB33940	Vignafuran	3.44	C16H14O4
425.3446	11.94	424.3341	HMDB06327	Alpha-Tocotrienol	7.62	C29H44O2
324.0443	9.33	301.0563	HMDB01062	N-Acetyl-D-Glucosamine 6-Phosphate	−3.86	C8H16NO9P

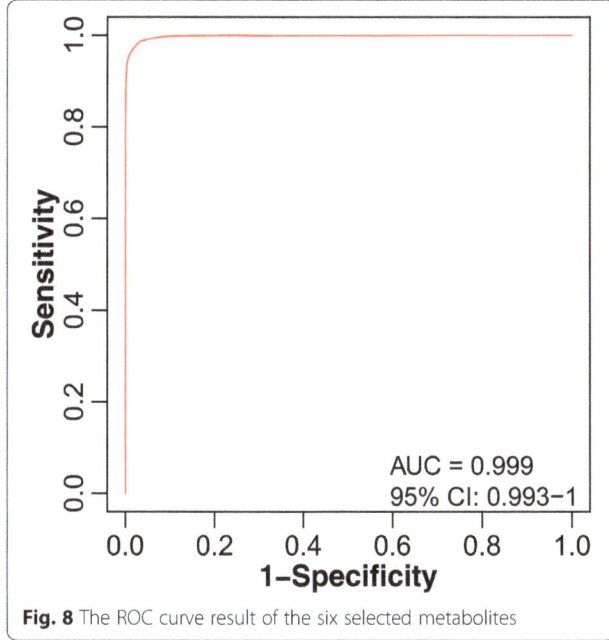

Fig. 8 The ROC curve result of the six selected metabolites

Biomarker analysis, metabolite identification and pathway analysis

To create the classification model between the disease and health groups, the functions implemented in metaX are used to conduct the biomarker selection, model creation and performance evaluation. A recursive feature elimination algorithm with the random forest model is used to select the best feature set. During the treatment, 5-fold cross-validation is used to optimize the model and reduce overfitting. As shown in Table 3, 8 features were selected. To further evaluate the performance of the 8 selected features, the 102 samples were randomly split into two sample sets. One sample set (Disease: 29, Health 29) was for model building and the other (Disease: 14, Health 30) was for testing. Based on the two data sets, the 8 features were used to build a random forest model, and a receiver operating characteristic (ROC) curve of this model was plotted and is shown in Fig. 8. The result indicated that the model based on the 8 features had a good result with an area under the ROC (AUROC) curve of 0.999. The 8 features were then identified based on the HMDB (version 3.6) database

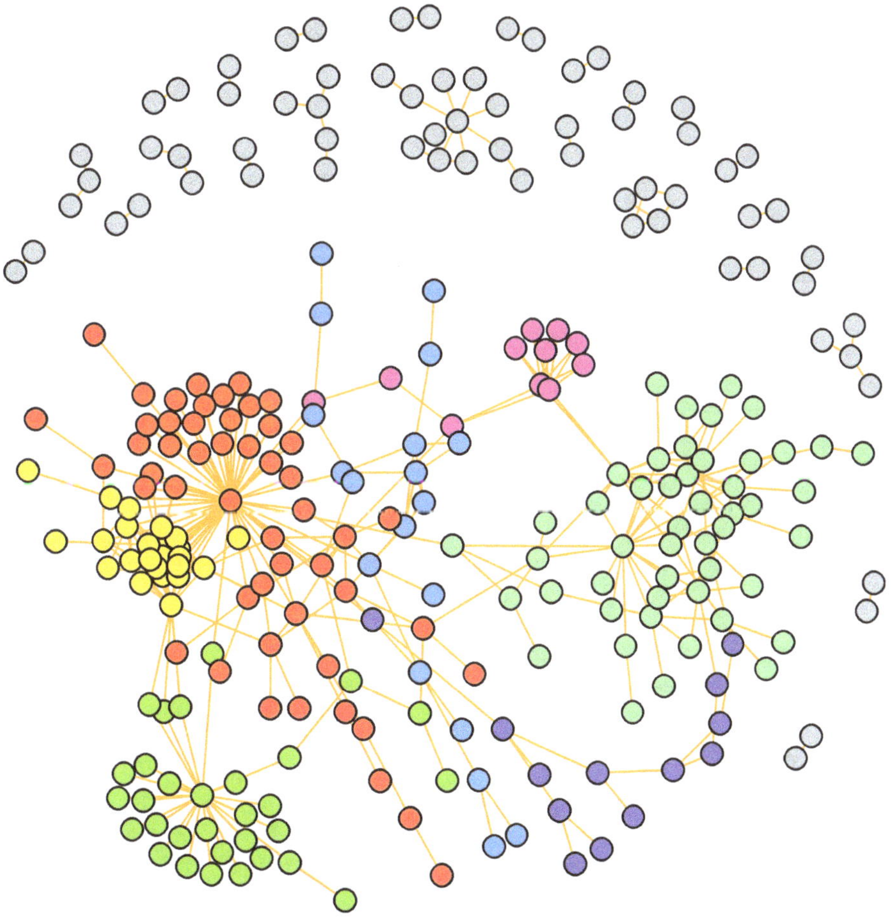

Fig. 9 The differential correction network. The top six largest numbers of nodes communities were color-coded. Detailed information about the samples and their communities are presented in Table S3

through metaX. Seven out of the 8 features were identified with a mass accuracy of < 10 ppm (parts per million). The putative identified metabolites were then submitted to the IMPaLA website (version 9) through metaX to perform the pathway analysis, and the results are presented in Additional file 2: Table S2.

Correlation network analysis

Network-based correlation analysis is a complementary method to the traditional univariate and multivariate statistics that is taken in metabolomics analysis to identify metabolite changes in response to variable status of physiology. All of the features with the normalized intensity described above were used to perform the differential correction network analysis. This analysis can be used to detect the interconnection of metabolite pairs whose relationships are significantly altered due to the disease process. In this study, only the metabolite pairs that had significant differential correlations (q-value < = 0.01) between the disease and health populations were used to build the network. As shown in Fig. 9, of the network with 266 nodes and 444 edges, a giant component (198/266, 74.44%) was found and the community detection analysis using the fast greedy modularity optimization algorithm against this component resulted in seven communities, in which each one has equal to or greater than 10 nodes detected. In addition, metaX can estimate three centrality metrics (degree, closeness and betweenness) for each node, and they reflect the importance of the node in the entire network (Additional file 3: Table S3). Differentially correlation network analysis is expected to provide useful insights into the underlying biological processes of the clinical development of CHD.

Conclusions

metaX presents a complete data processing software that is easy to operate and capable of dealing with large-scale metabolomics datasets. A metaX user can customize the pipeline according to the research requirements. Compared to software for metabolomics datasets that requires high-manual interaction, metaX requires much less manual interaction and can be used in a command line or web-based user-friendly interface. Based upon the fast process and the optimized workflow, therefore, metaX would greatly improve the interpretation of metabolomics data.

Abbreviations
FDR: False discovery rate; GUI: Graphical user interface; HMDB: Human metabolome database; MS/MS: Tandem mass spectrometry; OPLS-DA: Orthogonal partial least squares discriminant analysis; PCA: Principal component analysis; PLS-DA: Partial least square discriminant analysis; QC: Quality control; ROC: Receiver operating characteristic

Acknowledgements
We thank Dr. David Broadhurst for the implementation of QC-RSC method.

Funding
This study was supported in part by the International Science & Technology Cooperation Program of China (2014DFB30020), Chinese National Basic Research Programs (2014CBA02002-A, 2014CBA02005) and the National High-Tech Research and Development Program of China (2012AA020202). The funding body was not involved in the design of the study and collection, analysis, and interpretation of data or in writing the manuscript.

Authors' contributions
BW conceived of and designed the project. BW and CWZ wrote the R package. BW, MZL and CWZ developed the website of metaX. BW, MZL and CWZ performed the data analysis and tested the software. BW and SQL wrote the paper and all authors revised and approved.

Competing interests
The authors declare that they have no competing interests.

References
1. Rochfort S. Metabolomics Reviewed: A New "Omics" Platform Technology for Systems Biology and Implications for Natural Products Research. J Nat Prod. 2005;68(12):1813–20.
2. Gowda GAN, Djukovic D. Overview of Mass Spectrometry-Based Metabolomics. Oppor Challenges. 2014;1198:3–12.
3. Frédérich M, Pirotte B, Fillet M, de Tullio P. Metabolomics as a Challenging Approach for Medicinal Chemistry and Personalized Medicine. J Med Chem. 2016;59(19):8649–66.
4. Smith CA, Want EJ, O'Maille G, Abagyan R, Siuzdak G. XCMS: processing mass spectrometry data for metabolite profiling using nonlinear peak alignment, matching, and identification. Anal Chem. 2006;78(3):779–87.
5. Kuhl C, Tautenhahn R, Bottcher C, Larson TR, Neumann S. CAMERA: an integrated strategy for compound spectra extraction and annotation of liquid chromatography/mass spectrometry data sets. Anal Chem. 2012;84(1):283–9.
6. Fernandez-Albert F, Llorach R, Andres-Lacueva C, Perera A. An R package to analyse LC/MS metabolomic data: MAIT (Metabolite Automatic Identification Toolkit). Bioinformatics. 2014;30(13):1937–9.
7. Xia J, Psychogios N, Young N, Wishart DS. MetaboAnalyst: a web server for metabolomic data analysis and interpretation. Nucleic Acids Res. 2009;37(Web Server issue):W652–60.
8. Giacomoni F, Le Corguille G, Monsoor M, Landi M, Pericard P, Petera M, Duperier C, Tremblay-Franco M, Martin JF, Jacob D, et al. Workflow4Metabolomics: a collaborative research infrastructure for computational metabolomics. Bioinformatics. 2015;31(9):1493–5.
9. Di Guida R, Engel J, Allwood JW, Weber RJ, Jones MR, Sommer U, Viant MR, Dunn WB. Non-targeted UHPLC-MS metabolomic data processing methods: a comparative investigation of normalisation, missing value imputation, transformation and scaling. Metabolomics. 2016;12:93.
10. Engskog MKR, Haglöf J, Arvidsson T, Pettersson C. LC–MS based global metabolite profiling: the necessity of high data quality. Metabolomics. 2016;12(7):1–9.
11. Team RC. R: A Language and Environment for Statistical Computing. Vienna: R Foundation for Statistical Computing; 2016.
12. Ernest B, Gooding JR, Campagna SR, Saxton AM, Voy BH. MetabR: an R script for linear model analysis of quantitative metabolomic data. BMC Res Notes. 2012;5:596.
13. Huber W, Carey VJ, Gentleman R, Anders S, Carlson M, Carvalho BS, Bravo HC, Davis S, Gatto L, Girke T, et al. Orchestrating high-throughput genomic analysis with Bioconductor. Nat Methods. 2015;12(2):115–21.
14. Dunn WB, Broadhurst D, Begley P, Zelena E, Francis-McIntyre S, Anderson N, Brown M, Knowles JD, Halsall A, Haselden JN, et al. Procedures for large-scale metabolic profiling of serum and plasma using gas chromatography and liquid chromatography coupled to mass spectrometry. Nat Protoc. 2011;6(7):1060–83.
15. Gromski PS, Xu Y, Kotze HL, Correa E, Ellis DI, Armitage EG, Turner ML, Goodacre R. Influence of missing values substitutes on multivariate analysis of metabolomics data. Metabolites. 2014;4(2):433–52.
16. van den Berg RA, Hoefsloot HC, Westerhuis JA, Smilde AK, van der Werf MJ. Centering, scaling, and transformations: improving the biological information content of metabolomics data. BMC Genomics. 2006;7:142.

17. Edmands WM, Barupal DK, Scalbert A. MetMSLine: an automated and fully integrated pipeline for rapid processing of high-resolution LC-MS metabolomic datasets. Bioinformatics. 2015;31(5):788–90.

18. Dunn WB, Wilson ID, Nicholls AW, Broadhurst D. The importance of experimental design and QC samples in large-scale and MS-driven untargeted metabolomic studies of humans. Bioanalysis. 2012;4(18):2249–64.

19. Kirwan JA, Broadhurst DI, Davidson RL, Viant MR. Characterising and correcting batch variation in an automated direct infusion mass spectrometry (DIMS) metabolomics workflow. Anal Bioanal Chem. 2013; 405(15):5147–57.

20. Shen X, Gong X, Cai Y, Guo Y, Tu J, Li H, Zhang T, Wang J, Xue F, Zhu Z-J. Normalization and integration of large-scale metabolomics data using support vector regression. Metabolomics. 2016;12(5):1–12.

21. Johnson WE, Li C, Rabinovic A. Adjusting batch effects in microarray expression data using empirical Bayes methods. Biostatistics. 2006;8(1):118–27.

22. Szymanska E, Saccenti E, Smilde AK, Westerhuis JA. Double-check: validation of diagnostic statistics for PLS-DA models in metabolomics studies. Metabolomics. 2012;8 Suppl 1:3–16.

23. Benjamini Y, Hochberg Y. Controlling the false discovery rate: a practical and powerful approach to multiple testing. J R Stat Soc Ser B Methodol. 1995;1:289–300.

24. Mevik B-H, Wehrens R, Liland KH: pls: Partial least squares and principal component regression. R package version 2015;2(5). https://cran.r-project.org/web/packages/pls/index.html

25. Thévenot EA, Roux A, Xu Y, Ezan E, Junot C. Analysis of the Human Adult Urinary Metabolome Variations with Age, Body Mass Index, and Gender by Implementing a Comprehensive Workflow for Univariate and OPLS Statistical Analyses. J Proteome Res. 2015;14(8):3322–35.

26. van Iterson M, t Hoen PA, Pedotti P, Hooiveld GJ, den Dunnen JT, van Ommen GJ, Boer JM, Menezes RX. Relative power and sample size analysis on gene expression profiling data. BMC Genomics. 2009;10:439.

27. Fukushima A. DiffCorr: An R package to analyze and visualize differential correlations in biological networks. Gene. 2013;518(1):209–14.

28. Csardi G, Nepusz T. The igraph software package for complex network research. Inter J Complex Syst. 2006;1695(5):1–9.

29. Shannon P, Markiel A, Ozier O, Baliga NS, Wang JT, Ramage D, Amin N, Schwikowski B, Ideker T. Cytoscape: a software environment for integrated models of biomolecular interaction networks. Genome Res. 2003;13(11):2498–504.

30. Bastian M, Heymann S, Jacomy M. Gephi: an open source software for exploring and manipulating networks. ICWSM. 2009;8:361–2.

31. Wishart DS, Jewison T, Guo AC, Wilson M, Knox C, Liu Y, Djoumbou Y, Mandal R, Aziat F, Dong E, et al. HMDB 3.0-The Human Metabolome Database in 2013. Nucleic Acids Res. 2013;41(Database issue):D801–7.

32. Kanehisa M, Goto S. KEGG: kyoto encyclopedia of genes and genomes. Nucleic Acids Res. 2000;28(1):27–30.

33. Kanehisa M, Goto S, Sato Y, Furumichi M, Tanabe M. KEGG for integration and interpretation of large-scale molecular data sets. Nucleic Acids Res. 2011;40(D1):D109–14.

34. Horai H, Arita M, Kanaya S, Nihei Y, Ikeda T, Suwa K, Ojima Y, Tanaka K, Tanaka S, Aoshima K, et al. MassBank: a public repository for sharing mass spectral data for life sciences. J Mass Spectrom. 2010;45(7):703–14.

35. Wang Y, Xiao J, Suzek TO, Zhang J, Wang J, Bryant SH, Web S. PubChem: a public information system for analyzing bioactivities of small molecules. Nucleic Acids Res. 2009;37(Web Server):W623–33.

36. Sud M, Fahy E, Cotter D, Brown A, Dennis EA, Glass CK, Merrill Jr AH, Murphy RC, Raetz CR, Russell DW, et al. LMSD: LIPID MAPS structure database. Nucleic Acids Res. 2007;35(Database issue):D527–32.

37. Caspi R, Altman T, Billington R, Dreher K, Foerster H, Fulcher CA, Holland TA, Keseler IM, Kothari A, Kubo A, et al. The MetaCyc database of metabolic pathways and enzymes and the BioCyc collection of Pathway/Genome Databases. Nucleic Acids Res. 2014;42(D1):D459–71.

38. Kamburov A, Cavill R, Ebbels TM, Herwig R, Keun HC. Integrated pathway-level analysis of transcriptomics and metabolomics data with IMPaLA. Bioinformatics. 2011;27(20):2917–8.

39. Kuhn M. Building predictive models in R using the caret package. J Stat Softw. 2008;28(5):1–26.

40. Breiman L. Random forests. Mach Learn. 2001;45(1):5–32.

41. Gehlenborg N, Noble MS, Getz G, Chin L, Park PJ. Nozzle: a report generation toolkit for data analysis pipelines. Bioinformatics. 2013;29(8):1089–91.

42. Feng Q, Liu Z, Zhong S, Li R, Xia H, Jie Z, Wen B, Chen X, Yan W, Fan Y, et al. Integrated metabolomics and metagenomics analysis of plasma and urine identified microbial metabolites associated with coronary heart disease. Sci Rep. 2016;6:22525.

43. Feng Q, Liu Z, Zhong S, Li R, Xia H, Jie Z, Wen B, Chen X, Yan W, Fan Y, Dryad Data R, et al. Data from: Integrated metabolomics and metagenomics analysis of plasma and urine identified microbial metabolites associated with coronary heart disease, Dryad Data Repository. 2016.

44. Kessner D, Chambers M, Burke R, Agus D, Mallick P. ProteoWizard: open source software for rapid proteomics tools development. Bioinformatics. 2008;24(21):2534–6.

45. Chawade A, Alexandersson E, Levander F. Normalyzer: a tool for rapid evaluation of normalization methods for omics data sets. J Proteome Res. 2014;13(6):3114–20.

17

WebGIVI: a web-based gene enrichment analysis and visualization tool

Liang Sun[1,6], Yongnan Zhu[2,3], A. S. M. Ashique Mahmood[4], Catalina O. Tudor[4], Jia Ren[5], K. Vijay-Shanker[4], Jian Chen[2] and Carl J. Schmidt[1*]

Abstract

Background: A major challenge of high throughput transcriptome studies is presenting the data to researchers in an interpretable format. In many cases, the outputs of such studies are gene lists which are then examined for enriched biological concepts. One approach to help the researcher interpret large gene datasets is to associate genes and informative terms (iTerm) that are obtained from the biomedical literature using the eGIFT text-mining system. However, examining large lists of iTerm and gene pairs is a daunting task.

Results: We have developed WebGIVI, an interactive web-based visualization tool (http://raven.anr.udel.edu/webgivi/) to explore gene:iTerm pairs. WebGIVI was built via Cytoscape and Data Driven Document JavaScript libraries and can be used to relate genes to iTerms and then visualize gene and iTerm pairs. WebGIVI can accept a gene list that is used to retrieve the gene symbols and corresponding iTerm list. This list can be submitted to visualize the gene iTerm pairs using two distinct methods: a Concept Map or a Cytoscape Network Map. In addition, WebGIVI also supports uploading and visualization of any two-column tab separated data.

Conclusions: WebGIVI provides an interactive and integrated network graph of gene and iTerms that allows filtering, sorting, and grouping, which can aid biologists in developing hypothesis based on the input gene lists. In addition, WebGIVI can visualize hundreds of nodes and generate a high-resolution image that is important for most of research publications. The source code can be freely downloaded at https://github.com/sunliang3361/WebGIVI. The WebGIVI tutorial is available at http://raven.anr.udel.edu/webgivi/tutorial.php.

Keywords: Visualization, eGIFT, Gene iTerm, Gene enrichment, Web development

Background

High-throughput technologies provide biologists with large lists of genes or proteins when they compare expression data between two biological states (e.g., normal tissue vs. cancer tissue). Grouping enriched genes to known biological processes and pathways is a common strategy for understanding the biology that underlies the differences between the two states. Approaches include GO enrichment analysis such as DAVID [1, 2], GOEAST [3] and Gorilla [4], and pathway analysis such as KEGG [5] and Reactome [6].

eGIFT

eGIFT [7] uses a text-mining method to identify informative terms (iTerms) for individual genes. iTerms are not limited to gene ontology (GO) terms; they also capture more detailed biological knowledge. Consequently, eGIFT provides a finer grained interpretation of gene lists than GO analysis. The current gene analysis results of eGIFT provide users with a list of ranked iTerms and their associated genes in a tabular format. A graphic representation of these gene and iTerm relations would allow biologists to better interpret their input gene lists or gene-iTerm pair lists. This often captures the biological concept enriched in the input data.

Visualization tool

An effective visualization of large data sets can provide biologists with means to discover buried relationships in complex data sets. Currently, several different visualization

---

* Correspondence: schmidtc@udel.edu
[1]Department of Animal and Food Sciences, University of Delaware, Newark, DE, USA
Full list of author information is available at the end of the article

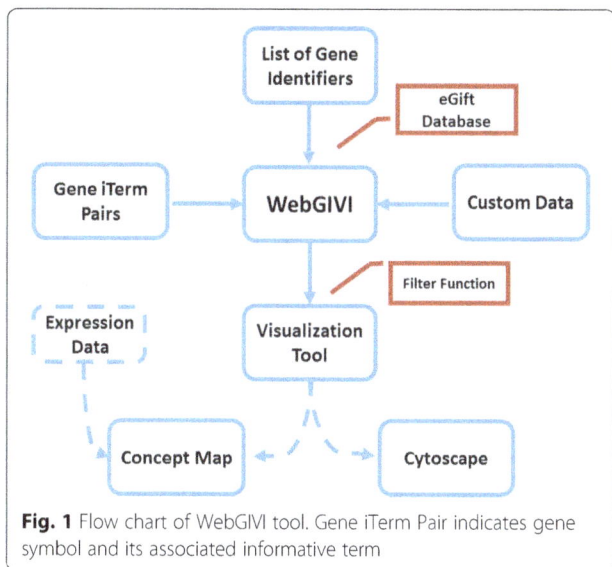

Fig. 1 Flow chart of WebGIVI tool. Gene iTerm Pair indicates gene symbol and its associated informative term

ability to visualize relationships without the overhead of having access to system administrators. Cytoscape.js [16] and D3.js [17] are the most popular visualization JavaScript libraries that can be applied to visualize network and biomolecule interactions. There are several successful web based visualization tools that use the Cytoscape and D3 JavaScript libraries in a biological context. For example, BNVC [18], a web-based visualization tool of biomolecular networks, can be used to compare two similar networks. PINV [19], also a web-based protein-protein interaction network visualizer, provides complex interaction networks with the ability to query, filter and group data of interest. However, complex graphs rendered by such tools are difficult to interpret when analyzing data with hundreds of nodes and edges. WebGIVI addresses the issue of visualizing large data sets by adopting a concept map method for visualization. The concept map aligns nodes on different layers and automatically calculates the distance between layers. This maximizes the amount of information that can be displayed despite limited physical size of the screen. In addition, a zoom-in feature of Concept Map allows the user to scale the graph for a better view. Now, a user can analyze hundreds of nodes in a high-resolution image produced by WebGIVI that can be saved and is suitable for publication.

Interactive WebGIVI provides an integrated graph to help users generate biological hypotheses. A database of rate-limiting genes, identified by text mining and manually

tools are used to capture the relationship between genes, protein and networks, such as Arena 3D [8], Medusa [9], Ondex [10], Osprey [11], Pajek [12], BioLayout Express^{3D} [13], Cytoscape [14] and ProViz [15]. However, most of these tools need to be installed on a local computer and require plugins or third party software such as Java Runtime Environment. Installation and maintenance of such tools can be difficult for those unfamiliar with computer system administration. In contrast, web-based tools offer the

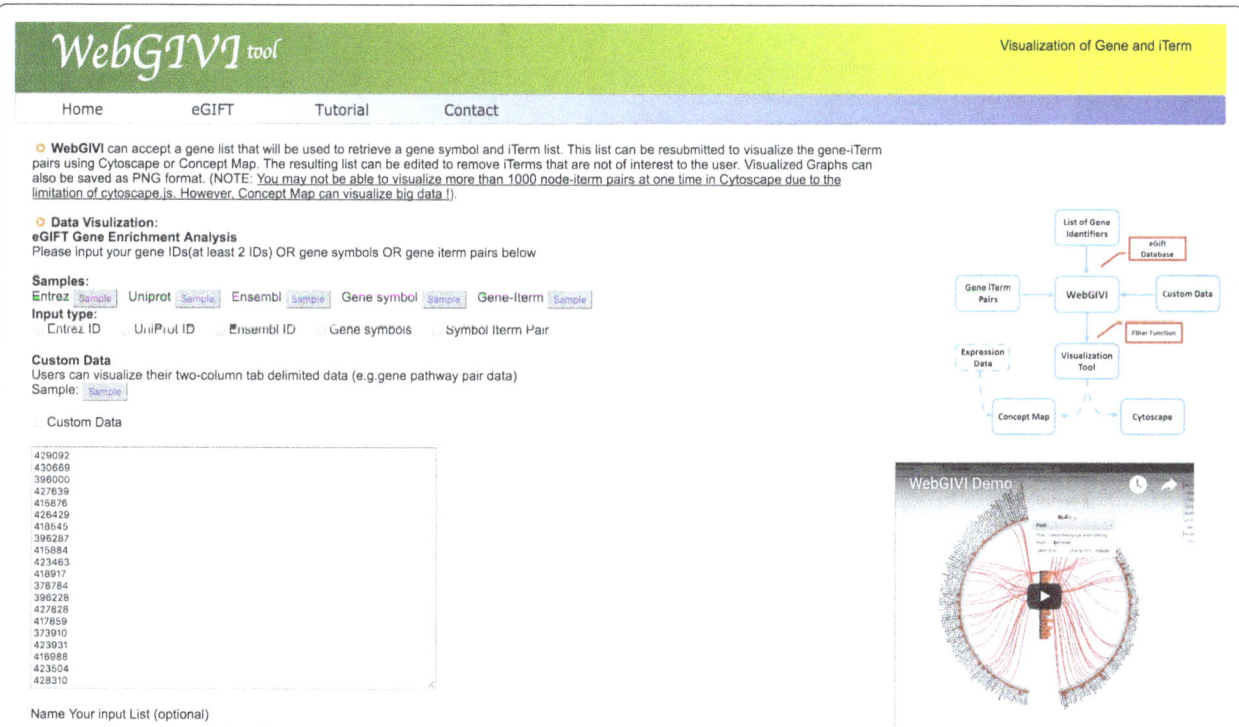

Fig. 2 Submission interface for WebGIVI. Gene lists can be input in several different formats and named in the text field above the submit button. To the right of the page is a short video demonstrating various WebGIVI functionalities

verified, was also integrated into WebGIVI. All genes encoding rate-limiting products are colored in the visualization. Expression data that is uploaded to WebGIVI can also be color-coded according to expression levels in two different biological states. These features are accomplished by using PHP, Cytoscape.js and D3.js to generate a powerful and interactive web based visualization tool that implements gene enrichment analysis and gene and iTerms visualization.

Implementation

The flow chart of the data processing is depicted in Fig. 1. Currently, WebGIVI accepts multiple input data formats including: NCBI Entrez gene ID, UniProt [20], or Ensemble [21] Gene IDs. Prior to accessing WebGIVI, the user will have identified genes of interest in their biological system. The WebGIVI input list is used to retrieve the gene symbols and iTerms through an eGIFT API. In the retrieved list of gene symbol-iTerm pairs, the first column contains the gene symbols, and the second column contains the gene-associated iTerms. Interactive WebGIVI also supports uploading and visualization of any two-column tab separated data. For

instance, gene symbols and related pathways data can be visualized.

WebGIVI has three major functions:

1. *Visualization.* Interactive table visualization includes sorting and deleting functions that help users reduce the graph to only iTerms and genes of interest. GUI functions in the Concept Map visualization include switching, filtering, sorting, searching, grouping and saving the data in tab delimited text files or as high-resolution images.
2. *Integration of Rate-limiting and Expression Data.* Rate limiting gene products determine the flow rate of metabolites or signal through pathways. We have identified 150 rate limiting gene products via text-mining of primary literature and biochemistry textbooks. This list was manually verified (CJS), and integrated into WebGIVI.
   When a user enters an expression data list, genes that encode a rate limiting gene product are colored when visualizing gene:iTerm pair in the concept map. If such genes are differentially regulated they are likely have a significant impact on the overall rate of metabolite or signal flow through their

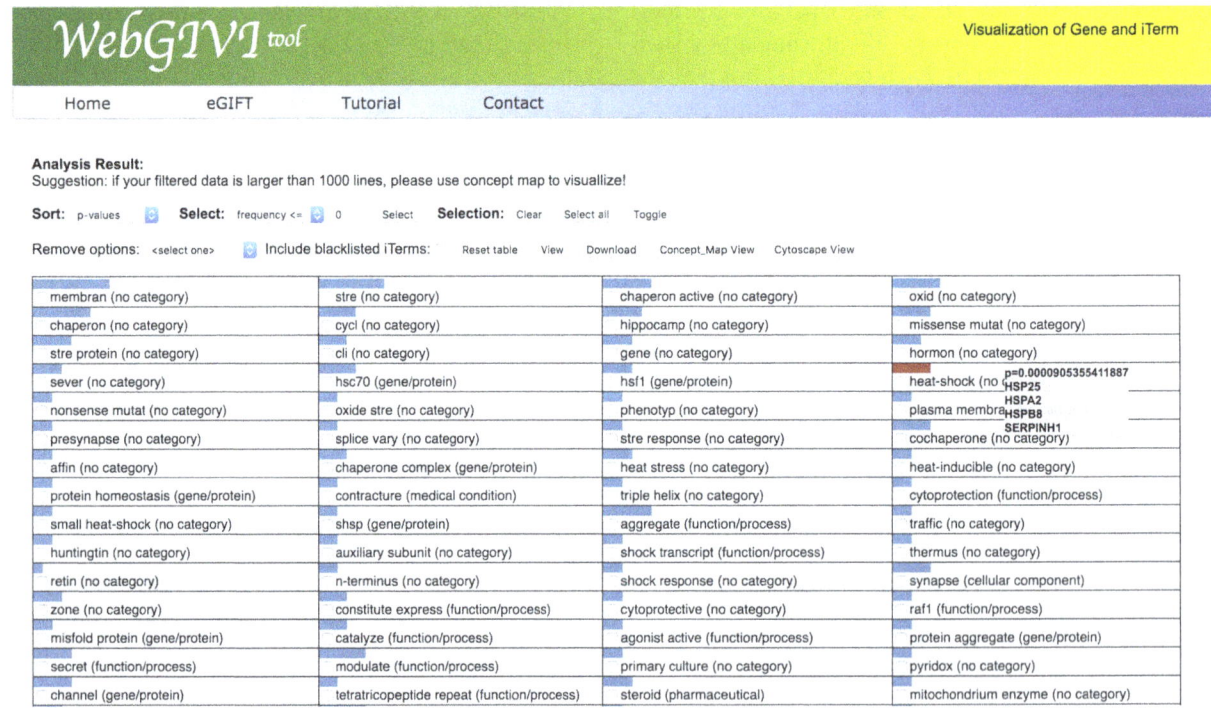

Fig. 3 Results pages containing iTerms enriched in the input gene list. In this case the list was sorted by *p*-values as determined by Fisher's exact test. One can use the Frequency text box to eliminate iTerms that have fewer genes associated with them based on the user's cutoff. The Selection buttons allow one to "Select All", Clear the selections or Toggle selected iTerms. Also, individual iTerms can be selected by clicking. The selection features are used to remove irrelevant iTerms based on the users knowledge. One can include blacklisted iTerms (see text) by selecting the check box. The Reset Table button can be used to recreate the starting table should errors be made while deleting iTerms thought to be irrelevant. The Concept Map View and Cytoscape View open new windows with the corresponding representations of the iTerm:gene pairs. Hovering over an iTerm box will display the genes associated with that iTerm from the input list along with the *p*-value of that iTerm

respective pathways. Furthermore, to emphasize certain features of their data, users can upload expression files that highlight nodes that they consider informative.

3. *Link-out Functions.* Nodes and edges provide link outs to gene specific, iTerms or gene:iTerm pairs via the NCBI, UniPROT, or eGIFT database. For example, gene-iTerm pairs link to eGIFT sentence web page, which contain sentences with the gene and iTerm highlighted in the text.

Results and discussion
Data filtering functions
Sorting Functions: eGIFT uses a precomputed text-mining database that has extracted all gene associated informative terms (iTerm) from PubMed abstracts. After submitting a gene list to WebGIVI, a table is returned to the user containing the iTerms associated with the input genes. By hovering over an iTerm, the user can see the genes associated with that iTerm. The default list is sorted based on the Fisher's exact test p-value, but the user can choose to sort based on alphabetical order, the gene ontology group (process, function, compartment or unclassified) to which the iTerm has been classified or the frequency of appearance of each iTerm.

Editing Functions: Not all iTerms are informative in all use cases, but could be important to others. For example, 'in situ hybridization' is an irrelevant iTerm to our use case scenario but will be interesting to researchers who might want to apply this experimental method to their own work. However, some iTerms are highly likely to be non-informative. To remove such iTerms a "blacklist" has been developed that includes terms such as "some cell" or "10 fold" that are typically non-informative to the general WebGIVI user. Since the developers of WebGIVI cannot be certain that a given iTerm is irrelevant to all users, the returned iTerm list includes the blacklisted terms; a checkbox is provided that allows the user to hide any terms that are included in the blacklist. It is also beneficial to the user to also be

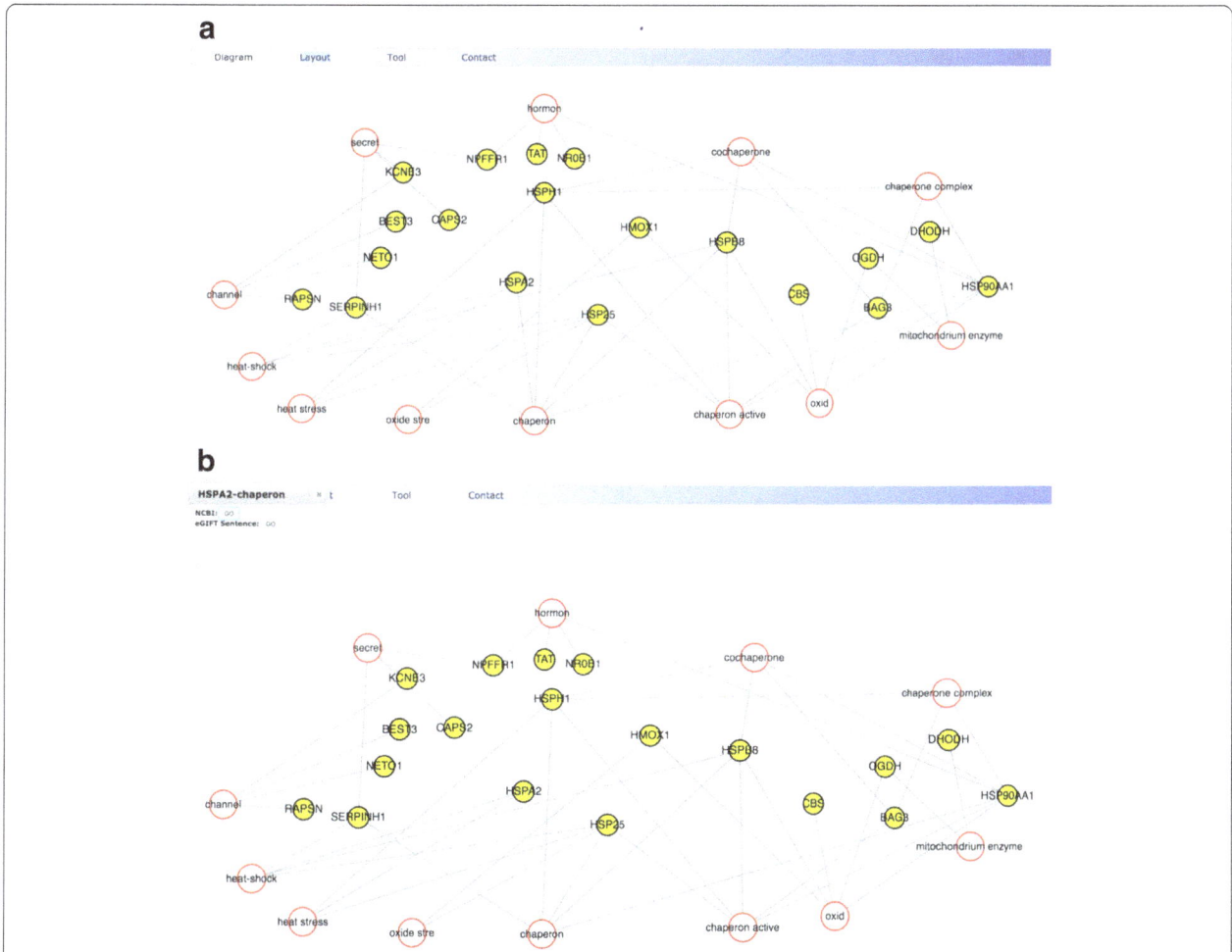

Fig. 4 a Cytoscape View of iTerm:gene pairs. **b** *Right clicking* on an edge connecting two nodes will activate a pop up window allowing the user to connect to either NCBI or eGIFT sentences (see text)

able to filter out irrelevant iTerms in the context of their study, and only save iTerms of direct relevance. Once data is submitted on WebGIVI's homepage, the returned list will allow the user to delete iTerms from the results table using deleting functions. If the user prefers, they can choose not to prefilter but visualize data in Concept Map or Cytoscape directly.

A biological use case scenario

We used the Sun et al. white-leghorn hepatocellular (LMH) cell heat stress dataset [22], which is a RNA-Seq study of LMH cells under heat stress. This study identified a total of 235 up-regulated and 578 down-regulated genes. Figure 2 shows a completed WebGIVI submission

page with a portion of the regulated genes from the LMH study (Additional file 1). In this case we used Entrez gene identification numbers. Following submission, an iTerm list (Fig. 3) is returned that can be sorted alphabetically, by frequency, Gene Ontology categories, or by p-value as determined by the Fisher's exact test. In this case the list is sorted by p-value. Hovering over an iTerm will show the corresponding p-value, along with the genes from the list associated with that iTerm. One can choose to display iTerms that have been blacklisted by checking the "Include blacklisted items". You can also select irrelevant iTerms by right clicking and delete them using the remove options. Users can view the output in either Cytoscape (Fig. 4) or as a Concept Map

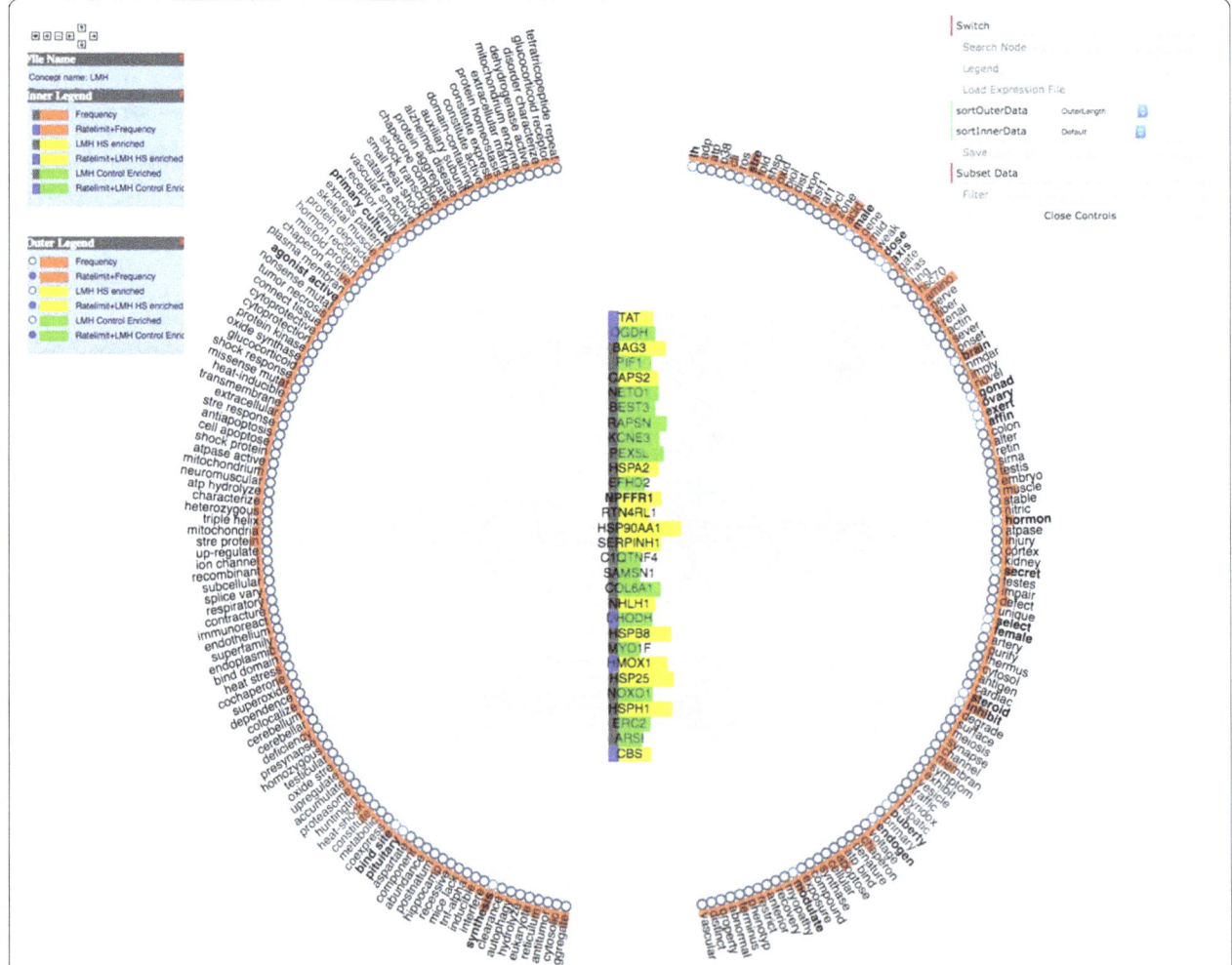

Fig. 5 Concept Map view of iTerm:gene pairs. Genes from the input list with iTerms are in the vertical column at the center and iTerms are around the outside. The panel to the left is the legend explaining the graph, while the panel to the *right* allows manipulation of the output. This allows the user to switch the positions of the genes and iTerms, Search for specific genes or iTerms and to toggle the legend window on or off. The Load Expression File allows the user to load a file that indicates if a gene was up or down regulated in the experiment. *Yellow* corresponds to up while *green* to down regulated genes. The Save button allows the user to save either the graph text as a tab delimited two-column files or save the image. The Filter button allows the user to choose the minimum number of edges that a node must have to be visualized. See Fig. 6 for explanation of Subset Data

(Fig. 5) by selecting the appropriate buttons. The default mode in Cytoscape generates a force graph (Fig. 4a) and clicking on an edge connecting a gene product to an iTerm pops up a window that allows the user to connect to either NCBI or eGIFT (Fig. 4b). Additional view modes include tree or circle that are accessible by the Layout button.

While Cytoscape can be useful for small graphs, the Concept map view is generally easier to use for larger input data sets. In the default view, gene symbols are displayed in a column at the center of the view, while iTerms are displayed as a wheel around the gene symbols (Fig. 5). If necessary, additional layers are automatically added to display more gene:iTerm relations. Several attributes are visible in the Concept Map view. In this case we have uploaded a file indicating how the gene was regulated by heat stress using the Load Expression File button. Genes highlighted in yellow are enriched under heat stress while those in green are

enriched in the control (thermoneutral) samples. In addition, genes encoding rate limiting gene products are indicated by a blue rectangle added to the symbol.

In the concept map view users can right click to select either genes or iTerms. Selecting one will create a gene:iTerm edge (Fig. 6) then clicking on the Subset Data button will create a new concept map with just the selected gene and iTerms (Fig. 7). This is useful to allow an investigator to link genes with similar iTerms for subsequent investigation. At any point, right clicking on an active edge will open a window that can be used to connect to PubMed, UniProt or eGIFT (Fig. 8). Linking out to NCBI database will search that database with the gene and iTerm, and retrieve links to abstracts that contain those two search terms. For example when searching for a gene:iTerm pair such as HSP90AA1 and the iTerm "chaperone" the search will be in the syntax "HSP90AA1 AND chaperone" and the results will

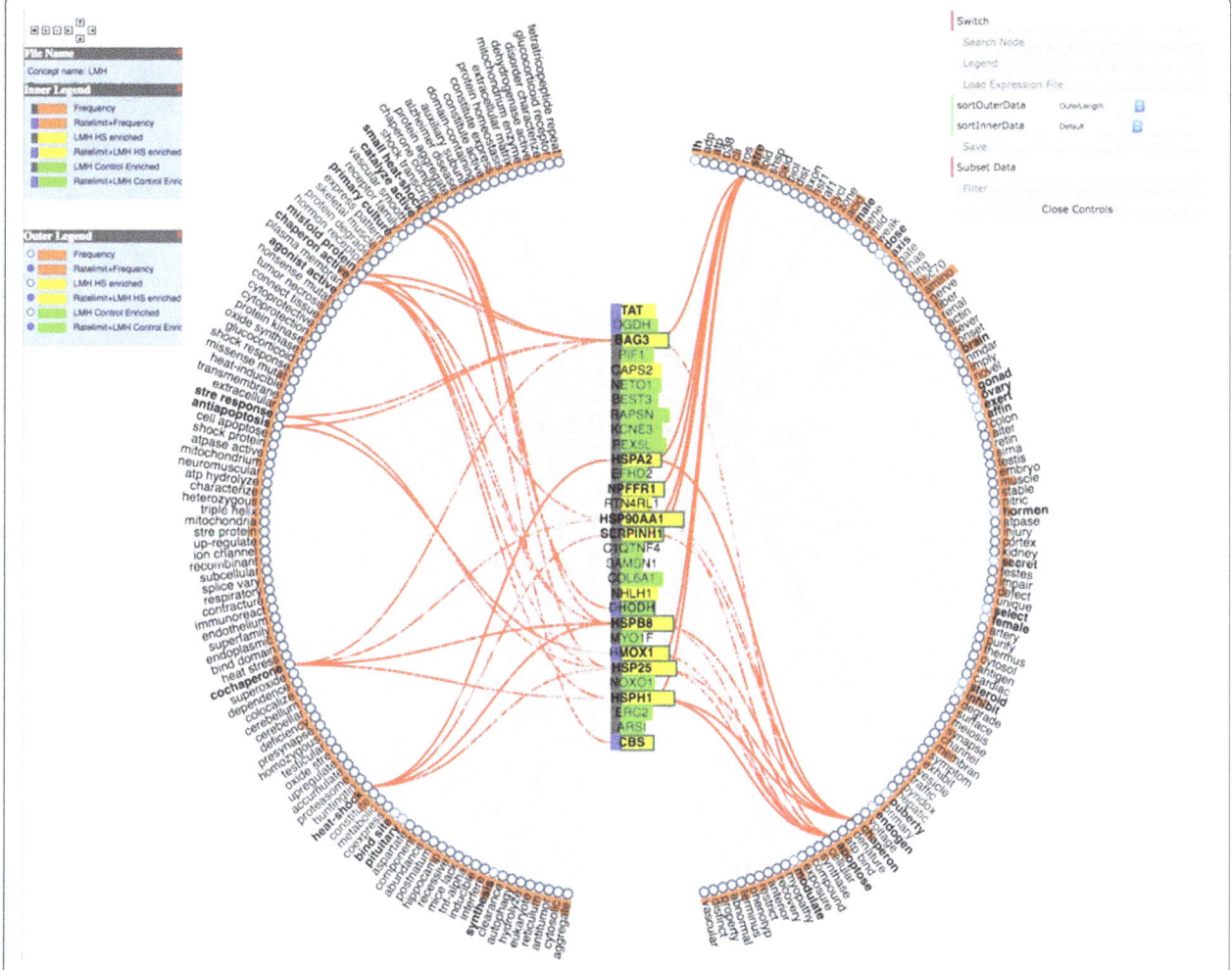

Fig. 6 Clicking on either iTerm or gene nodes will activate connecting edges. Pushing the Subset Data button will create a new graph (Fig. 7) containing only those selected nodes

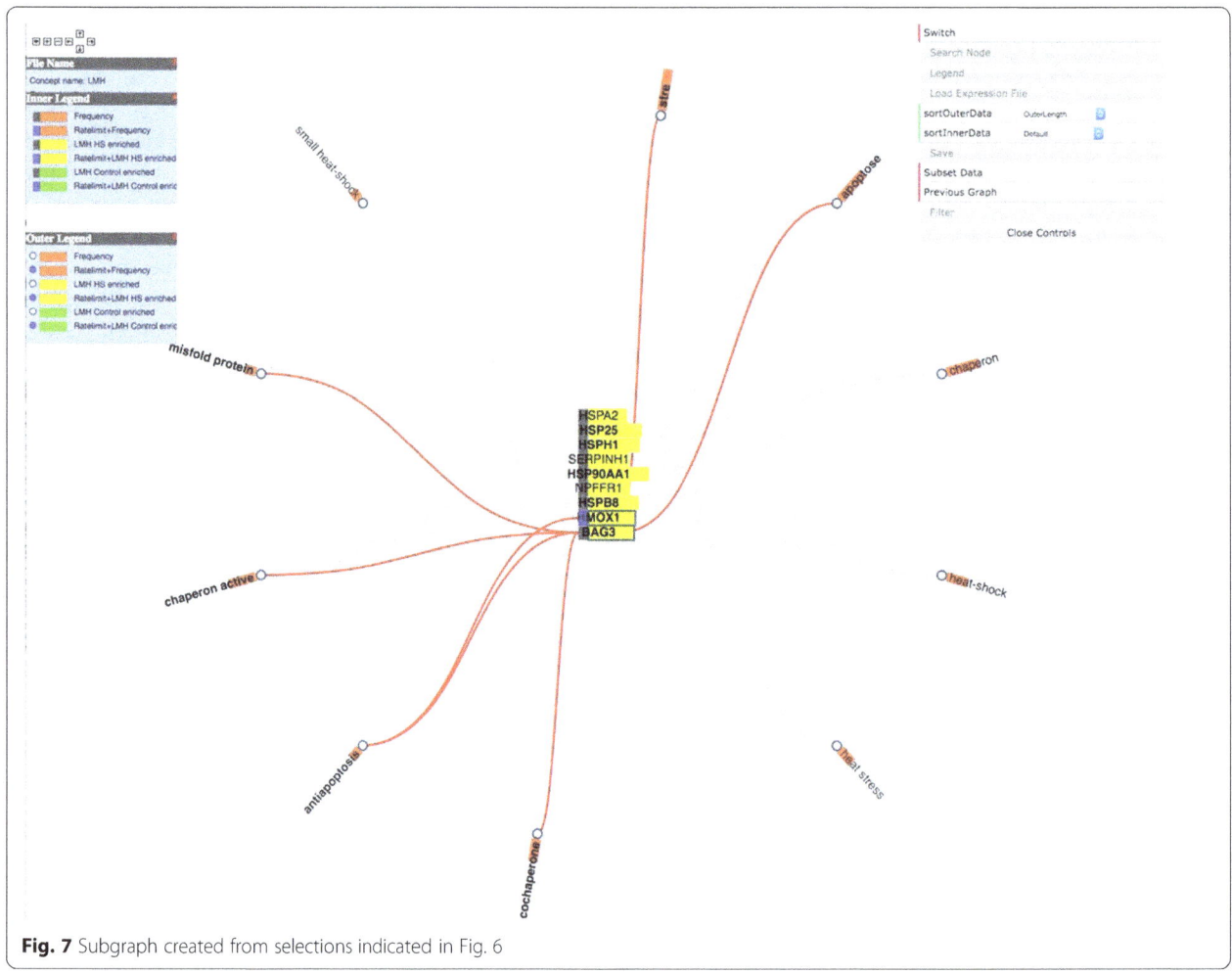

Fig. 7 Subgraph created from selections indicated in Fig. 6

include all abstracts that include both the HSP90AA1 and chaperone. Linking to UniProt will access a search page allowing the user to view the UniProt entry for the gene product. Linking to eGIFT web pages provides users with the sentences extracted from the literature that contain the gene and iTerm pair (Fig. 9). This feature of WebGIVI greatly facilitates the user's understanding of the gene product's function. In addition, the PMID under a sentence links to the PubMed abstract pages from which the sentence was extracted. This further aids in placing the gene product in biological context. The ability to provide users with these sentences is a unique facet of WebGIVI functionality.

Interactive WebGIVI also provides a way to save the current view: users can export all data on the Concept Map as a two-column tab separated file, which can be resubmitted in WebGIVI to obtain the same graph. This feature allows a user to readily share the data with collaborators. Users can also export the graph as a scalable vector graphics (SVG) format, which can be transformed to high resolution image types via readily accessible image conversion web sites.

Comparison of WebGIVI, DAVID and AmiGO2 analysis

Incorporating the text-mining tool, WebGIVI, into the analysis of high-throughput transcriptome experiments complements Functional Annotation clustering provided by DAVID and GO analysis provided by AmiGO2 [23]. DAVID is an online knowledgebase that can output a list of enriched biological concepts from an input list of gene identifiers. The site makes use of multiple resources including the Gene Ontology, the KEGG pathway database, Interpro [24], along with several others. Amigo2 is a product of the Gene Ontology consortium and allows users to submit gene lists and identify enriched GO terms. We chose to compare WebGIVI with DAVID and Amigo2 because these resources are easy to use and have been widely adopted by the scientific community.

Heat stress has been implicated in affecting cell cycle regulatory processes including DNA synthesis, DNA repair, cell cycle checkpoints, cell proliferation, and spindle formation. The objective was to compare the ability of DAVID, AmiGO2 and WebGIVI to identify genes affecting cell cycle regulation that are up-regulated by heat

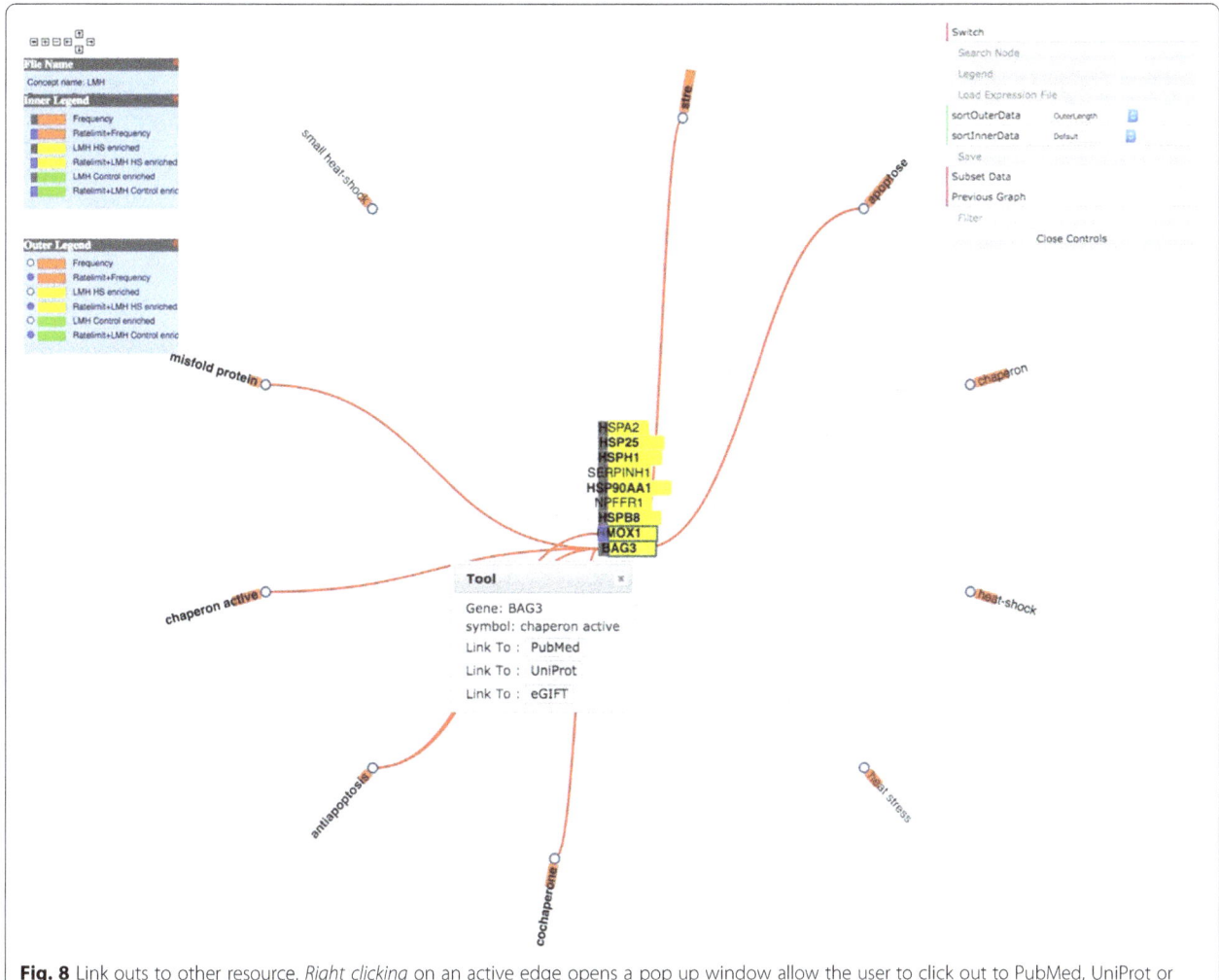

Fig. 8 Link outs to other resource. *Right clicking* on an active edge opens a pop up window allow the user to click out to PubMed, UniProt or eGIFT resources

stress in the liver of chickens (Jastrebski et al. manuscript submitted). Genes whose expression was increased by heat stress were identified and submitted to Amigo2, DAVID, and WebGIVI for comparative analysis. Genes recognized by each analytical method as associated with the concept of cell cycle regulation (including cell cycle, DNA replication, DNA repair, checkpoint) were considered in this comparison of methods (see Additional file 2 for complete list of identified genes).

In combination DAVID, AMIGO2 and WebGIVI identified a total of 214 genes affecting cell cycle regulation as enriched by heat stress in the liver. WebGIVI identified the largest percentage of total genes (80%) and uniquely identified the greatest percentage of genes (30%) not recognized by either DAVID or AMIGO2 (Table 1). However, WebGIVI missed 7 genes that were captured by either DAVID or Amigo2 (Fig. 10). Taken together, this analysis indicates that multiple approaches are best for categorizing the biology embodied in gene

lists, and that WebGIVI provides an important contribution to these analyses.

Future work
WebGIVI can be used to visualize not only eGIFT data, but also any two-column relationship data. For example, microRNA and target genes, kinase and substrate pairs, and protein-protein interactions could all be visualized with WebGIVI. Our tool can also be used as an extension for other tools. Compared to other visualization tools, our web-based gene iTerm visualization tool is highly customizable. Users can easily upload their own data and edit their data in the graph. No pre-installed or third party software such as Java Runtime Environment is required to visualize users' data. While we are developing a WebGIVI blacklist of iTerms we believe are not informative, users still need to examine all iTerms and remove ones they find uninformative manually. This inability of WebGIVI to learn as user's preferences and

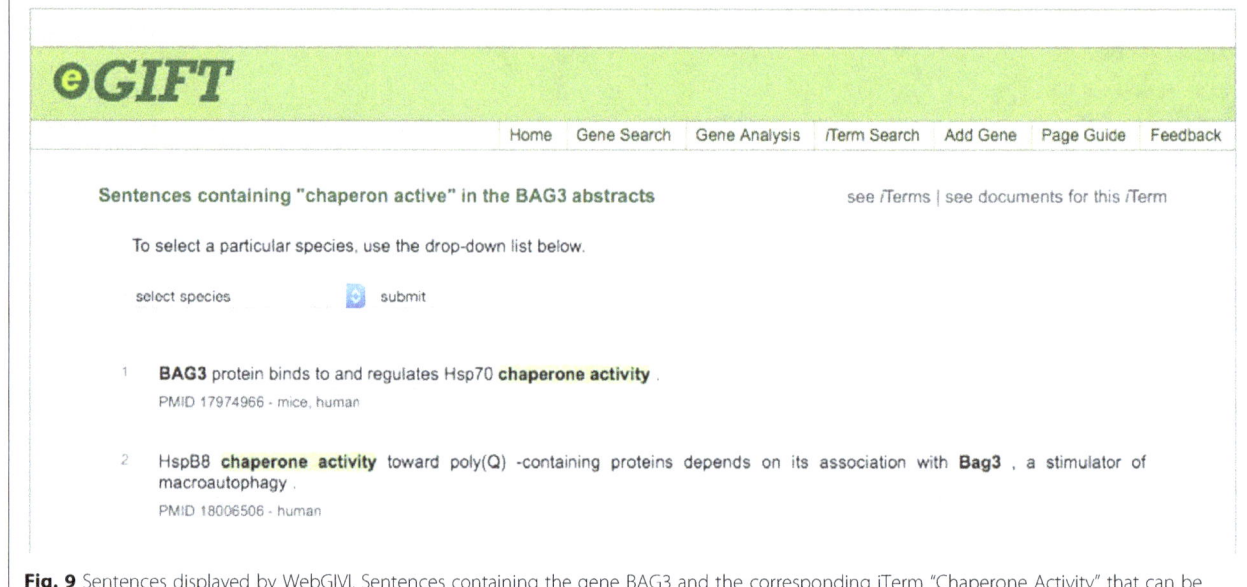

Fig. 9 Sentences displayed by WebGIVI. Sentences containing the gene BAG3 and the corresponding iTerm "Chaperone Activity" that can be accessed by clicking on the appropriate edge (see Fig. 9)

automatically remove iTerm provides room for future improvement. Potentially, machine-learning techniques could be applied to remove such iTerms. Future work will also integrate more bioinformatics databases such as protein kinase database and transcriptional factor database to enable the user to discover more interesting biological relationships and improve the usability of WebGIVI.

Conclusions

Interactive WebGIVI tool provides an integrated visualization and gene enrichment analysis tool. It helps biologists to visualize genes and iTerms online, makes sense of their biological data, and is useful to generating biological hypotheses from high throughput data.

Availability and requirements

Project name: WebGIVI: Web-based Gene and Iterm Visualization Tool

Table 1 Comparative analysis of WebGIVI, DAVID and AmiGO2

Analysis tool	Percent of total	Percent unique
WebGIVI	80%	30%
DAVID	42%	6%
AmiGO2	50%	11%

The three tools identified a total of 214 genes associated with the biological concept of cell cycle regulation" (see text). The numbers indicate the percentage of the 214 genes identified by the different tools along with the number of genes uniquely identified by the corresponding tools. In this analysis, WebGIVI identified 64 (30%) genes as associated with cell cycle regulation that were not associated with this concept by either DAVID or AmiGO2 (see Fig. 10)

Project home page: http://raven.anr.udel.edu/webgivi/
Source code: https://github.com/sunliang3361/WebGIVI
Operation system(s): Web based, Platform independent
Programming language: HTML, CSS, JavaScript, PHP
Other requirements: Modern Browser
License: BSD License
Any restrictions to use by non-academics: None.

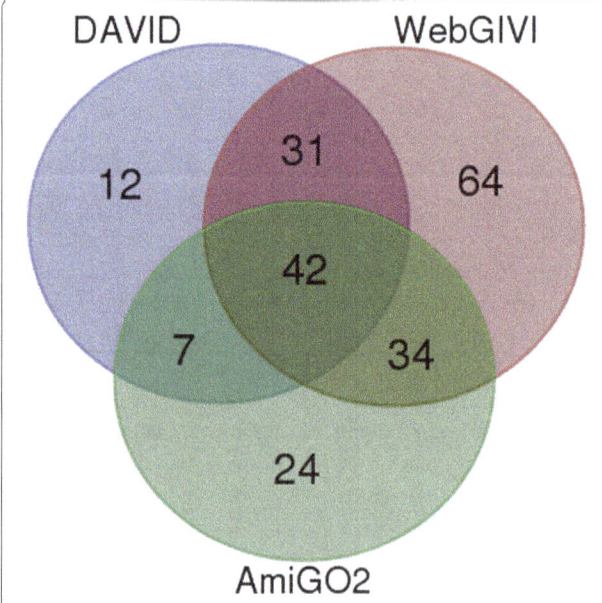

Fig. 10 Venn diagram depicting the number of genes associated with the biological concept of "cell cycle regulation" (see text) by DAVID, AmiGO2 and WebGIVI

Additional files

Additional file 1: NCBI Gene Entrez ID List for the Case Scenario. This file contains NCBI gene Entrez ID list. The first column is Entrez ID, and the second column is the gene symbol.

Additional file 2: NCBI Entrez IDs for Comparison of WebGIVI, DAVID and AmiGO2. This file contains NCBI gene Entrez ID list which is identifed by WebGIVI, DAVID and AmiGO2 to be associated with the concept of cell cycle regulation.

Abbreviations

GO: Gene ontology; iTerm: Informative term; LMH: White-leghorn hepatocellular cell; SVG: Scalable vector graphics

Acknowledgments

LS thanks Allen Hubbard for his help in editing this manuscript and Andrew Bancroft for his help in creating the tutorial video.

Funding

This work was supported by NSF grants NSF DBI-1147029 to CJS and NSF DBI-1260795 to JC.

Authors' contributions

LS, CJS, JC and YZ designed the software; LS, YZ and AM wrote the codes; COT, AM and KVS provided eGIFT data. JR helped design the Cytoscape network map. All authors helped revise the paper and implement the software. LS and CJS wrote the paper. All authors have read and approve of the final version of the manuscript.

Competing interests

The authors declare that they have no competing interests.

Author details

[1]Department of Animal and Food Sciences, University of Delaware, Newark, DE, USA. [2]Department of Computer Science and Electrical Engineering, University of Maryland Baltimore County, 1000 Hilltop Circle, Baltimore, MD, USA. [3]Department of Computer Science, Hangzhou Dianzi University, Hangzhou 310018, Zhejiang Province, People's Republic of China. [4]Department of Computer and Information Sciences, University of Delaware, Newark, DE 19716, USA. [5]Center for Bioinformatics and Computational Biology, University of Delaware, Newark, DE 19711, USA. [6]Current address: Computing Service, The Samuel Roberts Noble Foundation, Ardmore, OK 73401, USA.

References

1. Huang DW, Lempicki RA, Sherman BT. Systematic and integrative analysis of large gene lists using DAVID bioinformatics resources. Nat Protoc. 2009;4:44–57
2. Huang DW, Sherman BT, Lempicki RA. Bioinformatics enrichment tools: Paths toward the comprehensive functional analysis of large gene lists. Nucleic Acids Res. 2009;37:1–13.
3. Zheng Q, Wang XJ. GOEAST: a web-based software toolkit for Gene Ontology enrichment analysis. Nucleic Acids Res. 2008;36:358–63.
4. Eden E, Navon R, Steinfeld I, Lipson D, Yakhini Z. GOrilla: a tool for discovery and visualization of enriched GO terms in ranked gene lists. BMC Bioinformatics. 2009;10:48.
5. Ogata H, Goto S, Sato K, Fujibuchi W, Bono H, Kanehisa M. KEGG: Kyoto encyclopedia of genes and genomes. Nucleic Acids Res. 1999;27:29–34.
6. Croft D, O'Kelly G, Wu G, Haw R, Gillespie M, Matthews L, et al. Reactome: a database of reactions, pathways and biological processes. Nucleic Acids Res. 2011;39:691–7.
7. Tudor CO, Schmidt CJ, Vijay-Shanker K. eGIFT: mining gene information from the literature. BMC Bioinformatics. 2010;11:418.
8. Pavlopoulos GA, O'Donoghue SI, Satagopam VP, Soldatos TG, Pafilis E, Schneider R. Arena3D: visualization of biological networks in 3D. BMC Syst Biol. 2008;2:104.
9. Hooper SD, Bork P. Medusa: a simple tool for interaction graph analysis. Bioinformatics. 2005;21:4432–3.
10. Köhler J, Baumbach J, Taubert J, Specht M, Skusa A, Rüegg A, et al. Graph-based analysis and visualization of experimental results with ONDEX. Bioinformatics. 2006;22:1383–90.
11. Breitkreutz B, Stark C, Tyers M. Osprey: a network visualization system. Genome Biol. 2003;4:1–4.
12. Batagelj V, Mrvar A. Pajek – Program for Large Network Analysis. 1999;1–11 Downloaded May 2, 2017. http://vlado.fmf.uni-lj.si/pub/networks/doc/pajek.pdf.
13. Freeman TC, Goldovsky L, Brosch M, van Dongen S, Mazière P, Grocock RJ, et al. Construction, visualisation, and clustering of transcription networks from microarray expression data. PLoS Comput Biol. 2007;3:2032–42.
14. Shannon P, Markiel A, Ozier O, Baliga NS, Wang JT, Ramage D, et al. Cytoscape: a software environment for integrated models of biomolecular interaction networks. Genome Res. 2003;13:2498–504.
15. Iragne F, Nikolski M, Mathieu B, Auber D, Sherman D. ProViz: protein interaction visualization and exploration. Bioinformatics. 2005;21:272–4.
16. Cytoscape.js. Available from: http://cytoscape.github.io/cytoscape.js/. Accessed 2 May 2017.
17. Bostock M, Ogievetsky V, Heer J. D 3: Data-Driven Documents. IEEE Trans Vis Comput Graph. 2011;17:2301–9.
18. Jiang X, Ronggui Y, Huiran Z, Wu Z, Kawata S. Biomolecular network visualization based on CPSE-Bio. Comput. Converg. Technol. (ICCCT), 2012 7th Int. Conf. 2012; 1488–92
19. Salazar GA, Meintjes A, Mazandu GK, Rapanoël HA, Akinola RO, Mulder NJ. A web-based protein interaction network visualizer. BMC Bioinformatics. 2014;15:129.
20. Bateman A, Martin MJ, O'Donovan C, Magrane M, Apweiler R, Alpi E, et al. UniProt: A hub for protein information. Nucleic Acids Res. 2015;43:D204–12.
21. Kersey PJ, Allen JE, Armean I, Boddu S, Bolt BJ, Carvalho-Silva D, et al. Ensembl Genomes 2016: more genomes, more complexity. Nucleic Acids Res. 2015;44:D574–80.
22. Sun L, Lamont SJ, Cooksey AM, McCarthy F, Tudor CO, Vijay-Shanker K, et al. Transcriptome response to heat stress in a chicken hepatocellular carcinoma cell line. Cell Stress Chaperones. 2015;20:939–50.
23. Blake JA, Christie KR, Dolan ME, Drabkin HJ, Hill DP, Ni L, et al. Gene ontology consortium: going forward. Nucleic Acids Res. 2015;43:D1049–56.
24. Mitchell A, Chang HY, Daugherty L, Fraser M, Hunter S, Lopez R, et al. The InterPro protein families database: the classification resource after 15 years. Nucleic Acids Res. 2015;43:D213–21.

18

LASSIE: simulating large-scale models of biochemical systems on GPUs

Andrea Tangherloni[1], Marco S. Nobile[1,3], Daniela Besozzi[1], Giancarlo Mauri[1,3] and Paolo Cazzaniga[2,3*] (iD)

Abstract

Background: Mathematical modeling and in silico analysis are widely acknowledged as complementary tools to biological laboratory methods, to achieve a thorough understanding of emergent behaviors of cellular processes in both physiological and perturbed conditions. Though, the simulation of large-scale models—consisting in hundreds or thousands of reactions and molecular species—can rapidly overtake the capabilities of Central Processing Units (CPUs). The purpose of this work is to exploit alternative high-performance computing solutions, such as Graphics Processing Units (GPUs), to allow the investigation of these models at reduced computational costs.

Results: LASSIE is a "black-box" GPU-accelerated deterministic simulator, specifically designed for large-scale models and not requiring any expertise in mathematical modeling, simulation algorithms or GPU programming. Given a reaction-based model of a cellular process, LASSIE automatically generates the corresponding system of Ordinary Differential Equations (ODEs), assuming mass-action kinetics. The numerical solution of the ODEs is obtained by automatically switching between the Runge-Kutta-Fehlberg method in the absence of stiffness, and the Backward Differentiation Formulae of first order in presence of stiffness. The computational performance of LASSIE are assessed using a set of randomly generated synthetic reaction-based models of increasing size, ranging from 64 to 8192 reactions and species, and compared to a CPU-implementation of the LSODA numerical integration algorithm.

Conclusions: LASSIE adopts a novel fine-grained parallelization strategy to distribute on the GPU cores all the calculations required to solve the system of ODEs. By virtue of this implementation, LASSIE achieves up to 92× speed-up with respect to LSODA, therefore reducing the running time from approximately 1 month down to 8 h to simulate models consisting in, for instance, four thousands of reactions and species. Notably, thanks to its smaller memory footprint, LASSIE is able to perform fast simulations of even larger models, whereby the tested CPU-implementation of LSODA failed to reach termination. LASSIE is therefore expected to make an important breakthrough in Systems Biology applications, for the execution of faster and in-depth computational analyses of large-scale models of complex biological systems.

Keywords: Graphics Processing Unit, GPU computing, Reaction-based model, Deterministic simulation, Numerical integration method, LSODA, Nvidia CUDA, Fine-grained parallelization, Systems biology, Rule-based model

Background

Systems Biology is a multidisciplinary research field relying on the cross-talk between mathematical, computational and experimental tools to investigate the functioning of complex biological systems, and to predict how they might behave in both physiological and perturbed conditions. To this aim, different computational methods—e.g., parameter estimation, sensitivity analysis or reverse engineering [1, 2]—are usually exploited to define or calibrate the mathematical model that describes the system of interest. These methods require the execution of a large number of simulations, each one generally corresponding to a distinct model structure or parameterization, that is, to a different set of molecular interactions or to different initializations of the species amounts and/or reaction constants. As a result, the computational burden required by these computational analyses can rapidly overtake the capabilities of Central Processing Units (CPUs), therefore limiting

*Correspondence: paolo.cazzaniga@unibg.it
[2]Department of Human and Social Sciences, University of Bergamo, Piazzale Sant'Agostino 2, 24129 Bergamo, Italy
[3]SYSBIO.IT Centre of Systems Biology, Piazza della Scienza 2, 20126 Milano, Italy
Full list of author information is available at the end of the article

in-depth computational investigations to small-scale models consisting in a few tens of reactions and molecular species at most. General-purpose Graphics Processing Units (GPUs) can be exploited to overcome these drawbacks. Indeed, they are parallel multi-core co-processors that are drawing an ever-growing attention by the scientific community, since they give access to tera-scale performances on common workstations (and peta-scale performances on GPU-equipped supercomputers [3]). As such, they can markedly decrease the running times required by traditional CPU-based software, still maintaining low-costs and energetic efficiency. As a matter of fact, in the latter years GPUs have been widely adopted as an alternative approach to classic parallel architectures for the parallelization of computational methods in Systems Biology, Computational Biology and Bioinformatics [4].

In this work we propose LASSIE (LArge-Scale SImulator), a novel GPU-accelerated software designed to simulate large-scale reaction-based models of cellular processes, consisting in hundreds or thousands of reactions and molecular species. An example of killer-application of LASSIE would consist in the simulation of rule-based models according to the so-called indirect methods (see [5, 6] for more information), especially when some proteins are characterized by multiple phosphorylation sites or binding domains, a condition that yields a combinatorial explosion of intermediate chemical complexes and chemical reactions [7]. We designed LASSIE as a general "black-box" tool able to simulate, in principle, any large-scale reaction-based biochemical system based on mass-action kinetics (e.g., the ErbB signaling pathways modeled by Chen et al. [8]), given that the available GPU memory is sufficient to accommodate the necessary data structures. However, considering the difficulty in the manual definition of such massive models, LASSIE may be adopted as an efficient simulation engine for rule-based modeling tools (e.g., BioNetGen [9], PySB [10], Kappa [11]). As a matter of fact, rule-based modeling can generate extremely large-scale systems characterized by very long simulation times: LASSIE may represent an enabling tool to prevent the application of advanced computational investigations of such biological models.

In silico simulations allow to determine the quantitative variation of molecular species amount in time and/or in space, by exploiting either deterministic, stochastic or hybrid algorithms [12–14]. In particular, when the concentrations of molecular species is high and the effect of biological noise can be neglected [15], Ordinary Differential Equations (ODEs) represent the typical modeling approach for cellular processes. Given a model parameterization (i.e., the initial state of the system and the set of kinetic parameters), the temporal dynamics of the system can be simulated by solving the ODEs using some numerical integrator, such as Euler or Runge-Kutta methods [16].

Unfortunately, ODEs can be affected by a well-known phenomenon named stiffness [17], which occurs when the system of biochemical reactions is characterized by two well-separated dynamical modes, determined by fast and slow reactions, respectively [18]. Stiffness can cause the step-size of integration algorithms to reach extremely small values, thus increasing the overall running time. To solve this issue, advanced integration methods like LSODA [19] can be exploited, thanks to their capability of efficiently solving stiff systems. LSODA is able to recognize when a system is stiff and to dynamically select between the most appropriate integration algorithm: the Adams methods [16] in the absence of stiffness, and the Backward Differentiation Formulae (BDF) [20] otherwise. Despite the improvement of efficiency granted by LSODA, the numerical integration of the system of ODEs can become excessively burdensome when the numbers of reactions and molecular species increase. LASSIE overcomes this limitation by distributing over thousands of GPU cores all the calculations required by the numerical integration methods it embeds, therefore paving the way for fast simulations of large-scale and stiff models of cellular processes. One interesting feature of GPUs is that they can have different characteristics, both in terms of resources (e.g., amount of high performance memories, number of cores) and computing power (e.g., clock rate). Kernels' performances transparently scale on different GPUs, since they automatically leverage the additional resources offered by the latest architectures, a characteristic known as transparent scalability.

Notably, LASSIE was designed to be a "black-box" deterministic simulator, not requiring any expertise in mathematical modeling nor any GPU programming skill. More precisely, given the formalization of a cellular process as a reaction-based model [21, 22] and assuming mass-action kinetics [23, 24], LASSIE proceeds according to the following workflow: (1) it automatically generates the system of ODEs—one ODE for each molecular species occurring in the biochemical system—according to the biochemical reactions included in the model; (2) it automatically derives the Jacobian matrix, taking advantage of the symbolic derivation, to apply the BDF; (3) it executes the numerical integration of the ODEs by automatically switching between the Runge-Kutta-Fehlberg (RKF) [25] method in the absence of stiffness and first-order BDF (also known as Backward Euler method) [20] in presence of stiffness. We point out that LASSIE is a fully automatic simulator: the user does not need to enter the ODEs directly. On the contrary, the input consists in a set of (parameterized) chemical reactions, specified by means of text files. The corresponding system of ODEs is automatically determined according to the mass-action kinetics, making LASSIE usable without any prior knowledge about ODEs modeling and integration. In order to further simplify the execution of simulations, LASSIE is provided with a user-friendly Graphical User Interface (Fig. 1), whose functioning is described in Additional file 1. A comprehensive description of the input files is provided in Additional file 2.

The computational performances of LASSIE are assessed by measuring the running time required to simulate a set of randomly generated synthetic reaction-based models of increasing size—ranging from 64 to 8192 reactions and species—which is compared to the running time required by a CPU-implementation of LSODA. Moreover, we show the accuracy of LASSIE by comparing its outcome with LSODA outcome for the simulation of a model of the Ras/cAMP/PKA signal transduction pathway in *S. cerevisiae* [26], which is characterized by stiffness.

We highlight that, in general, the implementation of computational methods able to fully exploit the peculiar architecture of GPUs is challenging, since specific programming skills are required and a complete algorithm redesign is often necessary. For instance, the parallelization on the GPU cores can rely either on a coarse-grained or a fine-grained strategy. The first strategy allows to simultaneously run a massive number of independent simulations (each one characterized by, e.g., a different model parameterization); on the contrary, the second strategy consists in the parallelization of all the calculations required by a *single* simulation, an approach that is more suitable for large-scale models. By virtue of the novel *fine-grained* parallelization strategy used to implement LASSIE, our GPU-powered simulator achieves up to 92× speed-up with respect to LSODA.

Coarse-grained parallelizations of deterministic simulations were presented in [27–29]. The simulators proposed

in these works allow to reach a speed-up ranging from 28× to 86× with respect to the corresponding CPU-based simulators. Fine-grained parallelizations of stochastic simulations were presented in [30, 31]. Komarov and D'Souza proposed GPU-ODM [30], a fine-grained simulator of large-scale models based on the Stochastic Simulation Algorithm (SSA) [32]. This tool uses special data structures and functionalities to efficiently distribute all calculations over the multiple cores of GPUs. These optimizations allow GPU-ODM to outperform the most advanced CPU-based implementations of SSA. Komarov et al. also proposed a GPU-powered fine-grained implementation of τ-leaping [31], an approximate but accurate stochastic algorithm that is, in general, faster than SSA [33]. This tool was shown to be more efficient than its sequential counterpart in the case of extremely large biochemical networks (i.e., characterized by more than 10^5 reactions). Notably, to the best of our knowledge, no examples of fine-grained deterministic simulators, such as LASSIE, have been proposed so far.

LASSIE was developed using the most widespread GPU computing library, namely, Nvidia Compute Unified Device Architecture (CUDA). CUDA allows programmers to exploit the GPUs for general-purpose computational tasks (GPGPU computing). Nevertheless, the direct porting of an application to the GPU is usually unfeasible, so that the full exploitation of the computational power and of the massive parallelism of GPUs still represent the main challenges of GPGPU computing. To exploit the CUDA architecture, the programmer implements C/C++ functions (called kernels), which are loaded from the CPU (the host) to one or more GPUs (the devices), and replicated in many copies named threads. CUDA organizes threads in three-dimensional structures called blocks,

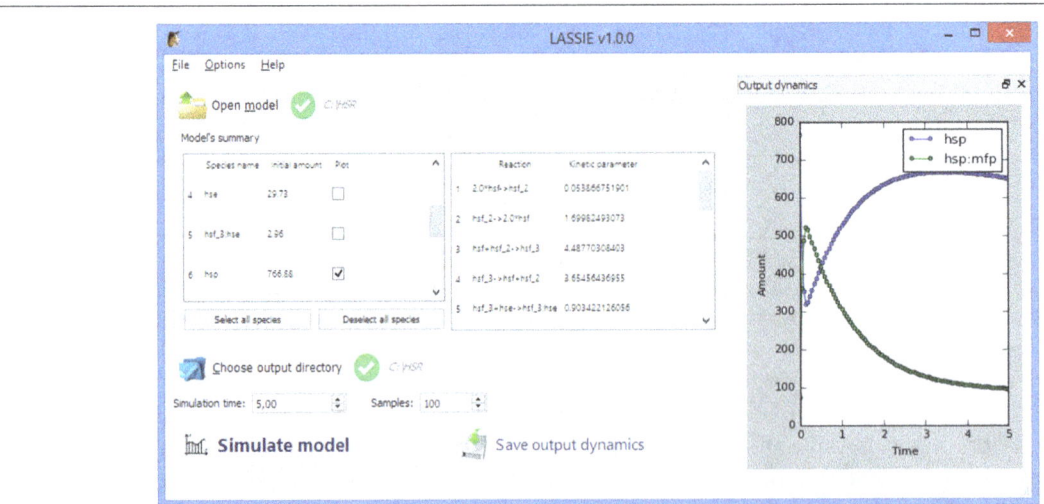

Fig. 1 LASSIE's Graphical User Interface that easily allows the user to (*i*) open a model, (*ii*) visualize its set of species, reactions and parameters, (*iii*) select the output directory, (*iv*) perform a simulation and (*v*) graphically represent the corresponding dynamics

which belong to three-dimensional structures named grids, as shown in Fig. 2 (left side). CUDA combines the Single Instruction Multiple Data (SIMD) architecture and a flexible multi-threading in order to handle any conditional divergence between threads. Figure 2 (right side) also shows a schematic representation of CUDA's memory hierarchy: the global memory (accessible from all threads), the shared memory (accessible from threads belonging to the same block), the local memory (each thread has its registers and arrays) and the constant memory (cached and read only). The global memory is large (a few GBs) but suffers from high access latencies; this problem, anyway, was mitigated thanks to the use of L1 cache since the introduction of the Fermi architecture. On the contrary, the constant memory is much smaller (i.e, up to 10 KB for each multi-processor) but faster than the global memory, as well as the shared memory (i.e., up to 112 KB for each multi-processor limited to 48 KB for each block); in particular, the latter should be exploited as much as possible in order to obtain the best performances. Though, the size of the shared memory and its scope restrict the possibility to use it, as only threads belonging to the same block can communicate through the shared memory.

Given the peculiar features of CUDA architecture, GPU programmers should be able to optimize both threads partitioning and memory usage, as well as to redesign the algorithm with appropriate kernels, in order to fully leverage the computational power of these multi-core devices. For instance, in the implementation of LASSIE, the shared memory is not used because several blocks are exploited to solve the system of ODEs, and threads do not communicate data with each other. Moreover, the data structures employed by LASSIE are larger than the total size of the shared memory, thus preventing the possibility to exploit it. In what follows, we show how GPU programming and CUDA features—including built-in support for vector types, which extend the standard C data types to vector—have been exploited to optimize the execution workflow of LASSIE.

The paper is structured as follows. In the next section we briefly introduce the formalism of reaction-based models and provide a general description of LASSIE's implementation. Then, we discuss the computational performance of LASSIE, showing the speed-up it achieves with respect to LSODA for the simulation of reaction-based models of different sizes. We also analyze how the number of reactions and the number of species affect the performances of LASSIE. We conclude the work with some final remarks about CUDA's architecture and LASSIE, proposing future improvements of the simulator. LASSIE is available on the GITHUB repository https://github.com/aresio/LASSIE.

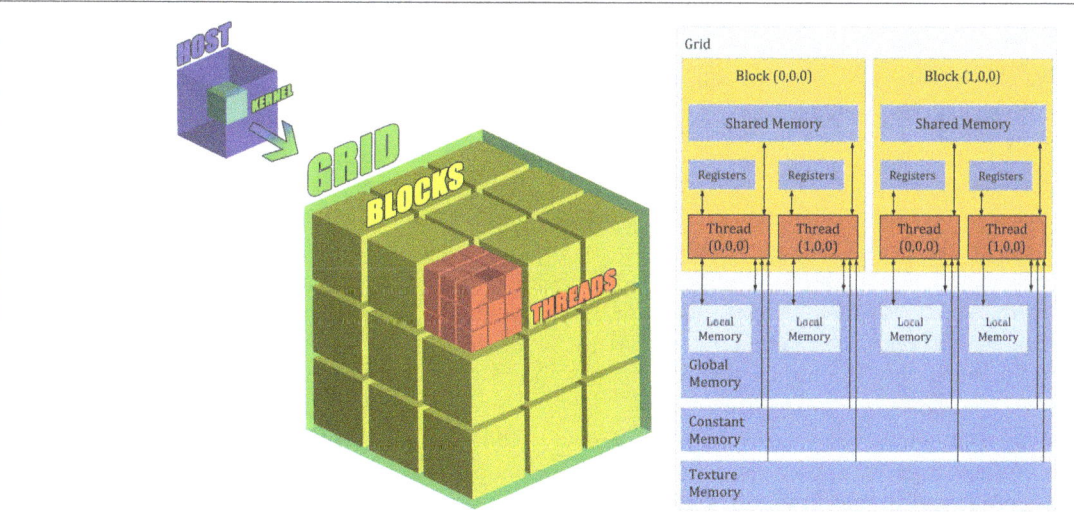

Fig. 2 Threads and memory hierarchy of CUDA's architecture. *Left side*. Thread organization: the host (CPU) launches a single kernel that is executed in multiple threads on the device (GPU). Threads (*red cubes*) are organized in three-dimensional structures called blocks (*yellow cubes*), which belong to three-dimensional grid (*green cube*). The programmer must explicitly define the dimensions of blocks and grids. Whenever a kernel is run by the host, the corresponding grid is created by the device which automatically schedules each block on one free streaming multiprocessor available. This solution allows a transparent scaling of performances on different devices. Moreover, if the machine is equipped with more than one GPU, it is also possible to distribute the workload by launching the kernel on each GPU. *Right side*. Memory hierarchy: in CUDA there are many different memories with different scopes. Each thread has two different kind of private memory: registers and local memories. Threads belonging to the same block can communicate through the shared memory, which has low access latency. The global memory suffers from high access latencies but it is accessible to all threads and it is cached since the introduction of the Fermi architecture. Also the texture and the constant memory are equipped with a cache as well, and all threads can read from these two memories

Methods

LASSIE is designed to be a "black-box" deterministic simulator, created to be easily used without any GPU programming or ODEs modeling skills. In this section we describe how LASSIE allows to perform deterministic simulations of large-scale biochemical models, distributing all required calculations on the cores of the GPU. It is worth noting that the parallelization strategy exploited by LASSIE represents one of the novelties of this work, and allowed to achieve the remarkable performance results presented in the next sections. In particular, LASSIE has been developed to solve systems of coupled ODEs specified in the form $\frac{d\mathbf{X}}{dt} = f(t, \mathbf{X})$, where $\mathbf{X} \equiv \mathbf{X}(t)$ represents the vector of concentration values at time t of all chemical species occurring in the system.

Reaction-based models and ODEs generation

Reaction-based modeling is a mechanistic, quantitative and parametric formalism to describe and simulate networks of biochemical reactions [22], which was exploited to analyze different signal transduction pathways (see, e.g., [34–39]). A reaction-based model is defined by specifying the set of N molecular species $\{S_1, \ldots, S_N\}$ and the set of M biochemical reactions $\{R_1, \ldots, R_M\}$ which appear in the cellular process under investigation [22]. A generic reaction is described as follows:

$$R_i : \sum_{j=1}^{N} a_{ij} S_j \xrightarrow{k_i} \sum_{j=1}^{N} b_{ij} S_j, \quad i = 1, \ldots, M, \qquad (1)$$

where $a_{ij}, b_{ij} \in \mathbb{N}$ are the stoichiometric coefficients and $k_i \in \mathbb{R}^+$ is the kinetic constant associated with R_i.

The set of reactions $\{R_1, \ldots, R_M\}$ can be written compactly in the matrix-vector form $\mathbf{AS} \xrightarrow{\mathbf{K}} \mathbf{BS}$, where $\mathbf{S} = [S_1 \cdots S_N]^T$ is the N-dimensional column vector of molecular species, $\mathbf{K} = [k_1 \cdots k_M]^T$ is the M-dimensional column vector of kinetic constants, and $\mathbf{A}, \mathbf{B} \in \mathbb{N}^{M \times N}$ are the so-called stoichiometric matrices whose (non-negative) elements $[A]_{i,j}$ and $[B]_{i,j}$ correspond to the stoichiometric coefficients a_{ij} and b_{ij} of the reactants and the products of all reactions, respectively. Since a reaction simultaneously involving more than two reactants has a probability to take place almost equal to zero, here we consider only first and second-order reactions (i.e., at most two reactant molecules of the same or different species can appear in the left hand side of Eq. 1). For this reason, the matrices $\mathbf{A}$ and $\mathbf{B}$ are sparse.

Given an arbitrary reaction-based model and assuming the law of mass-action [24, 40], it is possible to derive the corresponding system of coupled ODEs that describes the variation in time of the species concentrations. Specifically, by denoting the concentration of species S_j at time

t as X_j, where $X_j \in \mathbf{R}^{\geq 0}$ for $j = 1, \ldots, N$, the system of coupled ODEs can be obtained as follows:

$$\frac{d\mathbf{X}}{dt} = (\mathbf{B} - \mathbf{A})^T [\mathbf{K} \circ \mathbf{X}^\mathbf{A}], \qquad (2)$$

where $\mathbf{X}$ is the N-dimensional vector of concentration values at time t (representing the state of the system at time t), the symbol $\circ$ denotes the entry-by-entry matrix multiplication, and $\mathbf{X}^\mathbf{A}$ denotes the vector-matrix exponentiation form [40]. Formally, $\mathbf{X}^\mathbf{A}$ is a M-dimensional vector whose i-th component is given by $X_1^{Ai1} \cdots X_N^{AiN}$, for $i = 1, \ldots, M$.

We highlight that each ODE appearing in Eq. 2 is a polynomial function, consisting in at least one monomial that is associated with a specific kinetic constant.

Data structures and CUDA memory usage

Given a reaction-based model as input, LASSIE automatically generates the systems of ODEs according to Eq. 2 and encodes the matrices $\mathbf{A}$ and $\mathbf{H} = (\mathbf{B} - \mathbf{A})^T$ as two arrays of *short4* CUDA vector types, named $\mathbf{VA}$ and $\mathbf{VH}$, respectively.

CUDA vector types are multi-dimensional data ranging from 1 to 4 components, addressed by $.x$, $.y$, $.z$, and $.w$. Since the matrices $\mathbf{A}$ and $\mathbf{H}$ are sparse, LASSIE uses compressed data structures created by removing all zero elements from $\mathbf{A}$ and $\mathbf{H}$, in order to save memory and avoid unnecessary readings from the global memory. Namely, let h_{ji} be the element of $\mathbf{H}$ at row j and column i, and a_{ij} the element of $\mathbf{A}$ at row i and column j, for $i = 1, \ldots, M$ and $j = 1, \ldots, N$. For each non-zero element of $\mathbf{H}$, we store into the $.x$ and $.y$ components of $\mathbf{VH}$ the values j and i, respectively; the $.z$ component of $\mathbf{VH}$ is used to store the element h_{ji}, while the $.w$ component stores the index of the kinetic constant associated with that monomial. Similarly, for each non-zero element of $\mathbf{A}$, the $.x$ and $.y$ components of $\mathbf{VA}$ contain the values i and j, respectively. The value a_{ij} is stored into the $.z$ component of $\mathbf{VA}$, while the $.w$ component is left unused. Note that we exploited the *short4* CUDA vector type rather than the *short3* CUDA vector type, because the former is 8-aligned and requires a single instruction to fetch a whole entry, while the latter is 2-aligned and thus takes three memory operations to read each entry. In order to parse these arrays inside the GPU, we use two additional arrays of *short2* CUDA vector types, named $\mathbf{O_H}$ and $\mathbf{O_A}$, which store the offsets used to correctly read the entries of the $\mathbf{VH}$ and $\mathbf{VA}$ structures, respectively. The $.x$ and $.y$ components of each row of $\mathbf{O_H}$ contain, respectively, the first index and the last index to access the $\mathbf{VH}$ structure. Each thread uses its own pair of indexes to read the rows of the $\mathbf{VH}$ structure between the first index and the last one. Similarly, $\mathbf{O_A}$ stores the indexes that allow to correctly access the $\mathbf{VA}$ structure. Finally, the values of the kinetic constants are stored into an array of type *double*, named **K**. Figure 3 shows an example of the matrix encoding used in LASSIE.

Thanks to these CUDA structures, we obtain a twofold performance improvement: (*i*) at the instruction level, a single instruction is enough to either load or store a multiword vector. So doing, the total instruction latency for a particular memory transaction is lower and also the bytes per instruction ratio is higher; (*ii*) at the memory controller level, by using vector types a transfer request from a warp has a larger net memory throughput per transaction, yielding a higher bytes per transaction ratio. With a fewer number of transfer requests, the memory controller is able to reduce contentions producing a higher overall memory bandwidth utilization. The only limitation due to *short* data type is that indices are limited to $2^{2 \times 8} - 1$, which means that LASSIE cannot simulate systems larger than 65 536 chemical species and reactions.

Execution workflow and CUDA kernels

Once that the system of ODEs is generated by reading the input files (see Additional file 2) and appropriately stored according to the CUDA vector types, LASSIE solves it by automatically switching between the Runge-Kutta-Fehlberg (RKF) method [25] in the absence of stiffness, and the Backward Differentiation Formulae (BDF) methods [20] in presence of stiffness. The integration of the systems of ODEs is carried out from an initial time instant t_0, up to a given maximum simulation time t_{max}. In order to reproduce the dynamics of the cellular process described by the ODEs, the concentration values of the molecular species appearing in the reaction-based model are saved at specified time steps within the interval $[t_0, t_{max}]$ (such time steps might correspond, e.g., to the sampling times of laboratory experiments).

LASSIE's workflow consists in 6 distinct phases, as represented in Fig. 4. Note that phases P_1, P_4 and P_6 are

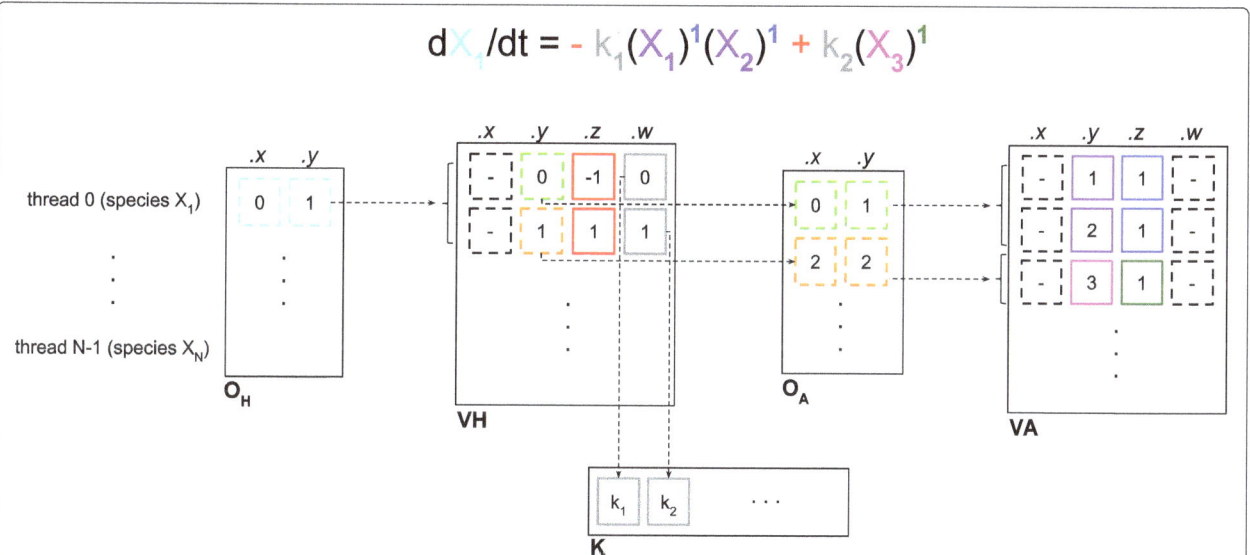

Fig. 3 Example of matrix encoding to automatically generate an ODE using LASSIE. All terms of the polynomial function describing the ODE of species X_1 given at the top of the figure are encoded in the components of the data structures $\mathbf{O_H}$, $\mathbf{VH}$, $\mathbf{O_A}$, $\mathbf{VA}$ and $\mathbf{K}$, as detailed hereby. Notice that only the data structures components with solid borders are used to automatically generate the ODE; the various terms appearing in the ODE are represented with corresponding colors in the data structure components. Matrix encoding starts from matrix $\mathbf{O_H}$. Each thread j, for $j = 0, \ldots, N - 1$, reads the values stored in the $.x$ and $.y$ components of $\mathbf{O_H}$ (denoted by the *lightblue borders*). In this example, we consider species X_1 that corresponds to thread 0. Each thread fetches the values in $\mathbf{VH}$, starting from the row indicated by the value stored in the $.x$ component of $\mathbf{O_H}$, up to the row corresponding to the value stored in the $.y$ component. In this example, thread 0 in matrix $\mathbf{O_H}$ reads the values contained in the first two rows—i.e., rows 0 and 1—in matrix $\mathbf{VH}$. Each row of $\mathbf{VH}$ encodes a monomial of an ODE: the $.x$ component is not used; the $.y$ components (denoted by *green* and *orange borders*) indicate the row numbers of the $\mathbf{O_A}$ structure that each thread must read; the $.z$ components (*red borders*) indicate the sign and the coefficient of the monomial; the $.w$ components (*gray borders*) indicate the positions of the array $\mathbf{K}$ containing the values of the kinetic constants corresponding to the reactions that the threads are parsing. In this example, the $.z$ and $.w$ components of $\mathbf{VH}$ allows to derive the coefficients $-1k_1$ and $+1k_2$ for the first and the second term of the ODE, respectively. Afterwards, as in the case of $\mathbf{O_H}$, each thread fetches the values in $\mathbf{VA}$, starting from the row indicated by the value stored in the $.x$ component of $\mathbf{O_A}$, up to the row corresponding to the value stored in the $.y$ component of $\mathbf{O_A}$. The values stored in the $.y$ (*violet* and *fuchsia borders*) and $.z$ (*blue* and *dark green borders*) components of $\mathbf{VA}$ correspond to the indexes of the species and the stoichiometric coefficients, respectively, while the $.x$ and $.w$ components of $\mathbf{VA}$ are left unused. In this example, row 0 in matrix $\mathbf{O_A}$ reads the values stored in rows 0 and 1 ($.y$ and $.z$ components) of matrix $\mathbf{VA}$, generating the factors $(X_1)^1 (X_2)^1$ in the first term of the ODE, while row 1 in matrix $\mathbf{O_A}$ reads the values stored in row 2 ($.y$ and $.z$ components) of matrix $\mathbf{VA}$, generating the factor $(X_3)^1$ in the second term of the ODE. Therefore, in this example, the matrix encoding overall generates the ODE of species X_1 consisting in the sum of two polynomial terms: $-k_1 (X_1)^1 (X_2)^1 + k_2 (X_3)^1$

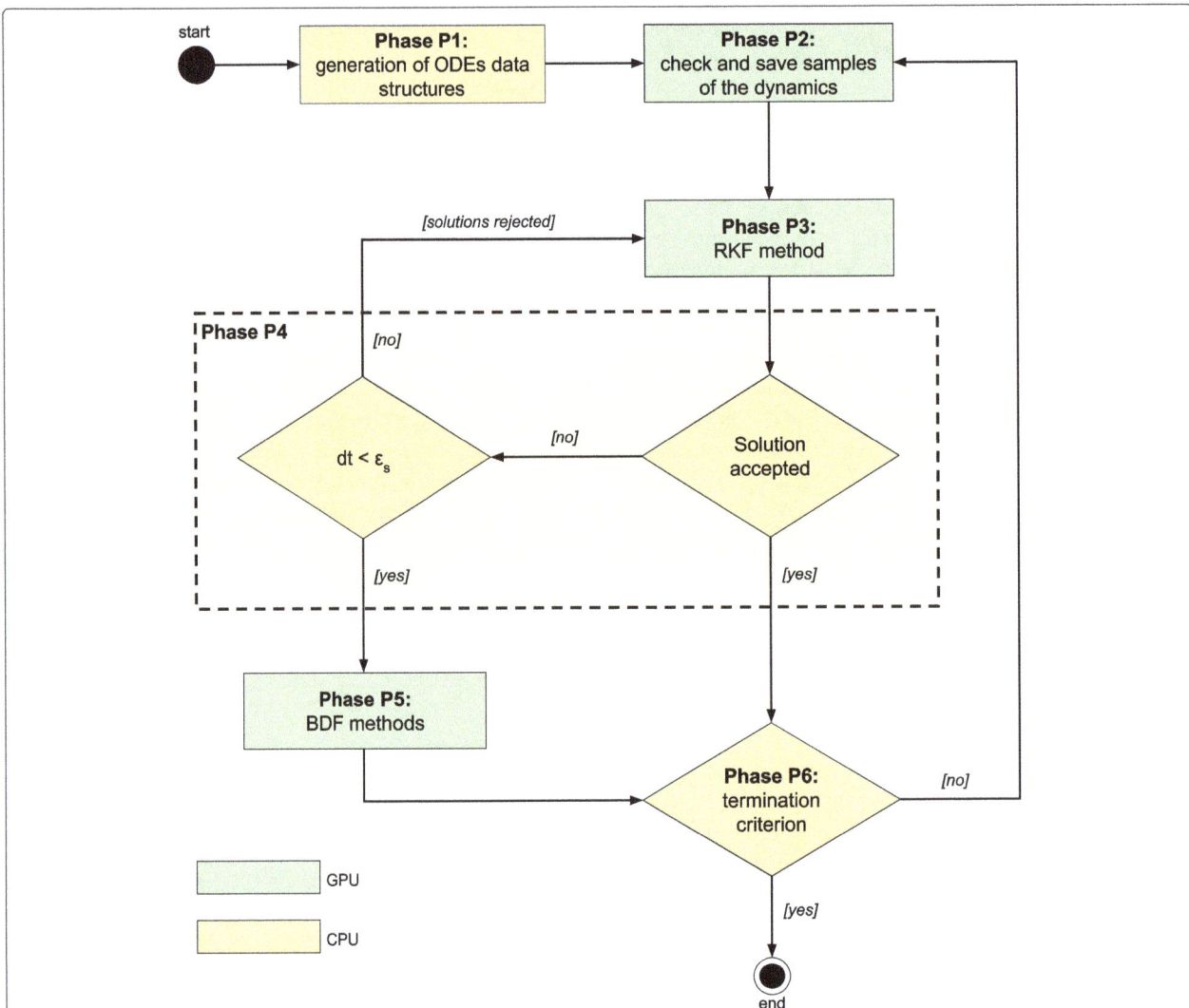

Fig. 4 Simplified scheme of LASSIE workflow. The data structures used to encode the system of ODEs are generated in phase P_1. In phase P_2, if the current simulation time t corresponds to a specified sampling time instant, then the current concentration values of all molecular species are saved; otherwise, the execution proceeds to the next phase. In phase P_3 each thread derives and solves the corresponding ODE by exploiting the RKF method, while in phase P_4 the RKF solutions are verified: (*i*) if the RKF solutions are rejected, then the integration step-size dt is reduced and phase P_3 is executed again; (*ii*) if RKF solutions are rejected but the integration step-size dt is too small, then phase P_5 is executed and the system of ODEs is solved using the BDF methods; (*iii*) if the RKF solutions are accepted, the termination criterion is verified during phase P_6 (all phases from P_2 on are iterated until the maximum simulation time t_{max} is reached)

executed by the host (yellow boxes in Fig. 4), while P_2, P_3 and P_5 are executed by the device (green boxes in Fig. 4). Overall, phases P_2, P_3 and P_5 rely on 25 different lightweight kernels, which were specifically developed to fully leverage the parallel architecture of the GPU for the implementation of the aforementioned numerical integration methods. We describe hereafter the main design and implementation choices of each phase and their related CUDA kernels, which result in a novel parallelization strategy with respect to state-of-the-art methodologies (see, e.g., [41]).

Phase P_1. It implements the generation of all data structures used to encode the ODEs, as described in the previous section. This phase is executed on the host.

Phase P_2. It is used to sample and save the system dynamics and it is implemented by means of a single CUDA kernel (**kernel K_1**). In particular, if the current simulation time t corresponds to one of the specified sampling time instants, LASSIE saves the concentration values of (possibly, a subset of) all molecular species into an array defined on the GPU. Otherwise, the execution proceeds to the next phase.

Phase P_3. It implements the RKF method [42], an explicit integration algorithm with variable step-size used by each thread j to solve the j-th ODE, for $j = 0, \ldots, N - 1$. This phase is implemented as 9 CUDA kernels.

During this phase, two different approximated states $\mathbf{u}(t + dt)$ and $\mathbf{w}(t + dt)$ of the state $\mathbf{X}(t + dt)$ of the system are generated at each step, thanks to the evaluation of six supplementary values $\mathbf{l}_1, \ldots, \mathbf{l}_6$ (see details in Additional file 3). To evaluate the accuracy of $\mathbf{u}$ and $\mathbf{w}$ at the current step-size dt, LASSIE exploits a user-defined vector tolerance $\boldsymbol{\varepsilon} \in \mathbf{R}^N$ (with $\varepsilon_j > 0$ for all $j = 1, \ldots, N$), and two additional arrays, $\mathbf{ER}, \boldsymbol{\delta} \in \mathbf{R}^N$, defined as follows:

$$\mathbf{ER} = \frac{|\mathbf{w}(t + dt) - \mathbf{u}(t + dt)|}{dt}, \quad \boldsymbol{\delta} = 0.84 \left(\frac{\boldsymbol{\varepsilon}}{\mathbf{ER}} \right)^{\frac{1}{4}}. \tag{3}$$

If $ER_j \leq \varepsilon_j$ for all $j = 1, \ldots, N$, then $\mathbf{u}$ is accepted as new state of the system, that is, $\mathbf{X}(t + dt) = \mathbf{u}(t + dt)$; otherwise, the solutions $\mathbf{u}$ and $\mathbf{w}$ are rejected and recalculated by using a new step-size. The new step-size is computed as $dt = dt \cdot \min\{\delta_1, \ldots, \delta_N\}$, being $\delta_1, \ldots, \delta_N$ the components of vector $\boldsymbol{\delta}$ (note that the new value of dt has to be chosen in order to satisfy the requested error tolerance for all ODEs).

Overall, phase P_3 is implemented by means of the following kernels:

- **kernel K_2**: used to evaluate each ODE at the current state $\mathbf{X}$ of the system;
- **kernels $K_3 - K_8$**: each thread j, for $j = 0, \ldots, N - 1$, computes the components $l_{1j}, \ldots, l_{6j}$ of $\mathbf{l}_1, \ldots, \mathbf{l}_6$, by invoking **kernel K_2**;
- **kernel K_9**: each thread j, for $j = 0, \ldots, N - 1$, computes the components w_j and u_j of the approximated states $\mathbf{u}$ and $\mathbf{w}$, respectively;
- **kernel K_{10}**: each thread j, for $j = 0, \ldots, N - 1$, calculates the components ER_j and δ_j of $\mathbf{ER}$ and $\boldsymbol{\delta}$, respectively.

Phase P_4. It is used to verify the RKF solutions calculated during phase P_3 and, accordingly, to choose the next phase to be executed: (*i*) if the solutions are rejected and the new step-size dt is acceptable (that is, $dt \geq \varepsilon_s$, for some $\varepsilon_s > 0$, e.g., $\varepsilon_s = 10^{-6}$), phase P_3 is executed again exploiting a smaller step-size dt; (*ii*) if the solutions are rejected and the new step-size dt becomes too small (that is, $dt < \varepsilon_s$), LASSIE executes phase P_5; (*iii*) if all solutions do not violate the specified RKF-tolerance vector $\boldsymbol{\varepsilon}$, then LASSIE executes phase P_6.

Note that point (*ii*) implicitly states that the system of ODEs is considered to be stiff, so that LASSIE automatically switches to phase P_5, where BDF methods are used for the numerical integration. Phase P_4 is executed on the host.

Phase P_5. It implements the BDF methods, the most widely used implicit multi-step numerical integration algorithms [43]. LASSIE switches to this phase if and only if the RKF solutions $\mathbf{u}$ and $\mathbf{w}$ evaluated during phase P_4 are rejected, and the RKF step-size dt becomes smaller than ε_s.

The general formula for a BDF can be written as

$$\sum_{i=0}^{q} \alpha_i \mathbf{X}(t - t_i) = dt\beta_0 f(t, \mathbf{X}(t)), \tag{4}$$

where the coefficients α_i (with $\alpha_0 = 1$) and β_0 are chosen according to the order q of BDF [43], and dt is user-defined. Note that, for $q > 6$, the absolute stability region of the resulting BDF methods is too small and such BDFs are numerical unstable [44]. Therefore, BDFs with an order q greater than 6 are not used. Since each BDF is an implicit method, at each time step it requires the solution of a nonlinear system of equations, which can be solved by using the iterative Newton–Raphson method [45]. This algorithm allows to find successively better approximations z of the zeros of a real-valued function $f(z) = 0$ by using the derivative of $f(z)$, and it is repeated until a sufficiently accurate value is reached. This idea can be extended to a system of nonlinear equations, by using the Jacobian matrix $\mathbf{J}(t, \mathbf{X}(t))$ of $f(t, \mathbf{X}(t))$, which is the matrix of all first-order partial derivatives. Since the evaluation of the Jacobian matrix at each iteration is computationally expensive, LASSIE actually exploits: (*i*) a modified Newton–Raphson method [46]; (*ii*) the LU factorization method [47] (we refer the interested reader to Additional file 3 for technical details). During phase P_5, the Newton-Raphson method is iterated until a user-defined maximum number of iterations max_{it} is reached, or a sufficiently accurate value is achieved (i.e., smaller than a user-defined tolerance value ε_{NR}).

Overall, phase P_5 is implemented by means of the following kernels:

- **kernel K_{11}**: each thread j, for $j = 0, \ldots, N - 1$, derives the j-th row of the Jacobian matrix and evaluates it on the current state of the system $\mathbf{X}$;
- **kernel K_{12}**: the Jacobian matrix is transposed in order to exploit the LU factorization method (accelerated on GPU by the cuBLAS library [48]);
- **kernels $K_{13} - K_{18}$**: based on the order q of the BDF, LASSIE invokes one of these kernels (i.e., **kernel K_{13}** for $q = 1$, **kernel K_{14}** for $q = 2$, ..., **kernel K_{18}** for

$q = 6$) to calculate the known terms of the linear system;

- **kernels K_{19} – K_{24}**: each **kernel $K_{(18+q)}$**, $q = 1, \ldots, 6$, performs the calculations of the q-th order BDF;
- **kernel K_{25}**: it updates the iteration vector needed to execute the Newton-Raphson method (see Additional file 3).

Phase P_6. It is used to verify the termination criterion: if the maximum time t_{max} is reached, then the simulation ends. On the contrary, the execution iterates from phase P_2. This phase is executed on the host.

All the temporary results computed by LASSIE are stored on the GPU, since data transfers between the host and the device are very time consuming. For the same reason, the output data (i.e., the concentration values of molecular species sampled at fixed time instants) are transferred to the host as soon as the whole simulation is completed.

Results and discussion

In this section we compare the computational performance of LASSIE against LSODA [19], which is generally considered one of the best numerical integration algorithms for deterministic simulations of biological systems, thanks to its capability of dealing with stiff and non-stiff systems. In particular, we exploited the LSODA implementation provided by SciPy library [49] (version 0.15.1), written in C language. LASSIE was run on a machine with a GPU Nvidia GeForce Titan GTX, based on the Kepler architecture and equipped with 2×15 streaming multiprocessors for a total of 5760 cores (clock 837 MHz) and a theoretical peak processing power of 1.3 TFLOPS in double precision. Instead, LSODA was run on GALILEO, a supercomputer created by the Italian consortium CINECA. GALILEO consists of 516 compute nodes, each one equipped with 2 CPUs octa-core Intel Xeon Haswell E5-2630 v3 (clock 2.40 GHz) for a total of 8256 cores, and 128 GB of RAM. Each CPU is capable of about 300 GFLOPS in double precision. In our tests, we exploited one node with 120 GB of RAM distributed over 5 cores.

The computational performance was evaluated by simulating a set of synthetic reaction-based models of increasing size, that is, having a number of reactions and species $M \times N$ arbitrarily chosen in the range from 64×64 to 8192×8192. The models were generated considering the methodology used in [30, 50], which was modified in order to randomly sample the initial concentration of each species with a uniform distribution in the range $[0, 1)$, and the kinetic constant of each reaction with a logarithmic distribution in the range $[10^{-8}, 1)$.

For each model size $M \times N$, we generated and simulated 30 different synthetic reaction-based models to the aim of

measuring the average running time of both LASSIE and LSODA. The simulation of each reaction-based model was performed multiple times, using different settings for the sampling of the time-series. Specifically, in each repetition, we saved either $10, 50, 100, 500$ or 1000 samples of the system dynamics of all chemical species, at regular intervals. All simulations were halted at time $t_{max} = 50$ (arbitrary units).

All simulations were executed—independently from the size of the model and the number of samples saved—by setting the following parameters of LASSIE:

- tolerance of RKF method $\varepsilon_j = 10^{-12}, j = 1, \ldots, N$;
- first–order BDF method ($q = 1$);
- BDF integration step $dt = 0.1$;
- tolerance of Newton-Raphson method $\varepsilon_{NR} = 10^{-6}$;
- maximum number of iterations allowed during each call of the Newton-Raphson method $max_{it} = 10^4$;
- initial integration step of RKF method equal to 10^{-3};
- tolerance value to switch between RKF and Backward Euler methods $\varepsilon_s = 10^{-6}$.

The following parameters of LSODA were used to run the simulations:

- relative tolerance equal to 10^{-6};
- absolute tolerance equal to 10^{-12};
- maximum number of internal steps equal to 10^4.

Table 1 reports the values of the average running times (given in seconds) of LSODA and LASSIE, required for the execution of each set of 30 different synthetic reaction-based models of size $M \times N$, each time considering $10, 50, 100, 500, 1000$ samples of the system dynamics of all chemical species. The speed-up values achieved by LASSIE with respect to LSODA are given in Table 1 and graphically represented in Fig. 5, for each tested case; note that when the speed-up value is greater than one, LASSIE is faster than LSODA, and vice versa. The break-even (blue line in Fig. 5) between the performances of LASSIE and LSODA is observed when the number of reactions and chemical species is between 128 and 256. Specifically, in the case of 256×256 model size and 10 samples, the running time of LSODA is almost twice with respect to LASSIE: 1.28 s vs. 0.67 s. In particular, we emphasize that the execution of the simulations for models characterized by 4096 reactions and 4096 species with 10 samples takes, on average, 249.8 s with LSODA and just 2.71 s with LASSIE, resulting in around 92× speed-up. Furthermore, LASSIE allows the simulation of large-scale models (e.g., 8192×8192) thanks to its smaller memory footprint with respect to LSODA, taking just 14.13 s to simulate the model characterized by 8192 reactions and 8192 species with 10 samples. Conversely, the version of LSODA implemented in SciPy library has a high memory footprint

Table 1 Average running time (in seconds) of LSODA and LASSIE – and corresponding speed-up value – required for the execution of the set of 30 synthetic reaction-based models of size $M \times N$ (with $M = N$), considering 10, 50, 100, 500, 1000 samples of the system dynamics of all chemical species

$M \times N$	10 samples			50 samples			100 samples			500 samples			1000 samples		
	LSODA	LASSIE	Speed-up	LSODA	LASSIE	Speed-up	LSODA	LASSIE	Speed-up	LSODA	LASSIE	Speed-up	LSODA	LASSIE	Speed-up
64 × 64	0.257	0.519	0.495	0.288	0.541	0.532	0.220	0.557	0.395	0.307	0.665	0.462	0.303	0.839	0.361
128 × 128	0.393	0.613	0.641	0.473	0.635	0.745	0.507	0.644	0.787	0.674	0.786	0.858	0.498	0.958	0.520
256 × 256	1.277	0.669	1.909	1.486	0.696	2.135	1.293	0.727	1.779	1.319	0.905	1.456	1.277	1.122	1.138
512 × 512	4.313	0.792	5.446	4.629	0.841	5.504	4.559	0.915	4.982	4.300	1.215	3.539	4.669	1.526	3.060
1024 × 1024	15.753	0.955	16.495	15.707	1.056	14.874	16.201	1.201	13.490	15.982	1.809	8.835	16.647	2.407	6.916
2048 × 2048	61.824	1.662	37.199	51.748	1.987	31.076	61.762	2.397	25.766	62.307	3.721	14.745	62.742	5.479	11.451
4096 × 4096	249.839	2.713	92.090*	248.234	4.571	54.306	249.422	5.665	44.029	249.546	12.407	20.113	254.416	17.393	14.627
8192 × 8192	NA	14.134	NA	NA	26.051	NA	NA	38.058	NA	NA	101.91	NA	NA	129.755	NA

*Maximum speed-up value

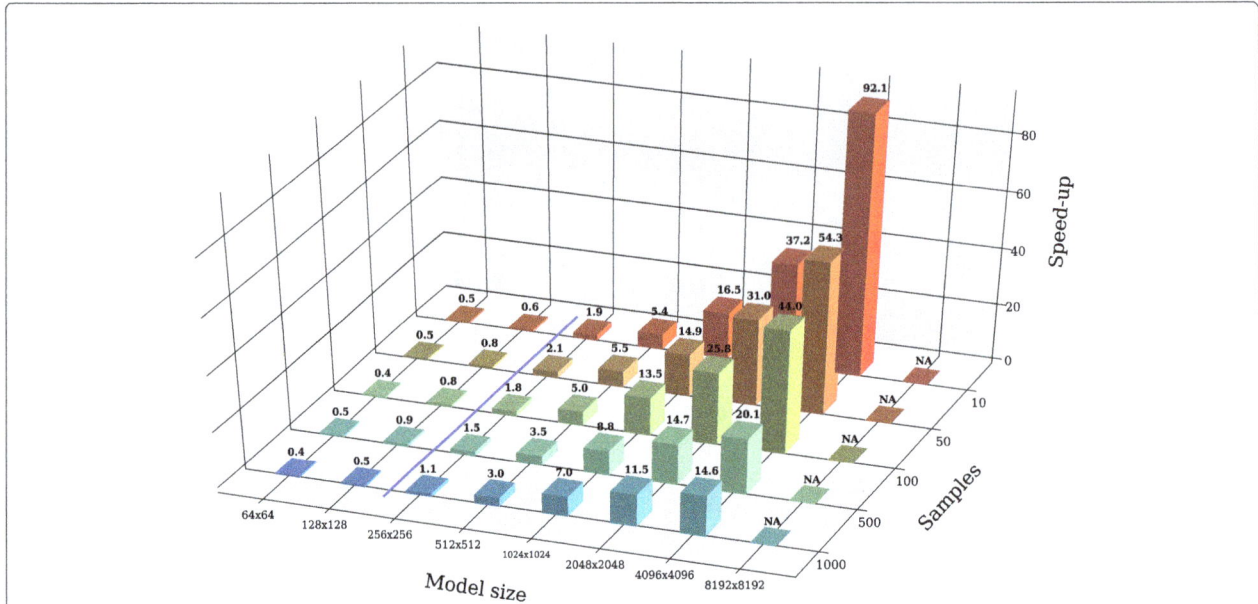

Fig. 5 Speed-up values (z-axis) achieved by LASSIE with respect to LSODA for the simulation of synthetic models of increasing size, having a number of reactions and of species $M \times N$ (x-axis) and characterized by an increasing number of sampling time instants of the system dynamics (y-axis). When the value of the speed-up is greater than one, LASSIE is faster than LSODA and vice versa

that does not allow to simulate models of this size on GALILEO, the supercomputer employed to perform the simulations.

Figure 5 also points out how the number of samples of the dynamics affects the performances of LASSIE, due to the different number of accesses to the high-latency global memory. For instance, the speed-up achieved with the model characterized by 4096 reactions and 4096 species decreases to 14.6× with 1000 samples, meaning that the simulations with 1000 samples are around 6× slower than the simulations with 10 samples. In models characterized by 2048 reactions and 2048 species, the speed-up obtained with 10 samples (37.2×) is around 3× larger compared to the one achieved with 1000 samples (11.4×), while in models characterized by 1024 reactions and 1024 species, the speed-up obtained with 10 samples (16.5×) is around 2× larger compared to the one achieved with 1000 samples (6.9×). Finally, Fig. 6 shows that the running time of LASSIE increases with the number of samples, while LSODA is characterized by an almost constant running time, irrespective of the number of samples. It is worth noting that CPU-bound integration methods like LSODA can be more efficient in the case of small-scale models. This is due to two concomitant circumstances. On the one hand, the clock frequency of CPUs is higher than the clock frequency of GPU (2.4 GHz with respect to 837 MHz, in the case of the hardware used to execute our tests). On the other hand, the communication and synchronization between threads can introduce a significant overhead, which is mitigated only when the calculations

are distributed over a relevant number of threads; therefore, LASSIE becomes profitable for medium/large-scale models characterized by hundreds of species. Notably, the bigger the model, the greater the speed-up.

As an additional test, we investigated whether the relationship between the number of reactions and the number of species could affect the overall performances of LASSIE. As the number of chemical species corresponds to the number of ODEs, the length of each ODE is roughly proportional to the number of reactions. Since GPUs have a lower clock frequency than CPUs (e.g., in the case of the hardware used for the tests, 837 MHz with respect to 2.4 GHz, respectively), each GPU core is slower than the CPU core to perform a single instruction[1]. For this reason, in order to obtain the highest performances, the calculations on the GPU should be spread across threads as much as possible, while the number of operations performed by each thread should be reduced.

Indeed, as reported in Table 2 and shown in Fig. 7, when the number of chemical species involved in a model is greater than the number of reactions, LASSIE achieves better performances than those obtained in the case of models with a number of chemical species smaller than the number of reactions. For instance, considering the models with $M \times N$ equal to 171×512, the running time of LASSIE is smaller than in the case of the models with size 512×171, irrespective of the number of samples of the system dynamics, thanks to the higher number of threads that are concurrently launched on the GPU in the first case. This is in general valid in all cases with the exception

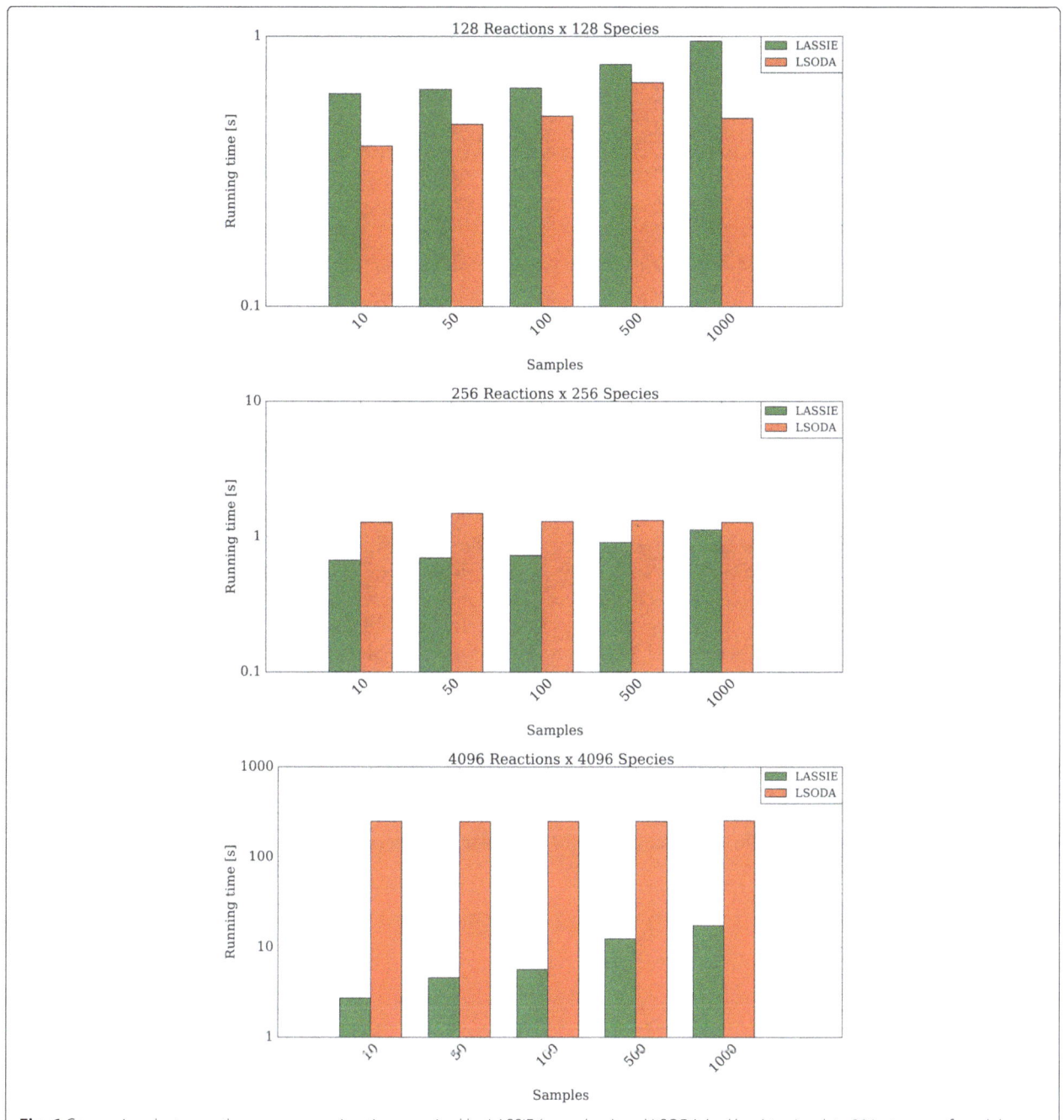

Fig. 6 Comparison between the average running time required by LASSIE (*green bars*) and LSODA (*red bars*) to simulate 30 instances of models characterized by 128 reactions and 128 species (*top*), 256 reactions and 256 species (*middle*), 4096 reactions and 4096 species (*bottom*), saving different numbers of sampling time instants of the dynamics. Note that the *y*-axes are in logarithmic scale

of the models characterized by 2048 chemical species with 500 and 1000 samples of the system dynamics. Here, the average running time of LASSIE is greater than in the case of models with 2048 reactions, since the required number of accesses to the high-latency global memory of the GPU impairs the performances of the simulations.

In order to assess the scalability of LASSIE, and of CUDA applications in general, we executed additional tests on different GPUs. Figure 8 shows a comparison of LASSIE's performance using three different GPU models (Table 3): a notebook video card (Nvidia GeForce 960M, red bars), the Nvidia GeForce Titan Z used throughout the paper (green bars), and a Tesla-class GPU (the Nvidia K20c, blue bars). To compare the speed-up provided by these GPUs we generated 30 different synthetic models (characterized by size $M \times N$ equal to 1024×1024,

Table 2 Average running time (in seconds) of LASSIE required for the execution of a set of 30 synthetic reaction-based models of size $M \times N$ (with $M \neq N$), considering 10, 50, 100, 500, 1000 samples of the system dynamics of all chemical species

$M \times N$	10 samples	50 samples	100 samples	500 samples	1000 samples
	LASSIE	LASSIE	LASSIE	LASSIE	LASSIE
171 × 512	0.589	0.623	0.662	0.900	1.149
512 × 171	1.032	1.062	1.087	1.318	1.540
341 × 1024	0.682	0.751	0.828	1.285	1.726
1024 × 341	1.119	1.179	1.220	1.597	1.895
512 × 1024	0.759	0.828	0.941	1.419	1.949
1024 × 512	1.053	1.112	1.192	1.624	1.999
683 × 2048	0.876	1.064	1.310	2.442	3.533
2048 × 683	1.389	1.512	1.664	2.406	3.029
1024 × 2048	1.002	1.201	1.508	2.858	3.889
2048 × 1024	1.317	1.444	1.613	2.614	3.281

2048×2048 and 4096×4096) and calculated the average running time.

Our results highlight the importance of two distinct factors on LASSIE's performances: the GPU's clock frequency and the amount of available resources (in this case, the cores). As a matter of fact, despite the lower amount of CUDA cores, the GeForce 960M turns out to be competitive on models of moderately large size thanks to its higher clock rate, with respect to the Titan Z and the K20c. When the ODEs largely outnumber the available cores (e.g., for 4096 reactions and chemical species), the GeForce 960M is no longer competitive. This is an example of transparent scalability of CUDA applications: the threads are automatically distributed over the available cores, improving the overall performances, without any user intervention. Moreover, as described in the Background section, threads are organized in blocks that are scheduled on the available multi-processors. Thanks to this characteristic, when the overall number of threads outnumbers the available cores, CUDA automatically creates a queue of blocks that are scheduled on the streaming multi-processors as soon as they become available for computation. Thus, LASSIE can, in principle, simulate any model on any GPU, as long as there is enough memory to store the data structures.

The Tesla K20c is characterized by a large amount of cores that, in the case of 1024×1024 models, are fully exploited only during the simulation of the stiff parts of

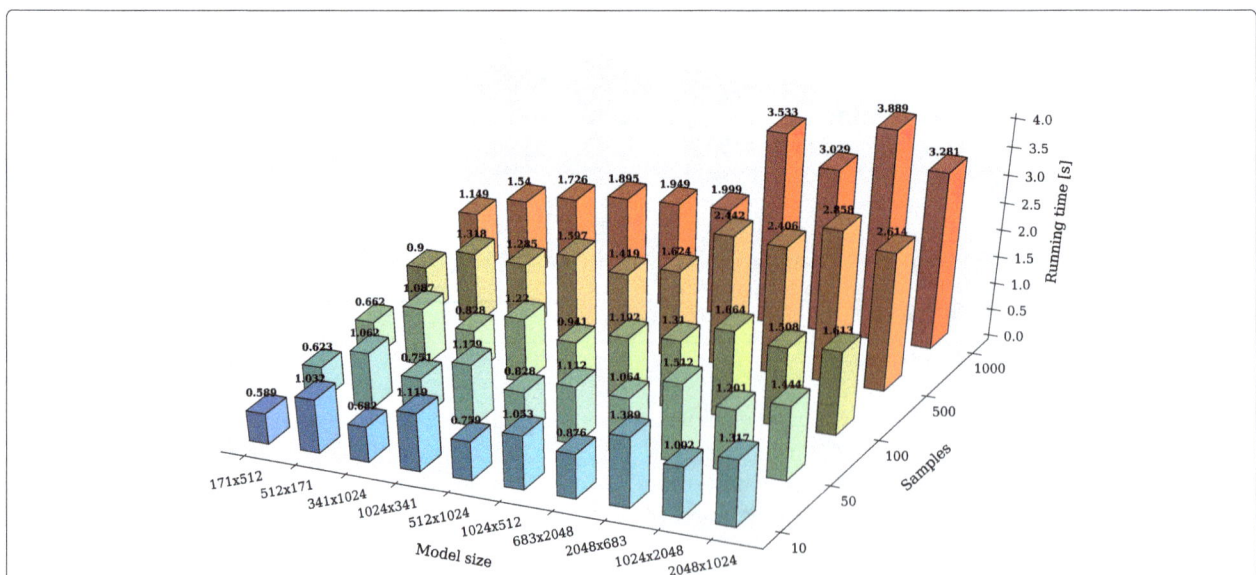

Fig. 7 Running time (z-axis) of LASSIE for the simulation of synthetic models of increasing size, having a number of reactions and of species $M \times N$ (x-axis), with $M \neq N$, and characterized by an increasing number of sampling time instants of the system dynamics (y-axis)

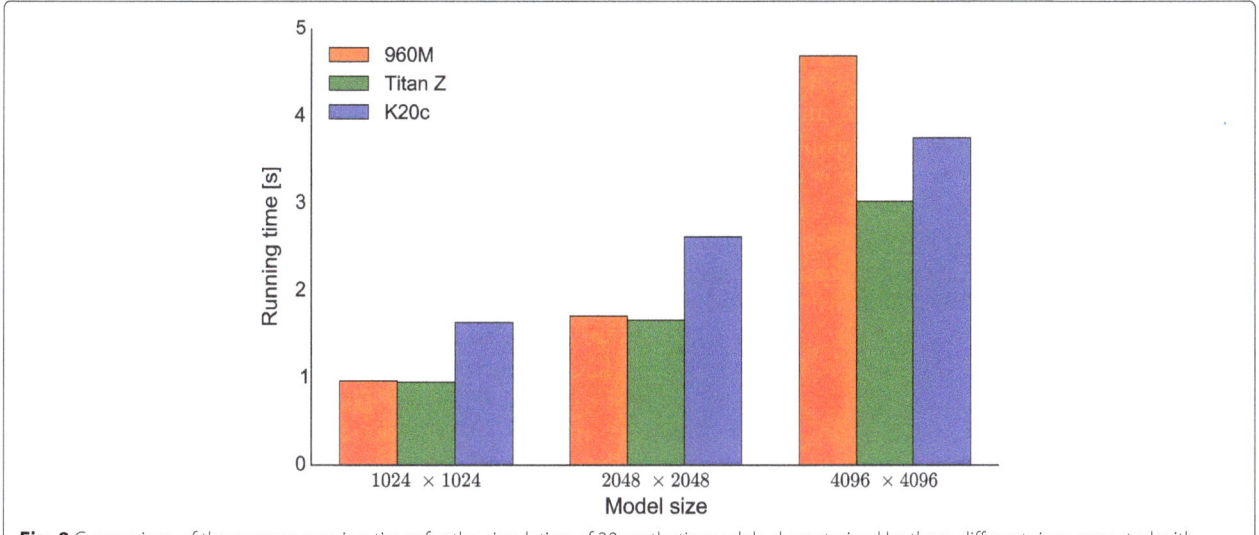

Fig. 8 Comparison of the average running times for the simulation of 30 synthetic models characterized by three different sizes, executed with different GPUs: a notebook GPU Nvidia GeForce 960M (*red bars*); a Nvidia GeForce Titan Z (*green bars*); a Tesla-class GPU Nvidia K20c (*blue bars*)

the dynamics. For the remaining parts of the simulation, half of its cores are actually used for computation with a slower clock rate with respect to the clock rate of the GeForce GPUs. Moreover, Tesla cards exploit Error Correcting Codes (ECC) on memories, ensuring additional checks of correctness to the data against potential corruption from electrical or magnetic interference, at the price of a significant overhead [51]. The ECC was enabled during all tests, partly explaining the reduced performance of the Tesla K20c on very large-scale models with respect to the Titan Z.

We assessed the accuracy of LASSIE by simulating the dynamics of the model of the Ras/cAMP/PKA signaling pathway in yeast presented in [26], and comparing the outcome of LASSIE with the result of the simulation performed with LSODA. We also investigated the influence of LASSIE parameters (e.g., tolerance values) on the running times and quality of the simulated solutions, by exploiting a model representing a chain of isomerizations. The accuracy results—which show an identical dynamics with respect to LSODA using default settings—are presented in Additional file 4.

As a final remark, we highlight that a fair comparison of GPUs and CPUs is a difficult task, in general, due to their deep architectural differences. The theoretical peak performances of both architectures are difficult to achieve: indeed, developers must implement code to the aim of maximizing the parallelism and the occupancy of the multi-processors, adhering as much as possible to the underlying SIMD computational model in the case of the GPU and exploiting vector instructions in the case of the CPU. However, GPUs allow the temporary divergence of the execution flow of threads, that is, a part of the threads can execute different portions of the code (e.g., the branches of an IF/THEN/ELSE statement). When this situation occurs, some threads get stalled waiting for reconvergence. This mechanism provides the programmer with a certain degree of freedom to abandon the SIMD paradigm, but at the same time it can potentially lead to the complete serialization of the execution affecting the overall performances. Hence, conditional branches should be avoided as much as possible. We also highlight that the usage of registers and shared memory influences the occupancy of the GPU, as these resources are scarce on each streaming multiprocessor. All these circumstances can

Table 3 Nvidia GPUs used to assess the scalability of LASSIE

	GeForce GTX 960M	GeForce GTX Titan Z	Tesla K20c
Global memory	4 GB	6 GB	5 GB
Number of streaming multiprocessors	5	15	13
CUDA cores per streaming multiprocessor	128	192	192
Total number of CUDA cores	640	2880	2496
Base clock	1.2 GHz	876 MHz	706 MHz

prevent the achievement of the peak computational power of a GPU.

To this aim, we developed kernels that maximize the parallelism and the occupancy of the multi-processors avoiding threads divergence as much as possible. Moreover, we optimized data structures to store the matrices **A** and **H** that encode the system of ODEs, and CUDA vector types that allow to increase the memory throughput and to reduce the number of memory accesses, all precautions that explain the performance boost achieved with LASSIE.

Conclusions

In this work we presented LASSIE, a GPU-powered simulator of large-scale biochemical systems based on mass-action kinetics. LASSIE is a "black-box" simulator able to automatically convert reaction-based models of biological systems into the corresponding systems of ODEs. Reaction-based models defined according to the law of mass-action do not hinge upon the use of any approximate kinetics functions (e.g., Michaelis-Menten rate law for enzymatic processes [23], Hill functions for cooperative binding [52], etc.), which are frequently used in Systems Biology for the definition of mathematical models based on differential equations. Although Michaelis-Menten kinetics or Hill functions can be useful in biological modeling, they rely on chemical assumptions that are valid only in certain conditions [53]. Therefore, the reason why we rely on mass-action based models is manifold. On the one hand, since the biological function and biochemical kinetics of all molecular species and all reactions appearing in the model are not approximated nor lumped together in any way, they can be analyzed independently from each other. As a consequence, this allows to determine the influence of every single species and reaction on the overall functioning of the system. On the other hand, the law of mass-action allows to derive a first order ODE for each species appearing in the model: it is worth noting that such ODE is a *polynomial function* that describes how the concentration of that species changes in time, according to all the reactions where it appears either as reactant or product [24]. The presence of polynomial functions simplifies the symbolic derivation that is needed to calculate the Jacobian matrix associated with the ODEs and exploited by the BDF. In addition, as described in the Methods section, polynomials can be efficiently encoded in the memory and parsed GPU-side. As a result, all GPU threads can perform the same task (i.e., polynomial decoding and evaluation), strongly reducing warps' divergence and the consequent stalling of threads due to serialization, a circumstance that would instead happen if each thread calculated an ODE characterized by an arbitrary kinetics. In order to solve systems of ODEs characterized by stiffness, LASSIE automatically

switches between the RKF and the BDF integration methods. LASSIE's execution flow is partitioned into 25 CUDA kernels, overall distributing the calculations over the available cores in order to fully exploit the massive parallel capabilities of modern GPUs, therefore achieving a relevant reduction of the running time in case of large-scale models.

In order to assess the computational performance of LASSIE, we performed a set of simulation tests using synthetic reaction-based models of increasing size, and we compared LASSIE's running time with respect to the LSODA numerical integration algorithm implemented in the SciPy library. The break-even between the performances of LASSIE and LSODA was observed when both the numbers of reactions and chemical species is in between 128 and 256. This result indicates that, for biological systems consisting in more than 256 reactions and 256 species, the GPU-powered simulator becomes more convenient than the LSODA algorithm running on CPU. Indeed, in the case of large-scale models, characterized by 4096 reactions and 4096 species, we obtained a considerable 92× speed-up. Moreover, thanks to its smaller memory footprint with respect to LSODA, LASSIE allows the simulation of even larger models, taking just an average of 14.13 s to simulate models characterized by 8192 reactions and 8192 species. On the contrary, LSODA did not allow the simulation of models of this size on the computer we used for the tests, as it crashed because of its very high memory footprint. We also highlight that COPASI [54], one of the most used software in Systems Biology, requires in general longer execution times with respect to the SciPy implementation of LSODA exploited in this work. In addition, COPASI fails when trying to simulate large-scale models. We provide an example of such model—characterized by 4096 reactions and 4096 species—as SBML file in the GITHUB repository.

BDFs are the most used integration algorithms to solve systems of ODEs in case of stiffness. The first–order BDF is a single-step implicit integration method, meaning that the next state of the system depends only on the current state of the system. Higher–order BDFs are multi-step methods, so that the next state of the system relies on multiple previous states of the system (i.e., the number of previous states is equal to the BDF order). This implies that the integration step-size should be the same for all previous states to ensure the correctness of the solution. For this reason, LSODA uses the multi-step Adams methods as explicit methods in addition to BDFs. Conversely, LASSIE uses the RKF method, which is a single-step explicit algorithm with variable step-size, and the Backward Euler method. Other single-step implicit methods belonging to the family of Runge-Kutta methods exist [55], the most known being the families of Lobatto and Radau methods [56].

These methods have been proven to be suitable for stiff systems, thanks to their accuracy and stability [56]. As a future development of LASSIE, we will investigate the feasibility and efficacy of replacing the Backward Euler and, more generally, the BDFs with implicit Runge-Kutta methods [57], in particular Lobatto and Radau methods.

In order to fully exploit the CUDA architecture, the memory hierarchy must be exploited as much as possible. Because of the peculiar sequential structure of both explicit and implicit integration algorithms, LASSIE's kernel are lightweight and rarely reuse any variables. For this reason, the current implementation only leverages the global memory (characterized by high latencies) and registers to manipulate the mutable data. The shared memory has not been exploited in any way, leaving room for potential future improvements of performances. However, on the GPUs where the L1 cache and the shared memory share the same resources, CUDA allows to express a preference to assign a larger amount of memory to the caching mechanisms. This functionality is enabled by default in LASSIE using the CUDA `cudaFuncSetCacheConfig` primitive, executed with the `cudaFuncCachePreferL1` argument. Also the constant memory has not been used, since all data structures are larger than the total size of this memory. In a future release, we plan to leverage these memories, for instance to store the array of the kinetic constants and the structures containing the offsets used to correctly decode the ODEs. LASSIE currently exploits only a single GPU, even on multi-GPU systems; as a future improvement of this work we plan to extend it in order to support multi-GPU systems, to further increase the size of the models to be simulated.

Finally, an additional goal in the development of LASSIE is to integrate and accelerate the investigation of rule-based models. In the next future, we plan to develop a set of tools that will leverage the functionalities offered by rule-based modeling frameworks [9–11], to convert a rule-based model into a set of reactions. Although rule-based tools already provide internal simulation methods (e.g., PySB allows to perform deterministic simulations using advanced integrators like LSODA), LASSIE can represent a valuable alternative for large-scale models, enabling the investigation of more detailed biological systems, paving the way to potential new discoveries in Systems Biology. LASSIE will be also integrated in COSYS, a free web-based platform for Systems Biology investigation available at http://www.sysbio.it/cosys [58].

Endnote

[1] The advances in GPU's technology will progressively reduce this gap. In middle 2016—with the introduction of the novel Pascal architecture and the 16 nm FinFET manufacturing process—Nvidia presented a GPU with a clock frequency of 1.7 GHz that, theoretically, is expected to double LASSIE's performances.

Abbreviations
BDF: Backward Differentiation Formulae; CPU: Central Processing Unit; CUDA: Compute Unified Device Architecture; ECC: Error Correcting Codes; GPGPU: General-purpose GPU; GPU: Graphics Processing Unit; LASSIE: LArge-Scale SImulator; ODE: Ordinary Differential Equation; RKF: Runge-Kutta-Fehlberg; SIMD: Single Instruction Multiple Data; SSA: Stochastic Simulation Algorithm

Acknowledgements
We acknowledge the CINECA award under the ISCRA initiative, for the availability of high performance computing resources and support. Authors would like to thank the SYSBIO.IT Centre of Systems Biology for the support.

Funding
Not applicable.

Authors' contributions
Conceived the idea: MSN. Designed the code: AT, MSN. Implemented the code and performed the experiments: AT. Analyzed the data: AT, MSN, DB, PC. Wrote the manuscript: AT, MSN, DB, PC. Critically read the manuscript and contributed to the discussion of the whole work: GM. All authors read and approved the final manuscript.

Competing interests
The authors declare that they have no competing interests.

Author details
[1] Department of Informatics, Systems and Communication, University of Milano-Bicocca, Viale Sarca 336, 20126 Milano, Italy. [2] Department of Human and Social Sciences, University of Bergamo, Piazzale Sant'Agostino 2, 24129 Bergamo, Italy. [3] SYSBIO.IT Centre of Systems Biology, Piazza della Scienza 2, 20126 Milano, Italy.

References

1. Aldridge BB, Burke JM, Lauffenburger DA, Sorger PK. Physicochemical modelling of cell signalling pathways. Nat Cell Biol. 2006;8(11): 1195–203.
2. Chou IC, Voit EO. Recent developments in parameter estimation and structure identification of biochemical and genomic systems. Math Biosci. 2009;219(2):57–83.
3. Joubert W, Archibald R, Berrill M, Brown WM, Eisenbach M, Grout R, Larkin J, Levesque J, Messer B, Norman M, Philip B, Sankaran R, Tharrington A, Turner J. Accelerated application development: The ORNL Titan experience. Comput Electr Eng. 2015;46:123–38.
4. Nobile MS, Cazzaniga P, Tangherloni A, Besozzi D. Graphics processing units in bioinformatics, computational biology and systems biology. Brief Bioinform. 2016;2016(bbw058).
5. Chylek LA, Harris LA, Tung CS, Faeder JR, Lopez CF, Hlavacek WS. Rule-based modeling: a computational approach for studying biomolecular site dynamics in cell signaling systems. Wiley Interdisci Rev Syst Biol Med. 2014;6(1):13–36.
6. Chylek LA, Stites EC, Posner RG, Hlavacek WS In: Prokop A, Csukás B, editors. Innovations of the rule-based modeling approach. Dordrecht: Springer; 2013. p. 273–300.
7. Blinov ML, Faeder JR, Goldstein B, Hlavacek WS. A network model of early events in epidermal growth factor receptor signaling that accounts for combinatorial complexity. Biosystems. 2006;83(2):136–51.
8. Chen WW, Schoeberl B, Jasper PJ, Niepel M, Nielsen UB, Lauffenburger DA, Sorger PK. Input–output behavior of ErbB signaling pathways as revealed by a mass action model trained against dynamic data. Mol Syst Biol. 2009;5(1):239.

9. Blinov ML, Faeder JR, Goldstein B, Hlavacek WS. BioNetGen: software for rule-based modeling of signal transduction based on the interactions of molecular domains. Bioinformatics. 2004;20(17):3289–91.

10. Lopez CF, Muhlich JL, Bachman JA, Sorger PK. Programming biological models in Python using PySB. Mol Syst Biol. 2013;9(1):646.

11. Feret J, Danos V, Krivine J, Harmer R, Fontana W. Internal coarse-graining of molecular systems. Proc Natl Acad Sci USA. 2009;106(16):6453–458.

12. Wilkinson D. Stochastic modelling for quantitative description of heterogeneous biological systems. Nat Rev Genet. 2009;10(2):122–33.

13. Székely Jr T, Burrage K. Stochastic simulation in systems biology. Comput Struct Biotechnol J. 2014;12(20–21):14–25.

14. Harris LA, Clancy P. A "partitioned leaping" approach for multiscale modeling of chemical reaction dynamics. J Chem Phys. 2006;125(14): 144107.

15. Eldar A, Elowitz MB. Functional roles for noise in genetic circuits. Nature. 2010;467(7312):167–73.

16. Butcher JC. Numerical Methods for Ordinary Differential Equations. Chichester West Sussex: Wiley; 2008.

17. Higham DJ, Trefethen LN. Stiffness of ODEs. BIT Numer Math. 1993;33(2): 285–303.

18. Gillespie DT. Stochastic simulation of chemical kinetics. Annu Rev Phys Chem. 2007;58:35–55.

19. Petzold LR. Automatic selection of methods for solving stiff and nonstiff systems of ordinary differential equations. SIAM J Sci Stat Comp. 1983;4: 136–48.

20. Cash JR. Backward Differentiation Formulae In: Engquist B, editor. Encyclopedia of Applied and Computational Mathematics. Berlin Heidelberg: Springer; 2015. p. 97–101.

21. Gillespie DT. A general method for numerically simulating the stochastic time evolution of coupled chemical reactions. J Comput Phys. 1976;22: 403–34.

22. Besozzi D. Reaction-based models of biochemical networks In: Beckmann A, Bienvenu L, Jonoska N, editors. Pursuit of the Universal. 12th Conference on Computability in Europe, CiE 2016, Proceedings. LNCS, vol. 9709. Switzerland: Springer; 2016. p. 24–34.

23. Nelson DL, Cox MM. Lehninger Principles of Biochemistry. New York: W. H. Freeman Co; 2004.

24. Voit EO, Martens HA, Omholt SW. 150 years of the mass action law. PLoS Comput Biol. 2015;11(1):1004012.

25. Fehlberg E. Classical fifth-, sixth-, seventh-, and eighth-order Runge-Kutta formulas with stepsize. NASA Tech Rep R-287, NASA. 1968.

26. Cazzaniga P, Pescini D, Besozzi D, Mauri G, Colombo S, Martegani E. Modeling and stochastic simulation of the Ras/cAMP/PKA pathway in the yeast *Saccharomyces cerevisiae* evidences a key regulatory function for intracellular guanine nucleotides pools. J Biotechnol. 2008;133(3):377–85.

27. Ackermann J, Baecher P, Franzel T, Goesele M, Hamacher K. Massively-parallel simulation of biochemical systems. In: Proceedings of Massively Parallel Computational Biology on GPUs, Jahrestagung der Gesellschaft Für Informatik e.V; 2009. p. 739–50.

28. Nobile MS, Cazzaniga P, Besozzi D, Mauri G. GPU-accelerated simulations of mass-action kinetics models with cupSODA. J Supercomput. 2014;69(1):17–24.

29. Zhou Y, Liepe J, Sheng X, Stumpf MP, Barnes C. GPU accelerated biochemical network simulation. Bioinformatics. 2011;27(6):874–6.

30. Komarov I, D'Souza RM. Accelerating the Gillespie exact stochastic simulation algorithm using hybrid parallel execution on graphics processing units. PLoS ONE. 2012;7(11):46693.

31. Komarov I, D'Souza RM, Tapia J. Accelerating the Gillespie τ-leaping method using graphics processing units. PLoS ONE. 2012;7(6):37370.

32. Gillespie DT. Exact stochastic simulation of coupled chemical reactions. J Phys Chem. 1977;81(25):2340–361.

33. Gillespie DT, Petzold LR. Improved leap-size selection for accelerated stochastic simulation. J Chem Phys. 2003;119:8229–234.

34. Amara F, Colombo R, Cazzaniga P, Pescini D, Csikász-Nagy A, Muzi Falconi M, Besozzi D, Plevani P. In vivo and in silico analysis of PCNA ubiquitylation in the activation of the Post Replication Repair pathway in *S. cerevisiae*. BMC Syst Biol. 2013;7(1):24.

35. Besozzi D, Cazzaniga P, Pescini D, Mauri G, Colombo S, Martegani E. The role of feedback control mechanisms on the establishment of oscillatory regimes in the Ras/cAMP/PKA pathway in *S. cerevisiae*. EURASIP J Bioinform Syst Biol. 2012;2012(10).

36. Cazzaniga P, Nobile MS, Besozzi D, Bellini M, Mauri G. Massive exploration of perturbed conditions of the blood coagulation cascade through GPU parallelization. BioMed Res Int. 2014;2014. Article ID 863298.

37. Intosalmi J, Manninen T, Ruohonen K, Linne ML. Computational study of noise in a large signal transduction network. BMC Bioinforma. 2011;12(1): 1–12.

38. Pescini D, Cazzaniga P, Besozzi D, Mauri G, Amigoni L, Colombo S, Martegani E. Simulation of the Ras/cAMP/PKA pathway in budding yeast highlights the establishment of stable oscillatory states. Biotechnol Adv. 2012;30:99–107.

39. Petre I, Mizera A, Hyder CL, Meinander A, Mikhailov A, Morimoto RI, Sistonen L, Eriksson JE, Back RJ. A simple mass-action model for the eukaryotic heat shock response and its mathematical validation. Nat Comput. 2011;10(1):595–612.

40. Chellaboina V, Bhat SP, Haddad WM, Bernstein DS. Modeling and analysis of mass-action kinetics. IEEE Control Syst. 2009;29(4):60–78.

41. Jackson KR. A survey of parallel numerical methods for initial value problems for ordinary differential equations. IEEE Trans Magn. 1991;27(5): 3792–797.

42. Mathews JH, Fink KD. Numerical Methods Using MATLAB. Upper Saddle River: Prentice-Hall Inc; 2004.

43. Thohura S, Rahman A. Numerical approach for solving stiff differential equations: A comparative study. J Sci Front Res Math Decision Sci. 2013;13:7–18.

44. Gear CW. The control of parameters in the automatic integration of ordinary differential equations. University of Illinois Urbana-Champaign. Int Rep File 757. 1968.

45. Ben-Israel A. A Newton-Raphson method for the solution of systems of equations. J Math Anal Appl. 1966;15(2):243–52.

46. Smooke MD. Error estimate for the modified Newton method with applications to the solution of nonlinear, two-point boundary-value problems. J Optim Theory Appl. 1983;39(4):489–511.

47. Bartels RH, Golub GH. The simplex method of linear programming using LU decomposition. Commun ACM. 1969;12(5):266–8.

48. Nvidia: cuBLAS library 7.5. 2015.

49. Jones E, Oliphant T, Peterson P, et al. SciPy: Open source scientific tools for Python. 2001. http://www.scipy.org/.

50. Nobile MS, Cazzaniga P, Besozzi D, Pescini D, Mauri G. cuTauLeaping: A GPU-powered tau-leaping stochastic simulator for massive parallel analyses of biological systems. PLoS ONE. 2014;9(3):91963.

51. Wilt N. The CUDA Handbook: A Comprehensive Guide to GPU Programming. Upper Saddle River: Addison-Wesley; 2013.

52. Weiss JN. The Hill equation revisited: uses and misuses. FASEB J. 1997;11(11):835–41.

53. Le Novère N. Quantitative and logic modelling of molecular and gene networks. Nat Rev Genet. 2015;16(3):146–58.

54. Hoops S, Sahle S, Gauges R, Lee C, Pahle J, Simus N, Singhal M, Xu L, Mendes P, Kummer U. COPASI - a COmplex PAthway SImulator. Bioinformatics. 2006;22(24):3067–074.

55. Butcher JC. Implicit Runge-Kutta processes. Math Comput. 1964;18(85): 50–64.

56. Prothero A, Robinson A. On the stability and accuracy of one-step methods for solving stiff systems of ordinary differential equations. Math Comput. 1974;28(125):145–62.

57. Butcher JC. On the implementation of implicit Runge-Kutta methods. BIT Numer Math. 1976;16(3):237–40.

58. Cumbo F, Nobile MS, Damiani C, Colombo R, Mauri G, Cazzaniga P. COSYS: A Computational Infrastructure for Systems Biology. LNCS, Springer, in press.

PERMISSIONS

The contributors of this book come from diverse backgrounds, making this book a truly international effort. This book will bring forth new frontiers with its revolutionizing research information and detailed analysis of the nascent developments around the world.

We would like to thank all the contributing authors for lending their expertise to make the book truly unique. They have played a crucial role in the development of this book. Without their invaluable contributions this book wouldn't have been possible. They have made vital efforts to compile up to date information on the varied aspects of this subject to make this book a valuable addition to the collection of many professionals and students.

This book was conceptualized with the vision of imparting up-to-date information and advanced data in this field. To ensure the same, a matchless editorial board was set up. Every individual on the board went through rigorous rounds of assessment to prove their worth. After which they invested a large part of their time researching and compiling the most relevant data for our readers.

The editorial board has been involved in producing this book since its inception. They have spent rigorous hours researching and exploring the diverse topics which have resulted in the successful publishing of this book. They have passed on their knowledge of decades through this book. To expedite this challenging task, the publisher supported the team at every step. A small team of assistant editors was also appointed to further simplify the editing procedure and attain best results for the readers.

Apart from the editorial board, the designing team has also invested a significant amount of their time in understanding the subject and creating the most relevant covers. They scrutinized every image to scout for the most suitable representation of the subject and create an appropriate cover for the book.

The publishing team has been an ardent support to the editorial, designing and production team. Their endless efforts to recruit the best for this project, has resulted in the accomplishment of this book. They are a veteran in the field of academics and their pool of knowledge is as vast as their experience in printing. Their expertise and guidance has proved useful at every step. Their uncompromising quality standards have made this book an exceptional effort. Their encouragement from time to time has been an inspiration for everyone.

The publisher and the editorial board hope that this book will prove to be a valuable piece of knowledge for researchers, students, practitioners and scholars across the globe.

LIST OF CONTRIBUTORS

Hamed Hassani Saadi and Reza Sameni
School of Electrical and Computer Engineering, Shiraz University, Shiraz, Iran

Amin Zollanvari
Department of Electrical and Electronic Engineering, Nazarbayev University, Astana, Kazakhstan

Jochen Kruppa and Klaus Jung
Institute for Animal Breeding and Genetics, University of Veterinary Medicine Hannover, Foundation, Bünteweg 17p, D-30559 Hannover, Germany

Andrey Smelter
School of Interdisciplinary and Graduate Studies, University of Louisville, Louisville, KY 40292, USA
Department of Computer Engineering and Computer Science, University of Louisville, Louisville, KY 40292, USA

Hunter N. B. Moseley
Department of Molecular and Cellular Biochemistry, University of Kentucky, Lexington, KY 40356, USA
Markey Cancer Center, University of Kentucky, Lexington, KY 40356, USA
Center for Environmental and Systems Biochemistry, University of Kentucky, Lexington, KY 40356, USA
Institute for Biomedical Informatics, University of Kentucky, Lexington, KY 40356, USA

Morgan Astra
Center for Environmental and Systems Biochemistry, University of Kentucky, Lexington, KY 40356, USA

Zhijiang Wan and Ning Zhong
Beijing Advanced Innovation Center for Future Internet Technology, Beijing University of Technology, Beijing, China
Department of Life Science and Informatics, Maebashi Institute of Technology, Maebashi, Japan
International WIC Institute, Beijing University of Technology, Beijing, China
Beijing Key Laboratory of MRI and Brain Informatics, Beijing, China
Beijing InternationalCollaboration Base on Brain Informatics and Wisdom Services, Beijing, China

Yishan He, Ming Hao and Jian Yang
Department of Life Science and Informatics, Maebashi Institute of Technology, Maebashi, Japan
International WIC Institute, Beijing University of Technology, Beijing, China
Beijing Key Laboratory of MRI and Brain Informatics, Beijing, China
Beijing InternationalCollaboration Base on Brain Informatics and Wisdom Services, Beijing, China

Vinay S. Kumar, Daniel Weaver and David Ando
Biological Systems and Engineering Division, Lawrence Berkeley National Laboratory, Berkeley, CA, USA
Joint BioEnergy Institute, Emeryville, CA, USA

Amit Ghosh
Biological Systems and Engineering Division, Lawrence Berkeley National Laboratory, Berkeley, CA, USA
Joint BioEnergy Institute, Emeryville, CA, USA
School of Energy Science and Engineering, Indian Institute of Technology (IIT), Kharagpur, India

Garrett W. Birkel and Tyler W. H. Backman
Biological Systems and Engineering Division, Lawrence Berkeley National Laboratory, Berkeley, CA, USA
Joint BioEnergy Institute, Emeryville, CA, USA
DOE Agile BioFoundry, Emeryville, CA, USA

Adam P. Arkin
Biological Systems and Engineering Division, Lawrence Berkeley National Laboratory, Berkeley, CA, USA
Department of Bioengineering, University of California, Berkeley, CA, USA
Environmental Genomics and Systems Biology Division, Lawrence Berkeley National Laboratory, Berkeley, CA, USA

Jay D. Keasling
Biological Systems and Engineering Division, Lawrence Berkeley National Laboratory, Berkeley, CA, USA

Joint BioEnergy Institute, Emeryville, CA, USA
Department of Chemical and Biomolecular Engineering, University of California, Berkeley, CA, USA
Department of Bioengineering, University of California, Berkeley, CA, USA
Novo Nordisk Foundation Center for Biosustainability, Technical University of Denmark, DK2970 Hørsholm, Denmark

Héctor García Martín
Biological Systems and Engineering Division, Lawrence Berkeley National Laboratory, Berkeley, CA, USA
Joint BioEnergy Institute, Emeryville, CA, USA
DOE Agile BioFoundry, Emeryville, CA, USA
BCAM, Basque Center for Applied Mathematics, Bilbao, Spain

Raquel Dias and Bryan Kolaczkowski
Department of Microbiology and Cell Science, University of Florida, Gainesville, FL, USA

Thales F. M. Carvalho
Departamento de Informática, Universidade Federal de Viçosa, Viçosa, Brazil

Jose Cleydson F. Silva
Departamento de Informática, Universidade Federal de Viçosa, Viçosa, Brazil
National Institute of Science and Technology in Plant-Pest Interactions/ BIOAGRO, Universidade Federal de Viçosa, Viçosa, Brazil

Fabio R. Cerqueira
Departamento de Informática, Universidade Federal de Viçosa, Viçosa, Brazil
Departamento de Engenharia de Produção, Universidade Federal Fluminense, Petrópolis, Rio de Janeiro, Brazil

Marcos F. Basso, Michihito Deguchi, Welison A. Pereira, Roberto R. Sobrinho, Otávio J. B. Brustolini and Maximiller Dal-Bianco
National Institute of Science and Technology in Plant-Pest Interactions/ BIOAGRO, Universidade Federal de Viçosa, Viçosa, Brazil

Elizabeth P. B. Fontes
National Institute of Science and Technology in Plant-Pest Interactions/ BIOAGRO, Universidade Federal de Viçosa, Viçosa, Brazil

Departamento de Bioquímica e Biologia Molecular, Universidade Federal de Viçosa, Viçosa, Brazil

Anésia A. Santos
National Institute of Science and Technology in Plant-Pest Interactions/ BIOAGRO, Universidade Federal de Viçosa, Viçosa, Brazil
Departamento de Biologia Geral, Universidade Federal de Viçosa, Viçosa, Brazil

Francisco Murilo Zerbini
National Institute of Science and Technology in Plant-Pest Interactions/ BIOAGRO, Universidade Federal de Viçosa, Viçosa, Brazil
Departamento de Fitopatologia, Universidade Federal de Viçosa, Viçosa, MG, Brazil

Pedro M. P. Vidigal
Núcleo de Biomoléculas, Universidade Federal de Viçosa, Viçosa, MG, Brazil

Fabyano F. Silva
Departamento de Zootecnia, Universidade Federal de Viçosa, Viçosa, Brazil

Renildes L. F. Fontes
Departamento de Solos, Universidade Federal de Viçosa, Viçosa, Brazil

José Santos and Ángel Monteagudo
Department of Computer Science, University of A Coruña, Campus de Elviña s/n, 15071 A Coruña, Spain

Lianbo Yu, Soledad Fernandez and Guy Brock
Center for Biostatistics, Department of Biomedical Informatics, The Ohio State University, 1800 Cannon Dr., 43210 Columbus, OH, USA

Zhi-An Huang, Zhenkun Wen, Qingjin Deng, Ying Chu and Zexuan Zhu
College of Computer Science and Software Engineering, Shenzhen University, Shenzhen 518060, China

Yiwen Sun
School of Medicine, Shenzhen University, Shenzhen 518060, China

Haijing Jin
Graduate Program in Structural and Computational Biology and Molecular Biophysics, Baylor College of Medicine, One Baylor Plaza, 77030 Houston, TX,USA

Ying-Wooi Wan
Department of Molecular and Human Genetics, Baylor College of Medicine, One Baylor Plaza, 77030 Houston, TX, USA

Zhandong Liu
Department of Pediatrics-Neurology, Jan and Dan Duncan Neurological Research Institute, Baylor College of Medicine, 1250 Moursund St., Suite 1325, 77030 Houston, TX, USA

Jing Liu, Yaxiong Chi and Chen Zhu
Key Laboratory of Intelligent Perception and Image Understanding of Ministry of Education, Xidian University, Xi'an 710071, China

Yaochu Jin
Department of Computer Science, University of Surrey, Guildford GU2 7XH, UK

Willy Vincent Bienvenut, Jean-Pierre Scarpelli, Johan Dumestier, Thierry Meinnel and Carmela Giglione
Institute for Integrative Biology of the Cell (I2BC), CEA, CNRS, Univ. Paris-Sud, Université Paris Saclay, 91198 Gif-sur-Yvette Cedex, France

Andrey Alexeyenko
National Bioinformatics Infrastructure Sweden, Science for Life Laboratory, Stockholm, Sweden
Department of Microbiology, Tumor and Cell biology, Karolinska Institutet, Stockholm, Sweden

Ashwini Jeggari
Department of Cell and Molecular Biology, Karolinska Institutet, Stockholm, Sweden

Namshik Han
Gurdon Institute, University of Cambridge, Cambridge, UK
School of Computer Science and School of Health Sciences, University of Manchester, Manchester, UK

Harry A. Noyes
School of Biological Sciences, University of Liverpool, Liverpool, UK

Andy Brass
School of Computer Science and School of Health Sciences, University of Manchester, Manchester, UK

Bo Wen, Zhanlong Mei, Chunwei Zeng and Siqi Liu
BGI-Shenzhen, Shenzhen 518083, China
China National GeneBank-Shenzhen, BGI-Shenzhen, Shenzhen, Guangdong 518083, China

Carl J. Schmidt
Department of Animal and Food Sciences, University of Delaware, Newark, DE, USA

Liang Sun
Department of Animal and Food Sciences, University of Delaware, Newark, DE, USA
Computing Service, The Samuel Roberts Noble Foundation, Ardmore, OK 73401, USA

Jian Chen
Department of Computer Science and Electrical Engineering, University of Maryland Baltimore County, 1000 Hilltop Circle, Baltimore, MD, USA

Yongnan Zhu
Department of Computer Science and Electrical Engineering, University of Maryland Baltimore County, 1000 Hilltop Circle, Baltimore, MD, USA
Department of Computer Science, Hangzhou Dianzi University, Hangzhou 310018, Zhejiang Province, People's Republic of China

A. S. M. Ashique Mahmood, Catalina O. Tudor and K. Vijay-Shanker
Department of Computer and Information Sciences, University of Delaware, Newark, DE 19716, USA

Jia Ren
Center for Bioinformatics and Computational Biology, University of Delaware, Newark, DE 19711, USA

Andrea Tangherloni and Daniela Besozzi
Department of Informatics, Systems and Communication, University of Milano-Bicocca, Viale Sarca 336, 20126 Milano, Italy

Marco S. Nobile and Giancarlo Mauri
Department of Informatics, Systems and Communication, University of Milano-Bicocca, Viale Sarca 336, 20126 Milano, Italy
SYSBIO.IT Centre of Systems Biology, Piazza della Scienza 2, 20126 Milano, Italy

Paolo Cazzaniga
Department of Human and Social Sciences, University of Bergamo, Piazzale Sant'Agostino 2, 24129 Bergamo, Italy

SYSBIO.IT Centre of Systems Biology, Piazza della Scienza 2, 20126 Milano, Italy

Index

www.ingramcontent.com/pod-product-compliance
Lightning Source LLC
Chambersburg PA
CBHW080405190526

45161CB00003B/143